D1809563

Texts and
Monographs
in Physics

A. Galindo   P. Pascual

Quantum Mechanics I

Translated by J. D. García and L. Alvarez-Gaumé

With 56 Figures

Springer-Verlag
Berlin Heidelberg New York
London Paris Tokyo
Hong Kong Barcelona

Professor Alberto Galindo

Universidad Complutense
Facultad de Ciencias Físicas
Departamento de Física Teórica
E-28040 Madrid, Spain

Professor Pedro Pascual

Universidad de Barcelona
Facultad de Física
Departamento de Física Teórica
E-08070 Barcelona, Spain

Translators:

Professor J. D. García

Department of Physics
University of Arizona
Tucson, AZ 85721, USA

Professor L. Alvarez-Gaumé

Theory Division
CERN
CH-1211 Genève 23, Switzerland

Editors

Wolf Beiglböck

Institut für Angewandte Mathematik
Universität Heidelberg
Im Neuenheimer Feld 294
D-6900 Heidelberg 1, FRG

Elliott H. Lieb

Department of Physics
Joseph Henry Laboratories
Princeton University
Princeton, NJ 08540, USA

Joseph L. Birman

Department of Physics, The City College
of the City University of New York
New York, NY 10031, USA

Tullio Regge

Istituto di Fisica Teorica
Università di Torino, C. so M. d'Azeglio, 46
I-10125 Torino, Italy

Robert P. Geroch

Enrico Fermi Institute
University of Chicago
5640 Ellis Ave.
Chicago, IL 60637, USA

Walter Thirring

Institut für Theoretische Physik
der Universität Wien, Boltzmanngasse 5
A-1090 Wien, Austria

Title of the original Spanish edition: Mecánica Cuántica (I)
© by the authors and Editorial Alhambra, Madrid 1978 (1st ed.)
© by the authors and EUDEMA, Madrid 1989 (2nd ed.)

ISBN 3-540-51406-6 Springer-Verlag Berlin Heidelberg New York
ISBN 0-387-51406-6 Springer-Verlag New York Berlin Heidelberg

Library of Congress Cataloging-in-Publication Data. Galindo, A. (Alberto), 1934– [Mecánica cuántica I. English] Quantum mechanics I / A. Galindo, P. Pascual ; translated by J. D. García and L. Alvarez-Gaumé. p. cm. – (Texts and monographs in physics) Translation of: Mecánica cuántica I. Includes bibliographical references. ISBN 0-387-51406-6 (U.S.) 1. Quantum theory. I. Pascual, Pedro. II. Title. III. Series. QC174.12.G3413 1990 530.1'2–dc20 90-9548

2155/3150-543210 – Printed on acid-free paper

Preface

The first edition of this book was published in 1978 and a new Spanish edition in 1989. When the first edition appeared, Professor A. Martin suggested that an English translation would meet with interest. Together with Professor A.S. Wightman, he tried to convince an American publisher to translate the book. Financial problems made this impossible. Later on, Professors E.H. Lieb and W. Thirring proposed to entrust Springer-Verlag with the translation of our book, and Professor W. Beiglböck accepted the plan. We are deeply grateful to all of them, since without their interest and enthusiasm this book would not have been translated.

In the twelve years that have passed since the first edition was published, beautiful experiments confirming some of the basic principles of quantum mechanics have been carried out, and the theory has been enriched with new, important developments. Due reference to all of this has been paid in this English edition, which implies that modifications have been made to several parts of the book. Instances of these modifications are, on the one hand, the neutron interferometry experiments on wave-particle duality and the 2π rotation for fermions, and the crucial experiments of Aspect et al. with laser technology on Bell's inequalities, and, on the other hand, some recent results on level ordering in central potentials, new techniques in the analysis of anharmonic oscillators, and perturbative expansions for the Stark and Zeeman effects.

Major changes have been introduced in presenting the path-integral formalism, owing to its increasing importance in the modern formulation of quantum field theory. Also, the existence of new and more rigorous results has led us to change our treatment of the W.B.K. method. A new section on the inverse scattering problem has been added, because of its relevance in quantum physics and in the theory of integrable systems. Finally, we have tried to repair some omissions in the first edition, such as lower bounds to the ground-state energy and perturbation expansions of higher order for two-electron atoms.

The material appears in a two-volume edition. Volume I contains Chaps. 1 to 7 and Appendices A to G. The physical foundations and basic principles of quantum mechanics are essentially covered in this part. States, observables, uncertainty relations, time evolution, quantum measurements, pictures, representations, path integrals, inverse scattering, angular momentum, symmetries, are among the topics dealt with. Simple dynamics under one-dimensional and central potentials provide useful illustrations. Pertinent mathematical complements are included in the Appendices as well as a brief introduction to some polemical issues as the collapse of quantum states and hidden variables. Volume II comprises Chaps. 8 to 15. Collision theory, W.B.K. approximation, stationary perturbation theory and variational method, time-dependent perturbations, identical particles, and the quantum theory of radiation are presented. As applications the Stark and Zeeman effects, the atomic fine structure, the van der Waals forces, the BCS theory of superconductivity and the interaction of radiation with matter are discussed.

To our acknowledgements we wish to add a special one to G. García Alcaine for providing us with highly valuable bibliographic material for the updating of this book. We also thank Professor L. Alvarez-Gaumé for the interest he showed in translating the manuscript of the second Spanish edition. We are delighted with his excellent work. Finally, we would like to thank Springer-Verlag for all their help.

Madrid, May 1990   *A. Galindo, P. Pascual*

Preface to the Spanish Edition

Quantum mechanics is one of the basic pillars of physics. It is therefore not surprising that new results have been continuously accumulating, not only in a wide variety of applications, but also in the very foundations, which were laid down almost entirely in the spectacular burst of creative activity of the 1920s. Above all else, these efforts were, and are, intended to establish its axiomatic scheme and to make more rigorous many of the methods commonly employed.

Following the classic texts by Dirac and von Neumann, a large number of books on quantum mechanics and its mathematical and physical foundations have appeared, many of which are excellent. Our reasons for adding another one to the list arise from long teaching experience in the subject. First was the necessity of collecting in one book the clarification of many points which, in our opinion, remain obscure in traditional treatments. Second was the desire to incorporate important material which in general can only be found dispersed among scientific journals or in specialized monographs.

Given the huge quantity of information available, we decided to confine ourselves to what is normally termed "nonrelativistic quantum mechanics". The omission of the relativistic aspects of physical systems is due to our conviction that, for reasons of physics rather than textbook space, developments of relativistic quantum theory should be more than simply the last chapter in a textbook on quantum mechanics. Additionally, Chaps. 1 and 14 were included since it seemed appropriate to include a historical perspective of the genesis of quantum mechanics and its physical basis (albeit a short one since our physics majors have already had a course in quantum physics). Also, given that the atom plays a central role in quantum mechanics, we decided to use it to illustrate (Chap. 14) how one deals with a complex system using the approximation techniques developed in previous chapters. In addition to restricting the scope of the book, we sometimes found it necessary (with the length of the work in mind) to summarize or omit altogether some points of unquestionable interest, e.g., in Chap. 8 the study of dispersion relations, Regge poles, the Glauber approximation, etc.

A major problem which presented itself at the outset was that of choosing the mathematical level of the work. Aiming essentially at students of physics and related sciences, we adopted as a base the knowledge which these students acquire in normal undergraduate training. Thus, we started with the supposition of a certain familiarity with the language and at least the elementary techniques of Hilbert spaces, groups, etc., although at times our desire to make concepts more precise or to include (generally without proof) rigorous results have forced us to use more advanced mathematical terminology. With the aim of helping the reader who wishes to pursue such aspects, in Appendix C we summarize the relevant points and provide a pertinent bibliography.

The postulational basis of quantum mechanics is developed in Chap. 2 in the spirit of the orthodox "Copenhagen" interpretation, and the content of (and necessity for) each postulate is discussed. This approach is completed in Chap. 13 with the symmetrization principle for systems of identical particles. Other more controversial aspects, such as the problem of measurement and hidden variable theory, are discussed briefly in Appendices E and F.

In Chaps. 3 to 8 we develop the quantum principles, introducing some basic concepts with applications to simple systems, the majority of which are exactly soluble. Chapter 3 is devoted to the study of the wave function and its time evolution, and ends with a short introduction to Feynman's alternative formulation of quantum mechanics.

The properties of bound states in essentially one-dimensional systems are treated in Chaps. 4 and 6, the collision problem being included in Chap. 4 (scattering states for one-dimensional problems) and comprising all of Chap. 8. In Chap. 5 angular momentum is studied in detail and, given its practical importance, this study is completed in Appendix B with a summary of the most common formulae and tables of Clebsch-Gordan and Racah coefficients. Finally, Chap. 7 contains a discussion of symmetry transformations, the most important invariances of physical systems, and the associated conservation laws.

The dynamical complexity of the majority of interesting physical problems makes it necessary to resort to approximation methods in order to understand their behaviour. In Chaps. 9 and 10 the techniques appropriate to a discussion of stationary states are presented, reserving for Chap. 11 the usual methods for time-dependent perturbations. In Chap. 12 we study charged particles moving in electromagnetic fields, discuss the gauge-invariance of their dynamics and apply the above-mentioned perturbative methods to calculate the fine structure and the Zeeman effect in hydrogenic atoms.

The symmetrization principle is developed in Chap. 13, introducing second quantization formalism and applying it to many-body systems of identical particles displaying quantum behaviour at the macroscopic level (the phenomena of superfluidity and superconductivity). In Chap. 15, the last one of the book, we present a treatment, relatively complete, of the problem of the interaction of radiation with matter, with the necessary quantization of the radiation field.

The book concludes with a collection of appendices which, in addition to those mentioned above, include a summary of the most important special functions (Appendix A), elements of the theory of distributions and Fourier transforms (Appendix D), and properties of certain antiunitary operators (Appendix G). Contrary to the usual custom, we have not included problems or exercises. Our intent is to collect these in a forthcoming book, which will complement the present work in a practical sense.

The mathematical notation utilized in this work is the traditional one, although in some instances, to simplify the formulae, we have omitted symbols where there can be little room for confusion. Thus, for example, we use indistinguishably $\lambda I - A$ and $\lambda - A$. We also omit the limits or domains of integration on many definite integrals when they coincide with the natural ones. Concerning the numbering of equations, (3.37) or (X.37) indicates equation number 37 of Chap. 3 or Appendix X, respectively. Finally, [XY 57] is a reference to the article or work by a single author whose last name begins with XY, or of various authors whose last names start with X and Y, respectively; the number indicates the last two digits of the year in which the publication appeared. When necessary, we use [XY 57n] to distinguish between analogous cases.

To conclude, we wish to express, first and foremost, our gratitude to our families, whose understanding, sacrifice and support have permitted us to dedicate many hours to the realization of this book. We also wish to show our gratitude to various coworkers for their critical reviews of various chapters and appendices: L. Abellanas (Appendices C and D); R. F. Alvarez-Estrada (Chap. 8); R. Guardiola (Chap. 15); A. Morales (Chap. 7); A. F. Rañada (Chaps. 2 and 3); and C. Sánchez del Rio (Chaps. 1 and 14). Their suggestions were extremely valuable. G. García-Alcaine and M. A. Goñi read considerable portions of the original, and we also benefitted from their comments. Lastly, we appreciate Sra. C. Marcos' typing of a first draft of this work, and especially the patience and care with which Srta. M. A. Iglesias typed the text and formulae of the final version.

We do not wish to end this prologue without thanking the editorial staff of Alhambra, who were most cooperative in accepting our suggestions during the preparation of this book.

Contents

Physical Constants

Avogadro number	$N_A = 6.022\,045\,(31) \times 10^{23}$ particles/mol
Speed of light	$c = 2.997\,924\,58\,(1.2) \times 10^{10}$ cm s^{-1}
Proton charge	$|e| = 4.803\,242\,(14) \times 10^{-10}$ Fr
	$\quad = 1.602\,189\,2\,(46) \times 10^{-19}$ C
Reduced Planck constant	$\hbar = 6.582\,173\,(17) \times 10^{-22}$ MeV s
	$\quad = 1.054\,588\,7\,(57) \times 10^{-27}$ erg s
	$\hbar c = 1.973\,285\,8\,(51) \times 10^{-11}$ MeV cm
	$\quad = 197.328\,58\,(51)$ MeV fm
	$\quad = 0.624\,007\,8\,(16)$ GeV mb$^{1/2}$
Fine structure constant	$\alpha = e^2/\hbar c = 1/137.036\,04\,(11)$
Boltzmann constant	$k_B = 1.380\,662\,(44) \times 10^{-16}$ erg K^{-1}
	$\quad = 8.617\,35\,(28) \times 10^{-11}$ MeV K^{-1}
Electron mass	$m_e = 9.109\,534\,(47) \times 10^{-28}$ g
	$m_e c^2 = 0.511\,003\,4\,(14)$ MeV
Proton mass	$m_p = 1836.151\,52\,(70)\,m_e$
	$m_p c^2 = 938.279\,6\,(27)$ MeV
Electron Compton wavelength	$\lambda_e = h/m_e c = 2.426\,308\,9\,(40) \times 10^{-10}$ cm
Bohr radius	$a_{\infty Bohr} = 0.529\,177\,06\,(44)$ Å
Rydberg (energy)	$R_\infty = \frac{1}{2}m_e(\alpha c)^2 = 13.605\,804\,(36)$ eV
Rydberg (frequency)	$R_\infty = m_e(\alpha c)^2/4\pi\hbar = 3.289\,842\,(17) \times 10^{15}$ Hz
Rydberg (wave number)	$R_\infty = m_e(\alpha c)^2/4\pi\hbar c = 109\,737.32\,(56)$ cm^{-1}
Bohr magneton	$\mu_B = \hbar|e|/2m_e c = 0.578\,837\,85\,(95) \times 10^{-14}$ MeV G^{-1}
Nuclear magneton	$\mu_N = \hbar|e|/2m_p c = 3.152\,451\,5\,(53) \times 10^{-18}$ MeV G^{-1}
Gravitational constant	$G = 6.672\,0\,(41) \times 10^{-8}$ cm^3 g^{-1} s^{-2}
Gravitational fine structure constant	$\alpha_G = \dfrac{Gm_p^2}{\hbar c} = 5.904\,2\,(36) \times 10^{-39}$

Note: The values of these constants have been taken from [AC 84]. Numbers within brackets correspond to one standard deviation and affect the last digits. It should be remembered that in October 1983 the General Conference on Weights and Measures adopted a new definition for the meter, i.e. the length traversed by light in vacuum in a time $1/299\,792\,458$ s, and therefore the speed of light is now, by definition, $c = 299\,792.458$ km/s.

After the printing of the Spanish second edition new adjustments of the fundamental constants have appeared which incorporate additional measurements and the definition for the meter quoted above [PD88].

1. The Physical Basis of Quantum Mechanics

1.1 Introduction

Newtonian mechanics and Maxwell's theory of the electromagnetic (EM) field were the pillars of physics until the end of the nineteenth century. Subsequently however, a series of phenomena were discovered which were totally incomprehensible within the context of those theories. Their explanation resulted in a combination of new hypotheses which, over the years, crystallized into what now constitutes quantum mechanics (QM). In this chapter we will discuss some of these phenomena, with the aim of developing a feeling for the fundamental concepts of QM.

While in this chapter there are many references to original works, we do not claim to present a systematic historical study of these accomplishments. Readers interested in this aspect can consult the excellent book by *Jammer* [JA 66].

1.2 The Blackbody

The concept of a *blackbody* was introduced by *Kirchhoff* [KI 60], who defined it to be a system which absorbs all EM radiation incident upon it. By means of thermodynamic arguments, he proved that the energy density per unit volume and frequency, $u(\nu, T)$, of the EM radiation in the interior of an insulated cavity with adiathermic walls, maintained at a temperature T, is independent of the nature of the wall material; furthermore, its functional dependence on ν and T coincides with that of the spectral emissivity of a blackbody. *Wien* [WI 94], also using thermodynamic reasoning, was able to demonstrate the so-called *displacement law*:

$$u(\nu, T) = \nu^3 \phi(T/\nu) \ . \tag{1.1}$$

To determine the form of the function $\phi(T/\nu)$, *Wien* [WI 96] supposed, following *Michelson* [MI 87], that Maxwell's formula for the velocity distribution of molecules in an ideal gas was also applicable to the molecules of a solid which behaves as a blackbody; he further assumed that the wavelength and intensity of the radiation emitted by a molecule are completely determined by its velocity, thus enabling him to establish that

$$u(\nu, T) = \alpha \nu^3 e^{-\beta \nu/T} \ , \tag{1.2}$$

where α and β are universal constants. This formula is known as *Wien's law*. Even though we now know that the hypotheses which led to (1.2) are incorrect, the results obtained were able to account for the experimental data which existed at that time. In a series of new experiments, *Paschen* [PA 97] and *Paschen* and *Wanner* [PW 99] confirmed the validity of the above law for the range of frequencies corresponding to visible light ($\nu \sim 10^{15}$ s^{-1}) and for temperatures up to 4000 K. The first discrepancies between Wien's law and experimental results were found by *Lummer* and *Pringsheim* [LP 97], who, while studying radiation emission of wavelengths in the range 12–18 μm ($\nu \sim 10^{13}$ s^{-1}), discovered that Wien's law is violated as frequency is decreased.

A new attempt at obtaining $u(\nu, T)$ theoretically was carried out by *Lord Rayleigh* [RA 00]. He began by calculating the number $N(\nu)$ of normal modes per unit volume and frequency for EM radiation in a closed cavity, and found that $N(\nu) = 8\pi\nu^2/c^3$ (including the correction made later by *Jeans* [JE 05]). He further supposed that for each one of these normal modes, resembling harmonic oscillators, the average energy $\langle E(\nu, T) \rangle_{\text{cl}}$ could be calculated via Boltzmann statistics:

$$\langle E(\nu, T) \rangle_{\text{cl}} = \frac{\int_0^\infty dE\; E \exp\left(-E/k_\text{B}T\right)}{\int_0^\infty dE \exp\left(-E/k_\text{B}T\right)} = k_\text{B}T \,, \tag{1.3}$$

where k_B [$= 8.61735(28) \times 10^{-11}$ MeV K^{-1} $= 1.380662(44) \times 10^{-16}$ erg K^{-1}] is the Boltzmann constant. Combining both results one obtains the expression

$$u(\nu, T) = \frac{8\pi\nu^2 k_\text{B}T}{c^3} \,, \tag{1.4}$$

which we recognize today as the *Rayleigh-Jeans* law. This relation reproduced the experimental results well at low frequencies, and furthermore, it satisfactorily explained the new and careful measurements of *Rubens* and *Kurlbaum* [RK 01] in this regime. However, it is clear that the Rayleigh-Jeans law is inadequate for high frequencies, since it leads to an energy density per unit volume

$$u(T) = \int_0^\infty d\nu\; u(\nu, T) = \infty \,, \tag{1.5}$$

which is totally absurd (ultraviolet catastrophe!).

Thus at the beginning of this century, two expressions were available, one valid at high frequencies (Wien's law) and another for low frequencies (the Rayleigh-Jeans law); it was necessary to find an expression for $u(\nu, T)$ which contained both of these as limiting cases. The first step towards obtaining the correct formula was made by *Planck* [PL 00]. Making use of thermodynamic arguments and of an *ad hoc* assumption in order to come up with an interpolation formula, he arrived at the celebrated expression which is now known as *Planck's law*, and which in modern notation is written

$$\boxed{u(\nu, T) = \frac{8\pi\nu^2}{c^3} k_\text{B}T \frac{h\nu/k_\text{B}T}{\exp\left(h\nu/k_\text{B}T\right) - 1}} \tag{1.6}$$

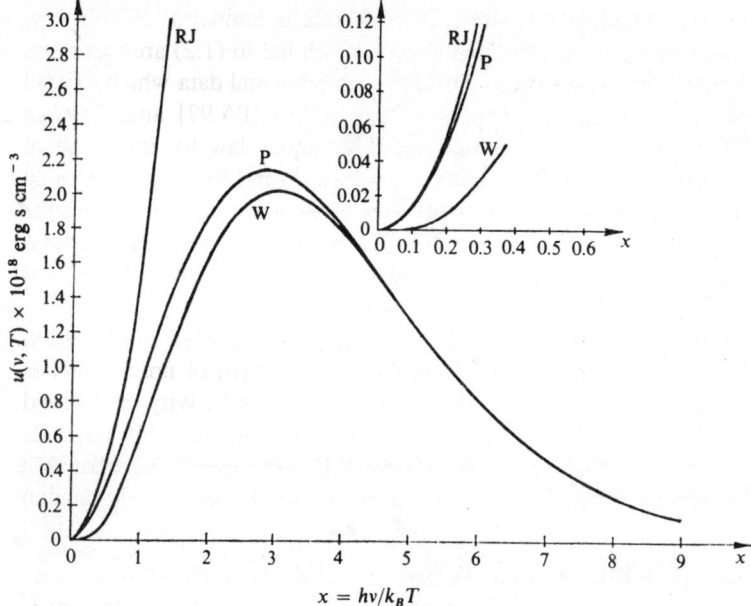

$$x = h\nu/k_B T$$

Fig. 1.1. A comparison of the Wien (W), Rayleigh-Jeans (RJ) and Planck (P) laws for $T = 300\,\mathrm{K}$

where h is *Planck's constant*. In practice it is more convenient to use the *reduced Planck's constant* $\hbar \equiv h/2\pi$ [= $6.582173(17) \times 10^{-22}\,\mathrm{MeV\,s} = 1.0545887(57) \times 10^{-27}\,\mathrm{erg\,s}$ experimentally], which plays a fundamental role throughout quantum mechanics. At appropriate values of T and ν, (1.6) approaches the Wien and Rayleigh-Jeans laws and a comparison of the three laws for $T = 300\,\mathrm{K}$ is given in Fig. 1.1.

The validity of Planck's law was firmly established by the measurements of *Rubens* [RU 00], so that it became imperative to find a justification for this fortunate interpolation formula ("*eine glückliche erratene Interpolationsformel*"). *Planck* [PL 01] himself provided an initial, tentative one, using the concept of entropy developed by Boltzmann. Later, in a communication [PL 12] to the Solvay Congress of 1911, he demonstrated how classical statistics should be modified to account for his law: It suffices to postulate that the energy E of each vibrational normal mode of the EM field in a cavity is not a continuous variable, but instead can only take on the values $E = n h\nu$, where ν is the frequency of the normal mode and n is a non-negative integer. In accord with this, the average energy of an oscillator would be

$$\langle E(\nu, T) \rangle_{\text{Planck}} = \frac{\sum_{n=0}^{\infty} n h\nu \, \exp\left(-n h\nu/k_B T\right)}{\sum_{n=0}^{\infty} \exp\left(-n h\nu/k_B T\right)}$$

$$= k_B T \frac{h\nu/k_B T}{\exp\left(h\nu/k_B T\right) - 1}, \tag{1.7}$$

instead of (1.3). Multiplying this quantity by $N(\nu)$, one obtains Planck's law.

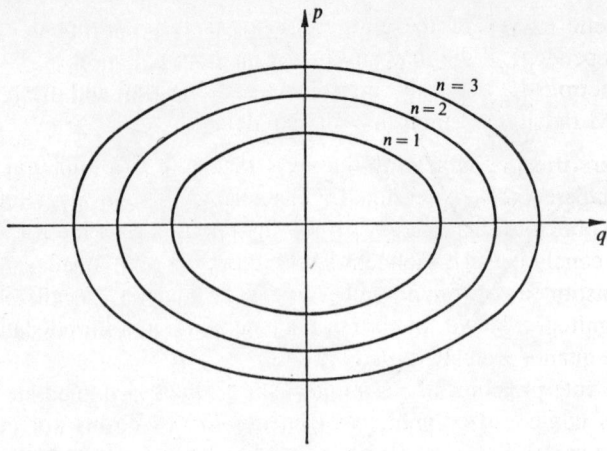

Fig. 1.2. Curves of constant energy $E = nh\nu$ in the (q, p) plane for the harmonic oscillator. The area between two consecutive ellipses is $h\nu$

In that same work [PL 12], *Planck* clearly established that classical dynamics, even together with relativity theory, constitutes a framework which is too narrow to explain microscopic phenomena. This led to an appeal to the effect that one should not seek any additional physical significance for h beyond that of an element of area in phase space. He observed that for a harmonic oscillator of frequency ν and energy $E = p^2/2m + 2\pi^2 m\nu^2 q^2$ the representative point in phase space (q, p) describes an ellipse centered at the origin with semi-axes $(E/2\pi^2 m\nu^2)^{1/2}$ and $(2mE)^{1/2}$, resulting in an area E/ν (Fig. 1.2). If one accepts that of all these ellipses, which form a continuum, the only ones which are physically allowed are those with area nh, then one obtains $E = nh\nu$. While this may seem to be an unimportant observation, we will see later that a generalization of this result led to the famous Sommerfeld-Wilson-Ishiwara quantization rules.

1.3 The Photoelectric Effect

During a period spanning the end of the last century and the beginning of the present one, the work of a large number of physicists, notably *Hertz* [HE 87], *Hallwachs* [HA 88], *Stoletow* [ST 88] and *Lenard* [LE 99], demonstrated the existence of the photoelectric effect: when a metal plate is illuminated by EM radiation, the metal emits electrons. These works empirically established the fundamental laws for this effect;

a) The number of electrons emitted is proportional to the intensity of the incident radiation.

b) For each metal there is a threshold frequency ν_0, such that for radiation of frequency $\nu < \nu_0$ no electrons are emitted.

c) The maximum kinetic energy of the emitted electrons is proportional to $(\nu - \nu_0)$ and is independent of the intensity of the incident radiation.

d) The emission of electrons is practically instantaneous, appearing and disappearing with the EM radiation without measurable delay.

Results (b) and (c) are irreconcilable with the wave theory of EM radiation.

Einstein [EI 05] compared the expression for the change of entropy with volume of EM radiation obeying Wien's law with the analogous expression for a system of particles. He concluded that monochromatic radiation of frequency ν behaves as if it were constituted of a finite number of energy "quanta", localized and independent, of magnitude $E = h\nu$. In current nomenclature, first introduced by *Lewis* [LE 26], these quanta are called *photons*.

With this result, the interpretation of the photoelectric effect is immediate. Light of frequency ν is composed of photons of energy $h\nu$. Electrons are in certain bound levels in a metal. Let $\phi_0(> 0)$ be the binding energy of the least-bound electrons, which evidently will depend on the kind of metal (it is customary to call ϕ_0 the work function of the metal; its value is typically a few electron volts). A photon incident in the metal can be absorbed by one of its electrons (only one, and total absorption if one accepts the indivisible character of the quanta). The energy of the photon is used, in part, to liberate the electron from its binding to the metal, and in part to provide kinetic energy. Thus, the maximum kinetic energy of the emitted electrons is $K_{max} = h\nu - \phi_0$, which corresponds to the removal of an electron from the least bound level. This makes evident the existence of a threshold frequency given by $h\nu_0 = \phi_0$.

Millikan [MI 16] was the first to confirm in detail this brilliant result of Einstein's, by using experiments on the photoelectric effect to measure Planck's constant directly, and finding for h a value coincident with that obtained from the analysis of blackbody radiation.

1.4 The Compton Effect

The phenomenon which perhaps played the most decisive role in convincing physicists at the turn of the century that light on some occasions displays corpuscular character was that which we now call the Compton effect, that is, the collision of X- or γ-rays with electrons. Even though the phenomenon was already known, *Compton* [CO 21] carried out the crucial experiments needed to understand it. In these, he demonstrated that upon illumination of a body with X-rays, some secondary radiation appeared in addition to the dispersed one. He proved that the wavelength of this secondary ray was independent of the material used in the experiment, depending only on the wavelength of the incident radiation and the angle between the incident and secondary rays. He also observed that the wavelength of the secondary radiation was always greater than or equal to that of the incident radiation.

Compton himself tried to explain this phenomenon within the framework of the wave theory of light but, finding this impossible, had recourse to a corpuscular

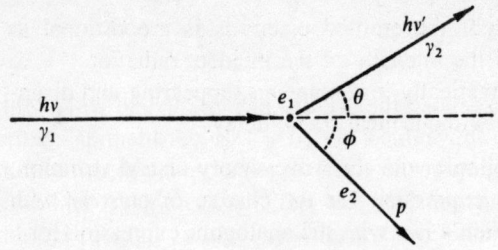

Fig. 1.3. A photon γ_1 incident on an electron e_1 at rest, creating a photon γ_2 and a recoil electron e_2

theory [CO 23]. Not only did he accept that light of frequency ν is composed of photons with energy E, but also that these photons have momentum p:

$$E = h\nu, \quad p = \frac{h\nu}{c} = \frac{h}{\lambda}, \tag{1.8}$$

the momentum being in the direction of the classical light ray. Inspection of the relativistic relation between energy and momentum, $E = (p^2c^2 + m^2c^4)^{1/2}$, shows that (1.8) is equivalent to supposing that photons are particles with zero rest-mass m_γ. (Experimental evidence confirms that $m_\gamma c^2 < 3 \times 10^{-33}$ MeV [AC 84].)

The energy of the photons in Compton's experiment was much larger than the binding energy of the electrons in the solid, so that the electrons can be considered to be at rest. Compton applied the energy-momentum conservation laws to the elastic electron-photon collision. With the notation of Fig. 1.3, and denoting the electron mass by m_e, these conservation laws lead to the relations

$$h\nu + m_e c^2 = h\nu' + \sqrt{m_e^2 c^4 + p^2 c^2}$$
$$\frac{h\nu}{c} = \frac{h\nu'}{c} \cos\theta + p \cos\phi \tag{1.9}$$
$$0 = \frac{h\nu'}{c} \sin\theta - p \sin\phi \,,$$

from which one can easily obtain

$$\Delta\lambda \equiv \lambda' - \lambda = \frac{h}{m_e c}(1 - \cos\theta) \,. \tag{1.10}$$

The constant $h/m_e c = 2.4263089(40) \times 10^{-10}$ cm is called the *Compton wavelength* of the electron. Excellent agreement with (1.10) was obtained in experiments using X-rays emitted by molybdenum and a graphite target.

1.5 Light: Particle or Wave?

For many years, physicists had associated the notions of particle and wave with being opposites; as a result, the above facts posed a serious dilemma in the conception of light. In order to interpret the classical experiments of diffraction and interference, it is necessary to consider EM radiation as waves, while an explanation of the photoelectric and Compton effects requires light to have a corpuscular character. Is light a wave or a particle? The answer to this question was first made possible in the second half of the 1920s, thanks primarily to the statistical interpretation of quantum mechanics developed from the fundamental work of *Born* [BO 26] and to the complementarity principle, announced in 1927 by *Bohr* [BO 28].

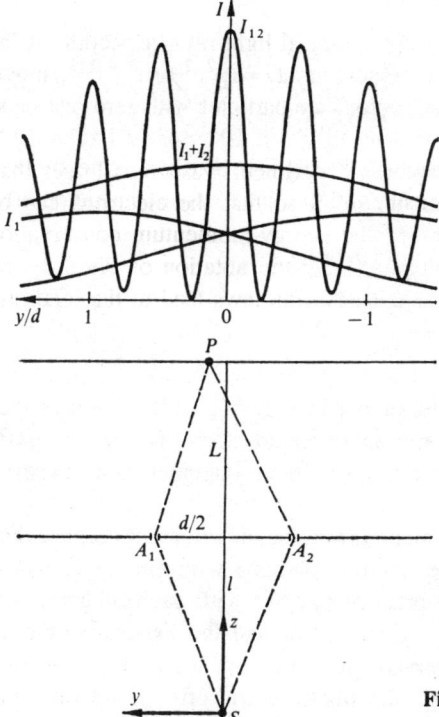

Fig. 1.4. Young's diffraction pattern (with $d^2/\lambda L = 2$, $a \ll d$)

In order to attempt to clarify the nature of light, let us consider the situation depicted in Fig. 1.4. A monochromatic light source, essentially a point source located at the origin of the coordinate system, emits light of wavelength λ which impinges on a metal screen located in the plane $z = l$. The screen has two circular openings, A_1 and A_2, each of radius a and with centers at the points $(0, d/2, l)$ and $(0, -d/2, l)$, respectively. The light travelling through the apertures is registered on a screen located at $z = L + l$. An arbitrary point P on this screen has the coordinates $(x, y, L+l)$. In order to simplify the discussion, let us suppose that $l = L \gg d$ and that $L\lambda \gg a^2$. With these assumptions [restricting our attention to those points P whose distance from the point $(0, 0, 2L)$ is at most of order d], it suffices to use the scalar theory of light (ignoring the polarization) and Kirchhoff's boundary conditions in the Fraunhofer approximation [BW 65]. The resulting amplitudes $\psi_1(P), \psi_2(P), \psi_{12}(P)$ of the wave at $P = (0, y, 2L)$ corresponding to the following three cases: (1), where only A_1 is open; (2), where only A_2 is open; and $(1, 2)$ where both are open, are given by the expressions

$$\psi_1(P) = Ae^{iks_1}\frac{J_1(x_1)}{x_1} \ , \quad \psi_2(P) = Ae^{iks_2}\frac{J_1(x_2)}{x_2} \ ,$$

$$\psi_{12}(P) = \psi_1(P) + \psi_2(P) \ , \tag{1.11}$$

where A is practically independent of the point P being considered. $J_1(x)$ is the Bessel function of order $1, k = 2\pi/\lambda$, and

$$s_1 = A_1 P = [L^2 + (y - d/2)^2]^{1/2} \ , \quad s_2 = A_2 P = [L^2 + (y + d/2)^2]^{1/2}$$

$$x_1 = \frac{2\pi\,da}{\lambda L}\left|1 - \frac{y}{d}\right| \ , \quad\quad\quad x_2 = \frac{2\pi\,da}{\lambda L}\left|1 + \frac{y}{d}\right| \ . \tag{1.12}$$

The intensity of the light which reaches P is given by the square of the modulus of the amplitude; thus, the intensities in the three cases considered are

$$I_1(P) = |A|^2\left|\frac{J_1(x_1)}{x_1}\right|^2 \ , \quad I_2(P) = |A|^2\left|\frac{J_1(x_2)}{x_2}\right|^2 \ ,$$

$$I_{12}(P) = I_1(P) + I_2(P) + 2\sqrt{I_1(P)I_2(P)}\ \cos\frac{2\pi\,dy}{\lambda L} \ . \tag{1.13}$$

Typical forms for these distributions are shown in Fig. 1.4, which displays the interference fringes that appear when both apertures are open. The results thus obtained using the wave theory of light are in excellent agreement with experimental results.

Let us now contemplate these phenomenon from the viewpoint of corpuscular theory; that is, suppose that the source S emits photons with energy $E = h\nu$ isotropically. These photons do not in practice interact with each other: if we reduce the intensity of the source by a given factor and the exposure time is increased by the same factor, the observed intensities do not change. On the other hand, experimental evidence (e.g., the photoelectric effect) supports the fact the the photon behaves as an indivisible particle.

Following classical reasoning, we should accept that photons have well-defined trajectories, and thus those reaching the recording screen have necessarily passed either through aperture A_1 or through aperture A_2. It is then completely impossible to explain how, when both apertures are open, there can be points on the recording screen for which fewer photons arrive than when only one is open, as happens experimentally.

On the other hand, individual photon counters are nowadays available, and it is possible, in principle, to substitute an array of such counters for the screen, and to make S sufficiently weak that, when a photon is emitted, the previous one has already arrived at the detectors. It can then be observed that the arrival of a photon is registered by a single counter, that successive arrivals are erratic (it being impossible to predict which counter will be triggered) and that the number of times which the counter located at P registers the arrival of a photon is proportional, for sufficiently large observation times, to the intensity $I_{12}(P)$, if both apertures are open. It will thus be observed that, while the photons arrive at the counters as localized entities, the probability distribution for the arrival of a photon at a point on the array is determined by the intensity distribution of the light wave. In this sense, one can say that a photon sometimes behaves as a wave and sometimes as a particle.

In this case the statistical interpretation of QM can be formulated by saying that the amplitude of the light wave, $\Psi(x; t)$, does not lead to a continuous spatial distribution of energy; instead, $|\Psi(x; t)|^2$ is a measure of the probability that at time t the photon can be found at the point x. This is how the unification of the wave and corpuscular aspects of light comes about. Neither alone furnishes a complete description of physical reality; instead, they are complementary aspects of a single reality, which become more or less evident according to the type of experiment being considered.

For another example of how experimental design can affect whether light manifests itself as a wave or as a particle, we return to Young's double slit experiment, this time adding two detectors to the experimental setup: one is located behind and close to A_1, the other similarly at A_2, and their purpose is to identify the slit through which a photon emitted from S passes. One finds in such a case that, upon forcing light to manifest its particle-like aspects in passing the slits, the experimental result is totally different. The number of photons arriving at the screen is then distributed in a fashion that more closely approximates $I_1 + I_2$ with increasing effectiveness of the detectors, and with the concomitant disappearance of the interference due to wave character.

From what we have said thus far, we can reach certain conclusions with respect to the behavior of light:

a) The probability of an event, such as the outcome of an experiment, is given by the square of the modulus of a complex number Ψ, which is called the "probability amplitude";
b) If an event can occur in various indistinguishable alternative forms, its probability amplitude is the sum of the probability amplitudes for each alternative;
c) If an experiment is performed in which it is possible to ascertain which of the alternatives has taken place, the probability of the event is the sum of the probabilities for these distinguishable alternatives.

This type of ideal "photon by photon" experiment is not easily realized. Nevertheless, a beautiful experiment has recently been carried out [GA 85], clearly demonstrating the interference between two different alternatives for a single photon, that is, the interference of a photon with itself.

1.6 Atomic Structure

The first atomic models appeared at the beginning of this century. Among them, one due to *Perrin* [PE 01] stands out, in which he imagined the atom as formed by a positively charged particle, around which enough electrons moved, planet-like, to compensate the total charge. A little later, *Nagaoka* [NA 04] proposed a more concrete model, in which the electrons were equidistant from the center of the atom and moved with a common angular velocity. Simultaneously, *Thomson's* model [TH 03] appeared, describing the atom as a positively charged sphere with the electrons oscillating around its center.

In 1911, *Rutherford* [RU 11] published his celebrated analysis of the collisions of alpha particles with thin gold and silver films. The experimental results favored, without any doubt, an atomic model in which all the positive charge, and practically all the mass, is concentrated in a very small volume of space constituting the nucleus, around which the electrons move in some fashion at distances large compared with the size of this nucleus.

The major difficulty of the Nagaoka-Rutherford model emerged when one tried to study its behaviour using Newtonian mechanics and Maxwell's EM theory. In the case of the hydrogen atom there is a nearly point-like nucleus of charge $|e|$, around which orbits an electron of charge $e = -|e|$. According to classical mechanics, if the electron's energy is $E < 0$, its relative motion with respect to the nucleus will describe an ellipse with the nucleus at one focus, and the frequency of motion is given by [GO 70] $\nu = (2|E|^3/\mu)^{1/2}(\pi e^2)^{-1}$, where μ is the reduced mass $(\mu^{-1} = m_e^{-1} + m_p^{-1})$ of the electron-proton pair. According to classical electrodynamics, the electron, being in accelerated motion, would emit radiation composed of a series of monochromatic waves of frequency equal to ν (or one of its harmonics). This process of radiating would be accompanied by a loss of energy by the electron and would result in the fall of the electron towards the nucleus in less than 10^{-10} seconds, producing a continuous spectrum of emission frequencies. In contrast, it is observed that, upon excitation of an atom, an emission spectrum of lines with well-defined frequencies appears, for which *Ritz* [RI 08] empirically discovered the *combination principle*, which can be stated as follows: the frequency of any spectral line of an element is expressible as a difference of two spectral terms. For hydrogen, the terms are of the form R/n^2, where R is a constant and n a positive integer; thus, the possible frequencies are

$$\nu_{m,n} = \frac{R}{n^2} - \frac{R}{m^2} , \quad m > n = 1, 2, \dots , \qquad (1.14)$$

a result which is impossible to understand classically. (In fact, Balmer had already proposed the relation (1.14) for hydrogen in 1885.)

Bohr [BO 13] was the first to give an explanation of the Ritz combination principle using the following hypotheses:

1) An atom has a discrete set of stationary states with energies $E_1, E_2, \dots$;

2) The emission and absorption of radiation by an atom does not occur in continuous form, as indicated by classical electrodynamics, taking place only when the atom goes from one stationary state to another;

3) In the transition from one stationary state E_i to another $E_j < E_i$ (or $E_j > E_i$), a photon of energy $h\nu = |E_i - E_j|$ is emitted (absorbed). (This formula, even though not exact, constitutes an excellent approximation in atoms, since the recoil of the atom absorbs only a negligible amount of that energy.);

4) An electron in a stationary state describes a circular orbit, with motion governed by the laws of classical mechanics; however, these laws are not valid in transitions from one stationary state to another;

5) The stationary orbits in an atom are determined by the condition that the orbital angular momentum of the electron should be a positive integer multiple of $\hbar = h/2\pi$.

For hydrogenic atoms, that is, those with nuclear charge $Z|e|$ and with only one electron of charge e, these postulates immediately yield the energies of the stationary states, since for circular orbits with angular momentum L the classical expression for the energy is $E = -\mu e^4 Z^2 / 2L^2$ [GO 70], where μ is the reduced mass of the system. As one assumes that $L = n\hbar$, $n = 1, 2, \ldots$, the energies of the stationary states are

$$\boxed{E_n = -\frac{1}{2}\mu(Z\alpha c)^2 \frac{1}{n^2}} \quad n = 1, 2, \ldots , \tag{1.15}$$

where $\alpha \equiv e^2/\hbar c = 1/137.03604(11)$ is the *fine structure constant*. The resulting radii of the orbits are

$$a_n = \frac{\hbar}{\mu(Z\alpha c)} n^2 . \tag{1.16}$$

Assuming an infinite mass for the proton, the quantity $a_{\infty Bohr}$, defined as $\hbar/m_e \alpha c$, would be the radius of the fundamental orbit ($n = 1$) of a hydrogen atom, with numerical value $a_{\infty Bohr} = 0.52917706(44)$ Å. Starting with (1.15) and making use of the third of Bohr's postulates, one obtains an expression for the frequencies:

$$\boxed{\nu_{m,n} = \frac{\mu(Z\alpha c)^2}{4\pi\hbar} \left(\frac{1}{n^2} - \frac{1}{m^2} \right)} \quad m > n = 1, 2, \ldots . \tag{1.17}$$

Quantitative agreement with the known data for the spectrum of the hydrogen atom was extraordinarily good, as it was for the few known lines of the ionized helium atom.

[Note that, although expression (1.15) continues to be valid in ordinary QM, the integer n, labelled the *principal quantum number*, has nothing to do with orbital angular momentum; in reality a state with principal quantum number n can have electrons with angular momenta $L/\hbar = 0, 1, 2, \ldots, n - 1$. This is a common abuse of terminology. Actually, what it means is that the modulus of the angular momentum is $[L(L + \hbar)]^{1/2}$, with $L/\hbar = 0, 1, 2, \ldots, n - 1$.]

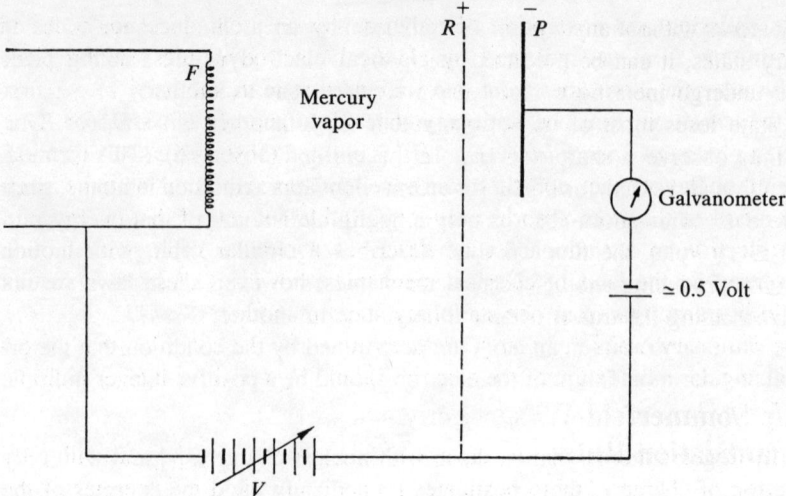

Fig. 1.5. The Franck-Hertz apparatus

This interpretation of atomic spectra established without a doubt, but in an indirect form, the existence in atoms of a series of stationary energy levels, in accord with Bohr's hypothesis. It was *Franck* and *Hertz* [FH 14] who, in a new set of experiments, provided a direct confirmation of their existence. The experimental layout used by Franck and Hertz is represented schematically in Fig. 1.5.

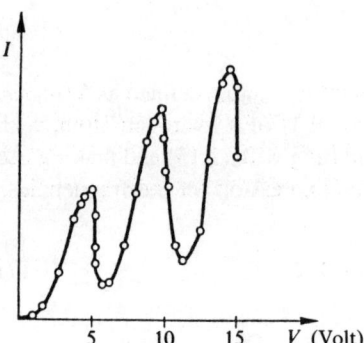

Fig. 1.6. The galvanometer signal as a function of V in the Franck-Hertz experiment

Mercury vapor at room temperature has virtually all of its atoms in the ground state: the average value of the kinetic energy is of the order of $k_B T \simeq (1/40)\,\text{eV}$, so that interatomic collisions cannot produce excitations, because the first excited state is 4.86 eV above the ground state. When V is near zero, the electrons coming from the hot filament F lack sufficient energy to reach the plate P, since they cannot overcome the potential difference between the grid and the plate. As V increases, more electrons arrive at P, many of which may have collided with mercury atoms without any significant transfer of energy. Given the existence of stationary states, it can be predicted that when $V \simeq 5$ volts, the electrons will begin to undergo inelastic collisions. An electron leaving an atom in the first excited state loses most of its energy, and is no longer able to reach P. One should thus observe a sharp decrease in the current in the vicinity of $V \simeq 5$ volts. If V is increased further, an electron will have enough energy to reach the plate even after an inelastic collision. Hence the current should rise again until $V \simeq 10$ volts, the first value at which the electrons could twice collide inelastically producing another decrease in the current. The experimental results thus corroborating the true existence of stationary states are shown in Fig. 1.6.

1.7 The Sommerfeld-Wilson-Ishiwara (SWI) Quantization Rules

One of the most important results of early quantum theory was the generalization of Bohr's theory carried out by *Sommerfeld* [SO 16], *Wilson* [WI 15] and *Ishiwara* [IS 15] for multi-periodic systems [GO 70]. A mechanical system described by a Hamiltonian is called multi-periodic if there exists a collection of generalized coordinates $(q_1, p_1, q_2, p_2, \ldots, q_n, p_n)$ such that, for each trajectory in phase space, p_i is a function only of q_i and in addition, the projection of the trajectory of the system onto each one of the $(q_i p_i)$ planes is a periodic curve (either in the sense of a libration or a rotation). One can then prove [GO 70] that the Hamiltonian for a multi-periodic system can be written as $H = H(J_1, J_2, \ldots, J_n)$, where the J_i are the so-called action variables defined as

$$J_i = \oint p_i dq_i \; . \tag{1.18}$$

The integral extends over a complete period of the projection of the motion of the system onto the $(q_i p_i)$ plane. The action variables are constants of motion and their conjugate variables w_i, called angle variables, obey the relations

$$\frac{dw_i}{dt} = \frac{\partial H(J_1, \ldots, J_n)}{\partial J_i} \equiv \nu_i(J_1, J_2, \ldots, J_n) \; , \tag{1.19}$$

ν_i being the frequency associated with the motion of the variables q_i and p_i.

For multi-periodic systems, SWI assumed that the action variables are quantized, and that the only possible values are

$$\boxed{J_i = n_i h} \; , \tag{1.20}$$

where the n_i are integers. In the theoretical justification of the quantization rules, the adiabatic invariance of the action variables played a prominent role [EH 13].

We now apply the SWI quantization rules to the problem of the hydrogenic atom. In spherical coordinates, the Hamiltonian of the system can be written as

$$H = \frac{1}{2\mu}\left[p_r^2 + \frac{1}{r^2}\left(p_\theta^2 + \frac{p_\phi^2}{\sin^2\theta}\right)\right] - \frac{Ze^2}{r} . \tag{1.21}$$

In this case, the constants of motion are the component L_z of angular momentum along the polar axis, usually chosen along the Oz axis; the square of the angular momentum vector $\boldsymbol{L}^2$; and the energy E. Thus

$$p_\phi = L_z , \quad p_\theta^2 + \frac{L_z^2}{\sin^2\theta} = L^2 , \quad \frac{1}{2\mu}\left(p_r^2 + \frac{L^2}{r^2}\right) - \frac{Ze^2}{r} = E , \tag{1.22}$$

thereby showing that each component of momentum is a function only of the corresponding coordinate. On the other hand, for the case $E < 0$ the trajectories are ellipses in the plane perpendicular to $\boldsymbol{L}$. The equation describing them is

$$\frac{1}{r} = \frac{\mu Ze^2}{L^2}(1 + \varepsilon \cos \psi) , \tag{1.23}$$

with eccentricity ε, semi-major axis a, and semi-minor axis b given by

$$\varepsilon = \sqrt{1 + \frac{2EL^2}{\mu Z^2 e^4}} , \quad a = \frac{Ze^2}{2|E|} , \quad b = \frac{L}{\sqrt{2\mu|E|}} \tag{1.24}$$

and the periodicity is

$$T = \pi Ze^2 \sqrt{\frac{\mu}{2|E|^3}} . \tag{1.25}$$

If the sense of circulation in the $(q_i p_i)$ plane is conveniently chosen, the action variables become [GO 70]

$$J_r = -2\pi\left(L - \frac{\mu Ze^2}{\sqrt{2\mu|E|}}\right) , \quad J_\theta = 2\pi(L - |L_z|) , \quad J_\phi = 2\pi|L_z| , \tag{1.26}$$

from which one obtains

$$|L_z| = \frac{1}{2\pi}J_\phi , \quad L = \frac{1}{2\pi}(J_\theta + J_\phi) , \quad E = -\frac{2\pi^2\mu Z^2 e^4}{(J_r + J_\theta + J_\phi)^2} . \tag{1.27}$$

Since the motion is periodic, the SWI quantization rules can be applied, and we get

$$J_r = n_r h , \quad J_\theta = n_\theta h , \quad J_\phi = n_\phi h , \tag{1.28}$$

where n_r, n_θ, n_ϕ are non-negative integers. The first, n_r, is called the *radial quantum number*. It is also customary to introduce the *principal quantum number* n, the *azimuthal quantum number* n_ψ, and the *magnetic quantum number* m, which are defined as

$$n = n_r + n_\theta + n_\phi , \quad n_\psi = n_\theta + n_\phi , \quad |m| = n_\phi \leq n_\psi . \tag{1.29}$$

In terms of these, the quantities characterizing the stationary orbits are

$$E = -\frac{1}{2}\mu(Z\alpha c)^2 \frac{1}{n^2} , \quad L = n_\psi \hbar , \quad L_z = m\hbar$$

$$\varepsilon = \sqrt{1 - \frac{n_\psi^2}{n^2}} , \quad a = \frac{\hbar}{\mu(Z\alpha c)}n^2 , \quad b = a\frac{n_\psi}{n} . \tag{1.30}$$

It should be noted first of all that the energy, as well as the angular momentum and its third component, end up being quantized. Classically, the total angular momentum cannot be zero if one wishes to avoid a head-on collision of the electron with the nucleus; thus, n_ψ and $n = n_r + n_\psi$ should be positive integers. In addition, the cosine of the angle α between L and the z-axis is given by $\cos\alpha = m/n_\psi$; it is thus quantized, a fact which is known as "space quantization". This space quantization is physically absurd, there being no external agent which singles out a privileged direction in space. The latter would occur, for example, if a uniform magnetic field were to act on the atom.

The lack of dependence of the energy on the magnetic quantum number m is an immediate consequence of the fact that the problem is invariant under the rotation group, i.e., that there is no preferred direction in space. This is of course true of all central potentials. On the other hand, the fact that the energy of a stationary state depends only on the principal quantum number and not on n_ψ, as might be expected in principle, is due to the particular form of the potential (*accidental degeneracy*).

The previous results, taken literally, imply that to each level associated with principal quantum number n there correspond n different orbits characterized by the values $n_\psi = 1, 2, \ldots, n$, and for each of these the orbit can be oriented in space in $(2n_\psi + 1)$ distinct forms corresponding to $m = -n_\psi, -n_\psi + 1, \ldots, n_\psi - 1, n_\psi$. Thus, the level under consideration is $n(n+2)$-fold degenerate. Nevertheless, we will see later that for this level the possible values of the angular momentum are in reality $L/\hbar = 0, 1, 2, \ldots, n-1$, and that for a given $L^* \equiv L/\hbar$ there are $(2L^* + 1)$ possible values of m, namely $m = -L^*, -L^* + 1, \ldots, L^* - 1, L^*$, so that the level is n^2-fold degenerate. (In fact, the electron possesses an internal degree of freedom, the spin, with two possible values, which when taken into account causes the level to be $2n^2$-fold degenerate.)

Actually, the small quantitative discrepancy between these results for large quantum numbers is not surprising when one keeps in mind that the SWI quantization rules lead to the *correspondence principle*, as we shall see in a moment. The latter is one of the concepts which played a prominent role in the development of early quantum theory and in the initiation of the modern theory.

The basis of the correspondence principle is the hypothesis that quantum theory, in its formalism, contains classical mechanics as a limiting case. The first manifestation of this principle appears in Planck's radiation theory, which converges to the classical Rayleigh-Jeans theory as the quantum of action h is made infinitesimally small. Although *Bohr* had already made veiled use of it in his theory of the hydrogen atom, his fundamental papers on the principle itself are much later works [BO 18], [BO 22], and his stance with respect to it changed during the course of his life [JA 66], [ME 65].

For a classical multi-periodic system, the frequencies ν_i are given by (1.19), and, according to classical electrodynamics, the system will radiate with frequencies

$$\nu_{cl} = \sum_{i=1}^{n} \tau_i \nu_i \ , \tag{1.31}$$

where τ_i are integers. Since the energy of the system is a continuous function $E(J_1, J_2, \ldots, J_n)$ of the action variables, the quantization of the energy levels comes about as a result of the fact that the action variables can only take on values belonging to a discrete set. If $E(J_1, J_2, \ldots, J_n) > E(J_1', J_2', \ldots, J_n')$ are the energies corresponding to two stationary states, the frequency associated with the transition from the first to the second is

$$\nu_{J,J'} = \frac{1}{h}[E(J_1, J_2, \ldots, J_n) - E(J_1', J_2', \ldots, J_n')] \ . \tag{1.32}$$

If the values of the J_i are very large and the action variables J, J' relatively close, the spacing between the energy levels is relatively small and one can write

$$\nu_{J,J'} \simeq \frac{1}{h} \sum_{i=1}^{n} \frac{\partial E(J_1, J_2, \ldots, J_n)}{\partial J_i}(J_i - J_i') = \frac{1}{h} \sum_{i=1}^{n} \nu_i(J_i - J_i') \ . \tag{1.33}$$

Now, using (1.20) we have $J_i - J_i' = \tau_i h$ (τ_i is an integer), and hence for large values of J_i the quantum frequency $\nu_{J,J'}$ acquires the classical form given in (1.31). This equality

$$\nu_{JJ'} \simeq \nu_{cl} \quad \text{if} \quad J_i, J_i' \gg h \ , \quad |J_i - J_i'| \ll J_i, J_i' \tag{1.34}$$

for the transition frequencies between large quantum numbers suggests that the quantum mechanical calculations of line intensities and polarizations should also coincide with their classical values in these same limiting situations. This is the essential content of the correspondence principle.

1.8 Fine Structure

One of the most spectacular results, and one which helped bring about the acceptance of the SWI quantization rules, was *Sommerfeld's* [SO 16], [SO 19] study of hydrogenic atoms within the framework of the theory of relativity. This study permitted him to give a theoretical explanation of the fine structure of the hydrogen atom discovered by *Michelson* [MI 91]. We do not know a Hamiltonian in relativistic mechanics which describes exactly the problem of two bodies interacting electromagnetically, and which depends uniquely on their variables. Nevertheless, an excellent approximation to the energy of relative motion of an electron of reduced mass μ and velocity v in the field of a nucleus of charge $Z|e|$, which incorporates the most important parts of the relativistic corrections and has an appropriate non-relativistic limit, is given by [BE 42]

$$E = \mu c^2 \left[\frac{1}{\sqrt{1 - v^2/c^2}} - 1 \right] - \frac{Ze^2}{r} . \tag{1.35}$$

If (r, ψ) are the polar coordinates in the plane of motion, their conjugate momenta are

$$p_r = \frac{\mu \dot{r}}{\sqrt{1 - v^2/c^2}} , \quad p_\psi = \frac{\mu r^2 \dot{\psi}}{\sqrt{1 - v^2/c^2}} . \tag{1.36}$$

Remembering that $v^2 = \dot{r}^2 + r^2 \dot{\psi}^2$, a little algebra permits us to write (1.35) as

$$p_r^2 + \frac{1}{r^2} p_\psi^2 = 2\mu \left(E + \frac{Ze^2}{r} \right) + \frac{1}{c^2} \left(E + \frac{Ze^2}{r} \right)^2 , \tag{1.37}$$

from which one can obtain the Hamilton-Jacobi equation via the substitutions $p_r = \partial W / \partial r$ and $p_\psi = \partial W / \partial \psi$. The equation thus obtained is separable, leads to orbits which are precessing ellipses, and the resulting action variables are

$$J_\psi = 2\pi L ,$$

$$J_r = \oint dr \sqrt{E \left(2\mu + \frac{E}{c^2} \right) + 2Ze^2 \left(\mu + \frac{E}{c^2} \right) \frac{1}{r} + \left(\frac{Z^2 e^4}{c^2} - L^2 \right) \frac{1}{r^2}} , \tag{1.38}$$

where L is the angular momentum. Evaluating this integral [GO 70], one obtains

$$J_\psi = 2\pi L , \quad J_r = -2\pi \left[\sqrt{L^2 - \frac{Z^2 e^4}{c^2}} - \frac{Ze^2(\mu + E/c^2)}{\sqrt{-E(2\mu + E/c^2)}} \right] . \tag{1.39}$$

The SWI quantization rules require $J_\psi = n_\psi h$ and $J_r = n_r h$, where $n_\psi > 0$ and $n_r \geq 0$ are integers. Thus, $L = n_\psi \hbar$ and

$$Z\alpha \left[\left(1 + \frac{E}{\mu c^2} \right)^{-2} - 1 \right]^{-1/2} = n_r + \sqrt{n_\psi^2 - Z^2 \alpha^2} . \tag{1.40}$$

Introducing the principal quantum number $n = n_r + n_\psi$ yields for the energies of the stationary states

$$E_{n n_\psi} = \mu c^2 \{ [1 + Z^2 \alpha^2 (n - n_\psi + \sqrt{n_\psi^2 - Z^2 \alpha^2})^{-2}]^{-1/2} - 1 \} . \tag{1.41}$$

Expanding in powers of $(Z\alpha)^2$, much less than unity in the case of interest, leads to

$$E_{n n_\psi} \simeq -\frac{1}{2} \mu (Z\alpha c)^2 \frac{1}{n^2} \left[1 + \frac{\alpha^2 Z^2}{n} \left(\frac{1}{n_\psi} - \frac{3}{4n} \right) \right] , \tag{1.42}$$

where n is a positive integer and $n_\psi = 1, 2, \ldots, n$. This equation not only permitted the explanation of the fine structure of hydrogen, but also agreed with the measurements of *Paschen* [PA 16] in ionized helium. The agreement of (1.42) with experiment is a bit fortuitous, since the correct formula is to be found using a relativistic wave equation (the Dirac equation) which correctly accounts for the spin of the electron. This leads to a formula identical to (1.42), but in which n_ψ has been replaced by $j + 1/2$, where j is the total angular momentum of the electron, with allowed values for level n given by $j = 1/2, 3/2, \ldots, n - 1/2$. Because of this, $j + 1/2$ takes on the same values as n_ψ, even if its physical significance is different. Moreover, the multiplicity of the levels in this relativistic model of Sommerfeld's is also not correct.

1.9 The Zeeman Effect

In 1897, *Zeeman* [ZE 97] discovered that the presence of external magnetic fields influences the process of emission of light by atoms. He observed that in a magnetic field of 32 kG the blue line of cadmium ($\lambda = 4800\,\text{Å}$) gave rise to a triplet of equidistant lines, the central one coinciding with the position of the original line. The same effect was subsequently observed in several other lines in zinc and cadmium, which showed no fine structure, and it was found that the separation between the multiplet components was proportional to B and independent of the atom under consideration. This effect, known as the "normal Zeeman effect", was explained classically by *Lorentz* [LO 97]. Near the end of 1897, *Preston* [PR 97] made it evident that there were cases for which the line splitting was different from that observed by Zeeman. Further experiments showed that in the presence of a magnetic field each one of the spectral lines due to fine structure gave rise to multiplets of various multiplicities; also, while the new lines were equidistant, the distance between them depended not only on the applied magnetic field, but on the line being considered as well. This new effect, labelled the "anomalous Zeeman effect", was not explainable via Lorentz's classical theory.

Let us now discuss the Zeeman effect for hydrogenic atoms using the SWI quantization rules. If B is a constant, uniform magnetic field directed along the Oz axis, we can take $A = (B \times r)/2$ as the vector potential. Ignoring the energy of interaction of the magnetic field with the nucleus ($m_{\text{nucleus}} \gg m_e$), as well as the effect of nuclear recoil on the magnetic interaction of the electron, the motion of the center of mass can be eliminated, and the Lagrangian which describes the relative motion is

$$L = \frac{\mu}{2}(\dot{r}^2 + r^2\dot{\theta}^2 + r^2\sin^2\theta\dot{\phi}^2) + \frac{Ze^2}{r} + \frac{eB}{2c}r^2\dot{\phi}\,\sin^2\theta\,. \tag{1.43}$$

From this we can deduce the conjugate momenta

$$p_r = \mu\dot{r}\,, \quad p_\theta = \mu r^2\dot{\theta}\,, \quad p_\phi = \mu r^2\sin^2\theta\left(\dot{\phi} + \frac{eB}{2\mu c}\right)\,, \tag{1.44}$$

and thus the Hamiltonian for the system is

$$H = \frac{1}{2\mu} \left(p_r^2 + \frac{1}{r^2} p_\theta^2 + \frac{1}{r^2 \sin^2 \theta} p_\phi^2 \right) - \frac{Ze^2}{r} - \frac{eB}{2\mu c} p_\phi + \frac{e^2 B^2}{8\mu c^2} r^2 \sin^2 \theta . \quad (1.45)$$

Taking into account that $r \sim a_{\infty\text{Bohr}}/Z$ and that $|p_\phi| \sim \hbar$, the relation between the magnitudes of the last two terms is approximately $|e|Ba_{\infty\text{Bohr}}^2/4\hbar c Z^2 \sim 10^{-10} B(\text{Gauss})/Z^2$. Since in practice $B \lesssim 10^5\,\text{G}$, the last term is negligible. (In quantum physics the angular momentum can be zero, in which case the last term of (1.45) is not negligible and contributes in an essential way to atomic diamagnetism.) As a result, the Hamilton-Jacobi equation is, to a very good approximation,

$$\frac{1}{2\mu} \left[\left(\frac{\partial W}{\partial r} \right)^2 + \frac{1}{r^2} \left(\frac{\partial W}{\partial \theta} \right)^2 + \frac{1}{r^2 \sin^2 \theta} \left(\frac{\partial W}{\partial \phi} \right)^2 - \frac{eB}{c} \left(\frac{\partial W}{\partial \phi} \right) \right]$$

$$- \frac{Ze^2}{r} = E . \quad (1.46)$$

This equation is separable, and the action variables are

$$J_r = -2\pi L + \frac{\pi \sqrt{2\mu} Z e^2}{\sqrt{-E - (eB/2\mu c)L_z}} , \quad (1.47)$$

$$J_\theta = 2\pi(L - |L_z|) , \quad J_\phi = 2\pi|L_z| ,$$

where L and L_z are the separation constants, viz., the magnitude of the angular momentum and its third component. Using the notation of (1.29), from the SWI quantization rules one obtains for the energy levels the expression

$$E_{nm} = -\frac{1}{2} \mu(Z\alpha c)^2 \frac{1}{n^2} - \frac{eB}{2m_e c} \hbar m , \quad (1.48)$$

where we have substituted m_e for μ in the last term. This is perfectly valid within the order of approximation required for (1.43). This term, which represents the correction to the energy levels due to the magnetic interaction, was to be expected physically. It is known that the electron has a magnetic moment due to its orbital motion, $\boldsymbol{\mu} = (e/2m_e c)\boldsymbol{\Lambda}$, with $\boldsymbol{\Lambda} \equiv \boldsymbol{r} \times (m_e \boldsymbol{v})$. Note that $\boldsymbol{\Lambda}$ is not necessarily the angular momentum $\boldsymbol{L} = \boldsymbol{r} \times \boldsymbol{p}$; instead they are related by $\boldsymbol{\Lambda} = \boldsymbol{L} - (e/2c)[r^2 \boldsymbol{B} - (\boldsymbol{r} \cdot \boldsymbol{B})\boldsymbol{r}]$. Because the projection of $\boldsymbol{L}$ on $\boldsymbol{B}$ is constant, the part $\boldsymbol{L}$ of $\boldsymbol{\Lambda}$ contributes only to the permanent orbital magnetic moment, while the corrective term in $\boldsymbol{\Lambda}$, proportional to $\boldsymbol{B}$, gives rise to an induced magnetic moment. We will see later (Chap. 12) how the interaction energy of this induced magnetic moment leads exactly to the diamagnetic term in (1.45). When $L \neq 0$, the induced magnetic moment is negligible compared to the permanent one, and the interaction energy is, to an excellent approximation, $-(e/2m_e c)\boldsymbol{L} \cdot \boldsymbol{B}$, thus justifying physically the last term in (1.48).

We should also note that the immersion of an atom in a magnetic field gives meaning to the spatial quantization of $\boldsymbol{L}$, since $\boldsymbol{B}$ determines a preferred direction with respect to which the orbital planes can acquire a quantized inclination. In fact, these planes are not motionless; L precesses around B with an angular frequency $\omega_L = |e|B/2m_ec$, called the Larmor frequency.

The frequency of the light emitted in a transition from the state (n, m) to the state (n', m') will be

$$\nu_{nm,n'm'} = \frac{\mu(Z\alpha c)^2}{4\pi\hbar}\left(\frac{1}{n'^2} - \frac{1}{n^2}\right) + \frac{eB}{4\pi m_ec}(m' - m)\,, \tag{1.49}$$

and the reduced frequency

$$\bar{\nu}_{nm,n'm'} \equiv \frac{\nu_{nm,n'm'}}{c}$$

$$\simeq \left[1.10 \times 10^5 Z^2 \left(\frac{1}{n'^2} - \frac{1}{n^2}\right) + 4.67 \times 10^{-5}(m' - m)B(\text{Gauss})\right]\ \text{cm}^{-1}\,. \tag{1.50}$$

While $(m' - m)$ can be any integer between $-(n+n')$ and $+(n+n')$, in practice the transitions are only important when $m' - m = 0, \pm1$. The reason for this is that the most intense spectral lines correspond to electric dipole transitions, for which it can be shown that there is a selection rule $|\Delta m| \leq 1$. This explains the triplets of the normal Zeeman effect. It is interesting to note that the term in (1.49) which depends upon B does not contain $\hbar$ and coincides with the expression obtained classically by *Lorentz* [LO 97].

No attempts within the framework of the old quantum theory were able to explain the existence of the anomalous Zeeman effect, until *Pauli* [PA 25] suggested the introduction of a fourth quantum number for describing the electron. This idea of Pauli's was immediately taken up by *Uhlenbeck* and *Goudsmit* [UG 25] in their theory of the electron spin.

Even though the Zeeman effect indirectly demonstrates the existence of a spatial quantization, it was convenient to have direct proof of this fact, which *Stern* and *Gerlach* [GS 22] succeeded in providing via the following experiment. A beam of silver atoms is produced by evaporating the metal in an oven, the beam being collimated by a series of apertures. This beam, which we assume to be directed along the Ox axis, enters a highly evacuated chamber in which an inhomogenous magnetic field B has been established. Because of the shape of the pole pieces which produce it, the field is basically directed along the Oz axis in the region traversed by the beam and also has a large gradient in the same direction. The silver atoms are paramagnetic and possess a permanent magnetic moment μ. In a magnetic field B, the center of mass of the atom is subjected to a force $F = \text{grad}(B \cdot \mu)$. Because of the precessional motion, μ_x and μ_y oscillate rapidly, while μ_z remains constant, so that, due to the oscillations, a force of magnitude $\mu_z \partial B/\partial z$ acts on the center of mass in the Oz direction on average. If L is the length of the magnetic field traversed by the beam, and K is the kinetic energy of the atoms, then at the end of the magnet an atom (whose trajectory is approximately parabolic) will have deviated from the direction Ox by an angle $\theta \simeq (\mu_z L/2K)\partial B/\partial z$. Thus, the deflection of the beam is proportional to μ_z.

If μ_z could take on any values in the interval $[-\mu, +\mu]$, the impacts on a screen perpendicular to the beam of atoms that had traversed the magnetic field would form a spot elongated in the Oz direction. In the case in question, two spots are observed precisely at the endpoints of the classically expected elongated spot. This demonstrates clearly the quantization of μ_z. Since, according to the Langevin theory of paramagnetic susceptibility, the permanent magnetic moment μ must be considered proportional to the angular moment L, the results of this experiment proved directly the quantization of the angular momentum.

It should be noted, however, that if the angular momentum were an integer multiple n_ψ of $\hbar$, as the SWI quantization rules indicate, then there would always be an odd number $(2n_\psi + 1)$ of spots and never an even number, as in the case of silver. This contradiction could only be overcome by introducing the spin.

1.10 Successes and Failures of the Old Quantum Theory

Up to this point, we have seen how the SWI quantization rules, applicable only to multi-periodic systems, allow the explanation of a number of experimental facts, such as the existence of stationary states, the discrete emission spectrum of atoms and the fine structure of hydrogen. Using these same quantization rules, *Schwarzschild* [SC 14] succeeded in explaining the Stark effect, that is, the influence of an external electric field on the energy levels of an atomic system. Finally, via the correspondence principle, one could obtain selection rules for the emission of radiation and methods for calculating the intensity of the spectral lines, as is described in *Sommerfeld*'s book [SO 19].

On the other hand, the old quantum theory predicted a quantization of the angular momentum which, while experimentally found, was not in quantitative agreement with the experimental data until the later introduction of spin permitted satisfactory explanation, not only of the anomalous Zeeman effect, but also of the fine structure of heavy atoms.

Even though the old quantum theory represented a large step forward in our understanding of the microcosmos, its structure was highly unsatisfactory; it was lacking in the two principles characteristic of all good theory: autonomy and self-consistency. In order to solve any problem, it was first necessary to solve it classically and then, using the correspondence principle in sometimes artificial ways, translate the result into quantum language. The theory was incapable of explaining the fundamental fact that there was no radiation from stationary states. Only multi-periodic systems could be quantized, and nothing could be done with systems in collision. It also contained typically classical concepts such as that of a trajectory. In the Franck-Hertz experiment, an electron, if it had a classical trajectory, should give up its energy continuously (not instantaneously in jumps). Furthermore, the old quantum theory shed no light whatsoever on the wave–particle character of EM radiation. Profound conceptual changes were necessary and these came during the years 1924–1930.

1.11 Matter Waves

In an attempt to clarify the wave–particle duality of EM radiation, *L. de Broglie* [BR 23] returned to a study of the analogies which exist between analytical mechanics and wave theories. Let us recall that the true trajectory of a particle of mass M, energy E, kinetic energy K, moving from point A to point B in the force field described in terms of a potential $V(r)$ is determined by Maupertuis' principle of least action [GO 70]

$$\delta \int_A^B 2K \, dt = \delta \int_A^B \sqrt{2M[E - V(r)]} ds = 0 \ . \tag{1.51}$$

The extremum refers to adjacent paths having the same total energy E, and where t and s are, respectively, the time and the path length along the particle trajectory. The magnitude of the local momentum which the particle would have at r if the energy were E is

$$p(r, E) = \sqrt{2M[E - V(r)]} \ ; \tag{1.52}$$

thus, the principle of minimum action can be written as

$$\delta \int_A^B p(r, E) ds = 0 \ , \quad E = \text{const} \ . \tag{1.53}$$

On the other hand, the path followed by a light ray of constant frequency ν passing through the points A and B is determined by Fermat's principle:

$$\delta \int_A^B \frac{ds}{v(r)} = 0 \ , \quad \nu = \text{const} \ , \tag{1.54}$$

where $v(r)$ is the velocity of light at the point r and is related to that in the vacuum, c, and to the index of refraction $n(r)$ by $v(r) = c/n(r)$. For this radiation of frequency ν, the wavelength is

$$\lambda(r, \nu) = \frac{v(r)}{\nu} \ , \tag{1.55}$$

and as a consequence another formulation of Fermat's principle is

$$\delta \int_A^B \frac{ds}{\lambda(r, \nu)} = 0 \ , \quad \nu = \text{const} \ . \tag{1.56}$$

Comparing the expressions (1.53) and (1.56), we see that the paths of particles in Newtonian dynamics can be identified with the rays of geometric optics if the wavelength is adjusted in such a way that the integrand in (1.56) is proportional to that of (1.53), i.e.,

$$p(r, E) = \frac{A}{\lambda(r, \nu)} \ , \tag{1.57}$$

where A is an arbitrary function of ν. As early as 1828, Hamilton noted this analogy between the two principles of Maupertuis and Fermat, and considered that this could be the foundation for a unifying formalism between optics and mechanics; however, the idea was then all but abandoned for nearly one hundred years.

We shall not pursue de Broglie's reasoning here; he used arguments based on relativity theory which are not to the point. In order to arrive at his fundamental results we need to merely repeat that an explanation of the Compton effect requires that a monochromatic light ray of wavelength λ is composed of photons of momentum $p = h/\lambda$. The comparison of this relation with (1.57) led him to hypothesize that any body with momentum p is accompanied by a wave which is inseparable from its motion and has a wavelength given by

$$\boxed{\lambda = \frac{h}{p}} \, . \tag{1.58}$$

As de Broglie himself remarked, this implies that under appropriate conditions any body can present wave behavior. A few years later, *Davisson* and *Germer* [DG 27] brilliantly confirmed de Broglie's prediction. With 54 eV electrons and a diffraction grating formed by a nickel crystal, they obtained a diffraction pattern corresponding to a wavelength $\lambda \simeq 1.65$ Å, in good agreement with (1.58), which gives $\lambda \simeq 1.67$ Å. An analogous experiment performed by *Thomson* and *Reid* [TR 27], in which they studied the diffraction of electrons by a crystalline film, also emphatically confirmed (1.58).

Various experiments have been carried out in the past decade using the new techniques of neutron interferometry. A perfect macroscopic silicon monocrystal is used as an interferometer; it is cut in such a way that the neutron beam encounters in its path three successive "ears", which are parallel to each other, and which stand up out of the crystal. The beam, upon impinging on the first ear, is divided through the Bragg effect into two beams, which after being focused by the second ear recombine coherently on the third. The separation between the two beams can be several centimeters; this facilitates external manipulation of each beam separately, if desired. The extent of the wave packet for each neutron is of order $1 \, \text{cm}^2$ in transverse cross-section and 0.01 cm in length, and the incident flux can be controlled in such a way that at any one time there is at most one neutron in the interferometer. Experiments formerly considered *gedanken* experiments have been made possible with this technique. The neutron double-slit experiments [SB 82] have confirmed beautifully the wave–particle duality for neutrons. The interested reader is urged to study the excellent review article by *Greenberger* [GR 83].

Returning to the hydrogen atom and considering circular orbits, the condition of quantization $2\pi r p = nh$ is equivalent to assuming that $2\pi r/\lambda = n$; namely, that the length of the circle should be an integer multiple of the wavelength of the electron, enabling the establishment of a stationary wave.

Relation (1.58) is the starting point for "wave mechanics". The first problem which presents itself is to determine which wave equation governs the evolution of the wave associated with the motion of a particle. This problem was solved by *Schrödinger* [SC 26], whose fundamental ideas we shall expound upon briefly. For simplicity, we consider the case of a particle of mass M upon which acts a force associated with a potential $V(r)$. Its Hamilton-Jacobi equation can be written as

$$\frac{\partial S(r;t)}{\partial t} + \frac{1}{2M}[\nabla S(r;t)]^2 + V(r) = 0 , \tag{1.59}$$

where $S(r;t)$ is called Hamilton's principal function. Since in this case the Hamiltonian does not depend explicitly on the time, this variable can be separated, and we can write $S(r;t) = W(r) - Et$, where the constant E is the total energy of the system, and where $W(r)$ is called Hamilton's characteristic function. Taking this into account,

$$|\nabla S(r;t)| = \sqrt{2M[E - V(r)]} . \tag{1.60}$$

Let us imagine the family of all surfaces of constant $S(r;t_0)$ in three dimensions at a given $t = t_0$, differentiated from each other by the value of S on the surface. As time elapses, the global properties of these surfaces do not change; the only change consists in having to provide a given surface with a different value of S. To be more precise, the surface which at t_0 has the value S_0 will, at the instant $t_0 + dt$, have the value $S_0 - E\, dt$. Alternatively, one can suppose that it is the surfaces themselves which are displaced, carrying the value of S; from this viewpoint, it is necessary that a point on the surface move in such a way that $\nabla S \cdot dn + (\partial S/\partial t)dt = 0$, where dn is the path element along the line described by a point on one of these surfaces, and which is chosen normal to the surface in question. Thus, the motion of these surfaces takes place with a normal velocity in the direction of ∇S given by

$$u(r) = \frac{E}{\sqrt{2M[E - V(r)]}} . \tag{1.61}$$

We thus see that the surfaces $S(r;t) = $ const can be considered as a collection of progressive wave surfaces in position space, and the phase velocity of this progressive motion satisfies (1.61) at each point.

Schrödinger adopted a wave amplitude, or *wave function*, associated with the motion of a particle and described precisely by

$$\Psi(r;t) = A(r) \exp[iS(r;t)/\hbar] = A(r) \exp[i(W(r) - Et)/\hbar] , \tag{1.62}$$

where $A(r)$ is a function to be determined. With this, supposing $A(r)$ to be real, one has achieved the coincidence of the wave fronts with the surfaces we have been discussing. Also, this guarantees that $\nu = E/h$, and that the wavelength is

$$\lambda(r) = \frac{u(r)}{\nu} = \frac{h}{\sqrt{2M[E - V(r)]}} , \tag{1.63}$$

in total accord with de Broglie's hypothesis. Taking into account that $E = h\nu$, the phase velocity of these waves can be written as

$$u(\boldsymbol{r}) = \frac{h\nu}{\sqrt{2M[h\nu - V(\boldsymbol{r})]}} \; , \qquad (1.64)$$

and thus the group velocity is

$$v(\boldsymbol{r}) = \frac{d\nu}{d(\nu/u(\boldsymbol{r}))} = \sqrt{\frac{2}{M}[E - V(\boldsymbol{r})]} \; , \qquad (1.65)$$

which coincides with the particle velocity.

This fact can be used to establish a more profound connection between the propagation of the wave and the motion of the representative point. One can attempt to construct a wave packet of small dimensions in each direction. Such a wave packet presumably would obey the same laws of motion as the moving representative point of the mechanical system. It would then be a substitute for the point image, insofar as its extent can be ignored, i.e., when its extent can be neglected compared to dimensions characteristic of the motion of the system and of its interaction. *Schrödinger* himself [SC 26] gave a method for linking his wave mechanics with ordinary mechanics. The true mechanical process is realized through – and is represented by – the wave process. The study of point motions, which is the object of classical mechanics, is therefore only an approximation. A macroscopic mechanical process may be represented by a wave–packet which (as time elapses) can for all practical purposes be considered a point compared to the natural dimensions of the system. This procedure ceases to have meaning when these dimensions are not large compared to the wavelength. In such a case, one must proceed in accord with wave theory, and not with the laws of ordinary classical mechanics, which would now be as useless as geometrical optics is for explaining an interference or diffraction pattern.

To determine the equation satisfied by $\psi(\boldsymbol{r})$, defined by

$$\Psi(\boldsymbol{r};t) = \psi(\boldsymbol{r})\mathrm{e}^{-\mathrm{i}Et/\hbar} \; , \qquad (1.66)$$

Schrödinger assumed that $\Psi(\boldsymbol{r};t)$ satisfied a typical wave equation

$$\Delta\Psi(\boldsymbol{r};t) - \frac{1}{u^2(\boldsymbol{r})}\frac{\partial^2}{\partial t^2}\Psi(\boldsymbol{r};t) = 0 \; . \qquad (1.67)$$

From this one immediately obtains

$$\boxed{-\frac{\hbar^2}{2M}\Delta\psi(\boldsymbol{r}) + V(\boldsymbol{r})\psi(\boldsymbol{r}) = E\psi(\boldsymbol{r})} \; , \qquad (1.68)$$

which today we call the *Schrödinger equation for stationary states*. The search for the quantum levels, which were previously obtained using the old SWI quantization, is now reduced to finding the eigenvalues of (1.68). To demonstrate the elegance and power of the new method, *Schrödinger* himself [SC 26] produced solutions for the hydrogen atom, the harmonic oscillator and various other problems. He also showed how one could develop a perturbation method for the above equation, when it became too complicated to be solved exactly.

In the last of his famous works, *Schrödinger* [SC 26] attacked the problem of generalizing (1.68). He observed that in this equation the value E of the energy of the stationary state appears explicitly, and thus this equation is not applicable when the states depend on the energy in a form different from that given in (1.66). If one multiplies (1.68) by $\exp(-iEt/\hbar)$ and uses the relation [implied by (1.66)]

$$E\Psi(r;t) = i\hbar\frac{\partial}{\partial t}\Psi(r;t) \tag{1.69}$$

to eliminate E, one obtains

$$-\frac{\hbar^2}{2M}\Delta\Psi(r;t) + V(r)\Psi(r;t) = i\hbar\frac{\partial}{\partial t}\Psi(r;t) \quad, \tag{1.70}$$

which he assumed to be also valid when $V = V(r;t)$; equation (1.70), with $V(r;t)$, is nowadays called the *time-dependent Schrödinger equation*. This equation has a form typical of a diffusion equation, but one in which the diffusion coefficient is imaginary. Thus, its solution requires the acceptance of a complex wave amplitude.

To obtain the Schrödinger equation for a system of N particles, one can proceed as follows: One starts with the classical Hamiltonian for the problem written in *cartesian coordinates* (the motivation for this restriction will be explained later in Sect. 2.12),

$$H = \sum_{j=1}^{N} \frac{1}{2M}p_j^2 + V(r_1,\ldots,r_N;t) \ . \tag{1.71}$$

Then we convert all terms of this expression into operators via the rules

$$p_j \rightarrow -i\hbar\nabla_j \ , \quad H \rightarrow +i\hbar\frac{\partial}{\partial t} \ , \quad r_j \rightarrow \text{ multiplication by } r_j \quad , \tag{1.72}$$

and we let these operators act on the wave function $\Psi(r_1,\ldots,r_N;t)$. The resulting Schrödinger equation is

$$i\hbar\frac{\partial}{\partial t}\Psi(r_1,\ldots,r_N;t)$$

$$= \left[-\sum_{j=1}^{N}\frac{\hbar^2}{2M_j}\Delta_j + V(r_1,\ldots,r_N;t)\right]\Psi(r_1,\ldots,r_N;t) \ . \tag{1.73}$$

Schrödinger [SC 26] also concerned himself with the physical meaning of the wave function, but at this point he failed, arguing that if e is the charge of the electron, then $e\Psi^*(r;t)\Psi(r;t)$ represents the charge density distribution of the electron. It was *Born* [BO 26] who correctly interpreted the physical meaning of the wave function by asserting that $|\Psi(r;t)|^2$ is the probability density that at the instant t the particle be found at the point r in space. In particular, if $\Psi_1(r;t)$ and $\Psi_2(r;t)$ are two wave functions which give rise to probability densities $P_1(r;t) = |\Psi_1(r;t)|^2$ and $P_2(r;t) = |\Psi_2(r;t)|^2$, respectively, then the probability density associated with a wave $\Psi(r;t) = \Psi_1(r;t) + \Psi_2(r;t)$, i.e., a superposition of the two, is given by

$$P(r;t) \propto |\Psi(r;t)|^2 = P_1(r;t) + P_2(r;t) + 2\text{Re}\{\Psi_1^*(r;t)\Psi_2(r;t)\} . \tag{1.74}$$

It is the presence of the final term which enables the explanation of interference and diffraction phenomena for photons as well as for massive particles.

1.12 Wave Packets

In order to see more clearly the relationship between wave mechanics and classical mechanics, we will analyze the particularly simple case of a free particle in one-dimensional motion. According to our previous assertions, the corresponding Schrödinger equation is

$$i\hbar \frac{\partial \Psi(x;t)}{\partial t} = -\frac{\hbar^2}{2M} \frac{\partial^2 \Psi(x;t)}{\partial x^2} . \tag{1.75}$$

The simplest solution of this equation is a plane monochromatic wave

$$\Psi(x;t) = A \exp[i(kx - \omega t)] , \tag{1.76}$$

where A is a constant and $\omega = \hbar k^2/2M$ gives the so-called dispersion relation for the medium. The frequency is given by $\nu = \omega/2\pi$, and the wavelength by $\lambda = 2\pi/|k|$. Its energy and momentum are $p = \hbar k$ and $E = \hbar\omega = p^2/2M$, in agreement with the classical relation between energy and momentum. Note that for this wave, whose momentum is well defined, $|\Psi(x;t)|^2 = |A|^2$, that is to say the particle has the same probability of being found at any point in space.

Packets of waves can be introduced to achieve better localization of the particle; these are defined by

$$\Psi(x;t) = \frac{1}{\sqrt{2\pi}} \int_{-\infty}^{+\infty} dk \, f(k) e^{i(kx - \omega t)} , \tag{1.77}$$

where $f(k)$ is a function which presents a sharp maximum at $k = k_0$, and which is concentrated within a small range Δk of this value. Because $\omega(k)$ is a quadratic function, we have

$$\omega(k) = \omega(k_0) + \left[\frac{d\omega(k)}{dk}\right]_{k=k_0} (k - k_0) + \frac{1}{2}\left[\frac{d^2\omega(k)}{dk^2}\right]_{k=k_0} (k - k_0)^2 . \tag{1.78}$$

So long as the times considered are such that

$$\left| \frac{t}{2} \left[\frac{d^2\omega(k)}{dk^2} \right]_{k=k_0} (k-k_0)^2 \right| = \left| \frac{\hbar t}{2M}(k-k_0)^2 \right| \ll 1 \tag{1.79}$$

for those values of k which are significant, the expression (1.77) can be approximated in the following way:

$$\Psi(x;t) \simeq \exp[\mathrm{i}(k_0 x - \omega_0 t)] \frac{1}{\sqrt{2\pi}} \int_{-\infty}^{+\infty} dk\, f(k) \exp[\mathrm{i}(k-k_0)(x-v_\mathrm{g}t)]$$

$$\equiv \exp[\mathrm{i}(k_0 x - \omega_0 t)]\phi(x - v_\mathrm{g}t) , \tag{1.80}$$

where $\omega_0 \equiv \omega(k_0)$ and the group velocity v_g is given by

$$v_\mathrm{g} = \left[\frac{d\omega(k)}{dk} \right]_{k=k_0} = \frac{\hbar k_0}{M} , \tag{1.81}$$

which in this case coincides with the velocity v_0 of a particle with momentum $p_0 = \hbar k_0$. The wave described by (1.80) corresponds to a plane monochromatic wave modulated by a function $\phi(x - v_\mathrm{g}t)$ which moves with velocity v_g; thus, in this approximation the wave packet moves without distortion.

To estimate the spatial extent of the packet, we reason in the following intuitive, if not rigorous, way: if $f(k) = |f(k)|\exp[\mathrm{i}\alpha(k)]$, then (1.77) can be written as

$$\Psi(x;t) = \frac{1}{\sqrt{2\pi}} \int_{-\infty}^{+\infty} dk\, |f(k)| e^{\mathrm{i}\phi(k;x,t)} , \tag{1.82}$$

where $\phi(k;x,t) \equiv kx - \omega t + \alpha(k)$. Those values of x and t which cause ϕ to vary substantially in the effective range of integration, i.e., in the interval $|k - k_0| \ll \Delta k/2$, contribute negligibly to $\Psi(x;t)$ due to the rapid oscillation of the exponential. On the other hand, at each instant t there exists a special point $x = x_0$ determined by the condition that $\phi(k;x,t)$ be stationary at k_0; that is,

$$x_0 = v_\mathrm{g}t - \left[\frac{d\alpha(k)}{dk} \right]_{k=k_0} . \tag{1.83}$$

This equality defines the so-called center of the packet, a point at which the cancellations are presumably minimal, and which moves with velocity v_g. Its characterization by means of (1.83) is essentially equivalent to taking x_0 as the average value of the probability density distribution $|\Psi(x;t)|^2$. Around this center, $\Psi(x;t)$ will be non-negligible only for those values of x for which ϕ changes by only a small amount (of order unity at most) in the interval Δk; this occurs when

$$\left| \frac{d\phi(k;x,t)}{dk} \right|_{k=k_0} \Delta k \lesssim 1 . \tag{1.84}$$

Since $[d\phi(k; x, t)dk]_{k=k_0} = x - v_g t + [d\alpha(k)/dk]_{k=k_0} = x - x_0$, the width of the packet satisfies the relation $\Delta x \cdot \Delta k \simeq 1$, or equivalently

$$\Delta x \cdot \Delta p \simeq \hbar \,. \tag{1.85}$$

This implies that if one attempts to define the position of a particle with a precision Δx, then its momentum can only be known with a precision $\Delta p \simeq \hbar/\Delta x$. This result is just a special case of the uncertainty principle, first stated by *Heisenberg* [HE 27]. Moreover, the order of magnitude of the time in which the packet traverses its own length is $\Delta t \simeq \Delta x/v_g \simeq M \Delta x/p_0$, and, the uncertainty in the energy of the particle being $\Delta E \simeq p_0 \Delta p_0/M$, we get

$$\boxed{\Delta E \cdot \Delta t \simeq \hbar} \,. \tag{1.86}$$

It is important to point out that, so long as the inequality (1.79) is obeyed, the wave packet does not deform as time elapses (save for an inessential phase); it simply accompanies its center, moving like a particle under the laws of classical mechanics. When (1.79) is violated, the packet deforms as it evolves, spreading asymptotically without limit (see Sect. 3.7), and the concept of the associated classical particle becomes devoid of meaning. [A proton with speed $v = c/30$ and with $\Delta k \simeq 10 \, \text{cm}^{-1}$, having an uncertainty in position of $\Delta x = 0.1 \, \text{cm}$, can travel a distance of 3.2×10^{10} cm before $|\hbar t|/2M(\Delta x)^2 \simeq 1$.]

1.13 Uncertainty Relations

Even though we will later give a rigorous proof of the uncertainty relations starting from first principles, this is a convenient point to comment on their physical meaning. We consider only the position-momentum uncertainty relations for now, given their importance. These are

$$\boxed{\Delta x \cdot \Delta p_x \simeq \hbar \,, \quad \Delta y \cdot \Delta p_y \simeq \hbar \,, \quad \Delta z \cdot \Delta p_z \simeq \hbar} \,, \tag{1.87}$$

which can be obtained in the same manner as was (1.85), generalizing the reasoning presented there to three-dimensional space; these imply that the more exactly one determines the position in a particular direction, the less well-defined will be the momentum in this direction, and vice versa.

Notice that the above relations refer to simultaneous knowledge of the position and momentum in a predetermined direction, but they do not at all restrict the precision with which position or momentum can be measured. If, for example, p_x for an electron is known exactly, its x position, due to (1.87), is totally undetermined. If it is then desired to localize its x position through some experimental procedure with a precision Δx, the uncertainty relations indicate that its momentum would be altered by an amount Δp_x which is unknown and not predictable, but such that $\Delta x \cdot \Delta p_x \simeq \hbar$.

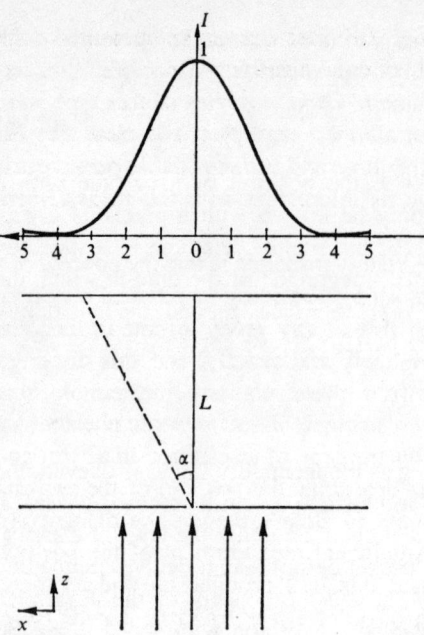

Fig. 1.7. Photon diffraction experiment

An experiment which would clarify this point can be imagined as follows: consider a screen in the $z = 0$ plane having a circular hole of radius $a/2$ that is centered at the origin. A beam of photons with momentum $\boldsymbol{p} = (0, 0, p_z)$ is incident on this screen (Fig. 1.7). Then, according to the laws of optics [BW 65], the intensity which is registered on a plate located at $z = L$ is given (in convenient units), under the assumption that $\lambda L \gg a^2$, by

$$I(x, y) = \left| \frac{2J_1(s)}{s} \right|^2 , \qquad (1.88)$$

where $s = \pi a \varrho / \lambda L$ and $\varrho \equiv (x^2 + y^2)^{1/2}$. This intensity distribution is also represented in Fig. 1.7. The photons which arrive at the recording plate had $\Delta x \simeq a$ when they passed through the hole and, as with all diffraction phenomena, they have some nonzero momentum p_x. According to (1.88), these photons are concentrated in a region of the photographic plate such that $\varrho \lesssim \lambda L/a$; this indicates that the momenta of the photons upon leaving the aperture were distributed within a cone whose opening angle has $\sin \alpha \simeq \varrho / L \simeq \lambda / a$. With this, the uncertainty in momentum is $\Delta p_x \simeq (h/\lambda)(\lambda/a) = h/a$, and so $\Delta x \cdot \Delta p_x \simeq h$. It is thus impossible, by this method, to measure the position and momentum with a precision greater than that permitted by the uncertainty relations.

Objections could be raised to this form of measurement: the uncertainty is due to the fact that the path taken by the photon is not known; one could think about determining it by making the screen mobile and measuring its recoil when the photon deflects. However, an analysis of this problem raises the question of how to determine exactly the position and momentum of the screen; one again encounters the condition that these variables cannot be measured with a precision greater than that permitted by the uncertainty principle, thus re-establishing that $\Delta x \cdot \Delta p_x \simeq h$ for the photon. Other analyses of this type were discussed by *Heisenberg* [HE 30], and in all such examples it is clear that the origin of the uncertainty relations lies in the finite and uncontrollable perturbation of the system being observed, caused by its interaction with the measurement apparatus.

An immediate consequence of the uncertainty principle is that the concept of a trajectory loses its meaning in an absolute sense; according to classical concepts, the determination of a trajectory implies that at any given instant in time the position and momentum are known completely and exactly, and this disagrees with the uncertainty relations. It follows from these relations, furthermore, that many concepts of classical physics have no analogue in microscopic phenomena. Thus, for example, it is impossible to follow the orbit of an electron in a hydrogen atom. To do this it would be necessary to determine the position of the electron at a number of successive instants in time. In Bohr's theory, the diameter of the orbit is of the order of 10^{-8} cm; a significant measurement of the position would require, say, $\Delta x \simeq 10^{-10}$ cm, and this will produce an undetermined change in the momentum of the electron of order $\Delta p \simeq \hbar/\Delta x \simeq 2 \times 10^5$ eV$/c$. Since the momentum of the electron in the atom is of the order of $p \simeq m_e \alpha c$, where α is the fine structure constant, the uncertainty caused in the energy would be $\Delta E \simeq p\Delta p/m_e \simeq 1 \times 10^3$ eV, which is the of the order of 100 times the ionization energy of the hydrogen atom. Thus, the first measurement would most likely ionize the atom, making it impossible to follow the electron's orbit.

The enormous impact which the uncertainty principle has had among physicists and nonphysicists alike is due to its connections with the concept of determinism. Classical determinism is the doctrine which affirms that if the present state of an isolated system is known, it is possible to predict exactly the results of any measurement on this system. Our foregoing results deny the possibility of such a prediction and therefore the basis for determinism.

Some physicists do not accept the nondeterministic character of QM and believe that the photon's arrival point on the plate in the experiment discussed above is totally determined by a certain set of conditions, even though we may not know exactly the conditions involved. In his discussion of this point, *Eisenbud* [EI 71] notes that macroscopic systems also display nondeterministic behavior when their internal structure is not taken into account. As an example of this, he considers a set of identical cubes; inside each one has been placed a gyroscope of different orientation and angular momentum. If the same torque were applied to each cube, it would be found that different cubes produce different responses to the applied force, and an observer who cannot examine the interior of the boxes could conclude that their behavior is nondeterministic.

It could be thought that the origin of the uncertainty principle in QM is similar to that encountered in Eisenbud's example above, and that, distinct from the currently recognized variables, new variables exist, "hidden variables", whose measurement would enable us to formulate a totally deterministic QM. In Appendix F we summarize some aspects of this question; for further details, consult [BE 73a], [CS 78], [PI 78], [ST 81], [AG 84]. The pioneering work of *Freedman* and *Clauser* [FC 72] deserves special attention in connection with this problem; strong experimental evidence against the existence of local hidden variables was deduced from its analysis. During the current decade, members of the Institute of Optics at Orsay, under the direction of *Aspect*, have carried out various elegant experiments with laser technology [AG 81], [AG 82], [AD 82]. These lead to spectacular agreement with quantum principles and to disagreements with respect to the predictions of local hidden variable models, which in one case exceeded 40 standard deviations. There being no experimental evidence in favor of hidden variables, we will ignore their possible existence and, following the "Copenhagen interpretation", we will accept that the wave function provides the maximum possible information about physical systems.

2. The Postulates of Quantum Mechanics

2.1 Introduction

In the previous chapter, we illustrated the failure of classical mechanics and the classical theory of the electromagnetic field to explain a variety of physical phenomena satisfactorily. In order to do so, a new theory emerged which is today known as quantum mechanics (QM).

We also noted that the new theory must possess some totally new characteristics, among which the following stand out: the introduction of a wave function having a probabilistic interpretation; the existence of uncertainty relations among certain dynamical variables which limit the precision of their simultaneous determination; and the discrete character of the values of some quantities, such as angular momenta or the energies of the bound states of the hydrogen atom.

In assessing early quantum theory, we have already indicated that it is absolutely necessary to have a complete revision of the basic concepts used in describing these newly discovered phenomena. Thanks to the prodigious contributions of scientists like Schrödinger, Heisenberg, Born, Dirac and many others, who established the physical basis of the new theory with such clarity, and also to the important mathematical foundations contributed by *von Neumann* [NE 32], we nowadays have at our disposal a framework within which it is possible, in principle, to pose and frequently to answer physical questions motivated by, e.g., atomic, molecular and nuclear systems, as well as by those macroscopic systems with ostensibly quantum behavior. This conceptual framework, which we will simply designate by QM, will be developed in this chapter from a set of fundamental postulates.

In attempting to carry out this task, two major types of difficulties appear. The first are of a mathematical nature; a rigorous development of QM implies the use of complicated mathematical methods (theory of distributions, spectral theory of unbounded operators, ...) which we intend to avoid as much as possible, even if this forces us at times (and to our regret) to be mathematically imprecise. Proceeding with complete mathematical rigor would most likely serve only to obscure the physical content of QM, requiring as it would substantial lengthening of the book, not to mention the fact that there exist important physical circumstances whose rigorous mathematical formulation has not yet been developed. The reader interested in the mathematical foundations of the discipline should consult one or more of the references [NE 32], [MA 63], [PR 71], [SI 71], [RS 72], [RS 75], [RS 78], [RS 79], [TH 81], [TH 83]. Appendix C contains a summary and a bibliography which should be adequate for most mathematical aspects of this, and subsequent, chapters.

The second type of difficulty is more delicate given that to date there does not exist a postulational basis for QM which is universally accepted, and that the most serious attempts along these lines are at a level which is beyond the level of this book. The reader interested in these aspects should consult, among other works, the references [JA 68], [ES 71], [ES 71a], [JA 74], [PI 76] and [WZ 83]. The difficulties are intimately related to the possible interpretation of QM. Here we shall adopt, in general, the so-called Copenhagen interpretation (due primarily to Niels Bohr), which is currently the most widely accepted.

2.2 Pure States

Whatever information we have about a given physical system arises from measurements made upon that system. We define a physical system to be a small part of the universe which is either isolated for all practical purposes, or has known and controllable external influences, and which is susceptible to experimental manipulation. After submitting the system to an appropriate series of experimental operations, it will be left in what we shall call a *state* of the system. Suppose we have at our disposal a collection of N identical systems, prepared under the same *relevant* experimental operations. After measuring a given physical quantity A of each of them, we obtain a set of values of A, not necessarily coincident, which give rise, in the limit as $N \to \infty$, to a probability distribution. It is precisely the collection of such probability distributions, one for each observable, which characterize the predictive aspects of the state of the system. Thus, two states, even if prepared by different operations, are considered the same if they lead to the same probability distributions for each observable.

Classically, it is in principle possible to prepare a system in which the positions and momenta of its particles are known with unlimited precision. The state of such a system is represented by a single point in phase space. In such a case, we can say that the state of the system is *pure*, our knowledge of the state being maximal. In particular, for such pure, identically prepared systems, the measurement of any observable always gives the same results. Nevertheless, for classical many-particle systems, limitations of a practical nature restrict such total control of the state. Only a few attributes of the system are generally fixed, and knowledge of its state suffices only to specify some extended region in phase space. We then speak of a *mixed state* (one for which we have incomplete information), and measurements of an observable on identically prepared systems will, in general, present some dispersion.

In attempting to do the same for a quantum system, we encounter difficulties: First, position and momentum are not simultaneously measurable with unlimited precision, due to the uncertainty relations; Second, in the measurement of one observable, as we saw in the previous chapter, the system-apparatus interaction can perturb the system uncontrollably in such a way that the measured value

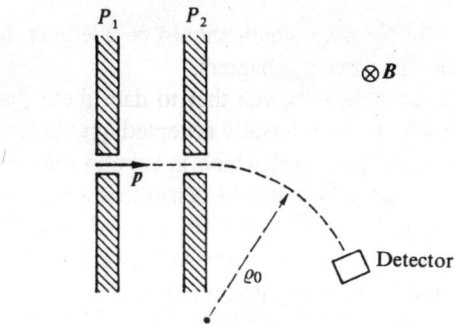

P_1 P_2

$\otimes B$

A

p

ϱ_0

Detector

Fig. 2.1. Experimental device for a measurement of the second kind of the momentum of a charged particle

is no longer solely attributable to the system, nor useful in inferring what the value is after the measurement. These considerations require us to distinguish clearly between two types of measurements: measurements of the *first kind*, or *preparations*, and those of the *second kind*. In order to clarify these ideas, we examine first how we could measure the momentum of a charged particle, and Fig. 2.1 shows schematically typical apparatus for doing this.

In a region A, a particle of known charge e (< 0) is produced which, after passing through the apertures in the screens P_1 and P_2, is deflected by a magnetic field B perpendicular to the plane of the diagram and is registered by the detector. The momentum of the particle upon leaving the region A is given by $|p| = |e|B\varrho_0/c$, and its direction is that indicated. Note that the result of the measurement gives us the possibility of calculating what the momentum of the particle was before its interaction with the detector. This, however, tells us nothing about the value of the momentum after detection has taken place, because at the instant of detection the momentum may change; it is even possible that the particle ceases to exist. This example illustrates a measurement of the *second kind*.

In contrast, we now consider a *preparation* which purports to produce a physical system in a definite state. Apparatus for preparing the same particles with definite momentum is shown schematically in Fig. 2.2. Particles proceeding from a region A pass through the apertures in screens P_1 and P_2. Between the screens P_2 and P_3 there is a magnetic field B perpendicular to the plane of the drawing. Particles which emerge from the aperture in P_3 have, at that point, a momentum p which can be determined from the geometry and the intensity B. Thus, particles can be prepared with any given momentum p by changing the position of the aperture in screen P_3 in a given way. Evidently, the preparation considered here is an idealization, because in practice the best that can be done is the preparation of a state having a momentum in some region $|p' - p| \leq \Delta p$, where Δp depends on the geometry of the system. Substantial gain in resolution can be achieved by using new screen collimators beyond P_3. In the limiting case in which $\Delta p/|p| \to 0$, with p fixed, we will have prepared the particle in a state with completely defined momentum.

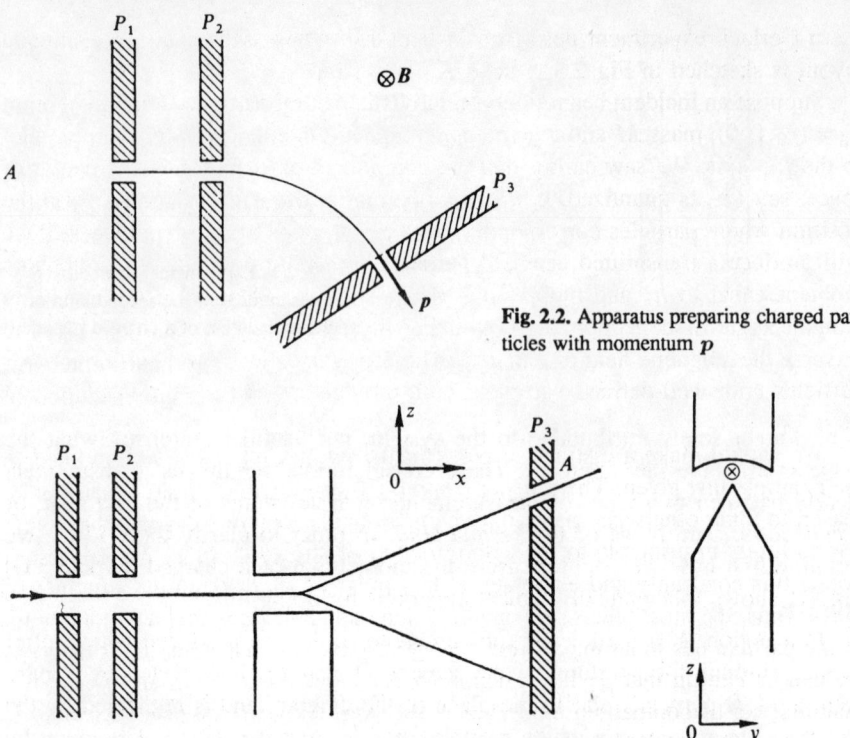

Fig. 2.2. Apparatus preparing charged particles with momentum p

Fig. 2.3. Geometry of the Stern-Gerlach experiment. The drawing on the right pictures the approximate form of the pole pieces for the electromagnet and the direction of the beam $\otimes$

Note that in QM not all physical attributes are *compatible*, that is, simultaneously measurable, which could result in perturbations of one set of measurements upon others. For example, it was shown in Chap. 1 that the position and momentum of one particle in one given direction of space are not compatible attributes; here we see that the three components of momentum for one particle are compatible. If, as we did above with p, we prepare a quantum system by specifying the largest number of independent compatible observables possible (*maximal preparation*), we obtain a state of the system about which we have the maximum amount of quantum information which it is possible to have. Such a quantum state is given the name *pure state*. Otherwise, when the preparation is not maximal, it is said to be a *mixed state*. But note that since not all quantum observables are compatible, unlike the classical case, quantum observables which necessarily have dispersion in their values exist, even when they are measured on pure quantum states. We cannot choose or enumerate *a priori* a maximal set of independent compatible attributes for a given system, since it is experiment which must supply this information. We shall return to this problem later. It is interesting, nevertheless, to note that under certain circumstances there are some quantities compatible with momentum. To this end, let us consider again the

Stern-Gerlach experiment described in Sect. 1.9, whose schematic experimental layout is sketched in Fig. 2.3.

Suppose an incident beam of non-relativistic neutral particles with momentum $p_0 = (p_0, 0, 0)$, mass M and a permanent magnetic moment μ is incident parallel to the Ox axis. We saw earlier that the component of μ in a given direction of space, say Oz, is quantized. If we now open an aperture in the screen P_3 at the position where particles corresponding to a given value of μ_z were detected, we will produce a transmitted beam of particles with third component of magnetic moment equal to μ_z and momentum $p = (p_0, 0, (\mu_z LM/p_0)\partial B/\partial z)$, assuming a small angle of deflection, LM/p_0 being the time which the particle takes to traverse the magnetic field of length L. Therefore, the above mechanism prepares particles with well-defined values of both momentum and the third component of μ.

We should make a distinction between the two meanings of well-defined in the example just given. There being theoretical justification and no experimental reason to think otherwise, p can only be regarded as a continuous variable (there are no limits in principle to the determination of any value of p; we will incorporate this continuity and elaborate on this point as we develop the structure of QM). Thus, the most precise statement which can be made is that the momentum of the particle lies in an interval $|p' - p| \leq \Delta p$. By contrast, since μ_z is discrete, we can be certain that μ_z has a definite value. Assuming that p and μ_z form a maximal set of compatible attributes of the particle in question, a pure state is the result of preparing the particle with given values of p and μ_z; these states are in fact idealizations of real situations in the limiting case $\Delta p/|p| \to 0$ and p constant.

In the construction of a physical theory, it is necessary to introduce a mathematical formalism, which generally consists of a set of primitive concepts, relations among these concepts, and dynamical laws, as well as a series of correspondence rules which relate the theoretical concepts of the mathematical formalism with the world of experience [PR 67], [TI 63]. Following this pattern, we begin the study of QM with the first of the basic postulates:

> *Postulate I.* The description of a physical system in QM is accomplished in terms of the elements of a separable complex Hilbert space associated with the physical system. At a given instant of time t, a pure state for the physical system is represented by a unit ray $|\Psi(t)\rangle_R$ in the corresponding Hilbert space. An element $|\Psi(t)\rangle$ of the ray $|\Psi(t)\rangle_R$ is called a *state vector* or *ket*.

(Readers unfamiliar with the mathematical concepts associated with Hilbert spaces and related topics of rays should consult Appendix C and the references given there. For heuristic purposes, and in the simplest of cases, we can think of a ray as the logical extension of the concept of a unit vector to include arbitrary phases, i.e., a bundle of unit vectors differing from each other only by a phase factor form a ray.)

Several questions naturally arise from this postulate, such as: Why the appearance of Hilbert spaces in QM? Why do these have to be separable and complex? Why do pure states correspond to unit rays? etc. In introducing Schrödinger's equation and Born's probabilistic interpretation of the wave function in Chap. 1, we saw that the structure of this equation leads to the use of a complex amplitude whose squared modulus $|\Psi(\boldsymbol{x};t)|^2$ should yield unity when integrated over all space since it represents a probability density. In other words, the wave function representing a state of a particle should be a unit vector in the separable complex Hilbert space $L^2(\mathbf{R}^3)$. On the other hand, there exist generalized mathematical schemes which analyze the measurement process and which lead logically to a Hilbert space structure over a rather arbitrary number field [JA 68], [JA 73], [PI 76]. Under certain plausible conditions, one can prove [GP 71] that this field should contain the real numbers, and in addition, if it is to be determined in a finite number of steps, it must consist of either the reals, the complex numbers or the quaternions, according to a classic theorem of Frobenius.

Quantum schemes based upon the reals and on quaternions have both been studied extensively and appear to be excluded by certain considerations ([ST 60], [SG 61], [SG 61a], [SG 62] and [EM 63]) since they lead to schemes equivalent to those based on the complex number field. Moreover, a hypothetical non-commutativity of the base number field leads to predictions which have been recently accessible to experimental verification [KG 84] and found conclusively wrong. A possible dynamics for quaternionic quantum systems is presented in [AD 85]. Finally, in Postulate III below, which details the statistical interpretation of the measurement process, we explain how two state-vectors from a given ray lead to the same probability distributions and therefore are equivalent representations of the state of the system. These considerations justify the content of Postulate I.

When the variables involved in the preparation of a state are discrete, selecting their values presents no difficulties, and if the set of such compatible discrete values is maximal, obtaining a pure state likewise presents no difficulty. The difficulty arises when one or more of them are of a continuous type, as occurs with the momentum of an electron. In this latter case, assurance that a given momentum lies in the neighborhood $|\boldsymbol{p} - \boldsymbol{p}_0| \leq \Delta p$ about the central value $\boldsymbol{p}_0$ still leaves much arbitrariness, since the Hilbert space of wave packets

$$\Psi(\boldsymbol{x};t) = (2\pi\hbar)^{-3/2} \int_{\mathbf{R}^3} d^3p \, \hat{\psi}(\boldsymbol{p}) \exp[\mathrm{i}(\boldsymbol{p} \cdot \boldsymbol{x} - E(\boldsymbol{p})t)/\hbar] , \qquad (2.1)$$

with $\boldsymbol{p} = \hbar\boldsymbol{k}$, $E(\boldsymbol{p}) = \hbar\omega(\boldsymbol{k})$, generated by functions $\hat{\psi}(\boldsymbol{p})$ localized in the region $|\boldsymbol{p} - \boldsymbol{p}_0| \leq \Delta p$ is of infinite dimension. Nevertheless, the process of repeated collimation, mentioned above in connection with Fig. 2.2, has an enormously smoothing influence on the details of the function $\hat{\psi}(\boldsymbol{p})$ associated with the resulting wave function, so that often in practical applications the explicit form of this function becomes irrelevant if Δp is sufficiently small (see Chap. 8). It is in this context that we may consider a state thus prepared as a pure state for all practical purposes. It is, in analogy to (1.80), described by the wave function

$$\Psi(\boldsymbol{x};t) = \exp[\mathrm{i}(\boldsymbol{p}_0 \cdot \boldsymbol{x} - E_0 t)/\hbar]\phi(\boldsymbol{x} - \boldsymbol{v}_g t) \,, \tag{2.2}$$

in which the modulating function ϕ is practically a constant in the space-time region of interest in a given problem. Outside this zone, ϕ must vary, since the square of $\Psi(\boldsymbol{x};t)$ must be integrable. But this variation is in general irrelevant in applications, and it is sometimes more useful in a given calculation to replace (2.2) by

$$\Psi(\boldsymbol{x};t) = \mathrm{const} \times \exp[\mathrm{i}(\boldsymbol{p}_0 \cdot \boldsymbol{x} - E_0 t)/\hbar] \,. \tag{2.3}$$

This wave function corresponds to the idealization $\Delta p \to 0$ and does *not* belong to the Hilbert space $L^2(\mathbb{R}^3)$.

The inclusion of state vectors outside the Hilbert space was foreshadowed by *Dirac* in his famous 1930 book [DI 58], and is widely used today. We should not forget, however, its idealized character and the fact that such states do not represent physically realizable states. There exist mathematically rigorous frameworks (the theory of rigged Hilbert spaces [GC 62]) in which these kets are allowed, though not with the operational ease which is frequently desired in practical applications.

By fixing a maximal set of compatible magnitudes, one can prepare the system in certain pure states which, as we shall see later, are represented by mutually orthogonal kets forming a complete orthonormal basis for the Hilbert space of the system. If any other maximal set is chosen for the preparation of pure states, these latter will be represented by a linear combination of the previous kets. This does not invariably mean that any unit vector in this Hilbert space represents a pure state of the system. This is only the case if there do not exist superselection rules which restrict the validity of the superposition principle.

Finally, we establish our notation for some operations with elements of the Hilbert space: if $|\psi_1\rangle$, $|\psi_2\rangle \in \mathcal{H}$, ($\mathcal{H}$ is the Hilbert space of the system), we will write their scalar product as $\langle\psi_2|\psi_1\rangle$, with the convention that it is linear in the first (right) entry and antilinear in the second (left) one. That is,

$$\langle\psi_2|\alpha\psi + \beta\psi'\rangle = \alpha\langle\psi_2|\psi\rangle + \beta\langle\psi_2|\psi'\rangle \,, \tag{2.4a}$$

where α and β are arbitrary complex numbers, while antilinearity results from

$$\langle\psi_2|\psi_1\rangle = \langle\psi_1|\psi_2\rangle^* \,. \tag{2.4b}$$

The (norm)^2, $\langle\psi|\psi\rangle$, of the ket $|\psi\rangle$ will frequently be denoted by $\|\psi\|^2$. If A is a linear operator, we will alternately denote the scalar product of $A|\psi_1\rangle$ with another ket $|\psi_2\rangle$ by $\langle\psi_2|A|\psi_1\rangle$ or by $\langle\psi_2, A\psi_1\rangle$. When the adjoint of A exists, we will denote it by $A^\dagger$. The symbol $\langle\psi|$ is usually called the *bra* associated with the ket $|\psi\rangle$ and represents a continuous linear functional of the kets:

$$\langle\psi| : |\psi_1\rangle \to \langle\psi|\psi_1\rangle \,. \tag{2.5}$$

That every such continuous functional can be associated with a ket is a consequence of the Riesz-Fréchet theorem, which states that the map of $|\psi\rangle$ onto

$\langle\psi|$ is one-to-one and antilinear. We will later formally extend the preceding notation and the ket $\leftrightarrow$ bra correspondence to those idealized kets discussed above which lie outside the Hilbert space.

2.3 Observables

In the mathematical formalism of QM it is necessary to introduce elements representing the possibles states of the physical system and its observable quantities (also simply called *observables*), these latter being physical attributes of the system that are measurable with an apparatus. (Note, for example, that in this sense the entropy, which is a measure of the missing information on the state of the system, is not an observable.) This correspondence for the states was established in Postulate I, and now Postulate II contains the mathematical characterization of the observables.

> *Postulate II.* Every observable of a physical system is represented in the mathematical formalism of QM by a linear self-adjoint operator which acts in the Hilbert space (specified in Postulate I) associated with the physical system being considered.

The justification of this postulate will be delayed until we discuss the statistical interpretation of measurement in the context of the formalism. Note that Postulate II does not assert that all self-adjoint operators represent some observable. We will discuss this point in connection with superselection rules.

This is an appropriate point to pause for a mathematical interlude, in order to display a little of the sophistication required in finding appropriate mathematical models for the behavior of physical systems, before continuing the analysis of the various postulates. Examples abound in physics in which small deviations from existing models have forced an entire reformulation of the conceptual structure of a sub-discipline. Some readers may wish just to skim through the material in this section. It should still be possible to appreciate that for important operators in quantum mechanics, the question of determining their self-adjointness is rather delicate. Furthermore, this interlude also gives us an opportunity to explore von Neumann's spectral decomposition of self-adjoint operators, a powerful tool which extends the well-known diagonalization procedure for Hermitian matrices in linear algebra to the general Hilbert space setting. From a mathematical point of view, it can be said that the spectral method lies at the very heart of QM, corresponding directly to the mathematical meaning of "the measured values of an observable". The reader should be warned, however, that it is important to have simple finite-dimensional linear algebra examples solidly in hand from previous elementary studies of quantum physics in order to see where the formal development is going.

Postulate II does not require that an operator A representing an observable be bounded; it will be seen that in general this property is not satisfied. As shown in Appendix C, for unbounded operators it is essential to distinguish between symmetric operators, $A \subseteq A^\dagger$, and self-adjoint operators $A = A^\dagger$. To clarify this distinction practically, we will discuss two operators which play important roles in QM. These also serve to introduce the formalism in a context with which the reader will have had some experience using ordinary methods.

Let the operator X be defined in $\mathcal{H} = L^2(\mathbb{R})$ by

$$(X\psi)(x) = x\psi(x) \tag{2.6}$$

with domain

$$D(X) = \left\{ \psi \in \mathcal{H} : \int_{-\infty}^{+\infty} dx \; x^2 |\psi(x)|^2 < \infty \right\} , \tag{2.7}$$

which is dense in $\mathcal{H}$. The operator X is symmetric ($X \subseteq X^\dagger$), since

$$\langle \psi | X | \varphi \rangle = \int_{-\infty}^{+\infty} dx \; x\psi^*(x)\varphi(x) = \langle X\psi | \varphi \rangle \tag{2.8}$$

is satisfied for any pair of functions $\psi, \varphi \in D(X)$. One way of establishing that X is also self-adjoint is to show that $D(X^\dagger) \subseteq D(X)$. To do this, consider a function $\psi \in D(X^\dagger)$, and define $\psi_1 = X^\dagger \psi$. Then

$$\langle \psi | X | \varphi \rangle = \langle \psi_1 | \varphi \rangle , \quad \forall \varphi \in D(X) ; \tag{2.9}$$

therefore

$$\int_{-\infty}^{+\infty} dx \; [x\psi^*(x) - \psi_1^*(x)]\varphi(x) = 0 , \quad \forall \varphi \in D(X) . \tag{2.10}$$

Since $D(X)$ is dense in $L^2(\mathbb{R})$, (2.10) requires that $\psi_1(x) = x\psi(x)$ almost everywhere. Hence $\psi \in D(X)$ and $\psi_1 = X\psi$, completing the demonstration of the self-adjointness of X.

We next consider the operator P, defined in $\mathcal{H} = L^2(\mathbb{R})$ by

$$(P\psi)(x) = -i\frac{d\psi(x)}{dx} \tag{2.11}$$

with domain

$$D(P) = \left\{ \psi \in \mathcal{H} : \psi \text{ absolutely continuous}, \int_{-\infty}^{+\infty} dx \left| \frac{d\psi(x)}{dx} \right|^2 < \infty \right\} . \tag{2.12}$$

Because $D(P)$ contains the set $C_0^\infty(\mathbb{R})$ of infinitely differentiable functions with compact support, it is dense in $\mathcal{H}$. To see that P is not bounded, it suffices to consider its effect on the functions $\psi_n(x) = \exp(-nx^2)$. P is symmetric ($P \subseteq P^\dagger$), since for all functions $\psi, \varphi \in D(P)$, integration by parts shows that

$$\langle \psi | P | \varphi \rangle = \int_{-\infty}^{+\infty} dx \ \psi^*(x) \left(-i \frac{d\varphi(x)}{dx} \right)$$

$$= \int_{-\infty}^{+\infty} dx \left(-i \frac{d\psi(x)}{dx} \right)^* \varphi(x) = \langle P\psi | \varphi \rangle , \qquad (2.13)$$

where we have used the vanishing of ψ, φ at the limits of the interval of integration $(-\infty, +\infty)$ implied by the definition of $D(P)$. Again, to show that P is self-adjoint we merely need to demonstrate that $D(P^\dagger) \subseteq D(P)$. To this end, consider a function $\psi \in D(P^\dagger)$ and define $\psi_1 = P^\dagger \psi$; then

$$\langle \psi | P | \varphi \rangle = \langle \psi_1 | \varphi \rangle , \quad \forall \varphi \in D(P) \qquad (2.14)$$

can be rewritten as

$$\langle \psi | P | \varphi \rangle = \int_{-\infty}^{+\infty} dx \ \psi_1^*(x) \varphi(x)$$

$$= i \int_{-\infty}^{+\infty} dx \left[\frac{d}{dx} \left(i \int_0^x dt \ \psi_1(t) + c \right)^* \right] \varphi(x) , \qquad (2.15)$$

where c is an arbitrary constant. Choosing $\varphi \in C_0^\infty$, integrating by parts, and taking into account that φ is zero outside a finite interval, we obtain

$$\int_{-\infty}^{+\infty} dx \left(\psi(x) - i \int_0^x dt \ \psi_1(t) - c \right)^* \left(-i \frac{d\varphi(x)}{dx} \right) = 0 , \ \forall \varphi \in C_0^\infty. \quad (2.16)$$

Since C_0^∞ is dense in $L^2(\mathbf{R})$, the first factor of the integrand in (2.16) must be a constant and hence, with a convenient choice c_0 for c, we can write almost everywhere

$$\psi(x) = c_0 + i \int_0^x dt \ \psi_1(t) , \qquad (2.17)$$

from which we identify $\psi_1(x) = -i \ d\psi(x)/dx$. We have thus established that $D(P^\dagger) \subseteq D(P)$ and that $P = P^\dagger$.

Note that the situation is totally different if we take the Hilbert space $\mathcal{H} = L^2(0, 2\pi)$. Then the definition (2.11) still holds, but the domain is now

$$D(P) = \left\{ \psi \in \mathcal{H} : \psi \text{ absolutely continuous}, \right.$$

$$\left. \int_0^{2\pi} dx \left| \frac{d\psi(x)}{dx} \right|^2 < \infty, \ \psi(0) = \psi(2\pi) = 0 \right\} , \qquad (2.18)$$

so that $D(P)$ is again dense and compact in $L^2(0, 2\pi)$ and the demonstration that $P \subseteq P^\dagger$ follows as before. Similarly one can prove that if $\psi \in D(P^\dagger)$, this function is absolutely continuous, but does not necessarily satisfy the boundary conditions $\psi(0) = \psi(2\pi) = 0$, so that $P \neq P^\dagger$. We see that even though P is symmetric for this case, it is not a self-adjoint operator. Furthermore one can show that P is not even essentially self-adjoint since it is already closed.

These considerations naturally bring up the question whether we may have restricted the initially selected domain $D(P)$ too much, resulting in P not being self-adjoint. This is in effect the case. The equation $P^\dagger \psi = \pm i\psi$ is equivalent to

$$-i\frac{d\psi(x)}{dx} = \pm i\psi(x) , \tag{2.19}$$

and its solutions are $\psi(x) = A\exp(\mp x) \in D(P^\dagger) \subset L^2(0, 2\pi)$. Consequently the deficiency indices (Appendix C) of P are $(1,1)$ and, being equal, P admits self-adjoint extensions. It is not difficult to establish [AG 61] that every one of these extensions, P_α, has the property of being a restriction of $P^\dagger$ to the set of functions ψ in $D(P^\dagger)$ which satisfy boundary conditions of the form

$$\psi(2\pi) = e^{i\alpha}\psi(0) , \quad \alpha \text{ real and fixed } \in [0, 2\pi) . \tag{2.20}$$

It is, of course, not always possible to find a self-adjoint extension of a given symmetric operator. The above operator P, defined according to (2.11), but in the space $\mathcal{H} = L^2(0, \infty)$, provides us with a good example. With the domain

$$D(P) = \left\{ \psi \in \mathcal{H} : \psi \text{ absolutely continuous },\right.$$
$$\left. \int_0^\infty dx \left|\frac{d\psi}{dx}(x)\right|^2 < \infty , \quad \psi(0) = 0 \right\} , \tag{2.21}$$

which is dense in $\mathcal{H}$, P is a closed, symmetric operator, and its adjoint $P^\dagger$ also acts like $-id/dx$; however, only the first two conditions in (2.21) hold for $D(P^\dagger)$ and in general not the third. Therefore, of the equations (2.19) only $P^\dagger \psi = i\psi$ has non-trivial solutions in $D(P^\dagger)$, since $\exp(x) \notin L^2(0, \infty)$. Thus, the deficiency indices of P are $(0, 1)$, and, being unequal, P has no self-adjoint extension.

The preceding examples, apart from indicating in an explicit way that one should distinguish carefully between symmetric operators and self-adjoint operators, also give rise to the suspicion that it may be very difficult to ensure in specific cases that the operators selected to mathematically represent observables are self-adjoint. One must keep in mind that the cases considered above, in addition to being extremely important, are among the simplest which appear in QM.

For the subsequent development of QM it is necessary to indicate that every self-adjoint operator A in Hilbert space $\mathcal{H}$ (which in this book will be assumed to be complex and separable) has a purely real *spectrum* $\sigma(A)$ and that furthermore its residual spectrum is empty. That is, its spectrum is the disjoint union of the *point spectrum* $\sigma_p(A)$ and the *continuous spectrum* $\sigma_c(A)$. Only when $\lambda \in \sigma_p(A)$, will

$$A|f\rangle = \lambda|f\rangle , \quad |f\rangle \in D(A) \tag{2.22}$$

have non-trivial solutions. The real number $\lambda \in \sigma_p(A)$ is called an *eigenvalue* of A, and the nonzero solutions of (2.22) are called the *eigenvectors* of A corresponding to the eigenvalue λ. The set of all such solutions for $\lambda \in \sigma_p(A)$ span a closed linear subspace $M_\lambda(A)$ corresponding to the eigenvalue λ. The dimension of $M_\lambda(A)$ is called the *multiplicity* of the eigenvalue λ. In $M_\lambda(A)$ we can choose an orthonormal basis with elements denoted by $|\lambda; \alpha\rangle$, where λ indicates the eigenvalue and α are those indices necessary to distinguish between different eigenvectors in the orthonormal basis. For self-adjoint operators one can prove that eigenvectors corresponding to different eigenvalues are orthogonal, and therefore a maximal set of mutually linearly independent eigenvectors of $A = A^\dagger$ can be chosen in such a way that it also constitutes an orthonormal set. This set though will not generally be complete in the Hilbert space $\mathcal{H}$.

Linear operators of particular interest due to the simplicity of their spectral analysis are the so-called *compact* (or completely continuous) operators. They are defined everywhere in the Hilbert space and they are bounded, being characterized by the fact that they transform any bounded set of elements of $\mathcal{H}$ into precompact sets (in the sense of strong convergence). For self-adjoint compact operators one can establish the following properties: all nonzero eigenvalues have only finite multiplicity; the point spectrum is discrete, except at most with zero as an accumulation point; and the orthonormal system constructed with the independent eigenvectors $|\lambda; \alpha\rangle$ is in this case complete. Thus any element $|f\rangle \in \mathcal{H}$ can be written

$$|f\rangle = \sum_\lambda \sum_\alpha f_{\lambda\alpha} |\lambda; \alpha\rangle , \quad f_{\lambda\alpha} \equiv \langle \lambda; \alpha | f \rangle , \tag{2.23}$$

where the first sum is over all different eigenvalues, and for a given λ the summation over α runs over the set of indices needed to characterize an orthonormal basis in $M_\lambda(A)$.

Unfortunately, expansions as simple as (2.23) do not exist for every self-adjoint operator. In fact, the equations corresponding to (2.22) for the self-adjoint operators X and P, defined in $L^2(\mathbb{R})$, are

$$\begin{aligned} (X\psi)(x) &= x\psi(x) = \lambda\psi(x) , \\ (P\psi)(x) &= -\mathrm{i}\frac{d\psi(x)}{dx} = \lambda\psi(x) \end{aligned} \tag{2.24}$$

with solutions, in the sense of distributions,

$$\psi_\lambda(x) = N\delta(x - \lambda) , \quad \psi_\lambda(x) = N\mathrm{e}^{\mathrm{i}\lambda x} , \tag{2.25}$$

where N is a nonzero arbitrary constant. These distributions do not belong to $L^2(\mathbb{R})$. Therefore, for these two operators, the point spectrum is empty, and the application of (2.23) is not possible within the Hilbert space framework.

Mathematicians have searched for weaker forms of the eigenvector expansion theorem (2.23). If A is a compact self-adjoint operator and E_{M_λ} denotes the orthogonal projection operator onto the eigensubspace $M_\lambda(A)$, we can show that

$$A = \sum_{\lambda \in \sigma_{\mathrm{p}}(A)} \lambda E_{M_\lambda} , \tag{2.26}$$

which can be immediately verified by having both members act on vectors of the complete orthonormal set $\{|\lambda; \alpha\rangle\}$. In particular, for $|f\rangle$ and $|g\rangle \in \mathcal{H}$, we obtain

$$\langle g|A|f\rangle = \sum_{\lambda \in \sigma_{\mathrm{p}}(A)} \lambda \langle g|E_{M_\lambda}|f\rangle . \tag{2.27}$$

These expressions allow a generalization for any self-adjoint operator, bounded or not.

For this generalization, it is necessary to introduce what is commonly understood as the *spectral resolution of the identity*. This consists of a family of projection operators E_λ, always understood to be orthogonal, which depend on the real parameter λ, and satisfy the following conditions:

1. $E_\lambda \leq E_\mu$, if $\lambda < \mu$
2. E_λ is a continuous function of λ from the right; i.e. $E_{\lambda+0} = E_\lambda$
3. $\lim_{\lambda \to -\infty} E_\lambda = 0$ and $\lim_{\lambda \to +\infty} E_\lambda = I$

One can then establish that there exists one – and only one – spectral resolution of the identity for each self-adjoint operator A, namely $\{E_\lambda\}$, frequently called the *spectral family* of A, which yields the spectral decomposition of this operator:

$$A = \int_{-\infty}^{+\infty} \lambda \, dE_\lambda , \tag{2.28}$$

where the integral is to be interpreted as a Lebesgue-Stieltjes (L-S) integral in the sense that for any $|f\rangle \in D(A)$, $|g\rangle \in \mathcal{H}$, the function $\langle g|E_\lambda|f\rangle$ is of bounded variation and

$$\langle g|A|f\rangle = \int_{-\infty}^{+\infty} \lambda \, d\langle g|E_\lambda|f\rangle \tag{2.29}$$

holds in the sense of an ordinary L-S integral. Formulae (2.28) and (2.29) are the generalizations of the expressions (2.26) and (2.27) to arbitrary self-adjoint operators. Conversely, to each spectral resolution of the identity $\{E_\lambda\}$, one can associate a self-adjoint operator by defining

$$A = \int_{-\infty}^{+\infty} \lambda \, dE_\lambda , \tag{2.30}$$

whose domain of definition $D(A)$ consists of all the elements $|f\rangle$ with the property

$$\int_{-\infty}^{+\infty} \lambda^2 \, d\langle f|E_\lambda|f\rangle < \infty . \tag{2.31}$$

The spectral family $\{E_\lambda\}$ of a self-adjoint operator A has the following distinguishing qualities:

1. A necessary and sufficient condition for $\lambda \in \sigma_p(A)$ is that $E_\lambda - E_{\lambda-0} > 0$, and then the projection $E_\lambda - E_{\lambda-0} = E_{M_\lambda}$ projects on the eigensubspace $M_\lambda(A)$.

2. A necessary and sufficient condition for a real value of λ *not* to belong to the spectrum of the operator A is the existence of an $\varepsilon > 0$ such that $E_{\lambda+\varepsilon} - E_{\lambda-\varepsilon} = 0$. These reals in union with those λ with Im $\{\lambda\} \neq 0$ make up the *resolvent set* $\varrho(A)$ of the operator A.

In view of these properties, one easily deduces that if A is also compact, its spectral family is given by

$$E_\lambda = \sum_{\substack{\mu \leq \lambda \\ \mu \in \sigma_p(A)}} E_{M_\lambda} , \tag{2.32}$$

whence (2.28) goes over into (2.26).

The spectral decomposition of A makes possible the development of a functional calculus for A. Given a complex or real Borel function φ defined on **R**, with finite values (continuity suffices, for example) then we can define an operator $\varphi(A)$ as

$$\varphi(A) = \int_{-\infty}^{+\infty} \varphi(\lambda) \, dE_\lambda . \tag{2.33}$$

This operator $\varphi(A)$ is densely defined and closed; its domain of definition $D(\varphi(A))$ consists of all elements $|f\rangle \in \mathcal{H}$ with

$$\int_{-\infty}^{+\infty} |\varphi(\lambda)|^2 \, d\langle f|E_\lambda|f\rangle < \infty . \tag{2.34}$$

This construction has the following properties [DS 64]:

$$\langle g|\varphi(A)|f\rangle = \int_{-\infty}^{+\infty} \varphi(\lambda) \, d\langle g|E_\lambda|f\rangle , \quad \forall |f\rangle \in D(\varphi(A)) ;$$

$$\|\varphi(A)f\|^2 = \int_{-\infty}^{+\infty} |\varphi(\lambda)|^2 \, d\langle f|E_\lambda|f\rangle , \forall |f\rangle \in D(\varphi(A)) ;$$

$$(\varphi(A))^\dagger = (\varphi^*)(A) ; \tag{2.35}$$

$$\alpha\varphi(A) = (\alpha\varphi)(A) , \quad \forall \text{ complex } \alpha ;$$

$$(\varphi_1 + \varphi_2)(A) \supseteq \varphi_1(A) + \varphi_2(A) ;$$

$$(\varphi_1\varphi_2)(A) \supseteq \varphi_1(A)\varphi_2(A) ;$$

$$\varphi(A)E_\lambda \supseteq E_\lambda\varphi(A) .$$

The calculus based on these rules is commonly called the *functional calculus for A*. When the functions φ_1, φ_2 and φ which appear above are bounded on $\sigma(A)$, the symbols $\supseteq$ can be replaced by equalities. Furthermore, as a consequence of these expressions, if φ is real in $\sigma(A)$, then $\varphi(A)$ is self-adjoint, and if φ has modulus 1 in $\sigma(A)$, then $\varphi(A)$ is unitary. Finally, the last relation in (2.35) expresses the fact that $\varphi(A)$ commutes with each projection operator E_λ, and therefore with A.

Using these techniques of spectral decomposition, *von Neumann* [NE 32] succeeded in producing a formalism in which the point spectrum and the continuous spectrum can be treated on the same footing; without a doubt, this constitutes one of the most substantial advances in the mathematical treatment of QM.

In terms of the spectral decomposition, an analog of (2.23) for the expansion in terms of eigenstates for any self-adjoint operator can be written in the form

$$|f\rangle = \int_{-\infty}^{+\infty} dE_\lambda |f\rangle , \tag{2.36}$$

where we have used the properties which define the spectral family in writing (2.36). Formally one would expect the elements of the "subspace" $(dE_\lambda)\mathcal{H}$ to be eigenvectors of A, with eigenvalue λ. But if $\lambda \in \sigma_c(A)$, then these elements cannot be in $\mathcal{H}$. Thus, for example, for the operator X in (2.6), formula (2.25) informs us that these elements, aside from normalizations, are of the form

$$|\lambda\rangle \rightarrow \psi_\lambda(x) = \delta(x - \lambda) , \quad \forall \text{ real } \lambda , \tag{2.37}$$

and similarly for P in (2.11),

$$|\lambda\rangle \rightarrow \psi_\lambda(x) = \frac{1}{\sqrt{2\pi}}e^{i\lambda x} , \quad \forall \text{ real } \lambda . \tag{2.38}$$

Using these "eigenfunctions" of X and P, respectively, we can write

$$
\begin{aligned}
\psi(x) &= \int_{-\infty}^{+\infty} d\lambda \, \psi(\lambda)\delta(x - \lambda) \\
\psi(x) &= \frac{1}{\sqrt{2\pi}} \int_{-\infty}^{+\infty} d\lambda \, \hat{\psi}(\lambda)e^{i\lambda x} , \quad \text{with} \\
\hat{\psi}(\lambda) &\equiv \frac{1}{\sqrt{2\pi}} \int_{-\infty}^{+\infty} dx \, \psi(x)e^{-i\lambda x} ,
\end{aligned}
\tag{2.39}
$$

relations which are valid $\forall \psi \in C_0^\infty(\mathbb{R})$. Thus, at least for a dense subspace in $\mathcal{H}$, the functions can be expanded [in the sense of (2.39)] as "linear combinations" of generalized eigenfunctions of these operators. Both expressions in (2.39) can be written as one equation, whose similarity to (2.23) is obvious:

$$|\psi\rangle = \int_{-\infty}^{+\infty} |\lambda\rangle \, d\lambda \, \langle\lambda|\psi\rangle \tag{2.40}$$

where $|\psi\rangle$ is the function ψ, $|\lambda\rangle$ denotes the "function" ψ_λ, and $\langle\lambda|\psi\rangle$ is the scalar product

$$\langle\lambda|\psi\rangle = \int_{-\infty}^{+\infty} dx \, \psi_\lambda^*(x)\psi(x) . \tag{2.41}$$

If we apply this scalar product formula to the calculation of $\langle \lambda' | \lambda \rangle$ using (2.37) and (2.38), we obtain in both cases

$$\langle \lambda' | \lambda \rangle = \delta(\lambda' - \lambda) . \tag{2.42}$$

This should be interpreted as a symbolic expression which only acquires meaning under an integration sign, and which expresses the orthogonal character of the generalized eigenstates corresponding to distinct points of the continuous spectrum.

In view of (2.23) and (2.40), the question naturally arises whether for every self-adjoint operator A and for every $|\psi\rangle$ we can always write

$$
\begin{aligned}
|\psi\rangle = &\sum_{\lambda_n \in \sigma_p(A)} \sum_{\alpha \in I(\lambda_n)} |\lambda_n; \alpha\rangle \langle \lambda_n; \alpha | \psi\rangle \\
&+ \int_{\sigma_c(A)} d\lambda \sum_{\alpha \in I(\lambda)} |\lambda; \alpha\rangle \langle \lambda; \alpha | \psi\rangle .
\end{aligned}
\tag{2.43}
$$

In (2.43) $|\lambda_n; \alpha\rangle$, $\alpha \in I(\lambda_n)$, is an orthonormal basis in the eigensubspace $M_{\lambda_n}(A) \subseteq \mathcal{H}$, of dimension Card I$(\lambda_n)$, and $|\lambda; \alpha\rangle$, $\alpha \in I(\lambda)$, is an orthonormal basis in the generalized "eigensubspace" $(dE_\lambda)\mathcal{H}$ corresponding to $\lambda \in \sigma_c(A)$, with dimension Card I(λ). The answer to this question is affirmative [if we interpret the integration of (2.43) as a direct integral] for any self-adjoint operator which does not have a continuous singular spectrum. This last condition allows us to write $d\lambda$ in (2.43), instead of having to write another, more general measure. The justification is provided by the spectral representation of A, in which the space $\mathcal{H}$ is decomposed as a direct integral of the Hilbert spaces $\mathcal{H}(\lambda)$, $\lambda \in \sigma(A)$. These are in a generalized sense eigenspaces of A, with eigenvalues λ, and are not subspaces of $\mathcal{H}$ when $\lambda \in \sigma_c(A)$ [DS 64]. With his theory of rigged Hilbert spaces, Gel'fand [GC 62] succeeded in giving a concrete functional characterization of the elements of $\mathcal{H}(\lambda)$ for a large class of self-adjoint operators which frequently appear in the applications of QM. Summarizing:

1) For any self-adjoint operator A in $\mathcal{H}$ not having a continuous singular spectrum, the expansions (2.43) in terms of (generalized) eigenvectors always exist either as direct integrals or, in some cases, in the distribution sense of Gel'fand.

2) If $|\psi_1\rangle$, $|\psi_2\rangle \in \mathcal{H}$, their scalar product in terms of the components $\langle \lambda; \alpha | \psi_1 \rangle$ and $\langle \lambda; \alpha | \psi_2 \rangle$ is expressed in the form

$$
\begin{aligned}
\langle \psi_2 | \psi_1 \rangle = &\sum_{\lambda_n \in \sigma_p(A)} \sum_{\alpha \in I(\lambda_n)} \langle \psi_2 | \lambda_n; \alpha\rangle \langle \lambda_n; \alpha | \psi_1 \rangle \\
&+ \int_{\sigma_c(A)} d\lambda \sum_{\alpha \in I(A)} \langle \psi_2 | \lambda; \alpha\rangle \langle \lambda; \alpha | \psi_1 \rangle .
\end{aligned}
\tag{2.44}
$$

3) If $\varphi(A)$ is a function of A, then

$$
\begin{aligned}
\varphi(A)|\psi\rangle = &\sum_{\lambda_n \in \sigma_p(A)} \varphi(\lambda_n) \sum_{\alpha \in I(\lambda_n)} |\lambda_n; \alpha\rangle\langle\lambda_n; \alpha|\psi\rangle \\
&+ \int_{\sigma_c(A)} \varphi(\lambda)\,d\lambda \sum_{\alpha \in I(\lambda)} |\lambda; \alpha\rangle\langle\lambda; \alpha|\psi\rangle ,
\end{aligned}
\tag{2.45}
$$

which follows readily from the observation that, formally,

$$
A|\lambda; \alpha\rangle = \lambda|\lambda; \alpha\rangle , \quad \varphi(A)|\lambda; \alpha\rangle = \varphi(\lambda)|\lambda; \alpha\rangle , \quad \forall \lambda \in \sigma(A) .
\tag{2.46}
$$

Although justifiable as a direct integral of operators, the expression (2.45) may also be symbolically written in the form

$$
\begin{aligned}
\varphi(A) = &\sum_{\lambda_n \in \sigma_p(A)} \varphi(\lambda_n) \sum_{\alpha \in I(\lambda_n)} |\lambda_n; \alpha\rangle\langle\lambda_n; \alpha| \\
&+ \int_{\sigma_c(A)} \varphi(\lambda)\,d\lambda \sum_{\alpha \in I(\lambda)} |\lambda; \alpha\rangle\langle\lambda; \alpha| .
\end{aligned}
\tag{2.47}
$$

4) For a Borel set $\Delta \subseteq \mathbf{R}$, with characteristic function φ_Δ, the self-adjoint operator $\varphi_\Delta(A)$ is a spectral projection operator $E_A(\Delta)$ [using (2.35) together with $\varphi_\Delta^2 = \varphi_\Delta$] which, according to (2.47), can be written

$$
\begin{aligned}
E_A(\Delta) = &\sum_{\lambda_n \in \sigma_p(A) \cap \Delta} \sum_{\alpha \in I(\lambda_n)} |\lambda_n; \alpha\rangle\langle\lambda_n; \alpha| \\
&+ \int_{\lambda \in \sigma_c(A) \cap \Delta} d\lambda \sum_{\alpha \in I(\lambda)} |\lambda; \alpha\rangle\langle\lambda; \alpha| .
\end{aligned}
\tag{2.48}
$$

In particular, if $\Delta = \mathbf{R}$, then $E_A(\mathbf{R}) = 1$, and the expression in (2.48) is called the *closure relation*, being nothing more than the equality

$$
I = \int_{-\infty}^{+\infty} dE_\lambda .
\tag{2.49}
$$

This explains why $\{E_\lambda\}$ is called a resolution of the identity.

5) If we formally take the scalar product of the expansions of $|\psi_1\rangle$ and $|\psi_2\rangle$ in terms of eigenvectors given by (2.43), we see that in order to get the expression (2.44) it suffices that the following symbolic orthonormalization relations be satisfied in a general sense:

$$
\begin{aligned}
&\langle\lambda_{n'}; \alpha'|\lambda_n; \alpha\rangle = \delta_{\lambda_{n'}\lambda_n}\delta_{\alpha'\alpha} \\
&\langle\lambda'; \alpha'|\lambda; \alpha\rangle = \delta(\lambda' - \lambda)\delta_{\alpha'\alpha} \\
&\langle\lambda; \beta|\lambda_n; \alpha\rangle = 0 .
\end{aligned}
\tag{2.50}
$$

Of course, only the first relation can be interpreted as the scalar product of two elements of $\mathcal{H}$.

2.4 Results of Measurements

The first two postulates express the correspondence between pure states and observables with mathematical elements of the quantum formalism; the third postulate will establish the connection between the possible values resulting from a measurement of an observable and these mathematical elements.

Postulate III. If a physical system is in a pure state described by a normalized vector $|\psi\rangle$, the probability of obtaining a value λ from a Borel set $\Delta \subseteq \mathbf{R}$ when measuring the observable A is given by

$$P_{A,\psi}(\Delta) = \|E_A(\Delta)\psi\|^2 . \tag{2.51}$$

Note that we have used the same symbol for the observable as for the self-adjoint operator which represents it.

Given the great importance of (2.51) we write it explicitly by using (2.48):

$$
\begin{aligned}
P_{A,\psi}(\Delta) = & \sum_{\lambda_n \in \sigma_p(A) \cap \Delta} \sum_{\alpha \in I(\lambda_n)} |\langle \lambda_n; \alpha | \psi \rangle|^2 \\
& + \int_{\lambda \in \sigma_c(A) \cap \Delta} d\lambda \sum_{\alpha \in I(\lambda)} |\langle \lambda; \alpha | \psi \rangle|^2 .
\end{aligned}
\tag{2.52}
$$

It is evident from (2.52) that if $\Delta_1 \subseteq \Delta_2$, then $P_{A,\psi}(\Delta_1) \leq P_{A,\psi}(\Delta_2)$ and, since $P_{A,\psi}(\mathbf{R}) = \|E_A(\mathbf{R})\psi\|^2 = \|\psi\|^2 = 1$, then $0 \leq P_{A,\psi}(\Delta) \leq 1$. On the other hand, if $\Delta \cap \sigma(A) = \emptyset$, it is clear from (2.52) that $P_{A,\psi}(\Delta) = 0$. Both results lead to the conclusion that the outcome of the measurement of an observable A is always a value belonging to the spectrum of the self-adjoint operator which represents it. Finally, it is trivial to see that the right-hand side in (2.52), as well as in (2.44), (2.45) and (2.48) are independent of the orthonormal basis $\{|\lambda; \alpha\rangle\}$ chosen in each generalized subspace.

In this postulate, one describes a pure state by using a normalized vector $|\psi\rangle$ belonging to a certain unit ray $|\psi\rangle_R$, but it is clear from (2.51) that the same probabilities would have been obtained using any other element of $|\psi\rangle_R$. In other words, all elements of $|\psi\rangle_R$ give rise to the same probability distribution with respect to a measurement of any observable A. Hence, in agreement with Postulate I, all these elements describe the same physical state.

If $\lambda_n \in \sigma_p(A)$, then it makes sense to speak of the probability that a measurement of the observable A yields the eigenvalue λ_n; this probability is given by

$$P_{A,\psi}(\lambda_n) \equiv P_{A,\psi}(\{\lambda_n\}) = \lim_{\varepsilon \downarrow 0} P_{A,\psi}(\{\lambda : |\lambda - \lambda_n| < \varepsilon\})$$

$$= \sum_{\alpha \in I(\lambda_n)} |\langle \lambda_n; \alpha | \psi \rangle|^2 . \tag{2.53}$$

By contrast, if λ is interior to $\sigma_c(A)$, then experimentally one can only ask for the probability that the measurement of the observable A yields a value that lies in the interval $(\lambda - \Delta\lambda/2, \ \lambda + \Delta\lambda/2) \subset \sigma_c(A)$. In the limit $\Delta\lambda \to 0$, one can generally speak of a probability density $\mathcal{P}_{A,\psi}(\lambda)$, defined by the relation

$$\mathcal{P}_{A,\psi}(\lambda) \equiv \lim_{\Delta\lambda \downarrow 0} \frac{1}{\Delta\lambda} P_{A,\psi}(\{\lambda' : |\lambda' - \lambda| < \Delta\lambda/2\})$$

$$= \sum_{\alpha \in I(\lambda)} |\langle \lambda; \alpha | \psi \rangle|^2 . \tag{2.54}$$

As stated before, the probabilities calculated from (2.53) and (2.54) do not depend on the orthonormal basis chosen in the eigensubspace.

In applications of QM, it often occurs that the state of the system must be represented by an idealized element $|\psi\rangle$ not belonging to the Hilbert space, and therefore not normalizable in the usual sense. We will formally use (2.52) in these cases as well to calculate *relative* probabilities, because $|\psi\rangle$ is still normalizable in a finite region of the space of those variables entering its wavefunction; we will explicitly point out when this is done.

Thus, for example, if $|\psi\rangle$ describes a spinless particle of well defined momentum, its wave function (2.3) is a plane wave and is therefore not absolutely normalizable as an element of $L^2(\mathbb{R}^3)$. However, it is normalizable in a relative sense, usually done in such a way that the "probability" of finding the particle in a unit volume is $(2\pi\hbar)^{-3}$, i.e.,

$$|\psi\rangle \to \frac{1}{(2\pi\hbar)^{3/2}} e^{i\boldsymbol{p}\cdot\boldsymbol{x}/\hbar} . \tag{2.55}$$

It is convenient to note that this "probability" per unit volume is really a probability density (per unit volume in both coordinate and momentum space), chosen so that in the wave packets (2.1), constructed by superposition of plane waves, $\Psi(\boldsymbol{x}; t)$ has dimensions of $L^{-3/2}$, and the probability amplitude $\hat{\psi}(\boldsymbol{p})$ has dimensions of (momentum)$^{-3/2}$, both being in this way absolutely normalizable.

The probabilities $P_{A,\psi}(\lambda_n)$ and the probability densities $\mathcal{P}_{A,\psi}(\lambda)$ introduced in (2.53) and (2.54) should be interpreted in the following way: if we prepare N identical physical systems, each in the same pure state, described by a normalized vector $|\psi\rangle$, and if we measure the observable A in each of them, we will find that if $\lambda_n \in \sigma_p(A)$, then this value will appear as the result of the measurement $n(\lambda_n)$ times. If λ is interior to $\sigma_c(A)$, we will find the result to be in the interval $(\lambda - \Delta\lambda/2, \ \lambda + \Delta /2) \subset \sigma_c(A)$ exactly $n(\lambda, \Delta\lambda)$ times. These repetition frequencies converge when $N \to \infty$ and $\Delta\lambda \to 0$ to the theoretical values given by (2.53) and (2.54):

$$P_{A,\psi}(\lambda_n) = \lim_{N \to \infty} \frac{n(\lambda_n)}{N} , \quad \mathcal{P}_{A,\psi}(\lambda) = \lim_{\substack{N \to \infty \\ \Delta\lambda \downarrow 0}} \frac{n(\lambda, \Delta\lambda)}{N\Delta\lambda} . \tag{2.56}$$

The *expectation value* or *mean value* of the observable A in the normalized state $|\psi\rangle$, which we will indicate by $\langle A \rangle_\psi$, is defined as the average of the results obtained when a large number N of measurements of this observable are carried out on identical systems, all in state $|\psi\rangle$. From its definition, we find that

$$\langle A \rangle_\psi = \sum_{\lambda_n \in \sigma_p(A)} \lambda_n P_{A,\psi}(\lambda_n) + \int_{\sigma_c(A)} d\lambda \, \lambda \mathcal{P}_{A,\psi}(\lambda) , \tag{2.57}$$

and, using (2.53) and (2.54), we obtain

$$\boxed{\begin{aligned} \langle A \rangle_\psi = &\sum_{\lambda_n \in \sigma_p(A)} \sum_{\alpha \in I(\lambda_n)} \lambda_n |\langle \lambda_n; \alpha | \psi \rangle|^2 \\ &+ \int_{\sigma_c(A)} d\lambda \sum_{\alpha \in I(\lambda)} \lambda |\langle \lambda; \alpha | \psi \rangle|^2 . \end{aligned}} \tag{2.58}$$

Since $|\lambda_n; \alpha\rangle$ and $|\lambda; \alpha\rangle$ are eigenvectors of the observable A,

$$\begin{aligned} \langle A \rangle_\psi = &\sum_{\lambda_n \in \sigma_p(A)} \sum_{\alpha \in I(\lambda_n)} \langle \psi | A | \lambda_n; \alpha \rangle \langle \lambda_n; \alpha | \psi \rangle \\ &+ \int_{\sigma_c(A)} d\lambda \sum_{\alpha \in I(\lambda)} \langle \psi | A | \lambda; \alpha \rangle \langle \lambda; \alpha | \psi \rangle , \end{aligned} \tag{2.59}$$

and, using (2.44), we finally get

$$\boxed{\langle A \rangle_\psi = \langle \psi | A | \psi \rangle} . \tag{2.60}$$

This expression is valid if $|\psi\rangle$ is normalized and belongs to the domain of definition $D(A)$ of the operator A [it actually suffices that $|\psi\rangle \in D(|A|^{1/2})$]. When $|\psi\rangle$ is normalizable, but is not necessarily normalized, the r.h.s. of (2.60) should be divided by $\langle \psi | \psi \rangle$. The name "expectation value" should not be interpreted as meaning that $\langle A \rangle_\psi$ is the most probable value, since it may well happen that $\langle A \rangle_\psi \notin \sigma(A)$.

To obtain a quantity which characterizes the dispersion of the results of a measurement, we introduce the so-called *uncertainty* or *standard deviation*, defined, if $|\psi\rangle$ is normalized, by

$$\boxed{\Delta_\psi A \equiv [\langle \psi | [A - \langle A \rangle_\psi]^2 | \psi \rangle]^{1/2}} . \tag{2.61}$$

Since

$$\begin{aligned} \langle \psi | [A - \langle A \rangle_\psi]^2 | \psi \rangle &= \langle \psi | [A^2 - 2\langle A \rangle_\psi A + \langle A \rangle_\psi^2] | \psi \rangle \\ &= \langle \psi | A^2 | \psi \rangle - 2\langle A \rangle_\psi^2 + \langle A \rangle_\psi^2 = \langle \psi | A^2 | \psi \rangle - \langle \psi | A | \psi \rangle^2 , \end{aligned} \tag{2.62}$$

an equivalent definition is

$$\Delta_\psi A = [\langle\psi|A^2|\psi\rangle - \langle\psi|A|\psi\rangle^2]^{1/2} .$$ (2.63)

From (2.61) one deduces that $\Delta_\psi A$ is nothing more than the norm of the vector $[A - \langle A\rangle_\psi]|\psi\rangle$, and therefore $\Delta_\psi A = 0$ if and only if $|\psi\rangle$ is an eigenstate of the operator A.

We can now comment on the content of Postulate II. On the one hand, we saw in Chap. 1 that, given a wave packet $\Psi(x;t)$ (for simplicity we consider only one dimension), $|\Psi(x;t)|^2$ is interpreted as the probability density for finding, at time t, the particle at the point x. Consequently the mean value of the position is $\int_{-\infty}^{+\infty} dx\, x|\Psi(x;t)|^2$, which is just the result of applying (2.60) if we identify the self-adjoint operator X given by (2.6) with the position observable. Similarly, taking a plane wave as representative of a particle with momentum p, this momentum appears as a generalized eigenvalue when the self-adjoint operator $P = -i\hbar d/dx$ is applied to this wave. Hence, this operator can be identified with the momentum observable in $L^2(\mathbb{R})$.

These simple cases, plus other analogous ones which were also well known in the early stages of the development of QM, led to the current form of Postulate II. It is conceivable, however, that some observables might have ended up being represented by symmetric operators having no self-adjoint extension, since the mean values calculated using (2.60) are real numbers for these as well. A complete and careful analysis, with special attention to physical meaning (and in which the measurement process is decomposed into the basic processes) leads to the conclusion that this cannot be the case [JA 68]. In fact, devices used for measurements have a bounded scale of values, and thus the associated symmetric operators are bounded; hence they are self-adjoint. The unbounded range of values for some observables is actually arrived at as an idealization via a limiting process, and the corresponding operators turn out to be self-adjoint. (It suffices to observe that if $\mathcal{H} = \oplus_1^\infty \mathcal{H}_n$, and $A = \oplus_1^\infty A_n$, where A_n is self-adjoint in $\mathcal{H}_n$, then A is also self-adjoint.)

2.5 Uncertainty Relations

Consider a physical system in a state characterized by the normalized vector $|\psi\rangle$ and let us assume that A and B are any two observables of the system. We will prove that if $|\psi\rangle$, $A|\psi\rangle$, $B|\psi\rangle \in D(A) \cap D(B)$, then

$$\boxed{\Delta_\psi A \cdot \Delta_\psi B \geq \tfrac{1}{2}|\langle\psi|[A,B]|\psi\rangle| ,}$$ (2.64)

where $[A,B] \equiv AB - BA$ is the *commutator* of A and B. This inequality is known as the *uncertainty relation* and is the mathematical expression for the Heisenberg principle. In fact, one can define the operators $A' \equiv A - \langle A\rangle_\psi$ and $B' \equiv B - \langle B\rangle_\psi$, which are self-adjoint, and from these construct a family of operators $A' + i\lambda B'$, where λ is any real parameter. Using the conditions imposed on $|\psi\rangle$,

$$y(\lambda) \equiv \langle \psi|(A' - i\lambda B')(A' + i\lambda B')|\psi\rangle = \|(A' + i\lambda B')\psi\|^2 \geq 0 , \tag{2.65}$$

and hence

$$y(\lambda) = \langle \psi|B'^2|\psi\rangle \lambda^2 + \langle \psi|i[A', B']|\psi\rangle \lambda + \langle \psi|A'^2|\psi\rangle \geq 0 . \tag{2.66}$$

Now, since $[A', B'] = [A, B]$, and bearing in mind (2.61), we obtain

$$y(\lambda) = (\Delta_\psi B)^2 \lambda^2 + \langle \psi|i[A, B]|\psi\rangle \lambda + (\Delta_\psi A)^2 \geq 0 . \tag{2.67}$$

Since $i[A, B]$ is symmetric in its domain of definition, the coefficients of the previous inequality are real, and (2.67) is satisfied for all real values of λ if and only if

$$\langle \psi|i[A, B]|\psi\rangle^2 - 4(\Delta_\psi A)^2(\Delta_\psi B)^2 \leq 0 , \tag{2.68}$$

leading directly to (2.64).

Before discussing the physical interpretation of the uncertainty relation, let us see what conditions $|\psi\rangle$ should satisfy for both sides of (2.64) to be equal. If $\Delta_\psi A \cdot \Delta_\psi B = 0$, then (2.64) is necessarily an equality and hence the r.h.s. must be zero. This is expected since in this case $|\psi\rangle$ must be an eigenvector either of A or of B and then $\langle \psi|[A, B]|\psi\rangle = 0$.

We now assume that $\Delta_\psi A \cdot \Delta_\psi B \neq 0$. If (2.64) is satisfied as an equality, then

$$y(\lambda) = [(\Delta_\psi B)\lambda \pm (\Delta_\psi A)]^2 \geq 0 , \tag{2.69}$$

and since $\Delta_\psi B \neq 0$, there exists a real $\lambda \neq 0$ for which $y(\lambda) = 0$. From the definition of $y(\lambda)$ in (2.65), this requires that $(A' + i\lambda B')|\psi\rangle = 0$, or equivalently, that $|\psi\rangle$ be a solution of

$$[A - a]|\psi\rangle = -i\lambda[B - b]|\psi\rangle , \tag{2.70}$$

where a, b, and $\lambda \neq 0$ are real. Obviously, $a = \langle A \rangle_\psi$, $b = \langle B \rangle_\psi$ and λ is the double root of $y(\lambda) = 0$; namely

$$\lambda = -\frac{\langle \psi|i[A, B]|\psi\rangle}{2(\Delta_\psi B)^2} . \tag{2.71}$$

In general, given a set of physical systems in the same state, measurement of the observable A in each one yields a dispersion of values, unless the system is in an eigenstate of A. In this case, and only in this case, $\Delta_\psi A = 0$; in other words, $|\psi\rangle$ is *dispersion-free* in A. The uncertainty relation tells us that, in general, it is impossible to prepare sytems in such a way that one can simultaneously measure A and B with arbitrarily small standard deviations, unless $[A, B] = 0$. If $[A, B] \neq 0$ and one prepares the system so that $\Delta_\psi A$ has a given value, then $\Delta_\psi B$ must satisfy relation (2.64); thus, in general, diminishing $\Delta_\psi A$ increases $\Delta_\psi B$

in such a way that (2.64) is always satisfied. We could say that the uncertainty relation expresses the alteration of the potentiality of the A values in $|\psi\rangle$ as a result of the process of actualizing the B values through measurement.

2.6 Complete Sets of Compatible Observables

With respect to the uncertainty relation, we can ask: Given two observables A and B, under what circumstances will there exist an orthonormal basis of dispersion-free vectors simultaneously for A and B? Since such vectors need to be eigenvectors of both A and B, it is easy to show that, if both operators are bounded and have only pure point spectra, such a basis exists if and only if this basis simultaneously diagonalizes both operators, i.e., if $[A, B] = 0$. We then say that A and B are compatible observables. In the general case, we define A and B to be *compatible* (making precise the concept utilized earlier for compatible attributes) when all arbitrary observables $\bar{A} = f(A)$ and $\bar{B} = g(B)$ [bounded functions of A and B, respectively, with purely discrete spectra] commute. (Note that in practice all experimental observables are of this type, given the finite resolution of measuring devices.) This is equivalent to asserting that the spectral projections of A and B commute:

$$[E_A(\Delta_1), \ E_B(\Delta_2)] = 0 \ . \tag{2.72}$$

Expression (2.72) is symbolized as

$$[A, B] = 0 \ . \tag{2.73}$$

When A and B are bounded, (2.72) and (2.73) are equivalent. If they are not, (2.72) implies that $[A, B] \subseteq 0$. since A and B only commute on a dense set. The reciprocal argument [that if A and B commute on a dense set, then (2.72) is satisfied] is not generally true [it suffices to consider the two extensions P_0 and P_π of the operator (2.11) in $L^2(0, 2\pi)$ with boundary conditions (2.20), where $\alpha = 0, \pi$]. Anyway, if A and B are compatible, one can demonstrate [SC 68] that there exists an orthonormal basis (which may be understood in a generalized sense if necessary) of eigenvectors common to A and B, and which we will continue to call dispersion-free, even when this is an abuse of terminology.

We now consider a physical system with an observable A; as stated above, the eigenvectors of A, $\{|a; \alpha\rangle\}$, form a complete orthonormal set, such that $A|a; \alpha\rangle = a|a; \alpha\rangle$. In general, some of the subspaces $M_a(A)$ spanned by these eigenvectors, for fixed a, will have dimension greater than one. As the previous discussion suggests, given another observable B which commutes with A, there will exist a common orthonormal basis $\{|a, b; \beta\rangle\}$ with $A|a, b; \beta\rangle = a|a, b; \beta\rangle$ and $B|a, b; \beta\rangle = b|a, b; \beta\rangle$. If all subspaces M_{ab}, spanned by $|a, b; \beta\rangle$ with a and b fixed, are all one-dimensional, the parameter β is superfluous and we can denote the basis simply by $\{|a, b\rangle\}$. When this is the case, we say that the observables A and B form a *complete set of compatible observables*, and in this case the set of simultaneous eigenvectors of A and B form a unique orthonormal basis. By "unique" we still understand here that each element of the basis is determined only up to a phase factor.

If the observables A and B have no common and unique orthonormal basis, i.e., if some of the subspaces M_{ab} are of dimension greater than one, then it is possible that there exists a third observable C, compatible with A and B, such that C distinguishes vectors with the same quantum numbers a and b. If this process is repeated as many times as necessary, we arrive at a set of compatible observables $A, B, C, \ldots$ such that the subspaces $M_{abc\ldots}$ are all one-dimensional.

A family of observables $A, B, C, \ldots$ are said to form a complete set of compatible observables for a physical system if: (1) the operators corresponding to these observables are compatible; (2) the basis formed by the simultaneous eigenvectors is unique up to phases; and (3) the system is not redundant – that is, (2) fails if we eliminate one of the observables from the set. Note that an eigenvector common to all of the observables in a complete set of compatible observables is completely characterized by the set of eigenvalues of each, except for normalization and phase. It is generally written as $|a, b, \ldots\rangle$, where $a, b, \ldots$ are the eigenvalues or quantum numbers of the observables $A, B, \ldots$, respectively. These vectors $|a, b, \ldots\rangle$, in which we have maximal information about a complete series of compatible attributes, are the typical representatives of systems prepared in pure states. One can prove that if an observable R is compatible with all elements of a complete set of compatible observables, then R must be a function of them: i.e., $R = R(A, B, \ldots)$ and $R|a, b, \ldots\rangle = R(a, b, \ldots)|a, b, \ldots\rangle$.

The number of observables in a complete compatible set for a given physical system depends on the nature of the system and on the experimental facilities. If, for example, one considers a free electron, the three components of linear momentum (P_x, P_y, P_z) are compatible; the theory embodies this fact by associating with them pairwise commuting operators which therefore could form part of a complete set of compatible observables. The problem then is to investigate whether this complete set of compatible observables contains some other observable which is not a function of P_x, P_y, P_z. In the early years of QM, it was believed that there was no other independent observable in the sense that, given the eigenvalues of P_x, P_y, and P_z, the state was completely determined. Experiment (Zeeman effect, fine strucutre, etc.) has amply demonstrated that this is not true, and it is now known that the subspaces M_p for electrons are two dimensional. Thus, a complete set of compatible observables is characterized, not only by P, but also by one of the components, say S_z, of a self-adjoint operator S, which has no classical analog and is called spin. As we will see later, the eigenvalues of S_z for electrons are $s_z = \pm\hbar/2$. Therefore, upon fixing p and s_z, the common eigenvector $|p_x, p_y, p_z, s_z\rangle$ is completely determined, except for normalization and phase. Only future experimental developments will permit us to assert whether the set $\{P_x, P_y, P_z, S_z\}$ forms a truly complete set of compatible observables.

All compatible sets of self-adjoint operators may be extended to a maximal set. A related problem arises, though, in that not all self-adjoint operators represent physical observables. However, the Hilbert space $\mathcal{H}$ associated with the system is not arbitrary, but admits of a concrete construction in terms of physical operations. It suffices to take an experimental set-up which prepares the system in states with quantum numbers determined by a series of compatible physical quantities, and which is complete within the experimental capabilities. The states obtained in this way are pure, and physically orthogonal in the sense that they differ in some of their quantum numbers. Thus, they form an orthogonal basis for a Hilbert space. This is the Hilbert space of the system, idealized conveniently in the limit of infinite precision. By its construction, the existence of a complete set of compatible observables is automatically assured. Another pure state, prepared by means of some other complete set of compatible observables, will necessarily be a linear combination of previous states. This will be evident if we accept, as in Postulate I, that there exists only one Hilbert space containing all such pure state vectors, and that, according to Postulate III, the sum of the probabilities of finding – in the second state – any one of the values of an observable which forms part of the first complete set of compatible observables is equal to one. Finally, the question of whether an observable always forms part of a complete set of compatible observables will be discussed (with an affirmative answer) when we analyze superselection rules.

Before continuing, we want to comment briefly on a famous theorem due to *von Neumann* [NE 32]: if $A_1, A_2, \ldots$ are pairwise commuting self-adjoint operators, there exists a self-adjoint operator R of which all of them are functions. In other words, $A_1 = A_1(R), A_2 = A_2(R), \ldots$. This theorem could in principle be of interest in the construction of a complete set of compatible observables, since it induces the thought that any such complete set $\{A_1, A_2, \ldots\}$ could be replaced by a single observable R. But even if this were true, the translation of readings on the scale of R to readings on the scales of $A_1, A_2, \ldots$ could be so complicated in general that the method loses its interest.

2.7 Density Matrix

We have already stated repeatedly that if one has maximal information on the state of a physical system, meaning that one has made a preparation such that the values of a complete set of compatible observables have been fixed, we can say that the system is in a pure state described by a normalized vector $|\psi\rangle$ which is completely determined up to a phase factor. To include in our theoretical development states about which we have only partial information, we introduce the concept of *mixed states*. An example of a mixed state is one about which we know that there are probabilities $p_1, p_2, \ldots, p_n, \ldots$ that the system can be found in the pure states represented by non-parallel normalized vectors $|\psi_1\rangle, |\psi_2\rangle, \ldots |\psi_n\rangle, \ldots$, respectively (which are not necessarily orthogonal). We will see at the end of this section that all mixed states are of this type. In the above, $\sum_i p_i = 1$, and, for all values i of the index, $0 \le p_i \le 1$. If some $p_n = 1$ and, therefore $p_i = 0$ for all $i \ne n$, then the mixed state is just a pure state.

In practice, it happens often that a state has definite values for some of the observables of a complete set of compatible observables, whereas for the remaining observables of the set the information is not complete. For example, in the case of an electron, an accelerator prepares states of momentum p, which are well defined in some idealized sense, but the information with respect to the Oz component of spin is not complete. In general, a beam which is prepared by a linear accelerator will have a probability $p_1 = 1/2$ that the electron be found in the state $|\psi_1 = |p; +\hbar/2\rangle$ and a probability $p_2 = 1/2$ that it be found in the state $|\psi_2\rangle = |p; -\hbar/2\rangle$. Evidently, in this case, the states $|\psi_1\rangle$ and $|\psi_2\rangle$ form a complete orthonormal set in the subspace of states with fixed momentum p.

In general, it is convenient to describe a mixed state by means of a *density matrix* or *density operator*, ϱ, defined by

$$\varrho \equiv \sum_i |\psi_i\rangle p_i \langle \psi_i| , \tag{2.74}$$

where p_i is the probability that the system be found in the state $|\psi_i\rangle$. This density operator has some interesting properties, which we will enumerate as we develop the subject. In the first place, ϱ is bounded and self-adjoint:

$$\varrho^\dagger = \varrho . \tag{2.75}$$

In fact, taking the adjoint of (2.74), we obtain formally for $\varrho^\dagger$ the same expression as for ϱ. But this does not prove (2.75), since the adjoint of a sum of infinite terms is not necessarily the sum of the adjoints. A rigorous demonstration is, however, straightforward. It suffices to keep in mind that ϱ is symmetric and defined in the entire Hilbert space, since if $|\psi\rangle \in \mathcal{H}$, then

$$\|\varrho\varphi\|^2 = \sum_{i,j} p_i p_j \langle\varphi|\psi_i\rangle\langle\psi_i|\psi_j\rangle\langle\psi_j|\varphi\rangle \le \sum_{i,j} p_i p_j \|\varphi\|^2 = \|\varphi\|^2 , \tag{2.76}$$

where we have used the Schwarz inequality. Since ϱ is symmetric and defined in all of $\mathcal{H}$, it is then necessarily self-adjoint, and (2.76) implies that $\|\varrho\| \le 1$.

On the other hand, given any element $|\psi\rangle$,

$$\langle\psi|\varrho^2|\psi\rangle = \sum_{i,j} p_i p_j \langle\psi|\psi_i\rangle\langle\psi_i|\psi_j\rangle\langle\psi_j|\psi\rangle \le \sum_{i,j} p_i p_j |\langle\psi|\psi_i\rangle\langle\psi_j|\psi\rangle|$$

$$= \left[\sum_i p_i |\langle\psi_i|\psi\rangle|\right]^2 \le \left(\sum_i p_i\right)\left(\sum_j p_j |\langle\psi_j|\psi\rangle|^2\right)$$

$$= \sum_j p_j |\langle\psi_j|\psi\rangle|^2 = \langle\psi|\varrho|\psi\rangle , \tag{2.77}$$

that is,

$$\varrho \ge \varrho^2 . \tag{2.78}$$

Since (2.75) permits writing $\varrho^2 = \varrho^\dagger \varrho$, and $\varrho^\dagger \varrho$ is a positive operator, then (2.78) implies that ϱ is positive:

$$\boxed{\varrho \geq 0}. \tag{2.79}$$

Moreover, given an orthonormal basis $\{|\varphi_j\rangle\}$,

$$\mathrm{Tr}\{\varrho\} = \sum_j \langle \varphi_j | \varrho | \varphi_j \rangle = \sum_j \sum_i p_i \langle \varphi_j | \psi_i \rangle \langle \psi_i | \varphi_j \rangle$$

$$= \sum_i p_i \|\psi_i\|^2 = \sum_i p_i , \tag{2.80}$$

and, keeping in mind the conditions imposed on the probabilities,

$$\boxed{\mathrm{Tr}\{\varrho\} = 1}. \tag{2.81}$$

Since ϱ is self-adjoint, positive and of finite trace $(= 1)$, it is a compact operator and its eigenvectors $|\phi_j\rangle$ form an orthonormal basis. If $\varrho|\phi_j\rangle = q_j|\phi_j\rangle$, then $q_j \geq 0$ and $\mathrm{Tr}\{\varrho\} = \sum_j q_j = 1$, so that one can write the decomposition

$$\varrho = \sum_j q_j |\phi_j\rangle\langle\phi_j| = \sum_{q \in \sigma(\varrho)} q E_{M_q} , \tag{2.82}$$

the first of which is completely analogous to (2.74), but with pairwise orthogonal projections. Such a decomposition of ϱ is unique except in the case of ambiguities due to degeneracy of its eigenvalues. The second is the spectral decomposition, and is therefore unique.

A simple and concrete example which illustrates the coexistence of distinct types of decompositions for a density matrix is the following: We assume that in a two-dimensional Hilbert space, a mixed state is prepared with weights $p_1 = 24/31$ and $p_2 = 7/31$, relative to the pure states $\frac{1}{\sqrt{2}}\binom{1}{1}$ and $\binom{1}{0}$, respectively. The corresponding density matrix according to (2.74) is

$$\varrho = \frac{24}{31}\binom{1/\sqrt{2}}{1/\sqrt{2}}\left(\frac{1}{\sqrt{2}}, \frac{1}{\sqrt{2}}\right) + \frac{7}{31}\binom{1}{0}(1,0) = \frac{1}{31}\begin{pmatrix} 19 & 12 \\ 12 & 12 \end{pmatrix} , \tag{2.83}$$

whose spectral decomposition is

$$\varrho = \frac{28}{31}\binom{4/5}{3/5}\left(\frac{4}{5}, \frac{3}{5}\right) + \frac{3}{31}\binom{3/5}{-4/5}\left(\frac{3}{5}, -\frac{4}{5}\right) , \tag{2.84}$$

which is immediately verified by looking for the eigenvalues and eigenvectors of ϱ.

Note that if A is an observable of the system, then the expectation value $\langle A \rangle_\varrho$ of A in the ϱ state is the statistical average, with weights p_i, of its mean values in the pure states $|\psi_i\rangle$, and hence

$$\langle A \rangle_\varrho = \sum_i p_i \langle \psi_i | A | \psi_i \rangle \, . \tag{2.85}$$

It is a useful observation that (2.85) is equivalent to

$$\boxed{\langle A \rangle_\varrho = \mathrm{Tr}\{\varrho A\}} \, . \tag{2.86}$$

In fact, if $\{|\varphi_i\rangle\}$ is an orthonormal basis,

$$\mathrm{Tr}\{\varrho A\} = \sum_i \langle \varphi_i | \varrho A | \varphi_i \rangle = \sum_i \sum_j \langle \varphi_i | \psi_j \rangle p_j \langle \psi_j | A | \varphi_i \rangle$$

$$= \sum_j p_j \langle \psi_j | A | \psi_j \rangle = \langle A \rangle_\varrho \, . \tag{2.87}$$

On the other hand, any operator ϱ which satisfies conditions (2.75), (2.79), and (2.81) can be considered as a candidate for a density operator: since ϱ is positive and self-adjoint, the condition $\mathrm{Tr}\{\varrho\} = 1$ guarantees that ϱ is a compact operator, and therefore its spectrum is pure point and discrete (except, at most, with zero as an accumulation point). Denoting its eigenvalues by $\omega_1, \omega_2, \ldots, \omega_n, \ldots$ (each one repeated as many times as its multiplicity) and corresponding orthonormal eigenvectors by $|1\rangle, |2\rangle, \ldots, |n\rangle, \ldots$ then

$$\varrho = \sum_n |n\rangle \omega_n \langle n| \, . \tag{2.88}$$

Since $\mathrm{Tr}\{\varrho\} = 1$, then $\sum_n \omega_n = 1$, and (ϱ being positive definite) $\omega_n \geq 0$. Consequently, ϱ may be a density operator, and indeed it will be one when, in addition to the above, it admits a decomposition (2.74) in which the pure states $|\psi_i\rangle$ are all physically realizable.

We will now see how to represent the density operator in a new orthonormal basis $\{|\varphi_i'\rangle\}$. These vectors are related to the eigenbasis $\{|\phi_i\rangle\}$ used in (2.82) by

$$|\varphi_i'\rangle = \sum_j c_{ji} |\phi_j\rangle \, , \tag{2.89}$$

where the transformation with elements c_{ji} must be unitary, that is to say,

$$\sum_k c_{ik} c_{jk}^* = \sum_k c_{ki} c_{kj}^* = \delta_{ij} \, , \tag{2.90}$$

and consequently,

$$|\phi_i\rangle = \sum_j c_{ij}^* |\varphi_j'\rangle \, . \tag{2.91}$$

Therefore, in the new basis the density matrix is given by

$$\varrho = \sum_{ij} |\varphi_i'\rangle p_{ij} \langle \varphi_j'| \, , \quad \text{with} \tag{2.92}$$

$$p_{ij} = \sum_k c_{ki}^* q_k c_{kj} = \langle \varphi_i' | \varrho | \varphi_j' \rangle \,. \tag{2.93}$$

Let us assume that the two bases are physically possible. Then the quantity p_{ii} (≤ 1, because $\|\varrho\| \leq 1$) represents the probability of finding the system in the state $|\varphi_i'\rangle$ and is called the *population* of the state $|\varphi_i'\rangle$. On the other hand, for $i \neq j$, p_{ij} is a sum of complex numbers like $q_k c_{ki}^* c_{kj}$, and contains the interference effects between states $|\varphi_i'\rangle$ and $|\varphi_j'\rangle$ appearing when $|\phi_k\rangle$ is expressed as a linear combination of the states $\{|\varphi_i'\rangle\}$. Immediately, one sees that $|p_{ij}|^2 \leq p_{ii} p_{jj}$, and in particular, $p_{ii} = 0$ implies $p_{ij} = 0$ for all j. The previous inequality is saturated if a pure state is being described. The complex numbers p_{ij} for $i \neq j$ are known as *coherences*, as they represent the residual coherence in ϱ of the vectors $|\varphi_i'\rangle$ and $|\varphi_j'\rangle$ when the statistical average is carried out [CD 73].

Obviously, pure states are special cases of mixed states. In a pure state it is known with certainty that the system is in a state represented by the normalized vector $|\psi\rangle$, so that it may be described by means of a density operator

$$\varrho = |\psi\rangle\langle\psi| \,, \tag{2.94}$$

which is just a projection on a one-dimensional space; hence

$$\varrho^2 = \varrho \,. \tag{2.95}$$

Conversely, if a mixed state satisfies (2.95), then it describes a pure state. In fact, keeping in mind (2.88), equality (2.95) can be written as

$$\sum_n \omega_n^2 |n\rangle\langle n| = \sum_n \omega_n |n\rangle\langle n| \,, \tag{2.96}$$

which implies that $\omega_n^2 = \omega_n$ for all values of n, so that either $\omega_n = 1$ or $\omega_n = 0$. Since $\sum_n \omega_n = 1$, some ω_i must be equal to one and $\omega_j = 0$ for all $j \neq i$. In other words such a system is in a pure state described by the vector $|i\rangle$.

As a consequence of Postulate III we have seen the important statement that when a measurement of an observable A is made upon a physical system in a pure state $|\psi\rangle$, the only possible results of this measurement are values corresponding to points in the spectrum of A, and we also get a procedure for calculating the probability of a particular outcome. For a system in a mixed state described by a density operator ϱ, expression (2.51), which gives the probability of obtaining a result in the Borel set $\Delta \subseteq \mathbf{R}$, has the obvious generalization

$$P_{A,\varrho}(\Delta) = \sum_i p_i P_{a,\psi_i}(\Delta) = \mathrm{Tr}\{\varrho E_A(\Delta)\} \tag{2.97}$$

as a consequence of the physical meaning of (2.74).

We claimed earlier, in connection with (2.74), that every mixed state of a system admits a representation of the type (2.74); that is, it could be represented by a density operator satisfying (2.75), (2.79) and (2.81). That this is so is a consequence of a deep theorem due to *Gleason* [GL 57], which we now explain.

From a physical viewpoint, any state ξ of the system is characterized by the probability distribution $P_{A,\xi}(\Delta)$ that, upon measurement of any observable A in ξ, the result lies in an arbitrary Borel set Δ. These probabilities should obviously satisfy

$$0 \leq P_{A,\xi}(\Delta) \leq 1 , \quad P_{A,\xi}(\emptyset) = 0 , \quad P_{A,\xi}(\mathbf{R}) = 1 , \tag{2.98}$$

and, if $\Delta_1, \Delta_2, \ldots$ are disjoint Borel sets, it is natural to require, at least for a finite number of sets, that (since we are performing measurements made with a single apparatus, and thus no question of compatibility arises)

$$P_{A,\xi}(\Delta_1 \cup \Delta_2 \cup \ldots) = P_{A,\xi}(\Delta_1) + P_{A,\xi}(\Delta_2) + \ldots . \tag{2.99}$$

Writing $P_{A,\xi}(\Delta) = \xi(E_A(\Delta))$, ξ maps the spectral projections of the observables into the interval $[0, 1]$, with properties (2.98) and (2.99). If one assumes that this mapping has a meaning for all the projections of the Hilbert space $\mathcal{H}$ (this extension will be commented upon in the discussion of superselection rules), assumed to be of dimension ≥ 3 and separable, and such that (2.98) and the extension of (2.99) are

$$0 \leq \xi(E) \leq 1 , \quad \xi(0) = 0 , \quad \xi(I) = 1$$
$$\xi(E_1 + E_2 + \ldots) = \xi(E_1) + \xi(E_2) + \ldots \tag{2.100}$$

whenever $E_i \perp E_j$, $i \neq j$, then Gleason's theorem guarantees that there exists a density matrix operator ϱ associated with ξ, such that

$$\xi(E) = \mathrm{Tr}\{\varrho E\} , \tag{2.101}$$

for all projections E in $\mathcal{H}$. In particular, if all self-adjoint operators represent observables, (2.98) and (2.99) imply that the assumptions for Gleason's theorem are satisfied, and therefore, if $\dim(\mathcal{H}) \geq 3$, all mixed states are described by a density operator. As an example of the fact that for $\dim(\mathcal{H}) = 2$ there exist mathematical states which satisfy (2.100), but are not of the form (2.101), it suffices to consider the mapping ξ, which, apart from satisfying $\xi(0) = 0$ and $\xi(I) = 1$, assigns to each one-dimensional projection $E_\phi = |\phi\rangle\langle\phi|$, with $\|\phi\| = 1$, the expectation value $\xi(E_\phi) \equiv (1 + \cos^3 \theta)/2$, where $\cos^2(\theta/2) = |\langle\phi_0|\phi\rangle|^2$, with ϕ_0 fixed and of unit norm. Note that if $E_\phi \perp E_{\phi'}$, then $\cos \theta + \cos \theta' = 0$, since $|\langle\phi_0|\phi\rangle|^2 + |\langle\phi_0|\phi'\rangle|^2 = \|\phi_0\|^2 = 1$. Furthermore, ξ has no dispersion in the one-dimensional projections E_{ϕ_0} and E_{ϕ_1}, $|\phi_1\rangle \perp |\phi_0\rangle$. (It is a peculiarity of two dimensions to allow the definition of dispersion-free states for *all* observables [BE 66].) These comments notwithstanding, given that for every real physical system the dimension of $\mathcal{H}$ is infinite and therefore its states satisfy (2.101), even in a practical case of restriction to finite dimension associated with relevant subsystems in a certain problem, it is expected that the states come from a reduction of a density operator and that they are therefore of the form (2.101) in every case. We henceforth assume this to be true.

Properties (2.100) characterize the quantum states ξ through the probabilities that, upon measuring a projection-like observable on ξ with spectrum $\{0, 1\}$, the outcome is 1. These particular observables are *yes-no experiments* or *propositions*, in terms of which any other observable can be analyzed. They also constitute the fundamental elements of the propositional calculus, which can be used as the basis for a rigorous formulation of QM [JA 68]. In terms of the basic properties (2.100), required only for projection observables, the distinction between pure states and strictly mixed states (i.e., non-pure states) is the following: ξ is said to be a strictly mixed state when there exist two distinct states, ξ_1 and ξ_2, and a $\lambda \in (0, 1)$, such that

$$\xi = \lambda\xi_1 + (1 - \lambda)\xi_2 , \tag{2.102}$$

this relation being understood in the sense that

$$\xi(E) = \lambda\xi_1(E) + (1 - \lambda)\xi_2(E) \tag{2.103}$$

for all yes-no observables. When such a decomposition does not exist, then ξ is said to be a pure state. If every self-adjoint operator represents an observable, and if $\dim(\mathcal{H}) \geq 3$, this definition of pure and mixed states coincides with the one previously established: pure state $\rightarrow$ unit vector, mixed state $\rightarrow$ density operator ϱ (with $\varrho^2 < \varrho$ if such a state is strictly mixed).

The reader interested on more details in the formalism of the density matrix may consult reference [FA 57].

2.8 Preparations and Measurements

We now assume that we want to measure an observable A in a certain physical system. If immediately before the measurement the system is known to be in a pure state represented by the vector $|\psi\rangle$, Postulate III permits us to calculate the probability of finding any value $a \in \Delta$ as a result of this measurement, using (2.51). When the measurement is carried out, one, and only one, of the possible results is obtained. It would seem natural for the state of the system after the measurement to contain in some way the information obtained in the measurement process, and that it should be, in general, different from the prior state $|\psi\rangle$. The next postulate specifies how the measurement alters the state of the system.

In Sect. 2.2, we distinguished measurements of the second kind, in which the measurement process destroys all information about the property being measured in our system, from the measurement of the first kind, which prepare the system in a state with definite values of some observables by means of an experimental device. For this latter kind, if we measure those observables immediately afterward, they will certainly have the same values with which we prepared them.

It is impossible to have definite information on the state of the system after an arbitrary measurement, since it might well happen that the system itself is destroyed. However, for measurements of the first kind, being at the same time preparations, the resulting state is not arbitrary. Let us assume first that we make such a measurement of an observable A, on a state ϱ and that the value obtained is a, with a nonzero statistical frequency. If the multiplicity of this eigenvalue of A is one, the state which results is completely determined: $|a\rangle$. That is, the measurement has transformed ϱ into the pure state $\varrho_{A,a}$:

$$\varrho \to \varrho_{A,a} \equiv \frac{E_{M_a} \varrho E_{M_a}}{\mathrm{Tr}\{E_{M_a} \varrho E_{M_a}\}} . \tag{2.104}$$

Note that the denominator is nonzero because it is the probability that the value a be the result of the measurement.

Nevertheless, if $\dim(M_a) \geq 2$, the measured value a does not determine the resulting state, since it depends on the type of apparatus used to measure the observable A. Whenever (2.104) is valid (here $\varrho_{A,a}$ will be a density matrix), we say that the measurement of the first kind leading to this is ideal, or we simply call it an *ideal measurement*, since the apparatus has affected the state in the least possible way. In other words, writing ϱ in a basis associated with a complete set of compatible observables containing A,

$$\varrho = \sum_{a_1 b_1 \dots, a_1' b_1' \dots} |a_1 b_1 \dots\rangle p_{a_1 b_1 \dots a_1' b_1' \dots} \langle a_1' b_1' \dots| , \tag{2.105}$$

a measurement upon this state will be called ideal if the state of the system afterwards is obtainable from (2.105) by fixing $a_1 = a_1' = a$, keeping the weights and coherences unchanged, and normalizing the result. Thus, for example, if ϱ describes electrons with $s_z = \hbar/2$, and arbitrary momenta, an ideal measurement of the momentum would be one which selects a fixed momentum and does not change the polarization of the state.

The transformation (2.104) refers to the state representing those systems initially prepared in the state ϱ, and for which the measurement of A has yielded the value a. In general, of course, of the original collection of systems some will produce the value a_1, others a_2, etc., for this observable A. If the apparatus is a *filter*, in the sense that only those systems are transmitted which have yielded values in the set Δ, the state representative of the resulting collection is a statistical mixture of $\varrho_{A,a}$ with $a \in \Delta$, and weights proportional to the respective probabilities of obtaining each of these values:

$$\varrho \to \varrho_{A,\Delta} = \frac{1}{\mathrm{Tr}\{\varrho E_A(\Delta)\}} \sum_{\alpha \in \Delta} E_{M_a} \varrho E_{M_a} . \tag{2.106}$$

This is due to the fact that the measurement destroys the coherences between the states corresponding to different values of A. Such disruptions of coherences are verified experimentally, and we incorporate them here in this way. However, the coherence between the different alternatives $a \in \Delta$ would remain if the measurement did not permit us to know which value in Δ the system actually has, that is, if the experimental device only eliminated the alternatives in $\mathbf{R} - \Delta$.

In this case, the sum in (2.106) should be replaced for an ideal measurement by $E_A(\Delta)\varrho E_A(\Delta)$. These considerations suggest:

Postulate IV. If a physical system is in the state ϱ, the state resulting after an ideal measurement of an observable A which acts as a filter for the values in the set Δ is described by the density matrix $\varrho_{A,\Delta}$:

$$\varrho_{A,\Delta} = \frac{1}{\mathrm{Tr}\{\varrho E_A(\Delta)\}} \sum_{\alpha \in \Delta} E_{M_a} \varrho E_{M_a} \; . \tag{2.107}$$

In reality, (2.104), (2.106), and (2.107) only make sense if the spectrum of A is pure point. However, one cannot pretend that an apparatus measures with infinite precision, and what is actually measured is a discretized approximation of the observable. It is in this approximation that the previous expressions make sense.

The change (2.106) in the state after an ideal measurement is called a *reduction of the wave packet* or a *collapse of the state*. The reconciliation between this type of stochastic, irreversible change and the deterministic change (2.123), which, as we will see, is associated with the time evolution of the system, is the critical problem for measurement theory in quantum mechanics. Some aspects of this problem are discussed in Appendix E.

We will now illustrate the above with a simple example: we consider a system with three observables A, B and C, which for simplicity are assumed to correspond to operators with discrete and non-degenerate spectra. We further assume that all the states which occur between consecutive measurements are stationary, which, as we will see later, means that such a state vector changes in time at most by a phase factor. Initially one prepares the system in a state $|a_0\rangle$ (in this example we assume all states are normalized) such that $A|a_0\rangle = a_0|a_0\rangle$, and one wishes to answer the following questions:

1) What is the probability, $P(a_0 \rightarrow b_j \rightarrow c_i)$, that the measurement of C yields the value c_i if one has previously made an ideal measurement of B which behaves as a filter for the value b_j?
2) What is the probability, $P(a_0 \rightarrow (b_j) \rightarrow c_i)$, that the measurement of C yields the value c_i if previously one has carried out a "non-filtering" ideal measurement of B?
3) What is the probability, $P(a_0 \rightarrow c_i)$, that the measurement of C yields the result c_i if no measurement of B has been made?

In the first case, after an ideal measurement of B, which resulted in the specific value b_j, the system is in the state $|b_j\rangle$; the probability for this to occur is $|\langle b_j|a_0\rangle|^2$ [see (2.104)], and hence, keeping in mind Postulate III, the desired probability is

$$P(a_0 \rightarrow b_j \rightarrow c_i) = |\langle c_i|b_j\rangle|^2 |\langle b_j|a_0\rangle|^2 \; . \tag{2.108}$$

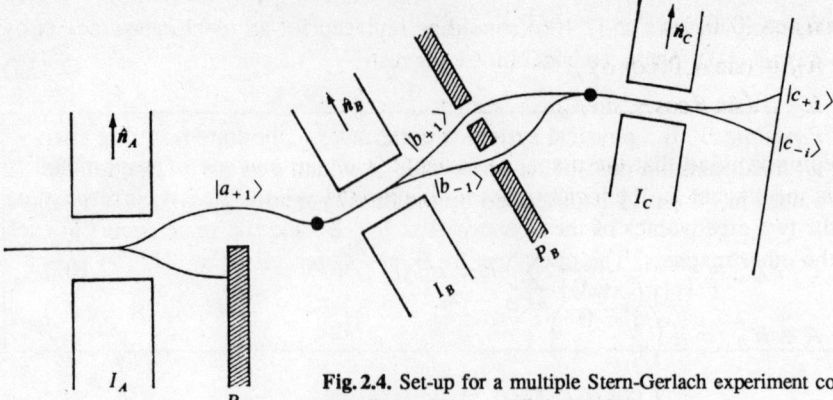

Fig. 2.4. Set-up for a multiple Stern-Gerlach experiment corresponding to $P(a_{+1} \to (b_j) \to c_j)$

In the second case, the probability of obtaining a specific b_j as a result is $|\langle b_j|a_0\rangle|^2$, and after the ideal measurement the system is in the mixed state $\sum_j |\langle b_j|a_0\rangle|^2 |b_j\rangle\langle b_j|$ [see (2.106)], so that the probability asked for is

$$P(a_0 \to (b_j) \to c_i) = \sum_j |\langle c_i|b_j\rangle|^2 |\langle b_j|a_0\rangle|^2 . \tag{2.109}$$

Finally, in the last case,

$$P(a_0 \to c_i) = |\langle c_i|a_0\rangle|^2 , \tag{2.110}$$

which may be written as

$$P(a_0 \to c_i) = \left| \sum_j \langle c_i|b_j\rangle\langle b_j|a_0\rangle \right|^2 . \tag{2.111}$$

In general, (2.109) and (2.111) give completely different results since (2.111) contains interference terms not present in (2.109).

An experiment of the type just described can be carried out by means of a sequence of three Stern-Gerlach devices, schematically shown in Fig. 2.4. An incident beam of unpolarized paramagnetic atoms (whose electrons have a total angular momentum J of magnitude $\hbar/2$) of well-defined momentum p are used. To say that the beam is unpolarized means that, given an arbitrary direction in space, the probabilities of obtaining $\pm\hbar/2$ when measuring the projection of J in this direction are equal. We will see later that the operator corresponding to J in this case is $J = (\hbar/2)\sigma$, where the σ are the so-called Pauli matrices, with their usual representation

$$\sigma_1 = \begin{pmatrix} 0 & 1 \\ 1 & 0 \end{pmatrix} , \quad \sigma_2 = \begin{pmatrix} 0 & -i \\ i & 0 \end{pmatrix} , \quad \sigma_3 = \begin{pmatrix} 1 & 0 \\ 0 & -1 \end{pmatrix} . \tag{2.112}$$

Assume that the axes of the three magnets, $I_A, I_B,$ and I_C are in the directions of the unit vectors $\hat{n}_A, \hat{n}_B,$ and $\hat{n}_C$, respectively, given by

$$\hat{n}_A \equiv (0, 0, 1) \,,$$
$$\hat{n}_B \equiv (\sin\alpha, 0, \cos\alpha) \,, \tag{2.113}$$
$$\hat{n}_C \equiv (\sin\beta\cos\gamma, \sin\beta\sin\gamma, \cos\beta) \,.$$

Keeping in mind that the magnetic moment μ of these atoms is proportional to σ, at the magnet I_A the incident beam divides into two sub-beams corresponding to the two eigenvalues of the operator $A \equiv \hat{n}_A \cdot \sigma$, and the same occurs at each of the other magnets. The operators A, B, and C are given by

$$A \equiv \hat{n}_A \cdot \sigma = \begin{pmatrix} 1 & 0 \\ 0 & -1 \end{pmatrix} \,,$$

$$B \equiv \hat{n}_B \cdot \sigma = \begin{pmatrix} \cos\alpha & \sin\alpha \\ \sin\alpha & -\cos\alpha \end{pmatrix} \,, \tag{2.114}$$

$$C \equiv \hat{n}_C \cdot \sigma = \begin{pmatrix} \cos\beta & \sin\beta e^{-i\gamma} \\ \sin\beta e^{+i\gamma} & -\cos\beta \end{pmatrix} \,.$$

These three operators have eigenvalues ± 1, and the corresponding eigenvectors are:

$$|a_{+1}\rangle = \begin{pmatrix} 1 \\ 0 \end{pmatrix} \,, \qquad\qquad |a_{-1}\rangle = \begin{pmatrix} 0 \\ 1 \end{pmatrix} \,,$$

$$|b_{+1}\rangle = \begin{pmatrix} \cos\frac{\alpha}{2} \\ \sin\frac{\alpha}{2} \end{pmatrix} \,, \qquad\qquad |b_{-1}\rangle = \begin{pmatrix} -\sin\frac{\alpha}{2} \\ \cos\frac{\alpha}{2} \end{pmatrix} \,, \tag{2.115}$$

$$|c_{+1}\rangle = \begin{pmatrix} \cos\frac{\beta}{2}e^{-i\gamma/2} \\ \sin\frac{\beta}{2}e^{+i\gamma/2} \end{pmatrix} \,, \qquad |c_{-1}\rangle = \begin{pmatrix} -\sin\frac{\beta}{2}e^{-i\gamma/2} \\ \cos\frac{\beta}{2}e^{+i\gamma/2} \end{pmatrix} \,.$$

I_A prepares the particles in the state $|a_{+1}\rangle$. The additional deflections of the trajectories are assumed to be produced by fields not considered in the figure, in such a way that they modify the beam's momentum, but not the angular momentum of the atoms, this latter being the observable of interest in this example [FL 77]. Then the probability $P(a_{+1} \to c_j)$ that measuring C yields the value c_j when the measurement of B has not taken place (i.e., if there is no detector in P_B to register the passage through the magnet I_B along paths $|b_{+1}\rangle$ or $|b_{-1}\rangle$) is

$$P(a_{+1} \to c_{+1}) = \tfrac{1}{2}[1 + \hat{n}_A \cdot \hat{n}_C] \,, \quad P(a_{+1} \to c_{-1}) = \tfrac{1}{2}[1 - \hat{n}_A \cdot \hat{n}_C] \,. \tag{2.116}$$

However, if such a detector is present and measures B, filtering the value $+1$ (blocking the path $|b_{-1}\rangle$), such that the state after the measurement is $|b_{+1}\rangle$, then one obtains the probabilities

$$P(a_{+1} \to b_{+1} \to c_{+1}) = \tfrac{1}{4}[1 + \hat{n}_B \cdot \hat{n}_C][1 + \hat{n}_A \cdot \hat{n}_B] \,,$$
$$P(a_{+1} \to b_{+1} \to c_{-1}) = \tfrac{1}{4}[1 - \hat{n}_B \cdot \hat{n}_C][1 + \hat{n}_A \cdot \hat{n}_B] \,. \tag{2.117}$$

Finally, if the I_B measurement is made (the detector at P_B is on), but this measurement does not filter, then the probabilities are

$$P(a_{+1} \to (b_i) \to c_{+1}) = \tfrac{1}{2}[1 + (\hat{n}_A \cdot \hat{n}_B)(\hat{n}_B \cdot \hat{n}_C)] \,,$$
$$P(a_{+1} \to (b_i) \to c_{-1}) = \tfrac{1}{2}[1 - (\hat{n}_A \cdot \hat{n}_B)(\hat{n}_B \cdot \hat{n}_C)] \,.$$
(2.118)

Comparing (2.116) and (2.118), one sees that the result of the experiment depends critically on whether or not the intermediate measurement is made.

2.9 Schrödinger Equation

With the previous postulates concerning the description of states and observables, the results of measurement and associated changes of these states, the "kinematics" of QM has been laid down. Notice that we have tacitly assumed that the measurements are accomplished at a given instant, disregarding their real duration. This idealization could be difficult to achieve in individual practical cases, but we continue to use it here. Nothing has been said, however, about how the states and observables vary with time between consecutive measurement processes. Equations (1.70) and (1.73) suggest that this evolution may be described by keeping the fundamental observables X and P fixed, while letting the states change according to the Schrödinger equation. The requirement of the invariance of the mathematical structure for closed systems as time varies leads, in basic formulations of quantum mechanics, to an identical conclusion [JA 68]. When the systems are not closed, but are interacting with classical external sources such that the latter's reaction to the interaction with the quantum mechanical system can be neglected (which is not the case in a measurement process), then the time evolution is assumed to be analogous to that of closed systems, but with a Hamiltonian which may vary in time. To summarize, we present the dynamical postulate:

Postulate V. In the time interval between two consecutive measurements, pure states of a physical system continue to be pure, and there exists in every unit ray $|\Psi(t)\rangle_{\mathrm{R}}$ some representative state vector $|\Psi(t)\rangle$ such that the evolution is given by the Schrödinger equation

$$\boxed{i\hbar \frac{d}{dt}|\Psi(t)\rangle = H(t)|\Psi(t)\rangle}\,,$$
(2.119)

where $H(t)$ is an observable called the *Hamiltonian* of the system. The observables of the system are represented by operators that are constant in time, unless the devices which measure those observables explicitly change in time, in which case the representative operators should also contain the change.

$H(t)$ is frequently referred to as the *energy* operator of the system, although this name is only appropriate when $H(t)$ does not explicitly depend on time.

From (2.119) and its adjoint, and from the definition (2.74) of the density operator $\varrho(t) \equiv \sum_n |\Psi_n(t)\rangle p_n \langle \Psi_n(t)|$, the equation of evolution for this operator is [using $H(t) = H^\dagger(t)$],

$$i\hbar \frac{d\varrho(t)}{dt} = \sum_n i\hbar \left[\frac{d|\Psi_n(t)\rangle}{dt} p_n \langle \Psi_n(t)| + |\Psi_n(t)\rangle p_n \frac{d\langle \Psi_n(t)|}{dt} \right]$$

$$= \sum_n [H(t)|\Psi_n(t)\rangle p_n \langle \Psi_n(t)| - |\Psi_n(t)\rangle p_n \langle \Psi_n(t)|H(t)] , \qquad (2.120)$$

and hence mixed states evolve according to

$$\boxed{i\hbar \frac{d\varrho(t)}{dt} = [H(t), \; \varrho(t)]} . \qquad (2.121)$$

The assumed time-independence of the statistical weights p_n expresses the fact that, during the time evolution, our information about the pure components of ϱ does not change. Also, the use of p_n varying with t would make impossible an equivalent description in the Heisenberg picture (see Sect. 2.14 below).

An important property of the Schrödinger equation is that during the evolution between two measurements, the norm of the state vector does not change. In fact,

$$i\hbar \frac{d}{dt} \langle \Psi(t)|\Psi(t)\rangle = \left[i\hbar \frac{d\langle \Psi(t)|}{dt} \right] |\Psi(t)\rangle + \langle \Psi(t)| \left[i\hbar \frac{d|\Psi(t)\rangle}{dt} \right]$$

$$= -\langle \Psi(t)|H(t)|\Psi(t)\rangle + \langle \Psi(t)|H(t)|\Psi(t)\rangle = 0 . \qquad (2.122)$$

On the other hand, the Schrödinger equation is linear, and its integration over the interval t_0 to t must yield an isometric linear operator (i.e., norm-preserving) $U(t, t_0)$ such that

$$\boxed{|\Psi(t)\rangle = U(t, t_0)|\Psi(t_0)\rangle} . \qquad (2.123)$$

We now examine the properties of this evolution operator $U(t, t_0)$. Since the Hilbert space is generated by pure states,

$$\boxed{U(t_0, t_0) = I} . \qquad (2.124)$$

In addition, substituting (2.123) into the Schrödinger equation gives

$$i\hbar \frac{dU(t, t_0)}{dt} |\Psi(t_0)\rangle = H(t)U(t, t_0)|\Psi(t_0)\rangle , \qquad (2.125)$$

and since this equation should hold for any pure state $|\Psi(t_0)\rangle$, one obtains

$$\boxed{i\hbar \frac{dU(t, t_0)}{dt} = H(t)U(t, t_0)} . \qquad (2.126)$$

Equation (2.126), together with the initial condition (2.124), are equivalent to the integral equation

$$U(t, t_0) = I - \frac{i}{\hbar} \int_{t_0}^{t} dt' H(t') U(t', t_0) \ . \tag{2.127}$$

From the relations

$$|\Psi(t_2)\rangle = U(t_2, t_1)|\Psi(t_1)\rangle \ , \quad |\Psi(t_3)\rangle = U(t_3, t_2)|\Psi(t_2)\rangle \ , \tag{2.128}$$

one deduces that $|\Psi(t_3)\rangle = U(t_3, t_2)U(t_2, t_1)|\Psi(t_1)\rangle$ and $|\Psi(t_3)\rangle = U(t_3, t_1)|\Psi(t_1)\rangle$, from which one gets

$$U(t_3, t_1) = U(t_3, t_2)U(t_2, t_1) \ . \tag{2.129}$$

In particular, taking $t_1 = t_3 \equiv t_0$ and $t_2 \equiv t$, and making use of (2.124), it follows that

$$U(t_0, t)U(t, t_0) = I \ , \tag{2.130}$$

and hence, since $U(t_2, t_1)$ is isometric, one has

$$U^{-1}(t_0, t) = U(t, t_0) \ . \tag{2.131}$$

In fact, it suffices to prove that $U(t, t_0)$ maps $\mathcal{H}$ onto $\mathcal{H}$. Otherwise, there would exist a nonzero $|\phi\rangle$ in $\mathcal{H}$ such that $\langle \phi|U(t, t_0)|\psi\rangle = 0, \forall |\psi\rangle \in \mathcal{H}$. Since $U(t_0, t)$ is isometric, it conserves the scalar products, and hence

$$0 = \langle \phi|U(t, t_0)|\psi\rangle = \langle U(t_0, t)\phi|U(t_0, t)U(t, t_0)|\psi\rangle$$
$$= \langle U(t_0, t)\phi|\psi\rangle \ , \tag{2.132}$$

using (2.130). As $|\psi\rangle$ is arbitrary, $U(t_0, t)|\phi\rangle = 0$ and since $\||\phi\| = \|U(t_0, t)\phi\| = 0$, this results in $|\phi\rangle = 0$, which contradicts the hypothesis. In summary, $U(t, t_0)$ is isometric and maps $\mathcal{H}$ onto $\mathcal{H}$; it is then necessarily unitary. Therefore

$$U(t, t_0) = U^{-1}(t_0, t) = U^{\dagger}(t_0, t) \ . \tag{2.133}$$

The formal structure of some of the previous demonstrations is clear. Under sufficiently general conditions on $H(t)$, [SI 71], the existence of a *unitary propagator* $U(t, t_0)$, satisfying (2.124), (2.129), and (2.133) can be demonstrated. This unitary propagator is a strongly continuous function of t_0, t and with the integral form of evolution (2.123), the state $|\Psi(t)\rangle$ satisfies the Schrödinger equation (2.119) whenever $|\Psi(t_0)\rangle \in D(H(t_0))$ for some t_0 (and consequently, for all t).

From (2.123) and the properties of $U(t, t_0)$, it follows that the mixed states evolve as

$$\varrho(t) = U(t, t_0)\varrho(t_0)U^{\dagger}(t, t_0) \ . \tag{2.134}$$

In the case of conservative systems, that is to say, those in which H does not explicitly depend on time, equation (2.126) can be formally integrated without difficulties. Imposing the initial conditions (2.124), one obtains the expression

$$U(t, t_0) = \exp[-i(t - t_0)H/\hbar] \quad, \tag{2.135}$$

which can be rigorously proved for all self-adjoint H.

For conservative systems, it can be seen from (2.135) that $U(t, t_0)$ depends only on the difference $t - t_0$, so that we may write $U(t, t_0) \equiv U(t - t_0)$. The physical reason is that for such systems there is no privileged origin of time, since H is constant, and the systems are invariant under time translations. In this case, the unitary time evolution operators $U(\tau)$ form a strongly continuous one-parameter group

$$U(\tau_1 + \tau_2) = U(\tau_1)U(\tau_2) , \quad U(0) = I , \quad U^\dagger(\tau) = U(-\tau) \quad. \tag{2.136}$$

These relations are immediate consequences of (2.135) and of functional calculus (2.35). On the other hand, a famous theorem by *Stone* [RS 55] asserts that any group $U(\tau)$ with the conditions given above is of the form $U(\tau) = \exp(-iH\tau/\hbar)$, where H is a self-adjoint operator called the group *generator*, and the Schrödinger equation holds whenever $|\Psi(t_0)\rangle \in D(H)$ for some t_0 (and therefore, for all t).

In the general case, one may obtain a formal solution of (2.127) by applying Neumann's iterative method:

$$U(t, t_0) = I + \sum_{n=1}^{\infty} \left(-\frac{i}{\hbar}\right)^n \int_{t_0}^{t} dt_1 H(t_1) \int_{t_0}^{t_1} dt_2 H(t_2) \int_{t_0}^{t_2} dt_3 H(t_3) \ldots$$

$$\int_{t_0}^{t_{n-1}} dt_n H(t_n) . \tag{2.137}$$

Letting $U^{(n)}(t, t_0)$ indicate the general term in this series, and introducing the step function $\theta(t)$, defined as usual by

$$\theta(t) = \begin{cases} 0, & t < 0 \\ 1, & t > 0 , \end{cases} \tag{2.138}$$

we have, for $t > t_0$,

$$U^{(n)}(t, t_0) = \int_{t_0}^{t} dt_1 \int_{t_0}^{t} dt_2 \ldots \int_{t_0}^{t} dt_n \theta(t_1 - t_2)\theta(t_2 - t_3) \ldots$$

$$\theta(t_{n-1} - t_n)H(t_1)H(t_2)\ldots H(t_n) . \tag{2.139}$$

This can be written as

$$U^{(n)}(t, t_0) = \frac{1}{n!} \int_{t_0}^{t} dt_1 \int_{t_0}^{t} dt_2 \ldots \int_{t_0}^{t} dt_n \sum_{\pi} [\theta(t_{\pi(1)} - t_{\pi(2)})\theta(t_{\pi(2)} - t_{\pi(3)}) \ldots$$

$$\theta(t_{\pi(n-1)} - t_{\pi(n)})H(t_{\pi(1)})H(t_{\pi(2)}) \ldots H(t_{\pi(n)})] , \tag{2.140}$$

where the sum runs over the $n!$ permutations of the indices $1, 2, \ldots, n$. If the order of the integrations is not important, then by changing the integration variables, it is immediately obvious that each one of the terms in (2.140) is equal to (2.139). We now define Dyson's time ordering operator P in the form

$$P[H(t_1)H(t_2)\ldots H(t_n)]$$
$$= \sum_\pi \theta(t_{\pi(1)} - t_{\pi(2)})\ldots$$
$$\theta(t_{\pi(n-1)} - t_{\pi(n)})H(t_{\pi(1)})H(t_{\pi(2)})\ldots H(t_{\pi(n)}),\quad (2.141)$$

which, bearing in mind (2.138), is equivalently written as

$$P[H(t_1)H(t_2)\ldots H(t_n)] = H(t_i)\ldots H(t_j)\ldots H(t_k),$$
$$t_i > \ldots > t_j > \ldots > t_k. \quad (2.142)$$

Therefore, for $t > t_0$, the solution (2.137) may be expressed as

$$U(t,t_0) = I + \sum_{n=1}^\infty \frac{1}{n!}\left(\frac{-i}{\hbar}\right)^n \int_{t_0}^t dt_1 \int_{t_0}^t dt_2 \ldots$$
$$\int_{t_0}^t dt_n P[H(t_1)H(t_2)\ldots H(t_n)], \quad (2.143)$$

and symbolically written as

$$U(t,t_0) = P\exp\left[-\frac{i}{\hbar}\int_{t_0}^t dt' H(t')\right], \quad t > t_0. \quad (2.144)$$

It is evident that if the system is conservative, the action of P is trivial and (2.135) is formally recovered. When $H(t)$ is bounded and continuous in the strong topology, each term in the series (2.137) and (2.143) is meaningful in that topology, and these series converge strongly. However, this boundedness of $H(t)$ is rarely met in practice.

2.10 Stationary States and Constants of the Motion

Let us now consider a conservative system, i.e., one for which the Hamiltonian operator, or energy observable, does not explicitly depend on time. If $|\Psi(t_0)\rangle$ is an eigenstate of H with eigenvalue E, then

$$\boxed{H|\Psi(t_0)\rangle = E|\Psi(t_0)\rangle}. \quad (2.145)$$

It is said that the state $|\Psi(t_0)\rangle$ is *stationary*, and in accord with (2.135), its time evolution is given by

$$\boxed{|\Psi(t)\rangle = e^{-iE(t-t_0)/\hbar}|\Psi(t_0)\rangle}. \quad (2.146)$$

In other words, the state varies only in its phase in the course of time; and at an arbitrary instant of time, $|\Psi(t)\rangle$ is an eigenstate of H with eigenvalue E.

If E belongs to the point spectrum of H, the solution $|\Psi(t_0)\rangle$ to (2.145) is called a *bound state* and its norm is finite. [This terminology only corresponds to relevant physical ideas in the cases in which E is less than the threshold energy of the essential spectrum, so that some energy, the binding energy, must be provided to break up a system in this state.] For bound states, the possible energies are therefore $E \in \sigma_p(H)$. If E belongs to the continuous spectrum of H, $|\Psi(t_0)\rangle$ is called a *scattering state* and has an infinite norm. Such a state lies in the generalized eigensubspace $\mathcal{H}(E)$ associated with the operator for the eigenvalue E. In many cases, H will be a differential operator, and when finding solutions of (2.145) as a differential equation, conditions will have to be imposed on these solutions to ensure that $|\Psi(t_0)\rangle$ belongs to $\mathcal{H}$ (bound states) or to $\mathcal{H}(E)$ (scattering states).

For an arbitrary system, either conservative or not, if $A(t)$ is an arbitrary observable and $|\Psi(t)\rangle$ describes the state of the system, the evolution of the expectation value $\langle \Psi(t)|A(t)|\Psi(t)\rangle$ is due in part to the evolution of the state vector and in part to A, which may explicitly depend on time. Thus, proceeding formally,

$$\frac{d}{dt}\langle \Psi(t)|A(t)|\Psi(t)\rangle$$

$$= \left[\frac{d}{dt}\langle \Psi(t)|\right] A(t)|\Psi(t)\rangle + \langle \Psi(t)|A(t) \left[\frac{d}{dt}|\Psi(t)\rangle\right]$$

$$+ \langle \Psi(t)|\frac{dA(t)}{dt}|\Psi(t)\rangle \,, \tag{2.147}$$

and from the Schrödinger equation, we obtain the time evolution law for expectation values:

$$i\hbar\frac{d}{dt}\langle \Psi(t)|A(t)|\Psi(t)\rangle = \langle \Psi(t)|[A(t),\ H(t)]|\Psi(t)\rangle$$

$$+ i\hbar\langle \Psi(t)|\frac{dA(t)}{dt}|\Psi(t)\rangle \,. \tag{2.148}$$

In conservative systems, if $|\Psi(t)\rangle$ is stationary,

$$\langle \Psi(t)|[A(t),\ H]|\Psi(t)\rangle$$

$$= \langle \Psi(t)|A(t)H|\Psi(t)\rangle - \langle \Psi(t)|HA(t)|\Psi(t)\rangle = 0 \,, \tag{2.149}$$

and hence

$$\frac{d}{dt}\langle \Psi(t)|A(t)|\Psi(t)\rangle = \langle \Psi(t)|\frac{dA(t)}{dt}|\Psi(t)\rangle \,. \tag{2.150}$$

In particular, if A does not explicitly depend on t, then its mean value in a stationary state is constant in time.

In the general case, we say that an observable $A(t)$ is a *constant of motion* if it satisfies

$$i\hbar \frac{d}{dt} A(t) + [A(t), \ H(t)] = 0 \ . \tag{2.151}$$

Evidently, for conservative systems the Hamiltonian itself is a constant of motion. From (2.151), it is immediately verified that the constants of motion of a physical system constitute a real Lie algebra, at least formally: if $A(t)$ and $B(t)$ are two such constants, then so are $\alpha A(t) + \beta B(t)$, where α and β are real, as is $i[A(t), B(t)]$, assuming of course, that these operations lead to observables. For the formal proof, it suffices to use the Jacobi identity,

$$[A,[B,C]] + [B,[C,A]] + [C,[A,B]] = 0 \tag{2.152}$$

for commutators of operators.

It is very important to realize that some of these results are close analogs of the corresponding classical mechanics results. If $A(q, p; t)$ is a function of generalized coordinates q_j and their conjugate momenta p_j, it is easily seen from Hamilton's equations that

$$\frac{dA}{dt} = [A, H]_P + \frac{\partial A}{\partial t} \ , \tag{2.153}$$

where $[.\,,.]_P$ denotes the Poisson bracket. This expression is completely analogous to (2.148), and suggests the transcription

$$[A, H]_P \overset{\text{cl}\rightarrow\text{QM}}{\longrightarrow} \frac{1}{i\hbar}[A, H] \ , \tag{2.154}$$

which the principle of canonical quantization will use and extend. Actually, the transcription does not have to be univalent, since QM is a generalization of classical mechanics and only the limit QM $\rightarrow$ classical mechanics as $\hbar \rightarrow 0$ provides unambiguous results.

We will now examine some characteristic properties of these observables which justify calling them constants of motion. In the first place, it is evident from (2.148) that the expectation value of an observable $A(t)$ satisfying (2.151) is time-independent, for any state $|\Psi(t)\rangle$. In the case of a conservative system, it is clear that a constant of motion A which does not explicitly depend on time commutes with the operator H; consequently, H and A can be chosen as members of a complete set of compatible observables for such a system. To simplify the notation, we assume that the spectra are pure point; then the basis consisting of the orthonormal eigenvectors common to all operators of the complete set of compatible observables can be written as $\{|E_i, a_j; \alpha\rangle\}$, where the index α indicates the eigenvalues of the remaining observables of the set. From the previous discussion, it is evident that $|E_i, a_j; \alpha\rangle$ is a stationary state of the system. On the other hand, if at the instant t the state is described by the vector $|\Psi(t)\rangle$, the probability of finding the eigenvalue $a_n \in \sigma_p(A)$ when measuring the observable A at this instant is

$$P_{A,\Psi}(a_n; t) = \sum_{E_i} \sum_{\alpha} |\langle E_i, a_n; \alpha|\Psi(t)\rangle|^2 \ . \tag{2.155}$$

This probability is independent of t. In fact, given the evolution equation for conservative systems, at another instant of time t we will have

$$P_{A,\Psi}(a_n; t') = \sum_{E_j} \sum_{\alpha} |\langle E_j, a_n; \alpha | e^{-i(t'-t)H/\hbar} |\Psi(t)\rangle|^2$$

$$= \sum_{E_j} \sum_{\alpha} |e^{-i(t'-t)E_j/\hbar} \langle E_j, a_n; \alpha | \Psi(t)\rangle|^2$$

$$= P_{A,\Psi}(a_n; t) , \qquad (2.156)$$

where we have used the fact that $|E_j, a_n; \alpha\rangle$ is an eigenstate of H with E_j as its eigenvalue.

In summary, for a conservative system, an observable not having explicit time dependence and which is a constant of motion will have a probability distribution which is also independent of time. Although we have proved it assuming pure point spectra, it is a completely general statement, as can be seen from the fact that $[E_A(\Delta), U(t, t_0)] = 0$, together with (2.51).

2.11 The Time-Energy Uncertainty Relation

An important consequence of (2.148) is the time-energy uncertainty relation, which we now study. A multitude of interpretations of this relation have been written, some of them erroneous. [AL 69] contains an accurate and detailed critique. Here we will concentrate on the *Mandelstam-Tamm* version [MT 45], although others, such as *Wigner*'s [WI 72], are also of interest. Consider a conservative system with an observable A which does not explicitly depend on time. For an arbitrary state $|\Psi(t)\rangle$, from Sect. 2.5 the relation

$$\Delta_\Psi A \cdot \Delta_\Psi H \geq \tfrac{1}{2} |\langle \Psi(t)|[A, H]|\Psi(t)\rangle| \qquad (2.157)$$

holds, which bearing in mind (2.148), can be rewritten as

$$\Delta_\Psi A \cdot \Delta_\Psi H \geq \frac{\hbar}{2} \left| \frac{d}{dt} \langle \Psi(t)|A|\Psi(t)\rangle \right| . \qquad (2.158)$$

The quantity $\Delta_\Psi A$ is the uncertainty in the measurement of A in the state $|\Psi(t)\rangle$, whereas $d\langle \Psi(t)|A|\Psi(t)\rangle/dt$ is the time rate of change of the expectation value of A, i.e., the velocity with which the mean value of the probability distribution for A in the state $|\Psi(t)\rangle$ moves. The time

$$\tau_\Psi(A) \equiv \frac{\Delta_\Psi A}{|d\langle \Psi(t)|A|\Psi(t)\rangle/dt|} \qquad (2.159)$$

is thus an estimate of the time interval which needs to elapse after a time t so that the mean of the distribution of values of A is displaced by the quantity $\Delta_\Psi A$. In this sense, one may consider $\tau_\Psi(A)$ as a characteristic time interval at the instant t for the evolution of the system with respect to the observable A. If

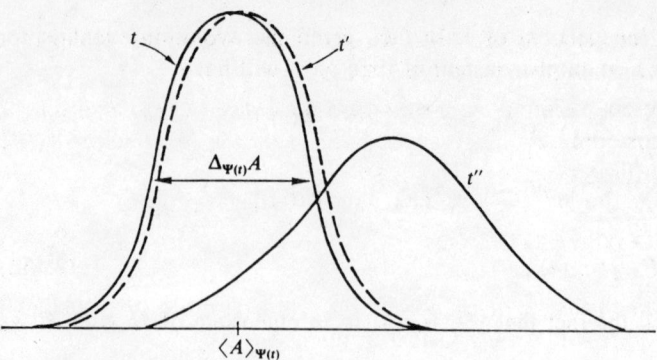

Fig. 2.5. Probability distribution for A at different times

one measures A at t in many identically prepared systems, all in the same state $|\Psi(t)\rangle$, then the statistical distribution of the results of such measurements are essentially the same as those obtained by making the measurement at the instant t', provided that $|t - t'| \ll \tau_\Psi(A)$. Examples of probability distributions of A at instants of time t, t', and t'', such that $|t' - t| \ll \tau_{\Psi(t)}(A)$, and $|t'' - t| \gtrsim \tau_{\Psi(t)}(A)$, are shown in Fig. 2.5.

It is evident that if A is a constant of motion, then the denominator of (2.159) is zero, as we have shown in Sect. 2.10, and since in general $\Delta_\Psi A \neq 0$, we have that $\tau_\Psi(A) = \infty$. The numerator of (2.159) is also zero if $|\Psi(t)\rangle$ is an eigenstate of A; in this case, the r.h.s. of (2.159) is not defined, but, bearing in mind the time evolution meaning of $\tau_\Psi(A)$, it is clear that one should take $\tau_\Psi(A) = \infty$.

If one considers the set of all observables $\{A\}$ which do not explicitly depend on time, we have, at a given instant of time t, a set of characteristic times of evolution $\{\tau_\Psi(A)\}$. Defining

$$\tau_\Psi \equiv \inf_{A \in \{A\}} \{\tau_\Psi(A)\}, \tag{2.160}$$

then $\tau_{\Psi(t)}$ may be considered as a time interval characteristic of the evolution of the system at t, in the sense that for any measured observable of the system, the statistical distribution of its value at times t and t' are essentially the same if $|t' - t| \ll \tau_{\Psi(t)}$; this ceases to be true if $|t' - t| \gtrsim \tau_{\Psi(t)}$. Recalling the inequality (2.158), one obtains

$$\boxed{\tau_\Psi \cdot \Delta_\Psi H \geq \frac{\hbar}{2}}. \tag{2.161}$$

This is called the *time-energy uncertainty relation*. For a physical system in an arbitrary state $|\Psi(t)\rangle$, the uncertainty in the energy and the time interval characteristic of the evolution of the system should be such that (2.161) is satisfied at all times. Note that if $|\Psi(t)\rangle$ is a stationary state, then $\Delta_\Psi H = 0$, and therefore (2.161) requires that $\tau_\Psi = \infty$, which is satisfied, since, according to what was discussed in Sect. 2.10, $\tau_\Psi(A) = \infty$ for stationary states, the probability distributions being constant in time for these states.

We will now present a simple example which illustrates (2.161). Consider a physical system whose time evolution is represented by the unit vector $|\Psi(t)\rangle$; let us assume that at time t_0,

$$|\Psi(t_0)\rangle = c_1|\psi_1\rangle + c_2|\psi_2\rangle , \tag{2.162}$$

where each $|\psi_i\rangle$ is an eigenstate of H, normalized and with energy E_i, such that $E_1 \neq E_2$ and $|c_1|^2 + |c_2|^2 = 1$. At any instant of time,

$$|\Psi(t)\rangle = c_1 e^{-iE_1(t-t_0)/\hbar}|\psi_1\rangle + c_2 e^{-iE_2(t-t_0)/\hbar}|\psi_2\rangle . \tag{2.163}$$

Measuring the energy of the system in this state, one finds either E_1 or E_2, and thus, $\Delta_\Psi H \simeq |E_1 - E_2|$ if the amplitudes $|c_1| \simeq |c_2|$ (an exact calculation yields $\Delta_\Psi H = |c_1 c_2||E_1 - E_2|$). If A is an arbitrary observable which does not depend on time explicitly, then the probability of obtaining the eigenvalue $a_n \in \sigma_p(A)$ (here assumed nondegenerate for the sake of simplicity) upon measurement of A is

$$\begin{aligned} P_{A,\Psi}&(a_n;t) \\ &= |\langle a_n|\Psi(t)\rangle|^2 = |c_1|^2|\langle a_n|\psi_1\rangle|^2 + |c_2|^2|\langle a_n|\psi_2\rangle|^2 \\ &\quad + 2\mathrm{Re}\{c_1 c_2^* \exp[-i(E_1 - E_2)(t - t_0)/\hbar]\langle a_n|\psi_1\rangle\langle a_n|\psi_2\rangle^*\} . \end{aligned} \tag{2.164}$$

Consequently, $P_{A,\Psi}(a_n, t)$ oscillates between two extreme values with angular frequency $|E_1 - E_2|/\hbar$. The characteristic time interval for the evolution with respect to the observable A is $\tau_\Psi(A) \gtrsim \hbar/|E_1 - E_2|$ whence $\tau_\Psi \gtrsim \hbar/|E_1 - E_2|$ and $\tau_\Psi \Delta_\Psi H \gtrsim \hbar$.

Actually, with the precise definition of $\tau_\Psi \equiv \int_0^\infty dt |\langle\Psi(0)|\Psi(t)\rangle|^2$ one can prove that $\tau_\Psi \Delta_\Psi H \geq (3\pi/5\sqrt{5})\hbar$ [GS 85].

2.12 Quantization Rules

Postulate II associates with every observable A of a physical system a self-adjoint operator, which we have also denoted by A, but we have said nothing yet about how one constructs these operators. For observables having a classical analog, a general law can be given for their construction, and thus we state Postule VI on *canonical quantization*:

Postulate VI. For a physical system in which the cartesian coordinates are $q_1, q_2 \dots, q_N$, with corresponding conjugate momenta $p_1, p_2, \dots, p_N$, the operators X_r and P_s, which represent these observables in QM, must satisfy the commutation relations

$$\boxed{[X_r, X_s] = 0 \,, \quad [P_r, P_s] = 0 \,, \quad [X_r, P_s] = i\hbar\delta_{rs}I} \,. \qquad (2.165)$$

If the system has an observable whose classical expression is $A(q_1, \dots, q_N, p_1, \dots p_N; t)$, in usual applications of QM the corresponding operator is obtained from this expression, conveniently written, by substituting the operators X_r and P_s for the variables q_r and p_s, respectively.

We have already given, in (1.72), a method for obtaining the Schrödinger equation by associating the operators X (multiplication by x) and $P = -i\hbar\nabla$ with the position and momentum variables. These operators satisfy (2.165). On the other hand, it is known that in classical mechanics

$$[q_r, q_s]_P = [p_r, p_s]_P = 0 \,, \quad [q_r, p_s]_P = \delta_{rs} \,, \qquad (2.166)$$

so that the transcription suggested by (2.154), namely

$$[A, B]_P \xrightarrow{\text{cl} \rightarrow \text{QM}} \frac{1}{i\hbar}[A, B] \,, \qquad (2.167)$$

will also lead to (2.165). Furthermore, a rigorous analysis of localizability in quantum mechanical systems [WI 62], [JA 66], in which the isotropy and homogeneity of space (conditions which have an especially simple formulation in cartesian coordinates, explaining their particular relevance to the previous postulate) play a fundamental role, implies that the unitary abelian groups

$$U_\alpha \equiv e^{-i\alpha \cdot X/\hbar} \,, \quad V_\beta \equiv e^{-i\beta \cdot P/\hbar} \,, \qquad (2.168)$$

where α and β are real (and where we have assumed compatibility between the X_r's and also among the P_s's), must satisfy *Weyl's form of the commutation relations*:

$$\boxed{U_\alpha V_\beta = e^{-i\alpha \cdot \beta/\hbar} V_\beta U_\alpha} \,. \qquad (2.169)$$

Their infinitesimal form leads immediately to $[X_r, P_s] = i\hbar\delta_{rs}I$. The interpretation of these groups as realizations of displacements of the conjugate variables will be discussed in Chap. 7. These considerations render plausible the contents of Postulate VI.

Note that in the second part of Postulate VI, there is little point in looking for a general rule for associating operators with arbitrary classical observables. QM is a generalization of classical mechanics, and many different quantum observables whose classical limits ($\hbar \rightarrow 0$) coincide may exist. Postulate VI only specifies the commutators between the basic observables, position and momentum. We will see using Weyl's form (2.169) that for irreducible systems these observables are completely determined. For other derived variables, such as angular momentum, energy, etc., which are also of great physical importance, very general arguments can often be found for determining their associated operators. These are summarized in the following practical description, justified ultimately by experimental evidence: Assume that such an observable has a classical expression of the form

$$A(q_1, \ldots, p_N; t) = f(q_1, \ldots, q_N; t) + g(p_1, \ldots, p_N; t)$$

$$+ \sum_{r=1}^{N} h_r(q_1, \ldots, q_N; t) p_r \qquad (2.170)$$

where f, g and h are real, arbitrary functions of their arguments. The third term can, in classical mechanics, be written in different forms, e.g.,

$$\sum_{r=1}^{N} h_r(q_1, \ldots, q_N; t) p_r = \sum_{r=1}^{N} p_r h_r(q_1, \ldots, q_N; t)$$

$$= \sum_{r=1}^{N} \frac{1}{2} [h_r(q_1, \ldots, q_N; t) p_r + p_r h_r(q_1, \ldots, q_N; t)] , \qquad (2.171)$$

and each one of these forms will, upon substitution of the corresponding operators for q_r and p_s, give rise to operators which are, in general, different. When they differ, the ambiguity can be resolved by prescribing the operator associated with the quantity to be

$$A = f(X_1, \ldots, X_N; t) + g(P_1, \ldots, P_N; t) + \sum_{r=1}^{N} \frac{1}{2} [h_r(X_1, \ldots, X_N; t) P_r$$

$$+ P_r h_r(X_1, \ldots, X_N; t)] . \qquad (2.172)$$

Evidently, this method of constructing operators that correspond to observables having classical analogs leads us to formally self-adjoint operators. In general, the condition that A be actually self-adjoint imposes a series of restrictions on the arbitrary functions which appear in (2.172), as we will see in some particular cases. There exist in the literature [SH 59], [AN 69] some methods of functional calculus which assign self-adjoint operators to very general classical quantities. All these present serious practical drawbacks, e.g., the fact that they do not satisfy the multiplicative property.

We move on to an important point associated with the commutation relations of the position and momentum operators given in (2.165). For simplicity, we assume a one-dimensional case. What can one say about two self-adjoint operators X and P obeying $[X, P] = i\hbar I$? In the first place it is clear that such a relation cannot be realized with finite matrices since in that case $\text{Tr}\{[X, P]\} = 0$. In an infinite-dimensional Hilbert space, operators exist that satisfy this commutator relation (except for multiplicative constants): for example, the X and P given by (2.6) and (2.11) in $L^2(\mathbb{R})$. But it is also satisfied in the same sense by X and

P_α in $L^2(0, 2\pi)$, though the structure of these two is very different. It suffices to observe that in the first case the spectra of X and P are **R**, while in the second case it is $[0, 2\pi]$ for X and $\{0, \pm 1, \pm 2, \ldots\}$ for P_α if $\alpha = 0$. On the other hand, at least one of these observables is unbounded in both cases. It can be shown [PU 67] that: first, $[X, P] = i\hbar I$ has no solutions if X and P are bounded; therefore, if the "=" in $[X, P] = i\hbar I$ is understood in a strict sense, this relation has no solutions, and we should understand it as being valid only in a subspace of $\mathcal{H}$ which we will take to be dense. Second, if the observables X and P satisfy $[X, P] = i\hbar I$ on a dense set M of vectors, stable under X and P, and if $r^2 X^2 + s^2 P^2$ (for some real r and s, $rs \neq 0$) is essentially self-adjoint in M, then the Hilbert space is a direct sum of orthogonal spaces stable under X and P, such that in each one of them the pair X and P is (unitarily equivalent to) a *Schrödinger pair*:

$$\left.\begin{array}{l} X = \text{multiplication by } x \\ P = -i\hbar\dfrac{d}{dx} \end{array}\right\} \text{ in } L^2(\mathbf{R}) . \tag{2.173}$$

In particular, if the algebra of observables generated by X and P is irreducible, then X and P form a Schrödinger pair. These results are generalizable to the case of $X_r, P_s, r, s = 1, 2, \ldots N$, with analogous conclusions.

An alternative way of introducing the commutation rules between operators X and P, in such a way that they are necessarily decomposible as direct sums of Schrödinger pairs, is that proposed by Weyl via (2.168) and (2.169). Thus let

$$U_\alpha = e^{-i\alpha X/\hbar} , \quad V_\beta = e^{-i\beta P/\hbar} , \tag{2.174}$$

where α and β are real arbitrary constants. The operator X appears as the generator of a continuous, unitary one-parameter group U_α, and is defined on all those vectors $|f\rangle$ for which the strong limit

$$\lim_{\alpha \to 0} \frac{i\hbar}{\alpha}[U_\alpha - 1]|f\rangle \equiv X|f\rangle \tag{2.175}$$

exists. Analogously, P is the generator of V_β, with a domain formed by the vectors $|g\rangle$ for which the strong limit

$$\lim_{\beta \to 0} \frac{i\hbar}{\beta}[V_\beta - 1]|g\rangle \equiv P|g\rangle \tag{2.176}$$

exists. The commutation relation in Weyl's form is now

$$U_\alpha V_\beta = e^{-i\alpha\beta/\hbar} V_\beta U_\alpha , \tag{2.177}$$

which presents no domain problem in its interpretation, since all operators which appear here are defined in the entire Hilbert space. Starting with (2.177), one can show [PU 67] that X and P are self-adjoint and that their commutator satisfies $[X, P] = i\hbar I$ on a set of vectors dense in $\mathcal{H}$. Formally, this commutation relation is obtained by expanding both members of (2.177) into power series.

Although the primary justification of Weyl's form (2.177) for the commutation relations is a study of localizability of quantum mechanical systems, it is illuminating to obtain them formally using only the commutation rule $[X, P] = i\hbar I$. The reasoning which follows is, of course, wrong in a strict sense; we have already remarked that this latter commutation relation has more solutions than those associated with (2.177).

We begin by "showing" that if A and B are two operators such that $[C, A] = [C, B] = 0$, where $C = [A, B]$, then

$$e^A e^B = e^{A+B+[A,B]/2} .$$
(2.178)

To prove this relation, we need to use the equality

$$[A, F(B)] = [A, B]\frac{dF(B)}{dB}$$
(2.179)

for any function F given as a power series. In fact, the relation

$$[A, B^n] = [A, B]nB^{n-1}$$
(2.180)

is true for $n = 0, 1$. Assuming that it is valid for general n,

$$[A, B^{n+1}] = [A, BB^n] = B[A, B^n] + [A, B]B^n$$
$$= B[A, B]nB^{n-1} + [A, B]B^n = [A, B](n + 1)B^n ,$$
(2.181)

so that it holds for all $n \geq 0$. Hence,

$$[A, B^n] = [A, B]\frac{dB^n}{dB} .$$
(2.182)

Then if $F(B) = \sum_{n=0}^{\infty} f_n B^n$,

$$[A, F(B)] = \sum_{n=0}^{\infty} f_n[A, B^n] = [A, B]\sum_{n=0}^{\infty} f_n\frac{dB^n}{dB} = [A, B]\frac{dF(B)}{dB} ,$$
(2.183)

which proves (2.179), and allows us to prove the relation (2.178). To this end, we introduce a family of operators $G(t)$, dependent on a real variable t, defined as

$$G(t) = e^{At}e^{Bt} .$$
(2.184)

Differentiating $G(t)$ results in

$$\frac{dG(t)}{dt} = Ae^{At}e^{Bt} + e^{At}Be^{Bt} = [A + e^{At}Be^{-At}]G(t) .$$
(2.185)

From (2.179) we obtain

$$[B, e^{At}] = [B, A]te^{At} ,$$
(2.186)

so that we can write (2.185) as

$$\frac{dG(t)}{dt} = (A + B + t[A, B])G(t) . \tag{2.187}$$

Since, by definition, $(A + B)$ commutes with $[A, B]$, we immediately obtain as the solution of this differential equation

$$G(t) = \exp\{(A + B)t + [A, B]t^2/2\}G(0) . \tag{2.188}$$

Recalling that $G(0) = I$, we have

$$G(t) = \exp\{(A + B)t + [A, B]t^2/2\} , \tag{2.189}$$

which for $t = 1$ leads to (2.178).

It can occasionally be useful to have a formula more general than (2.178), whose formal validity does not depend on any assumptions concerning the commutator of A and B. Such a formula was given by *Campbell-Hausdorff* [JA 62]:

$$e^A e^B = e^{\eta(A,B)} ,$$

$$\eta(A, B) \equiv \sum_{m \geq 1} \frac{(-1)^{m-1}}{m}$$

$$\times \sum_{\substack{p_i, q_i \\ p_i + q_i \geq 1}} \frac{[\overbrace{A \cdots A}^{p_1} \overbrace{B \cdots B}^{q_1} \overbrace{A \cdots A}^{p_2} \overbrace{B \cdots B}^{q_2} \cdots \overbrace{A \cdots A}^{p_m} \overbrace{B \cdots B}^{q_m}]}{[\sum_j (p_j + q_j)] p_1! q_1! p_2! q_2! \cdots p_m! q_m!} ,$$

$$[CDE \ldots J] \equiv [\ldots [[C, D], E], \ldots, J] , \tag{2.190}$$

and, explicitly displaying the first few terms of $\eta(A, B)$:

$$\eta(A, B) = A + B + \tfrac{1}{2}[A, B] + \tfrac{1}{12}[[A, B], B] + \tfrac{1}{12}[[B, A], A] + \cdots . \tag{2.191}$$

We repeat that (2.190), as well as the particular case (2.178), are purely formal relations, whose strict validity in each application must be studied with great care [WE 63].

Returning to (2.178) and taking $A = -i\alpha X/\hbar$ and $B = -i\beta P/\hbar$ we get

$$U_\alpha V_\beta = \exp[-i(\alpha X + \beta P)/\hbar] \exp[-i\alpha\beta/2\hbar] , \tag{2.192}$$

and similarly,

$$V_\beta U_\alpha = \exp[-i(\alpha X + \beta P)/\hbar] \exp[+i\alpha\beta/2\hbar] , \tag{2.193}$$

leading immediately to (2.177).

2.13 The Spectra of the Operators X and P

Given the fundamental importance of these observables, we continue to analyze some of their spectral properties. As already stated in Sect. 2.12, Weyl's relation (2.177) determines the structure of the generators X and P in such a way that the Hilbert space $\mathcal{H}$ is decomposed into a direct sum $\mathcal{H}_1 \oplus \mathcal{H}_2 \oplus \ldots$ of orthogonal subspaces, stable under U_α and V_β. In each one of these subspaces, the operators X and P are (unitarily equivalent to) a Schrödinger pair, i.e., such subspaces are isomorphic to $L^2(\mathbb{R})$, and under this isomorphism

$$X : \psi(x) \to x\psi(x) \,, \quad P : \psi(x) \to -i\hbar\frac{d}{dx}\psi(x) \,, \tag{2.194}$$

with the natural domains (2.7) and (2.12).

Equations (2.37), (2.38), and (2.39) tacitly incorporate $\sigma_c(X) = \sigma_c(P) = \mathbb{R}$ for (2.194), and although this is a consequence of the general theory involving rigged Hilbert spaces, it is interesting to verify this explicitly. We will only do this for X, since the Fourier transform in $L^2(\mathbb{R})$ interchanges X and P, and, this transformation being unitary, X and P are spectrally identical. We already know that X has no point spectrum; on the other hand,

$$(\lambda - X)^{-1} : \psi(x) \to \frac{1}{\lambda - x}\psi(x) \,, \tag{2.195}$$

and hence if $\lambda \in \mathbb{R}$, then $(\lambda - X)^{-1}$ is not defined in all of $L^2(\mathbb{R})$. Thus, $\lambda \notin \varrho(X)$, that is, $\lambda \in \sigma(X) = \sigma_c(X)$. Consequently, $\sigma_c(X) = \mathbb{R}$. Moreover, it is not difficult to prove that the spectrum is absolutely continuous and simple. (Analogous statements hold for P.) For the case in which X and P are sums of Schrödinger pairs (for example, if the particle has other degrees of freedom), the spectra of X and P are absolutely continuous, but their multiplicity is equal to the number of such pairs.

A formal proof that $\sigma_c(X) = \mathbb{R}$ and is of uniform multiplicity is as follows: Let $|\lambda\rangle$ be an eigenvector of U_α, that is,

$$U_\alpha|\lambda\rangle = \lambda|\lambda\rangle \,. \tag{2.196}$$

Using (2.177) gives

$$U_\alpha V_\beta|\lambda\rangle = e^{-i\alpha\beta/\hbar}\lambda V_\beta|\lambda\rangle \,, \tag{2.197}$$

and hence $\lambda \in \sigma(U_\alpha)$ implies that $\exp(-i\alpha\beta/\hbar)\lambda \in \sigma(U_\alpha)$. Consequently, if $\alpha \neq 0$, any complex number of unit modulus belongs to $\sigma(U_\alpha)$. Since $\sigma(U_\alpha) = \{\lambda : \lambda = \exp(-i\alpha\mu/\hbar), \mu \in \sigma(X)\}$, and α is arbitrary, it is clear that $\sigma(X) = \mathbb{R}$; and since V_β, which is unitary, permits continuous running over the spectrum of X, this spectrum is continuous and of uniform multiplicity.

It is necessary to point out that other formal demonstrations not employing Weyl's expression (2.177), relying instead only on $[X, P] = i\hbar I$ and "leading" to conclusions equivalent to the above, are untenable because, as we have discussed, this commutation relation has solutions for which one of the operators has a discrete spectrum.

2.14 Time Evolution Pictures

In the description of time evolution in QM given by Postulate V, the states carried all of the burden of the evolution, while the observables, if functionally static, were represented by operators constant in time. However, since what matters is the comparison between theory and experiment, and this always reduces to calculation of matrix elements $\langle \Psi_1(t)|A(t)|\Psi_2(t)\rangle$, the concrete choice of vectors $|\Psi(t)\rangle$ and of representative operators $A(t)$ is irrelevant, so long as such quantities do not change. The choice made in Postulate V is only one of many possibilities, and is called the *Schrödinger picture* for the description of the time evolution. All the quantities in this picture are denoted by the subscript S in this section. In the rest of the book we will omit this subscript, since in QM this picture is by far the most important and will always be used unless otherwise stated. Thus, the Schrödinger equation in this picture is

$$i\hbar\frac{d}{dt}|\Psi_S(t)\rangle = H_S(t)|\Psi_S(t)\rangle \ . \tag{2.198}$$

The invariance of the matrix elements $\langle \Psi_1(t)|A(t)|\Psi_2(t)\rangle$ under a transformation is guaranteed if representative vectors and operators are transformed as

$$|\Psi'(t)\rangle = V(t)|\Psi(t)\rangle \ , \quad A'(t) = V(t)A(t)V^\dagger(t) \ , \tag{2.199}$$

where $V(t)$ is an arbitrary isometric operator. If $V(t)$ is not unitary, only a part of the original Hilbert space $\mathcal{H}$ is physically realizable in the picture generated by $V(t)$. This is not a problem if we work only within this subspace at a given instant t, but in practice it is preferable not to change the Hilbert space associated with the system, and thus we shall take $V(t)$ to be unitary. Among the few pictures of practical use, the so-called *Heisenberg picture* stands out, and is used frequently in quantum field theory. This picture is obtained from the Schrödinger picture at fixed t_0 by means of the relations

$$\boxed{\begin{aligned}|\Psi_H(t)\rangle &= U^\dagger(t, t_0)|\Psi_S(t)\rangle \\ A_H(t) &= U^\dagger(t, t_0)A_S(t)U(t, t_0) \ ,\end{aligned}} \tag{2.200}$$

where $U(t, t_0)$ is the unitary operator solution of the integral equation

$$U(t, t_0) = I - \frac{i}{\hbar}\int_{t_0}^t dt' H_S(t')U(t', t_0) \tag{2.201}$$

which was studied in Sect. 2.9. It is evident from this that the Schrödinger and Heisenberg pictures coincide at $t = t_0$. On the other hand, from (2.123) and (2.134), we have

$$\boxed{|\Psi_H(t)\rangle = |\Psi_S(t_0)\rangle \equiv |\Psi_H\rangle \ , \quad \varrho_H(t) = \varrho_S(t_0) \equiv \varrho_H} \ , \tag{2.202}$$

i.e., in the Heisenberg picture the states do not change with time between measurements. As we remarked in Sect. 2.9, the constancy of the statistical weights p_n of ϱ in time makes possible the fact that $\mathrm{Tr}\{\varrho_H A_H(t)\} = \mathrm{Tr}\{\varrho_S(t)A_S(t)\}$.

The differential equation for the evolution of operators is obtained by differentiating (2.200) with respect to time:

$$i\hbar \frac{dA_H(t)}{dt} = i\hbar \frac{dU^\dagger(t,t_0)}{dt} A_S(t)U(t,t_0) + i\hbar U^\dagger(t,t_0)A_S(t)\frac{dU(t,t_0)}{dt}$$
$$+ i\hbar U^\dagger(t,t_0)\frac{dA_S(t)}{dt}U(t,t_0) \ , \tag{2.203}$$

which, using (2.126) and its adjoint, may be written as

$$i\hbar \frac{dA_H(t)}{dt} = U^\dagger(t,t_0)[A_S(t),H_S(t)]U(t,t_0)$$
$$+ i\hbar U^\dagger(t,t_0)\frac{dA_S(t)}{dt}U(t,t_0) \tag{2.204}$$

or, equivalently, as

$$\boxed{i\hbar \frac{dA_H(t)}{dt} = [A_H(t),H_H(t)] + i\hbar \frac{\partial A_H(t)}{\partial t}} \ . \tag{2.205}$$

The last term in the r.h.s. denotes the operator $dA_S(t)/dt$ in the Heisenberg picture and is zero if $A_S(t)$ is independent of time, as is the case for most of the usual observables. The analogy between (2.205) and (2.153) is evident.

For conservative systems, those in which H does not depend on time, $U(t,t_0)$ is given by

$$U(t,t_0) = \exp[-i(t-t_0)H_S/\hbar] \ , \tag{2.206}$$

which leads to $H_H(t) = H_S \equiv H$. Finally, the condition (2.151) for A to be a constant of motion in the Schrödinger picture is simply expressed in the Heisenberg picture, by virtue of (2.205), as $dA_H(t)/dt = 0$.

Another important picture, the *interaction* or *Dirac picture*, is studied in detail later (Sect. 11.4).

2.15 Superselection Rules

The reader may have noticed that in the preceding sections we have emphasized the distinction between arbitrary unit vectors of $\mathcal{H}$ and physically realizable state vectors representative of pure states. We have also pointed out on a number of occasions that not all self-adjoint operators are necessarily observable. These observations are important because of the existence of *superselection rules* [WW 52], which limit the possibility of superposing coherently two arbitrary state vectors. If $|\psi_1\rangle$ and $|\psi_2\rangle$ are two vectors representing two different pure states, and if we form a linear superposition $|\psi\rangle = \alpha_1|\psi_1\rangle + \alpha_2|\psi_2\rangle$ (where $|\alpha_1|^2 + |\alpha_2|^2 = 1$), we cannot guarantee *a priori* that $|\psi\rangle$ will represent a pure state if $\alpha_1\alpha_2 \neq 0$. For this to be so, it would be necessary to know that there exists a complete set of compatible observables which can be used to prepare the system in the pure state associated with this vector. Thus, for example, nobody knows how to prepare a pure state which is a nontrivial linear superposition of a pure state with electric charge q_1 and another with electric charge $q_2 \neq q_1$; or for systems having only electrons, one with an even number of electrons and the other with an odd number.

Experience seems to indicate that there exist a series of observables $\{Q_i\}$ compatible with all observables, such that they belong to any complete set of compatible observables. Such Q_i are called *superselection observables*, and they give rise to the superselection rules: if $|\psi_1\rangle$ and $|\psi_2\rangle$ are pure states which differ in some of the quantum numbers $\{q_i\}$ associated with $\{Q_i\}$, then there exists no observable which connects them $[\langle\psi_2|A|\psi_1\rangle = 0$ for any observable $A]$. Furthermore, no vector $|\psi\rangle = \alpha_1|\psi_1\rangle + \alpha_2|\psi_2\rangle$, where $\alpha_1\alpha_2 \neq 0$ represents a pure state, since it will not have well-defined quantum numbers $\{q_i\}$. Assuming that $|\psi\rangle$ describes some state of the system, the expectation value of any observable A in that state would be

$$\langle\psi|A|\psi\rangle = |\alpha_1|^2\langle\psi_1|A|\psi_1\rangle + |\alpha_2|^2\langle\psi_2|A|\psi_2\rangle \,, \tag{2.207}$$

thus being indistinguishable from the mixed state described by the density matrix $\varrho = |\alpha_1|^2|\psi_1\rangle\langle\psi_1| + |\alpha_2|^2|\psi_2\rangle\langle\psi_2|$. This tells us that the Hilbert space of the system decomposes into a direct sum (or integral) $\mathcal{H} = \oplus_{\{q_i\}} \mathcal{H}(\{q_i\})$, where $\mathcal{H}(\{q_i\})$ denotes the subspace of the states with superselection quantum numbers $\{q_i\}$. Any two of them (according to the above statements) are incoherent: the relative phases between a vector of one subspace and a vector of another subspace are not measurable, and each one of these subspaces is stable under the action of any observable.

There being no profound reasons to doubt the coherence between states of the same subspace $\mathcal{H}(\{q_i\})$, we admit that these subspaces are completely *coherent*: this means that any unit vector which is a linear superposition of state vectors represents a pure state (that is, the so-called *linear superposition principle* holds without restrictions). As a consequence of this hypothesis, it follows that any projection in $\mathcal{H}(\{q_i\})$ will represent an observable. This fact, together with the spectral decomposition of self-adjoint operators, guarantees (after an appropriate limiting process) that any such operator in a coherent subspace represents an observable. In particular, with such an abundance of observables, it is clear that two vectors corresponding to different unit rays in coherent subspaces are always physically distinguishable through the expectation values of some observable (it suffices to take the projection over one of these). Furthermore, every observable will belong to some complete set of compatible observables, as anticipated in Sect. 2.6.

All known superselection operators (electric charge and parity of fermion number, or univalence, and in an approximate sense, according to modern grand unification theories, baryon and lepton charges) have discrete spectra; these superselection operators do not change with time neither in the Schrödinger picture nor in the Heisenberg picture. Since time evolution is a continuous process, a state vector always remains in the same coherent subspace with passing time. Thus, the superselection quantum numbers are constants of the motion. The use of the prefix "super" signals the fact that, while it is possible to construct pure states with imprecisely determined quantum numbers for ordinary observables which are constants of the motion (selection rules), it is impossible to do so for superselection operators; hence, superselection rules.

From now on we will consider only systems with well-defined superselection quantum numbers, i.e., we will always stay inside a coherent Hilbert space; thus a one-to-one correspondence will exist between pure states and unit rays, and between observables and self-adjoint operators. Nevertheless, it must always be recognized that in reality it may be very difficult, or practically impossible, to construct an apparatus which measures the observable associated with an arbitrary self-adjoint operator. However, since to our knowledge there are no fundamental reasons which forbid such a correspondence, we will simply assume it.

3. The Wave Function

3.1 Introduction

In the previous chapter, the abstract formalism of Hilbert space was used to represent the states and observables of a physical system. On occasion, with clarity in mind, we have referred to concrete functional realizations of these spaces [for example, $L^2(\mathbf{R}^3)$ to describe electrons without spin] in which some basic observables have specific representations such as multiplication by the independent variable of the function or $-i\hbar\nabla$, etc.

The goal of the present chapter is the introduction of an appropriate realization of the Hilbert space in a general form for any complete set of compatible observables, in which the vectors are described in terms of their components with respect to an orthonormal eigenbasis of the given set. In this basis, the observables of this set all have diagonal representations. The remainder of the chapter is devoted to a discussion of some aspects of quantum mechanics which are more easily analyzed in terms of wave functions associated with particular representations.

3.2 Wave Functions

We consider a physical system and let $\{A, B, C \ldots\}$ be a complete set of compatible observables; the set of their eigenvectors, which we will indicate by $\{|a, b, c, \ldots\rangle\}$, forms an orthonormal basis. Given an element $|\psi\rangle$ in the Hilbert space $\mathcal{H}$ associated with the system, we introduce the function

$$\boxed{\psi(a, b, c, \ldots) \equiv \langle a, b, c, \ldots |\psi\rangle}\,, \tag{3.1}$$

whose variables run over the spectra of the operators of the complete set of compatible observables. This function is called the *wave function* of $|\psi\rangle$ in the *representation* associated with the set of observables and characterizes a pure state of the system in a manner totally equivalent to that of $|\psi\rangle$. It should nevertheless be noted that, even with a given complete set of compatible observables, it is necessary to specify the phases of the $\{|a, b, c, \ldots\rangle\}$ in order to be able to speak of *the* wave function. Since the closure relation for the chosen basis $\{|a, b, c, \ldots\rangle\}$ can be written symbolically as

$$\sum_{a,b,c,\dots} |a,b,c,\dots\rangle\langle a,b,c,\dots| = I \;, \tag{3.2}$$

where the summation sign indicates a sum over the point spectrum as well as an integration over the continuous spectrum for each observable, we have

$$|\psi\rangle = \sum_{a,b,c,\dots} \psi(a,b,c,\dots)|a,b,c,\dots\rangle \;. \tag{3.3}$$

This expression is nothing more than the expansion of $|\psi\rangle$ in a linear combination of vectors from the chosen basis. From (3.3) it is evident that $|\psi(a,b,c,\dots)|^2$ is the probability that upon simultaneous measurement of the observables $\{A,B,C,\dots\}$ on a system in a pure state represented by a unit vector $|\psi\rangle$, the results will be the values $\{a,b,c,\dots\}$, respectively (of course, if a were contained in $\sigma_c(A)$, for example, one should speak of a probability density with respect to this observable).

The complex number $\psi(a,b,c,\dots)$, whose modulus squared is a probability (or a probability density), represents a *probability amplitude*; from (3.1) it can be immediately deduced that the probability amplitude of a linear superposition of pure states is the same linear combination of the individual amplitudes.

On the other hand, if $|\psi\rangle$ is normalized, we get immediately from (3.3) that

$$\sum_{a,b,c,\dots} |\psi(a,b,c,\dots)|^2 = 1 \;, \tag{3.4}$$

[a special case of (2.44)], where the sums are to be understood, as before, in the sense of sums over the point spectra and as integrations over the continuous ones. The equality (3.4) therefore tells us that in this case the wave function is also normalized.

An arbitrary linear operator X will be completely specified, with respect to this complete set of compatible observables, by its matrix elements $\langle a',b',c',\dots|X|a,b,c,\dots\rangle$ and its action on the wave functions will be given by

$$(X\psi)(a,b,c,\dots) = \sum_{a',b',c',\dots} \langle a,b,c,\dots|X|a',b',c',\dots\rangle\psi(a',b',c',\dots) \;. \tag{3.5}$$

In the particular case that $X = f(A,B,C,\dots)$, it is clear that

$$(f(A,B,C,\dots)\psi)(a,b,c,\dots) = f(a,b,c,\dots)\psi(a,b,c,\dots) \;, \tag{3.6}$$

since

$$\begin{aligned} &\langle a',b',c',\dots|f(A,B,C,\dots)|a,b,c,\dots\rangle \\ &= f(a,b,c,\dots)\delta(a',a)\delta(b',b)\delta(c',c)\dots \;, \end{aligned} \tag{3.7}$$

where the symbol $\delta(x', x)$ represents a Kronecker symbol $\delta_{x', x}$ if x (or x') belongs to $\sigma_{\mathrm{p}}(X)$, and a Dirac symbol $\delta(x' - x)$ when x', $x \in \sigma_{\mathrm{c}}(X)$.

Given two complete sets of compatible observables $\{A_{(1)}, B_{(1)}, \ldots\}$ and $\{A_{(2)}, B_{(2)}, \ldots\}$, the relationship between the two associated representations is specified as soon as the matrix elements of the unitary transformation $\langle a'_{(2)}, b'_{(2)}, \ldots | a_{(1)}, b_{(1)}, \ldots \rangle$ which connects the two bases are known. Thus, the wave functions $\psi_{(1)}, \psi_{(2)}$ for a single state $|\psi\rangle$ in these two representations satisfy the relation

$$\psi_{(2)}(a'_{(2)}, b'_{(2)}, \ldots) = \sum_{a_{(1)}, b_{(1)}, \ldots} \langle a'_{(2)}, b'_{(2)}, \ldots | a_{(1)}, b_{(1)}, \ldots \rangle \psi_{(1)}(a_{(1)}, b_{(1)}, \ldots)$$.(3.8)

The rules for transformations among the matrix elements of an operator in the various bases are obtained in an analogous way.

Spectral representation theory associated with a maximal abelian algebra permits a rigorous development of the concept of a wave function, and thus a partial justification of the previous formal development. In each particular case, it would be necessary to study the conditions which would assure the validity of every step.

3.3 Position and Momentum Representations

We begin by considering the case of a single spinless particle. Experiment suggests in this case that its three position observables X form a complete set of compatible observables, since specifying the position of the particle determines its state. Therefore, the simultaneous eigenvectors $\{|x\rangle\}$, whose phases will be specified shortly, constitute an orthonormal basis. The wave function representing a pure state $|\Psi(t)\rangle$ in this basis, or *position representation*, as it is commonly referred to, is written as

$$\Psi(x; t) = \langle x | \Psi(t) \rangle ,$$ (3.9)

where x are points in the space $\mathbf{R}^3$ (we recall that the spectrum of each X_i is continuous and fills $\mathbf{R}$). According to the discussion of the previous section,

$$\varrho(x; t)\, d^3 x \equiv |\Psi(x; t)|^2 d^3 x$$ (3.10)

gives the probability that upon measurement of the position of the particle at time t, it will be found in a volume $d^3 x$ surrounding the point x. The quantity $\varrho(x; t)$ introduced in (3.10) is for this reason called the *probability density*.

For this case, the identity (3.4) implies

$$\int_{\mathbf{R}^3} d^3 x \, |\Psi(x; t)|^2 = 1 ,$$ (3.11)

so that in this position representation the Hilbert space is realized in terms of $L^2(\mathbb{R}^3)$.

According to (3.6), the action of the position operators $X_i (i = 1, 2, 3)$ in this representation is

$$(X_i \Psi)(\boldsymbol{x}; t) = x_i \Psi(\boldsymbol{x}; t) ; \tag{3.12}$$

that is to say, in the position representation the position operator X acts simply by multiplying by the variable $\boldsymbol{x}$. As a consequence, the action of $U_\alpha = \exp(-i\boldsymbol{\alpha} \cdot \boldsymbol{X}/\hbar)$ is given by

$$(U_\alpha \Psi)(\boldsymbol{x}; t) = e^{-i\boldsymbol{\alpha} \cdot \boldsymbol{x}/\hbar} \Psi(\boldsymbol{x}; t) . \tag{3.13}$$

One can easily derive from (2.196) and (2.197) that

$$U_\alpha V_\beta |\boldsymbol{x}\rangle = e^{-i\boldsymbol{\alpha} \cdot (\boldsymbol{x}+\boldsymbol{\beta})/\hbar} V_\beta |\boldsymbol{x}\rangle , \tag{3.14}$$

and thus

$$X V_\beta |\boldsymbol{x}\rangle = (\boldsymbol{x} + \boldsymbol{\beta}) V_\beta |\boldsymbol{x}\rangle . \tag{3.15}$$

The interpretation of (3.15) is clear: V_β translates the particle from a point $\boldsymbol{x}$ to the point $\boldsymbol{x} + \boldsymbol{\beta}$, and is thus called a translation operator. Given this, and taking into account that V_β is unitary, we can write

$$V_\beta |\boldsymbol{x}\rangle = e^{i\phi(\beta, \boldsymbol{x})} |\boldsymbol{x} + \boldsymbol{\beta}\rangle , \tag{3.16}$$

where ϕ is real. It is always possible to choose the relative phases among various $|\boldsymbol{x}\rangle$ in such a way that $\phi(\boldsymbol{\beta}, \boldsymbol{x}) \equiv 0$; it merely suffices to fix some $|\boldsymbol{x}_0\rangle$ and to define $|\boldsymbol{x}\rangle \equiv V_{\boldsymbol{x}-\boldsymbol{x}_0} |\boldsymbol{x}_0\rangle$. The relation $V_{\beta_1+\beta_2} = V_{\beta_1} V_{\beta_2}$ ensures the consistency of this choice with (3.16), and $\phi \equiv 0$. [Without this convention, the group law for the operators V_β would be guaranteed by $\phi(\boldsymbol{\beta}, \boldsymbol{x}) + \phi(\boldsymbol{\beta}', \boldsymbol{x}+\boldsymbol{\beta}) = \phi(\boldsymbol{\beta}+\boldsymbol{\beta}', \boldsymbol{x})$ (modulo 2π). For example, it would suffice to take $\phi(\boldsymbol{\beta}, \boldsymbol{x}) = \alpha(\boldsymbol{x}) - \alpha(\boldsymbol{x}+\boldsymbol{\beta})$, α real but otherwise arbitrary.] We will thus henceforth assume relative phases chosen so that

$$\boxed{V_\beta |\boldsymbol{x}\rangle = |\boldsymbol{x} + \boldsymbol{\beta}\rangle} , \tag{3.17}$$

which, in terms of wave functions is equivalent to

$$\boxed{(V_\beta \Psi)(\boldsymbol{x}; t) = \Psi(\boldsymbol{x} - \boldsymbol{\beta}; t)} , \tag{3.18}$$

as can be deduced from (3.17) and (3.9). The sign change in the combination $\boldsymbol{x} \pm \boldsymbol{\beta}$ between (3.17) and (3.18) reflects the fact that vectors and their corresponding components transform contragrediently.

To find the expression for the momentum operator $\boldsymbol{P}$, the generator of translations, in the position representation, we will assume that in (3.18) $\boldsymbol{\beta}$ is an infinitesimal quantity. In this case, (3.18) reduces to

$$\Psi(\boldsymbol{x};t) - i\frac{\beta}{\hbar} \cdot (P\Psi)(\boldsymbol{x};t) + O(\beta^2) = \Psi(\boldsymbol{x};t) - \beta \cdot (\nabla\Psi)(\boldsymbol{x};t) + O(\beta^2) \,, \quad (3.19)$$

and, since β is arbitrary,

$$\boxed{(P\Psi)(\boldsymbol{x};t) = \langle \boldsymbol{x}|P|\Psi(t)\rangle = -i\hbar\nabla\Psi(\boldsymbol{x};t)} \,. \qquad (3.20)$$

Thus, the expression for P in the position representation is $-i\hbar\nabla$. [Had we retained the general form (3.16), $P = -i\hbar\nabla - \hbar(\nabla\phi)(0,\boldsymbol{x})$.]

The domains upon which the operators X_i, P_i, are defined, given their prescriptions (3.12) and (3.20) in $L^2(\mathbf{R}^3)$, are those formed by the $\psi(\cdot) \in L^2(\mathbf{R}^3)$ such that, respectively,

$$D(X_i): \qquad \int_{\mathbf{R}^3} d^3x \, |x_i|^2 |\psi(\boldsymbol{x})|^2 < \infty \,; \qquad (3.21)$$

$D(P_i)$: the distribution $-i\hbar\nabla_i\psi(\cdot) \in L^2(\mathbf{R}^3)$, that is,

$$\int_{\mathbf{R}^3} d^3x \, |\nabla_i\psi(\boldsymbol{x})|^2 < \infty \,. \qquad (3.22)$$

One can examine the functions

$$\varphi_{n_1 n_2 n_3}(\boldsymbol{x}) = \prod_{i=1}^{3}[\sqrt{\pi}2^{n_i}n_i!]^{-1/2}\exp(-x_i^2/2)H_{n_i}(x_i) \,, \qquad (3.23)$$

where the n_i are non-negative integers and the $H_{n_i}(x_i)$ are Hermite polynomials (Appendix A), and show that they form a complete orthonormal system in $L^2(\mathbf{R}^3)$, whose finite linear combinations constitute a dense linear subspace $D \subseteq D(X) \cap D(P)$, stable under X and P and over which these operators are essentially self-adjoint. The set $C_0^\infty(\mathbf{R}^3)$ has the same properties.

We could have chosen, for a particle without spin, the set of momentum operators P as a complete set of compatible observables in the same manner in which we chose X as a complete set, since the state of a particle is also determined if we specify its momentum; this set's simultaneous eigenvectors $\{|\boldsymbol{p}\rangle\}$, whose phases we shall determine shortly, form an orthonormal basis. The wave function in this basis, or *momentum representation* as it is commonly called, is written

$$\boxed{\hat{\Psi}(\boldsymbol{p};t) = \langle \boldsymbol{p}|\Psi(t)\rangle} \,. \qquad (3.24)$$

The caret symbol ($\hat{}$) is used to indicate wave functions in this representation. The variable $\boldsymbol{p}$ runs over $\mathbf{R}^3$, since as we have seen, the spectrum of each P_i is $\mathbf{R}$.

In this case, $|\hat{\Psi}(\boldsymbol{p};t)|^2 d^3p$ is the probability that a measurement of the momentum of the particle at the instant t will find it to be within d^3p of the value $\boldsymbol{p}$. Carrying the reasoning which we have just completed for the position representation to this new representation, and choosing the relative phases of $\{|\boldsymbol{p}\rangle\}$

such that $U_\alpha |p\rangle = |p - \alpha\rangle$, we easily find that the expression for the position and momentum operators in the momentum representation are given by

$$\boxed{(X\Psi)\hat{}(p;t) = i\hbar \nabla_p \hat{\Psi}(p;t) \;, \quad (P\Psi)\hat{}(p;t) = p\hat{\Psi}(p;t)} \;, \tag{3.25}$$

where ∇_p indicates the gradient with respect to the momentum coordinates. There is a close analogy between the two representations, a fact which is not very surprising since the commutation relations in Weyl's form are not changed under the interchange $\alpha \cdot x \leftrightarrow \beta \cdot P$ if one simultaneously changes the sign of the imaginary unit i.

We have these two distinct representations for position and momentum, and we need to make explicit the relation (3.8) between the two wave functions for the same vector in both bases. For this we need the matrix elements $\langle x|p\rangle$ for the change of basis. But $\langle x|p\rangle \equiv \psi_p(x)$ is just the wave function for the state $|p\rangle$, which is an eigenstate of P, in the position representation, and thus (3.20) leads to the following equation:

$$-i\hbar \nabla \psi_p(x) = p\psi_p(x) \;, \tag{3.26}$$

whose solution is

$$\psi_p(x) = C(p)e^{ip\cdot x/\hbar} \;. \tag{3.27}$$

Since we have fixed the relative phases of $\{|p\rangle\}$ such that $U_\alpha|p\rangle = |p - \alpha\rangle$, then by projecting this relation on $|x\rangle$, which is an eigenstate of U_α, and using (3.27), one immediately concludes that $C(p) = C(p - \alpha) \equiv C$. Only the relative phase between $|x_0\rangle$ and $|p_0\rangle$ remains to be fixed; we choose it in such a way that $C > 0$. Finally, to determine the value of C we use the closure relations

$$\boxed{\int_{\mathbf{R}^3} d^3x \, |x\rangle\langle x| = \int_{\mathbf{R}^3} d^3p \, |p\rangle\langle p| = I} \;, \tag{3.28}$$

which permits us to write

$$\langle x|\psi\rangle = \int_{\mathbf{R}^3 \times \mathbf{R}^3} d^3p \, d^3x' \langle x|p\rangle\langle p|x'\rangle\langle x'|\psi\rangle$$

$$= |C|^2 \int_{\mathbf{R}^3 \times \mathbf{R}^3} d^3p \, d^3x' \, e^{ip\cdot(x-x')/\hbar} \langle x'|\psi\rangle$$

$$= |C|^2 (2\pi\hbar)^3 \int_{\mathbf{R}^3} d^3x' \, \delta(x - x')\langle x'|\psi\rangle = |C|^2 (2\pi\hbar)^3 \langle x|\psi\rangle \;, \tag{3.29}$$

from which follows

$$\boxed{\langle x|p\rangle = \frac{1}{(2\pi\hbar)^{3/2}} e^{ip\cdot x/\hbar}} \;. \tag{3.30}$$

Given the matrix element $\langle x|p\rangle$, the relation between $\psi(x)$ and $\hat{\psi}(p)$ is

$$
\boxed{
\begin{aligned}
\psi(x) &= \frac{1}{(2\pi\hbar)^{3/2}} \int_{\mathbf{R}^3} d^3p\, e^{ip\cdot x/\hbar}\hat{\psi}(p) \\
\hat{\psi}(p) &= \frac{1}{(2\pi\hbar)^{3/2}} \int_{\mathbf{R}^3} d^3x\, e^{-ip\cdot x/\hbar}\psi(x)\,,
\end{aligned}
}
\tag{3.31}
$$

which is just the expression for the Fourier transform (see Appendix D).

In the general case of a system of several particles with cartesian position observables $X_1, X_2, \ldots, X_N$, and conjugate momenta $P_1, P_2, \ldots, P_N$, the position representation is obtained by considering as the complete set of compatible observables the set $\{X_1, X_2, \ldots, X_N; A\}$, where A indicates a collection of operators necessary to make this set complete. The system of common eigenvectors $\{|x_1, x_2, \ldots, x_N; \alpha\rangle\}$ forms an orthonormal basis. In this representation, with phases chosen as indicated above, the expressions for the position and momentum operators are, respectively,

$$
\begin{aligned}
(X_r\psi)(x_1, x_2, \ldots, x_N; \alpha) &= x_r\psi(x_1, x_2, \ldots, x_N; \alpha)\,, \\
(P_r\psi)(x_1, x_2, \ldots, x_N; \alpha) &= -i\hbar\frac{\partial}{\partial x_r}\psi(x_1, x_2, \ldots, x_N; \alpha)\,,
\end{aligned}
\tag{3.32}
$$

this latter requiring that A and P_r be compatible.

In complete analogy, the momentum representation is obtained by considering as the complete set of compatible observables the set $\{P_1, P_2, \ldots, P_N; B\}$, where again B designates a collection of operators necessary to complete the set. The system of common eigenvectors $\{|p_1, p_2, \ldots, p_N; \beta\rangle\}$ is once more an orthonormal basis. In this representation, with appropriate choice of phases, the expressions for the position and momentum operators are

$$
\begin{aligned}
(X_r\psi)\hat{}\,(p_1, p_2, \ldots, p_N; \beta) &= +i\hbar\frac{\partial}{\partial p_r}\hat{\psi}(p_1, p_2, \ldots, p_N; \beta)\,, \\
(P_r\psi)\hat{}\,(p_1, p_2, \ldots, p_N; \beta) &= p_r\hat{\psi}(p_1, p_2, \ldots, p_N; \beta)\,,
\end{aligned}
\tag{3.33}
$$

assuming for the first of these that B and X_r are compatible.

When $A = B$ (a choice which is often possible since these operators are generally associated with the spin of the particles or other variables not related to ordinary space-time) we have

$$
\langle x_1, x_2, \ldots, x_N; \alpha'|p_1, p_2, \ldots, p_N; \beta\rangle
$$

$$
= (2\pi\hbar)^{-N/2}\exp\left[\frac{i}{\hbar}\sum_{j=1}^{N}p_ix_i\right]\delta(\alpha', \alpha)\,;
\tag{3.34}
$$

the wave functions in both representations are related in a form analogous to the case $N = 3$.

In the general case, in which the $X_1, X_2, \ldots, X_N$, or the $P_1, P_2, \ldots, P_N$ may not suffice to form a complete set of compatible observables, the Hilbert space is no longer compelled to be realized as an $L^2(\mathbf{R}^N)$. The norm square of a vector $|\psi\rangle$ with wave function $\psi(x_1, \ldots, x_N; \alpha)$ is

$$\|\psi\|^2 = \sum_\alpha \int_{\mathbf{R}^N} d^N x \, |\psi(x_1, \ldots, x_N; \alpha)|^2 \, , \tag{3.35}$$

where the sum runs over the spectra of the remainder of the observables of the complete set, and can in some cases be an integral. Thus, the Hilbert space is a direct sum (or direct integral) of $L^2(\mathbf{R}^N)$ spaces, or if one prefers, the tensor product of $L^2(\mathbf{R}^N)$ with the Hilbert space associated with the quantum numbers $\{\alpha\}$. Thus, for example, for a system of a single particle of spin $\hbar/2$, in the position representation, the wave function will be of the form $\psi(x; \alpha)$, with $\alpha = \pm\hbar/2$, if we choose as the complete set the operators $\{X, S_z\}$. One then expresses this wave function as the ordered pair or *spinor*

$$|\psi\rangle \rightarrow \begin{pmatrix} \psi_+(x) \\ \psi_-(x) \end{pmatrix} \, , \quad \psi_\pm(x) \equiv \psi(x; \pm\hbar/2) \, , \tag{3.36}$$

in which $\psi_\pm(x)$ represents the probability amplitude for finding the particle at the point x with spin $\pm\hbar/2$ in the Oz direction. Now the Hilbert space has the form $L^2(\mathbf{R}^3) \otimes \mathbf{C}^2 \simeq L^2(\mathbf{R}^3, \mathbf{C}^2) \simeq L^2(\mathbf{R}^3) \oplus L^2(\mathbf{R}^3)$.

3.4 Position-Momentum Uncertainty Relations

The position representation formalism which we have just developed will permit us to analyze more closely the position-momentum uncertainty relations. For the sake of simplicity, we frequently consider in this and subsequent sections the case of a spinless particle, even when the results we obtain are immediately generalizable to other situations. The commutation relations are

$$[X_i, P_j] = i\hbar\delta_{ij}I \, , \quad [X_i, X_j] = [P_i, P_j] = 0 \, , \tag{3.37}$$

so that the general uncertainty relation (2.64) implies

$$\boxed{\Delta_\psi X_i \cdot \Delta_\psi P_j \geq \frac{\hbar}{2}\delta_{ij}} \, . \tag{3.38}$$

That is, it is impossible to prepare a system in a pure state for which the uncertainties in X_i and P_i are made arbitrarily small, since (3.38) must always be satisfied.

We would like next to inquire whether there exists a *minimal packet*, that is, a state of the system with wave function $\psi(x)$ such that the equal sign in (3.38) holds for $i = j = 1, 2, 3$. This problem was discussed in complete generality in Sect. 2.5. The only thing remaining to be examined now is whether (2.70) has solutions, and if so, to find them. In the position representation the equations corresponding to (2.70) are

$$(x_j - a_j)\psi(\boldsymbol{x}) = -i\lambda_j \left[-i\hbar \frac{\partial \psi(\boldsymbol{x})}{\partial x_j} - b_j\psi(\boldsymbol{x}) \right] , \quad j = 1, 2, 3 , \tag{3.39}$$

where

$$a_j = \langle X_j \rangle_\psi , \quad b_j = \langle P_j \rangle_\psi , \quad \lambda_j = \frac{\hbar}{2(\Delta_\psi P_j)^2} = \frac{2(\Delta_\psi X_j)^2}{\hbar} , \tag{3.40}$$

which may be immediately seen from the earlier work [in particular from (2.71)]. For notational simplicity, and when the state is understood, we will omit the subscript ψ on the expectation values and the standard deviations. The equation (3.39) is separable, and its solutions are of the form

$$\psi(\boldsymbol{x}) = \prod_{j=1}^{3} N_j \exp\{-[x_j^2/2 - (a_j + i\lambda_j b_j)x_j]/\hbar\lambda_j\} , \tag{3.41}$$

where N_j are arbitrary constants. The normalized wave functions corresponding to minimal wave packets are thus, aside from an irrelevant global phase

$$\psi(\boldsymbol{x}) = \prod_{j=1}^{3} (\pi\hbar\lambda_j)^{-1/4} \exp[-(x_j - a_j)^2/2\hbar\lambda_j] \exp[ib_j(x_j - a_j/2)/\hbar] , \tag{3.42}$$

from which one can verify directly the validity of (3.40). The phase has been chosen such that the Fourier transform of $\psi(\boldsymbol{x})$ will be related to (3.42) in a simple way. In the momentum representation the minimal packet can be calculated using (3.31), that is, by carrying out the Fourier transform, which yields

$$\hat{\psi}(\boldsymbol{p}) = \prod_{j=1}^{3} \left(\frac{\lambda_j}{\pi\hbar} \right)^{1/4} \exp[-\lambda_j(p_j - b_j)^2/2\hbar] \exp[-ia_j(p_j - b_j/2)/\hbar] . \tag{3.43}$$

The symmetry between these two representations is made clearer if we use (3.40), since we can then write

$$\psi(\boldsymbol{x}) = \prod_{j=1}^{3} [2\pi(\Delta X_j)^2]^{-1/4} \exp\{-[(x_j - \langle X_j \rangle)/2\Delta X_j]^2\}$$
$$\times \exp[i\langle P_j \rangle(x_j - \langle X_i \rangle/2)/\hbar]$$

$$\hat{\psi}(\boldsymbol{p}) = \prod_{j=1}^{3} [2\pi(\Delta P_j)^2]^{-1/4} \exp\{-[(p_j - \langle P_j \rangle)/2\Delta P_j]^2\} \tag{3.44}$$
$$\times \exp[-i\langle X_j \rangle(p_j - \langle P_j \rangle/2)/\hbar] ,$$

which is a consequence of the nearly symmetric roles which X_i and P_i play in the commutation rules, a fact upon which we have already commented.

If one would like to have not only that $\Delta X_i \cdot \Delta P_i = \hbar/2$ for $i = 1, 2, 3$, but also that $\Delta(\hat{\boldsymbol{n}} \cdot \boldsymbol{P}) \cdot \Delta(\hat{\boldsymbol{n}} \cdot \boldsymbol{X}) = \hbar/2$ for any real unit vector $\boldsymbol{n}$, then since we have, for the previous wave function $\psi(\boldsymbol{x})$,

$$\Delta(\hat{\boldsymbol{n}} \cdot \boldsymbol{X}) = \sqrt{\sum_{i=1}^{3} n_i^2 (\Delta X_i)^2} \,, \quad \Delta(\hat{\boldsymbol{n}} \cdot \boldsymbol{P}) = \sqrt{\sum_{i=1}^{3} n_i^2 (\Delta P_i)^2} \,, \tag{3.45}$$

a simple application of the Schwarz inequality leads to $\Delta X_1 = \Delta X_2 = \Delta X_3 \equiv \Delta X$, $\Delta P_1 = \Delta P_2 = \Delta P_3 \equiv \Delta P$. Thus, such a minimal wave packet is

$$\begin{aligned}
\psi(\boldsymbol{x}) &= [2\pi(\Delta X)^2]^{-3/4} \exp[-(\boldsymbol{x} - \langle \boldsymbol{X} \rangle)^2/4(\Delta X)^2] \\
&\quad \times \exp[\mathrm{i}\langle \boldsymbol{P} \rangle \cdot (\boldsymbol{x} - \langle \boldsymbol{X} \rangle/2)/\hbar] \\
\hat{\psi}(\boldsymbol{p}) &= [2\pi(\Delta P)^2]^{-3/4} \exp[-(\boldsymbol{p} - \langle \boldsymbol{P} \rangle)^2/4(\Delta P)^2] \\
&\quad \times \exp[-\mathrm{i}\langle \boldsymbol{X} \rangle \cdot (\boldsymbol{p} - \langle \boldsymbol{P} \rangle/2)/\hbar] \,,
\end{aligned} \tag{3.46}$$

where $\Delta P = \hbar/2\Delta X$.

In summary, every plane wave modulated by a gaussian represents a minimal packet. We will see shortly, however, that it loses this minimal packet character in the course of its free-particle time evolution.

3.5 Probability Density and Probability Current Density

In Sect. 3.3 we have seen that if the wave function $\Psi(\boldsymbol{x}; t)$ for a spinless particle in the position representation is normalized, then

$$\varrho(\boldsymbol{x}; t) \equiv |\Psi(\boldsymbol{x}; t)|^2 \tag{3.47}$$

is the probability density for finding the particle at the point $\boldsymbol{x}$ at the instant t. We will see how $\varrho(\boldsymbol{x}; t)$ changes with time in the Schrödinger picture.

In the position representation, the Schrödinger equation is

$$\boxed{\mathrm{i}\hbar\frac{\partial\Psi(\boldsymbol{x}; t)}{\partial t} = H\Psi(\boldsymbol{x}; t)} \,, \tag{3.48}$$

where H is the Hamiltonian operator in this representation. For example, if the particle is non-relativistic (the only kind of situation we will consider) and if it is subjected to forces described by a potential $V(\boldsymbol{x}; t)$, the expression for the classical Hamiltonian is

$$H = \frac{1}{2M}\boldsymbol{p}^2 + V(\boldsymbol{x}; t) \,, \tag{3.49}$$

M being the mass of the particle. According to postulate VI, the corresponding operator is

$$H = \frac{1}{2M}\boldsymbol{P}^2 + V(\boldsymbol{X}; t) \,, \tag{3.50}$$

which is formally self-adjoint since we always assume V to be real. Later we will discuss conditions which will ensure its self-adjoint character. In the position representation, the above Hamiltonian becomes, using (3.12) and (3.20),

$$H = -\frac{\hbar^2}{2M}\Delta + V(\boldsymbol{x}; t) \quad . \tag{3.51}$$

If, instead of cartesian coordinates in the euclidean space under consideration, we use arbitrary curvilinear coordinates (ξ^i) such that the square of the distance between the points (ξ^i) and $(\xi^i + d\xi^i)$ is given by $ds^2 = g_{ij}(\xi)d\xi^i d\xi^j$, the change of variables in (3.51) can be shown to lead to the following differential operator for the Hamiltonian in the new coordinates:

$$H = -\frac{\hbar^2}{2M}\frac{1}{\sqrt{g(\xi)}}\frac{\partial}{\partial\xi^i}\left(\sqrt{g(\xi)}g^{ij}(\xi)\frac{\partial}{\partial\xi^j}\right) + V'(\xi; t) \quad , \tag{3.52}$$

where $V'(\xi; t) \equiv V(\boldsymbol{x}; t)$, $g(\xi) \equiv \det g_{ij}(\xi)$ and $g^{ij}(\xi)g_{jk}(\xi) = \delta_k^i$. One must be extremely careful not to obtain the operator H from the classical expression

$$H = \frac{1}{2M}g^{ij}(\xi)\eta_i\eta_j + V'(\xi; t) \quad , \tag{3.53}$$

where $\eta_i = Mg_{ij}(\xi)\dot{\xi}^j$ are the momenta conjugate to the variables ξ^i, by blindly applying the substitution $\eta_j \to -i\hbar\partial/\partial\xi^j$. In the first place, (3.53) can be written in many equivalent forms, which lead to different operators under this substitution. Second, this operator is not in general even formally self-adjoint since the volume element is now $\sqrt{g(\xi)}d^3\xi$. A formally self-adjoint operator which obeys the canonical commutation relations and which would be a candidate for representing η_j is

$$\eta_j \to -i\hbar\frac{1}{g^{1/4}}\frac{\partial}{\partial\xi^j}g^{1/4} \quad , \tag{3.54}$$

even though it may be necessary to add a real function of the ξ^i to the r.h.s. of (3.54) in order to ensure given relative phases in the new basis $|\xi^1, \xi^2, \xi^3\rangle$. The operator (3.54), together with the prescription of expressing (3.53) in the form

$$H = \frac{1}{2M}g^{-1/4}\eta_i g^{1/4}g^{ij}g^{1/4}\eta_j g^{-1/4} + V'(\xi; t) \quad , \tag{3.55}$$

leads to the result (3.52). Both (3.52) and (3.54) may be singular at some points, generally located on the boundaries of the ranges of the variables ξ_i. In such cases, one must choose boundary conditions pertinent to the appropriate self-adjoint extension of the operator [RH 63].

If $\Psi(\boldsymbol{x}; t)$ satisfies (3.48), then $\Psi^*(\boldsymbol{x}; t)$ will evolve according to

$$-i\hbar\frac{\partial\Psi^*(\boldsymbol{x}; t)}{\partial t} = [H\Psi(\boldsymbol{x}; t)]^* \quad , \tag{3.56}$$

and thus

$$\frac{\partial}{\partial t} \varrho(\boldsymbol{x}; t) = \Psi^*(\boldsymbol{x}; t) \frac{\partial \Psi(\boldsymbol{x}; t)}{\partial t} + \frac{\partial \Psi^*(\boldsymbol{x}; t)}{\partial t} \Psi(\boldsymbol{x}; t)$$

$$= \frac{1}{\mathrm{i}\hbar} [\Psi^* H \Psi - (H\Psi)^* \Psi] , \qquad (3.57)$$

from which we immediately obtain

$$\frac{d}{dt} \int_{\mathbf{R}^3} d^3x \; \varrho(\boldsymbol{x}; t) = \frac{1}{\mathrm{i}\hbar} \int_{\mathbf{R}^3} d^3x \; [\Psi^* H \Psi - (H\Psi)^* \Psi] = 0 , \qquad (3.58)$$

given that $H = H^\dagger$. The integral in the l.h.s. is the probability of finding the particle somewhere, and we see that H being self-adjoint guarantees that this probability does not change in time. That is, that there are no sources or sinks for this particle. Evidently, this is an immediate consequence of the fact that if $H = H^\dagger$, the operator $U(t, t_0)$ is unitary, and the norm of the state vector does not change in time:

$$\int_{\mathbf{R}^3} d^3x \; \varrho(\boldsymbol{x}; t_0) = \int_{\mathbf{R}^3} d^3x \; \Psi^*(\boldsymbol{x}; t_0) \Psi(\boldsymbol{x}; t_0)$$

$$= \int_{\mathbf{R}^3} \langle \Psi(t_0) | \boldsymbol{x} \rangle \, d^3x \, \langle \boldsymbol{x} | \Psi(t_0) \rangle$$

$$= \langle \Psi(t_0) | \Psi(t_0) \rangle = \langle \Psi(t) | \Psi(t) \rangle = \int_{\mathbf{R}^3} d^3x \; \varrho(\boldsymbol{x}; t) . \qquad (3.59)$$

On the other hand, if H is of the form (3.51), the relation (3.57) may be written as

$$\frac{\partial}{\partial t} \varrho(\boldsymbol{x}; t) = \frac{\mathrm{i}\hbar}{2M} \{ \Psi^*(\boldsymbol{x}; t) \Delta \Psi(\boldsymbol{x}; t) - [\Delta \Psi^*(\boldsymbol{x}; t)] \Psi(\boldsymbol{x}; t) \} , \qquad (3.60)$$

or, equivalently,

$$\boxed{\frac{\partial \varrho(\boldsymbol{x}; t)}{\partial t} + \nabla \cdot \boldsymbol{J}(\boldsymbol{x}; t) = 0} , \text{ where} \qquad (3.61)$$

$$\boxed{\boldsymbol{J}(\boldsymbol{x}; t) \equiv \frac{-\mathrm{i}\hbar}{2M} \{ \Psi^*(\boldsymbol{x}; t)[\nabla \Psi(\boldsymbol{x}; t)] - [\nabla \Psi^*(\boldsymbol{x}; t)] \Psi(\boldsymbol{x}; t) \}} . \qquad (3.62)$$

Equation (3.61) is formally identical to the continuity equation for a fluid of density $\varrho(\boldsymbol{x}; t)$ and current density $\boldsymbol{J}(\boldsymbol{x}; t)$ having no sources or sinks. In (3.60) $\varrho(\boldsymbol{x}; t)$ represents the *probability density* and, by analogy with what has just been said, $\boldsymbol{J}(\boldsymbol{x}; t)$ acquires the name *probability current density*. In the special case of $\Psi(\boldsymbol{x}; t_0)$ being real except for a global phase, (3.62) indicates that $\boldsymbol{J}(\boldsymbol{x}; t_0) \equiv 0$.

For stationary states of energy E, assuming that V is independent of time, we have

$$\Psi(\boldsymbol{x}; t) = \psi(\boldsymbol{x}) \mathrm{e}^{-\mathrm{i}Et/\hbar} , \qquad (3.63)$$

so that the explicit time dependence of $\varrho(\boldsymbol{x}; t)$ and $\boldsymbol{J}(\boldsymbol{x}; t)$ disappears and thus (3.61) reduces to

$$\nabla \cdot J(x) = 0 \, , \tag{3.64}$$

that is to say, the current is solenoidal.

In the case of a free particle, $V(x) \equiv 0$, the stationary solution of the Schrödinger equation with well-defined energy E and momentum p is

$$\Psi(x;t) = (2\pi\hbar)^{-3/2}e^{i(p\cdot x - Et)/\hbar} \, , \quad E = \frac{1}{2M}p^2 \, , \tag{3.65}$$

with the normalization discussed in Sect. 2.4. For such a function,

$$\varrho(x;t) = (2\pi\hbar)^{-3} \, , \quad J(x;t) = (2\pi\hbar)^{-3}\frac{1}{M}p \, , \tag{3.66}$$

and thus the relative probability current density is the relative probability density multiplied by the velocity of the particle p/M, in perfect agreement with the idea of a current.

The expressions (3.47) and (3.62) admit a simple generalization [JA 68]: if A is any observable, we can define probability densities and probability current densities associated with A at the point x for a particle in the state $|\Psi(t)\rangle$ as

$$\varrho_A(x;t) \equiv \left\langle \frac{1}{2}(E_x A + A E_x) \right\rangle_{\Psi(t)} \, ,$$

$$J_A(x;t) \equiv \left\langle \frac{1}{4M}[(E_x P + P E_x)A + A(E_x P + P E_x)] \right\rangle_{\Psi(t)} \, , \tag{3.67}$$

where $E_x \equiv |x\rangle\langle x|$. For the case $A = I$, we obtain (3.47) and (3.62). Note that these quantities are real and that the integral over $\mathbf{R}^3$ of $\varrho_A(x;t)$ yields the mean value of A in the state $|\Psi(t)\rangle$. If A does not depend explicitly on time, and is a constant of motion, the Schrödinger equation leads to the continuity equation

$$\frac{\partial \varrho_A(x;t)}{\partial t} + \nabla \cdot J_A(x;t) = 0 \, . \tag{3.68}$$

Finally, we should point out that knowledge of $\varrho(x;t_0)$ gives nothing more than the modulus of the wave function. Given the dual nature of the observables X and P, one could imagine that the phase of $\Psi(x;t_0)$ is essentially fixed if we also know the momentum probability distribution

$$\tilde{\varrho}(p;t_0) \equiv |\hat{\Psi}(p;t_0)|^2 \, . \tag{3.69}$$

This is, however, not necessarily so, as the following one-dimensional counter-example demonstrates: let

$$\varphi_n(x) \equiv [\sqrt{\pi}2^n n!]^{-1/2}e^{-x^2/2}H_n(x) \, , \quad n = 0, 1, 2, \dots \, , \tag{3.70}$$

where H_n are the Hermite polynomials. These functions form an orthonormal basis in $L^2(\mathbf{R})$. Their Fourier transform $\hat{\varphi}_n(p)$ satisfy the relation [see (4.51)]

$$\hat{\varphi}_n(p) = (-i)^n\varphi_n(p) \, . \tag{3.71}$$

Now consider the wave functions

$$\psi_1(x) \equiv \frac{1}{\sqrt{2}}[\varphi_0(x) + i\varphi_4(x)] ,$$

$$\psi_2(x) \equiv \frac{1}{\sqrt{2}}[\varphi_0(x) - i\varphi_4(x)] ,$$

(3.72)

which are mutually orthogonal and of unit norm. Since the φ_n are real, the probability densities $\varrho_1(x) \equiv |\psi_1(x)|^2$, $\varrho_2(x) \equiv |\psi_2(x)|^2$ are equal. On the other hand, from (3.71) we deduce

$$\hat{\psi}_1(p) = \frac{1}{\sqrt{2}}[\varphi_0(p) + i\varphi_4(p)] = \psi_1(p) ,$$

$$\hat{\psi}_2(p) = \frac{1}{\sqrt{2}}[\varphi_0(p) - i\varphi_4(p)] = \psi_2(p) ,$$

(3.73)

so that $\tilde{\varrho}_1(p) = \tilde{\varrho}_2(p)$. Therefore, ψ_1, ψ_2 have identical position and momentum distributions but are nonetheless orthogonal.

The question of whether or not $\varrho(\boldsymbol{x}; t_0)$ and $(\partial\varrho/\partial t)(\boldsymbol{x}; t_0)$ determine, within a global phase factor, the wave function $\Psi(\boldsymbol{x}; t_0)$ has been analyzed with an affirmative answer, provided certain very restrictive conditions hold [KE 37]. It is obvious that one should not expect $\varrho, \dot{\varrho} \equiv \partial\varrho/\partial t$ at t_0 to determine Ψ in the general case: it suffices to consider $\Psi = \Psi_1 + \Psi_2$, with $\Psi_1, \Psi_2 \in C_0^\infty(\mathbf{R}^3)$ and with disjoint supports at $t = t_0$. From this, $\varrho = |\Psi_1|^2 + |\Psi_2|^2$, and the continuity equation (3.61) permits one to see that at $t = t_0$, $\dot{\varrho}$ is also a sum of two analogous terms with disjoint supports. Thus, ϱ and $\dot{\varrho}$ at $t = t_0$ do not distinguish between $\Psi_1 + \Psi_2$ and $\Psi_1 + \exp(i\alpha)\Psi_2$, α real.

3.6 Ehrenfest's Theorem

In classical mechanics, the dynamical state of a system is determined by knowing the position and momentum coordinates of each of its particles at some initial instant. This is no longer possible in quantum mechanics due to the incompatibility of the position and momentum observables. Only in the limit $\hbar \to 0$, when this incompatibility disappears, can we expect to recover the classical description. In a quantal description, a pure state of a system is determined totally by its wave function $\Psi(x_1, \ldots, x_N; t)$. One can attempt to recover the classical picture by attributing position and momentum coordinates to each particle which are precisely the expectation values of the corresponding operators for that state $\langle X_i \rangle_\Psi$ and $\langle P_i \rangle_\Psi$, neglecting their fluctuations. In order for this procedure to succeed, it is necessary, first of all, that the expectation values of the position and momentum observables obey, to a good approximation, the classical laws of motion, and secondly, that the typical dimensions of the packet $\Psi(x_1, \ldots, x_N; t)$ associated with the system's motion be small compared to the natural dimensions of the system and remain so in the time interval of interest.

We now prove Ehrenfest's theorem, which will allow us to be more precise about the situations in which the first of the two conditions mentioned above will be satisfied, leaving until later a study of the second. For simplicity, we will restrict ourselves to the case of a spinless particle whose classical Hamiltionian is

$$H = \frac{1}{2M}p^2 + V(\boldsymbol{x}; t) \ . \tag{3.74}$$

The classical equations of motion are

$$\frac{d\boldsymbol{x}}{dt} = \frac{1}{M}\boldsymbol{P} \ , \quad \frac{d\boldsymbol{p}}{dt} = \boldsymbol{F}(\boldsymbol{x}; t) \ , \tag{3.75}$$

where the force is given by $\boldsymbol{F}(\boldsymbol{x}; t) = -\boldsymbol{\nabla}V(\boldsymbol{x}; t)$.

In quantum mechanics, the temporal variation of the expectation values is dictated by (2.148), which when written explicitly for the operators X and P reads

$$\frac{d}{dt}\langle X \rangle_\Psi = \frac{1}{i\hbar}\langle [X, H] \rangle_\Psi \ , \quad \frac{d}{dt}\langle P \rangle_\Psi = \frac{1}{i\hbar}\langle [P, H] \rangle_\Psi \ . \tag{3.76}$$

From the canonical commutation rules, and using as H the operator corresponding to (3.74), one obtains

$$[X, H] = \frac{i\hbar}{M}P \ , \quad [P, H] = -i\hbar\boldsymbol{\nabla}V(\boldsymbol{x}; t) = i\hbar\boldsymbol{F}(\boldsymbol{x}; t) \ , \tag{3.77}$$

and (3.76) can then be written as

$$\boxed{\frac{d}{dt}\langle X \rangle_\Psi = \frac{1}{M}\langle P \rangle_\Psi \ , \quad \frac{d}{dt}\langle P \rangle_\Psi = \langle \boldsymbol{F}(\boldsymbol{x}; t) \rangle_\Psi} \ , \tag{3.78}$$

which express the content of *Ehrenfest's theorem*. It should be noted that these equations do not let us affirm that $\langle X \rangle_\Psi$ and $\langle P \rangle_\Psi$ obey classical laws; for that, it is necessary that

$$\langle \boldsymbol{F}(\boldsymbol{x}; t) \rangle_\Psi = F(\langle X \rangle_\Psi; t) \ , \tag{3.79}$$

which is, in general, not true. What Ehrenfest's theorem states is that $\langle X \rangle_\Psi$ and $\langle P \rangle_\Psi$ "formally" satisfy the classical laws, but with "averaged forces" whose values depend upon the wave function. In order to see in which cases (3.79) can be considered satisfied as an approximation, we expand $\nabla_i V(\boldsymbol{x}; t)$ around the point $\langle X \rangle_{\Psi(t)}$:

$$\nabla_i V(\boldsymbol{x}; t) = (\nabla_i V)_{\text{cl}} + (\nabla_j \nabla_i V)_{\text{cl}}(x_j - \langle X_j \rangle_\Psi)$$
$$+ \tfrac{1}{2}(\nabla_k \nabla_j \nabla_i V)_{\text{cl}} \cdot (x_j - \langle X_j \rangle_\Psi)(x_k - \langle X_k \rangle_\Psi) + \dots \ , \tag{3.80}$$

where by "cl" we indicate that after calculating the term we have substituted $\langle X \rangle_\Psi$ for $\boldsymbol{x}$. With this, the second of equations (3.78) becomes

$$\frac{d}{dt}\langle P\rangle_{\Psi} = F(\langle X\rangle_{\Psi}; t) - \frac{1}{2}(\nabla_k\nabla_j\nabla V)_{\text{cl}}$$
$$\times [\langle X_j X_k\rangle_{\Psi} - \langle X_j\rangle_{\Psi}\langle X_k\rangle_{\Psi}] + \dots . \tag{3.81}$$

From (3.81) [or directly from (3.79)] one can see that Ehrenfest's theorem predicts that $\langle X\rangle_{\Psi}$ and $\langle P\rangle_{\Psi}$ move according to classical laws of motion whenever $V(\boldsymbol{x}; t)$ depends, at the most, quadratically on $\boldsymbol{x}$. This occurs, for example, for a free particle, for a particle subject to a uniform electric field, and for a particle subjected to a harmonic type potential. On the other hand, (3.81) allows us to state that the classical equations will be approximately valid whenever the anharmonicities of the potential are negligible in the region in which $\Psi(\boldsymbol{x}; t)$ is relevant.

The generalization of Ehrenfest's theorem to systems with several particles is obvious.

3.7 Propagation of Wave Packets (I)

For the motion of a wave packet to simulate that of a classical particle it is necessary, according to what we have said, that two conditions are met: first, that the average values of the position and momentum observables satisfy the classical laws to a good approximation, and second, that the dimensions of the packet be sufficiently small and remain so during the evolution. A glance at (3.79) shows that the first condition can hardly be fulfilled without the second being satisfied. We thus find ourselves drawn into an analysis of the evolution of a wave packet in time.

We consider the motion of a three-dimensional packet $\Psi(\boldsymbol{x}; t)$ which moves under the action of a force associated with a potential $V(\boldsymbol{x})$. We limit ourselves for the moment, following [ME 59], to the case in which $V(\boldsymbol{x})$ is at most quadratic in its variables:

$$V(\boldsymbol{x}) = V_0 - \sum_i F_i x_i + \frac{1}{2}\sum_{i,j} v_{ij} x_i x_j . \tag{3.82}$$

The results for a more general potential, which nevertheless can be approximated by (3.82) in the region covered by the wave packet during the time period of interest, will be well approximated by those obtained for this quadratic potential. In other words, the calculations will be correct except for terms of the order of the third (and higher) derivatives of the potential [AN 81].

Exactly as in the study of small oscillations in classical mechanics, it is advisable to introduce a set of principal coordinates which diagonalize v_{ij}. In this system of coordinates, also cartesian, $V(\boldsymbol{x})$ can be written as in (3.82), with $v_{ij} = v_i\delta_{ij}$. Thus H becomes the sum of three uncoupled Hamiltonians, so that in order to study the change with time of each of the quantities $\chi_i(t) \equiv (\Delta X_i)^2_{\Psi(t)}$, which contain the information about the evolution of the typical dimensions of the packet, it suffices to consider the one-dimensional case of each of the principal directions in turn.

The corresponding one-dimensional problem has as its Hamiltonian

$$H = \frac{1}{2M}P^2 + \alpha + \beta X + \frac{1}{2}\gamma X^2 . \tag{3.83}$$

Taking into account that

$$[X, H] = i\hbar \frac{1}{M} P , \quad [X^2, H] = i\hbar \frac{1}{M}(XP + PX) , \tag{3.84}$$

and using (2.148), one obtains

$$\frac{d\chi}{dt} = \frac{1}{M}[\langle XP + PX \rangle - 2\langle X \rangle \langle P \rangle] . \tag{3.85}$$

Since

$$[XP + PX, H] = 2i\hbar \left[\frac{1}{M}P^2 - X\frac{dV}{dX} \right] , \tag{3.86}$$

again (2.148) leads to

$$\frac{d^2\chi}{dt^2} = \frac{2}{M^2}\tilde{\chi} - \frac{2}{M}\left[\langle X\frac{dV}{dX} \rangle - \langle X \rangle \langle \frac{dV}{dX} \rangle \right] = \frac{2}{M^2}\tilde{\chi} - \frac{2}{M}\gamma\chi , \tag{3.87}$$

where $\tilde{\chi} \equiv (\Delta P)^2_{\Psi(t)}$. The r.h.s. in (3.87) is nothing more than $4[\langle T-V \rangle - L_{cl}]/M$ where $L_{cl} \equiv \langle P \rangle^2/2M - V_{cl}$. This suggests that in order to eliminate $\tilde{\chi}$ (whose time dependence we do not know) from (3.87), it suffices to show that the quantity

$$\varepsilon = \frac{\tilde{\chi}}{2M} + \frac{1}{2}\gamma\chi = \langle H \rangle - E_{cl} \tag{3.88}$$

is constant in time. This is a simple exercise. We can thus write

$$\frac{d^2\chi}{dt^2} = \frac{4}{M}[\varepsilon - \gamma\chi] , \tag{3.89}$$

an equation which indicates that the energy ε, which is the difference between the quantum energy of the packet and the classical energy of its center, is associated with the fluctuations of the position observable of the particle around its classical position.

The solution to (3.89), for constant ε, is

$$A > 0 : \quad \chi(t) = \alpha_0 e^{iA^{1/2}t} + \beta_0 e^{-iA^{1/2}t} + \frac{B}{A}$$

$$A = 0 : \quad \chi(t) = \alpha_0 + \beta_0 t + \frac{B}{2}t^2 \tag{3.90}$$

$$A < 0 : \quad \chi(t) = \alpha_0 e^{(-A)^{1/2}t} + \beta_0 e^{-(-A)^{1/2}t} + \frac{B}{A} ,$$

where $A \equiv 4\gamma/M$, $B \equiv 4\varepsilon/M$, and α_0, β_0 are constants of integration. These expressions permit us to calculate $\chi(t)$ in terms of $\chi_0, (d\chi/dt)_0$ and $\widetilde{\chi}_0$, where the subscript denotes the initial time t_0.

From (3.90) we learn that if $A \leq 0$, $\chi(t)$ can grow without limit, and thus even if the classical description were initially correct, it would cease to be so for sufficiently large times.

We next consider explicit cases with Hamiltonians of the type given by (3.83), for which the expressions (3.90) are exact:

1) **Particle Under the Action of a Constant Force.** Now $V(x) = V_0 - Fx$ and as a consequence

$$\chi(t) = \chi_0 + \left(\frac{d\chi}{dt}\right)_0 t + \frac{\widetilde{\chi}_0}{M^2} t^2 . \tag{3.91}$$

But $\widetilde{\chi}_0 \equiv (\Delta P)^2_{\Psi(t_0)}$ cannot be zero (we assume normalizable packets), so that (3.91) shows that, after a sufficiently large time, $\chi(t)$ surpasses any fixed quantity. If at the instant $t = t_0$ the packet were a minimal one,

$$\Psi(x; t_0) = (2\pi\chi_0)^{-1/4} \exp\{-[x - \langle X\rangle_0]^2/4\chi_0\} \exp(i\langle P\rangle_0 x/\hbar) , \tag{3.92}$$

it is easy to show that $\langle XP + PX\rangle_0 = 2\langle X\rangle_0\langle P\rangle_0$ and, according to (3.85), $(d\chi/dt)_0 = 0$. On the other hand, for this packet, $(\Delta X)_0(\Delta P)_0 = \hbar/2$ and as a consequence, the relation (3.91) reads

$$(\Delta X)_t = (\Delta X)_0 \sqrt{1 + \frac{\hbar^2 t^2}{4M^2(\Delta X)_0^4}} . \tag{3.93}$$

Thus, a minimal wave packet evolving under the action of a constant force always spreads as time elapses.

This spreading from initial time is not restricted to minimal packets. Any packet which, aside from an overall phase, is real at t_0 will have $(d\chi/dt)_0 = 0$, and thus has this same property. In this connection, it should be noted that it is always possible to construct packets which begin by contracting, since if $(d\chi/dt)_0 \neq 0$, this quantity being negative will produce such a packet. If this quantity is positive, it is easily shown that for the complex conjugate packet $\Psi^*(x; t_0)$ the corresponding value is $(d\chi^*/dt)_0 = -(d\chi/dt)_0$ and thus this packet will contract initially. This should not be surprising. As we shall see (Sect. 7.8) complex conjugation inverts the sense of time-evolution.

When we assume that $(d\chi/dt)_0 = 0$ initially in (3.91) (a relation which is evidently valid for a free particle; $F = 0$) and we use the uncertainty relation $\chi_0\widetilde{\chi}_0 \geq (\hbar/2)^2$, we immediately obtain the inequalities

$$\chi(t) = \chi_0 + \widetilde{\chi}_0 t^2/M^2 \geq 2(\chi_0\widetilde{\chi}_0)^{1/2}|t|/M \geq \hbar|t|/M . \tag{3.94}$$

The limitation $[\chi(t)]^{1/2} \geq (\hbar|t|/M)^{1/2}$ is known as the *standard quantum limit* for determining the position of a free particle, and it suggests that if two measurements of the position are carried out with a lapse of time $|t|$ separating them, the second one will necessarily have an uncertainty $\geq (\hbar|t|/M)^{1/2}$. The possibility of contracting packets $[(d\chi/dt)_0 < 0]$ after the first measurement could invalidate the previous conclusion. This problem remains unresolved [YU 84], [CA 85].

2) **Harmonic Oscillator**. Suppose that the particle is acted upon by a force which derives from the potential $V(x) = kx^2/2$. According to (3.78), the equations of motion for $\langle X \rangle$ and $\langle P \rangle$ are

$$\frac{d\langle X \rangle}{dt} = \frac{1}{M}\langle P \rangle , \qquad \frac{d\langle P \rangle}{dt} = -k\langle X \rangle , \tag{3.95}$$

which coincide with the classical equations. If $\omega \equiv (k/M)^{1/2}$, then

$$\langle X \rangle = \langle X \rangle_0 \cos \omega(t - t_0) + \frac{\langle P \rangle_0}{M\omega} \sin \omega(t - t_0) ,$$
$$\langle P \rangle = \langle P \rangle_0 \cos \omega(t - t_0) - M\omega\langle X \rangle_0 \sin \omega(t - t_0) . \tag{3.96}$$

Thus, both quantities $\langle X \rangle$ and $\langle P \rangle$ oscillate sinusoidally with angular frequency ω. It can be readily shown from the first of the formulae (3.90) that $\chi(t)$ describes a sinusoidal motion with frequency 2ω.

3.8 Wave Packet Propagation (II)

In the previous section we encountered equations which permitted us to examine, in some cases in exact form (or approximately by neglecting terms of the order of the third and higher derivatives of the potential), the distortion of a wave packet as time elapses. We now wish to attack the complete problem of its propagation; that is, the determination of $\Psi(x; t)$ starting from the wave function at initial time t_0, or, equivalently, solving the wave equation

$$i\hbar\frac{\partial\Psi(x; t)}{\partial t} = \left[-\frac{\hbar^2}{2M}\Delta + V(x) \right] \Psi(x; t) , \tag{3.97}$$

which is, in general, a difficult problem. For Hamiltonians which do not depend explicitly on time, such as the one considered here, we have already seen (in Sect. 2.9) that the evolution of the state vectors is given by

$$|\Psi(t_2)\rangle = e^{-iH(t_2-t_1)/\hbar}|\Psi(t_1)\rangle , \tag{3.98}$$

which, in the position representation, corresponds to

$$\Psi(x_2; t_2) = \int_{\mathbf{R}^3} d^3x_1\, G(x_2, t_2; x_1, t_1)\Psi(x_1; t_1)$$
$$G(x_2, t_2; x_1, t_1) \equiv \langle x_2|e^{-iH(t_2-t_1)/\hbar}|x_1\rangle . \tag{3.99}$$

It is evident that $\lim G(\boldsymbol{x}_2, t_2; \boldsymbol{x}_1, t_1) = \delta(\boldsymbol{x}_2 - \boldsymbol{x}_1)$ as $t_2 \to t_1$. We first consider the evolution into the future: $t_2 > t_1$. In this case, it is convenient to define a *retarded propagator*

$$G_+(\boldsymbol{x}_2, t_2; \boldsymbol{x}_1, t_1) = \theta(t_2 - t_1)\langle \boldsymbol{x}_2 | e^{-iH(t_2-t_1)/\hbar} | \boldsymbol{x}_1 \rangle \, , \qquad (3.100)$$

which is zero for $t_2 < t_1$, and which when substituted into (3.99) gives the evolution of the wave packets into the future. Then (3.99) shows that $\Psi(\boldsymbol{x}_2; t_2)$ is a superposition of all of the amplitudes at $(\boldsymbol{x}_2; t_2)$ produced by propagation of the waves emitted by $(\boldsymbol{x}_1; t_1)$, for all points $\boldsymbol{x}_1$ and at an instant $t_1 < t_2$; each of these points contributes with a probability amplitude $\Psi(\boldsymbol{x}_1; t_1)$, and $G_+(\boldsymbol{x}_2, t_2; \boldsymbol{x}_1, t_1)$ is the amplitude of the wave at $(\boldsymbol{x}_2; t_2)$ produced by a delta-function type source at $\boldsymbol{x}_1$ at the instant $t_1 < t_2$.

The expression for $G_+(\boldsymbol{x}_2, t_2; \boldsymbol{x}_1, t_1)$ presented in (3.100) is of little practical utility, since it is a purely formal expression in which the matrix element defining it is in general very difficult to evaluate. We thus look for a differential or integral equation for G_+, which might suggest an approximate calculation.

In order to derive the differential equation satisfied by the retarded propagator, we proceed in the following fashion: We assume for simplicity that H has a pure point and nondegenerate spectrum. Let $\{|E_n\rangle\}$ be the associated orthonormal basis. From (3.100) we find

$$G_+(\boldsymbol{x}_2, t_2; \boldsymbol{x}_1, t_1)$$
$$= \theta(t_2 - t_1) \sum_n \sum_m \langle \boldsymbol{x}_2 | E_n \rangle \langle E_n | e^{-iH(t_2-t_1)/\hbar} | E_m \rangle \langle E_m | \boldsymbol{x}_1 \rangle$$
$$= \theta(t_2 - t_1) \sum_n \langle \boldsymbol{x}_2 | E_n \rangle \langle E_n | \boldsymbol{x}_1 \rangle e^{-iE_n(t_2-t_1)/\hbar}$$
$$= \theta(t_2 - t_1) \sum_n \varphi_n(\boldsymbol{x}_2) \varphi_n^*(\boldsymbol{x}_1) e^{-iE_n(t_2-t_1)/\hbar} \, , \qquad (3.101)$$

where $\varphi_n(\boldsymbol{x}) \equiv \langle \boldsymbol{x} | E_n \rangle$ is the wave function for $|E_n\rangle$. Taking into account that $d\theta(t_2 - t_1)/dt_2 = \delta(t_2 - t_1)$, it is easily proved, using (3.101), that

$$\left(i\hbar \frac{\partial}{\partial t_2} - \left[-\frac{\hbar^2}{2M} \Delta_2 + V(\boldsymbol{x}_2) \right] \right) G_+(\boldsymbol{x}_2, t_2; \boldsymbol{x}_1, t_1)$$
$$= i\hbar \delta(t_2 - t_1) \sum_n \varphi_n(\boldsymbol{x}_2) \varphi_n^*(\boldsymbol{x}_1) \, . \qquad (3.102)$$

The closure relation now implies that

$$\left(i\hbar \frac{\partial}{\partial t_2} - \left[-\frac{\hbar^2}{2M} \Delta_2 + V(\boldsymbol{x}_2) \right] \right) G_+(\boldsymbol{x}_2, t_2; \boldsymbol{x}_1, t_1)$$
$$= i\hbar \delta(t_2 - t_1) \delta(\boldsymbol{x}_2 - \boldsymbol{x}_1) \, , \qquad (3.103)$$

which is formally valid for any H.

The differential equation (3.103) is of complicated structure, but it can happen that we are able to solve it for a Hamiltonian H_0 "near" H, in which case the

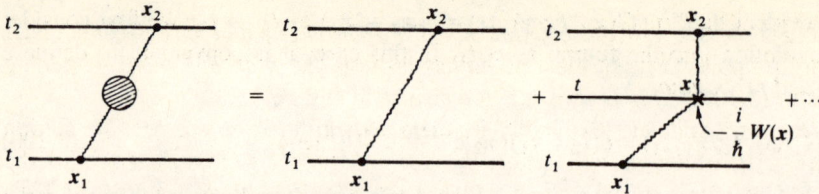

Fig. 3.1. Graphical representation of the expansion (3.109)

propagator for H might be found from that for H_0 by a perturbative calculation. The relation between both propagators can be formally established as follows: We consider the operator $\exp[iH_0(t-t_1)/\hbar]\exp[-iH(t-t_1)/\hbar]$. An easy calculation leads to

$$i\hbar\frac{d}{dt}e^{iH_0(t-t_1)/\hbar}e^{-iH(t-t_1)/\hbar} = e^{iH_0(t-t_1)/\hbar}(H-H_0)e^{-iH(t-t_1)/\hbar} , \quad (3.104)$$

which, upon integration with the appropriate boundary condition, gives

$$e^{iH_0(t_2-t_1)/\hbar}e^{-iH(t_2-t_1)/\hbar}$$
$$= I - \frac{i}{\hbar}\int_{t_1}^{t_2} dt\, e^{-iH_0(t-t_1)/\hbar}(H-H_0)e^{-iH(t-t_1)/\hbar} , \quad (3.105)$$

from which

$$e^{-iH(t_2-t_1)/\hbar}$$
$$= e^{-iH_0(t_2-t_1)/\hbar} - \frac{i}{\hbar}\int_{t_1}^{t_2} dt\, e^{-iH_0(t_2-t)/\hbar}(H-H_0)e^{-iH(t-t_1)/\hbar}. \quad (3.106)$$

Using $t_2 > t_1$, and taking matrix elements between $|x_2\rangle$ and $|x_1\rangle$, one obtains

$$G_+(x_2,t_2;x_1,t_1) = G_{+0}(x_2,t_2;x_1,t_1)$$
$$- \frac{i}{\hbar}\int_{\mathbf{R}^4} dt\, d^3x\, G_{+0}(x_2,t_2;x,t)W(x)G_+(x,t;x_1,t_1) , \quad (3.107)$$

where we assume that $H - H_0$ is local: $\langle x'|H-H_0|x\rangle = W(x)\delta(x'-x)$. We have been able to extend the integration over t to $(-\infty,+\infty)$ because we are considering retarded propagators. We condense (3.107) to

$$G_+ = G_{+0} - \frac{i}{\hbar}G_{+0}WG_+ , \quad (3.108)$$

which lends itself to a formal perturbative expansion:

$$G_+ = G_{+0} - \frac{i}{\hbar}G_{+0}WG_{+0} + \left(-\frac{i}{\hbar}\right)^2 G_{+0}WG_{+0}WG_{+0} + \dots . \quad (3.109)$$

The interpretation of (3.109) is clear: the probability amplitude $G_+(x_2,t_2;x_1,t_1)$ for finding a particle at $(x_2;t_2)$ given an initial position $(x_1;t_1)$ is the sum of various terms: the analogous amplitude for evolution under H_0; plus the amplitude that the particle propagates retardedly with H_0 up to some arbitrary point $(x; t)$ where it feels the action of $-i(H - H_0)/\hbar$, then proceeds, again under H_0, to the point $(x_2; t_2)$; plus, in similar fashion, the third and other remaining terms in (3.109). This is shown symbolically in Fig. 3.1.

We now calculate explicitly the retarded propagator for free evolution $(V(x) \equiv 0)$, which we will denote by $G_+^{(0)}(x_2, t_2; x_1, t_1)$. Taking as a basis the states of definite momentum $|p\rangle$, which diagonalize H, we derive from (3.101)

$$G_+^{(0)}(x_2, t_2; x_1, t_1) = \theta(t_2 - t_1)\frac{1}{(2\pi\hbar)^3}$$

$$\times \int_{\mathbf{R}^3} d^3p\, e^{i p \cdot (x_2 - x_1)/\hbar} e^{-ip^2(t_2 - t_1)/2M\hbar} . \qquad (3.110)$$

The angular integrals can be done immediately, giving $(p \equiv |p|)$

$$G_+^{(0)}(x_2, t_2; x_1, t_1) = \theta(t_2 - t_1)\frac{1}{2\pi^2\hbar^2|x_2 - x_1|}$$

$$\times \int_0^\infty dp\, p \, \sin\left[\frac{p|x_2 - x_1|}{\hbar}\right] e^{-ip^2(t_2 - t_1)/2M\hbar} . \qquad (3.111)$$

For the last integral, it suffices to make use of the formula

$$\int_{-\infty}^{+\infty} dx\, x\, e^{-i(ax^2 + 2bx)} = -\frac{b\sqrt{\pi}}{a^{3/2}}e^{-i\pi/4}e^{ib^2/a} , \qquad a > 0 , \qquad (3.112)$$

from which

$$\boxed{\begin{array}{c} G_+^{(0)}(x_2, t_2; x_1, t_1) = \theta(t_2 - t_1)e^{-i3\pi/4}\left[\dfrac{M}{2\pi\hbar(t_2 - t_1)}\right]^{3/2} \\[2mm] \times \exp[iM(x_2 - x_1)^2/2\hbar(t_2 - t_1)] , \end{array}} \qquad (3.113)$$

which is the explicit expression for the free retarded propagator.

The analogy of (3.113) with the classical propagator of diffusion theory is manifest: the free-particle Schrödinger equation is a diffusion equation with a purely imaginary diffusion coefficient.

On the other hand, (3.113) demonstrates that $G_+^{(0)}$ depends on the differences $x_2 - x_1$, $t_2 - t_1$, reflecting the homogeneity of space-time for free-particle evolution. In addition, since the support is entirely $\mathbf{R}^3$, free particle evolution presents the characteristics of an instantaneous diffusion: if $\Psi(x; t_1) \propto \delta(x - x_0)$, that is, if the particle is localized at x_0 at the instant t_1, then $\Psi(x; t_2) \propto G_+^{(0)}(x, t_2; x_0, t_1)$, and thus the probability density is nonzero at all points for $t_2 > t_1$.

Using (3.99) and (3.113), it is easy to show that if $\Psi(.; t_1) \in L^2(\mathbf{R}^3) \cap L^1(\mathbf{R}^3)$, then $\Psi(x; t_2)$ is essentially bounded for all x, when $t_2 \neq t_1$ and that this bound decreases as $|t_2 - t_1|^{-3/2}$ as $|t_2 - t_1|$ increases. In other words, the probability of finding the particle in a finite region of $\mathbf{R}^3$ decreases as $|t_2 - t_1|^{-3}$ in the course of time evolution of a free particle, and so the particle escapes to infinity.

It might be thought that the characteristics of the spreading described above are due to the non-relativistic form of the kinetic energy, which imposes no limits on the propagation velocity. This is not the case. It suffices that the energy $H_0(P)$ be a continuous, non-constant function bounded from below in order for any freely moving wave packet, initially localized in a bounded region, to have a noncompact support at any later time [GA 68], [HR 80]. If in addition, H_0 is spherically symmetric and strictly increasing with $|P|$, and $\Psi(x; t_0)$ presents a "hole", that is, a non-empty open set $\Omega \subset \mathbf{R}^3$ in which Ψ is zero, the function $\Psi(x; t)$ will "fill it" at some $t \neq t_0$ as close to t_0 as desired [HR 80]. For $H_0(P) = [c^2 P^2 + M^2 c^4]^{1/2} - Mc^2$, these results appear to be in conflict with einsteinian causality, were it possible to construct initial packets with compact support. However, the difficulty vanishes if one takes into account that the construction of such states requires a precise delimitation of their boundaries, whose observability implies the use of, for example, photons of arbitrarily high energy. This would lead to the subsequent production of particle-antiparticle pairs, and the impossibility of describing these processes in the framework of a quantum theory with a finite number of degrees of freedom. See also [HE 85].

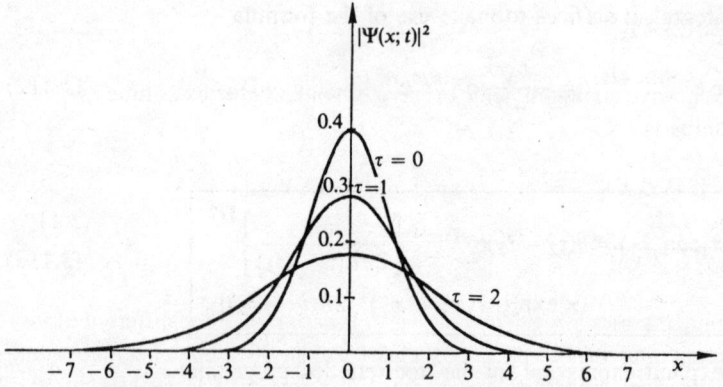

Fig. 3.2. Representation of $|\Psi(x; t)|^2$ as a function of x for different values of the parameter $\tau \equiv \hbar t / 2M(\Delta X)_0^2$, with $(\Delta X)_0 = 1$

Let us analyze in particular the free evolution of a packet which at $t = 0$ is isotropic and minimal [(3.46)]

$$\Psi(x; 0) = [2\pi(\Delta X)_0^2]^{-3/4} \exp[-(x - \langle X \rangle_0)^2 / 4(\Delta X)_0^2]$$
$$\times \exp[i\langle P \rangle_0 \cdot (x - \langle X \rangle_0 / 2)/\hbar] . \tag{3.114}$$

Using (3.99) and the propagator given in (3.113), one obtains after a long but elementary calculation that for $t \geq 0$

$$\Psi(x; t) = [2\pi(\Delta X)_0^2]^{-3/4} \left[\frac{(\Delta X)_0}{(\bar{\Delta} X)_t} \right]^3 \exp \frac{-[x - \langle X \rangle_0 - t\langle P \rangle_0 / M]^2}{4(\bar{\Delta} X)_t^2}$$

$$\times \exp\{\mathrm{i}\langle P\rangle_0 \cdot [\boldsymbol{x} - (\langle X\rangle_0 + t\langle P\rangle_0/M)/2]/\hbar\} \ , \tag{3.115}$$

where

$$(\bar{\Delta}X)_t^2 \equiv (\Delta X)_0^2 + \mathrm{i}\frac{\hbar t}{2M} \ , \quad \arg(\bar{\Delta}X)_t < \frac{\pi}{4} \ . \tag{3.116}$$

Notice here that the center of the wave packet, $\langle X\rangle_t = \langle X\rangle_0 + t\langle P\rangle_0/M$, moves with the velocity $\langle P\rangle_0/M$, the so-called group velocity. The average phase velocity is $\langle P\rangle_0/2M$. For the width of the wave packet, which is no longer minimal for $t \neq 0$, we find

$$(\Delta X_i)_t = \frac{|(\bar{\Delta}X)_t^2|}{(\Delta X)_0} = (\Delta X)_0\sqrt{1 + \frac{\hbar^2 t^2}{4M^2(\Delta X)_0^4}} \ , \quad i = 1, 2, 3 \ , \tag{3.117}$$

which demonstrates the spreading of the packet in the course of time and coincides with (3.93), found earlier.

For a one-dimensional problem under the assumption $\langle X\rangle_0 = \langle P\rangle_0 = 0$, the expression equivalent to (3.115) is

$$\Psi(x; t) = [2\pi(\Delta X)_0^2]^{-1/4}\frac{(\Delta X)_0}{(\bar{\Delta}X)_t} \exp\left[-x^2/4(\bar{\Delta}X)_t^2\right] \ , \tag{3.118}$$

which describes a wave anchored at the origin and which deforms as time evolves. Its square modulus is

$$|\Psi(x; t)|^2 = [2\pi(\Delta X)_0^2]^{-1/2}\varrho_t^{-1/2} \exp\left[-x^2/2(\Delta X)_0^2\varrho_t\right] \ ,$$
$$\varrho_t \equiv 1 + \frac{\hbar^2 t^2}{4M^2(\Delta X)_0^4} \ . \tag{3.119}$$

Shown in Fig. 3.2 are plots of $|\Psi(x; t)|^2$ as a function of x for different times; progressive broadening of the packet is clearly seen. It should be noted that in a typically classical situation, where $M = 1\,\mathrm{g}$ and $(\Delta X)_0 = 0.01\,\mathrm{cm}$, the uncertainty in position has increased by a factor of two in a time $t \simeq 3 \times 10^{23}\,\mathrm{s}$, while in the case of an electron with $M \simeq 9 \times 10^{-28}\,\mathrm{g}$, and $(\Delta X)_0 \simeq 10^{-8}\,\mathrm{cm}$, the required time is only $t \simeq 3 \times 10^{-16}\,\mathrm{s}$.

Graphs of $\mathrm{Re}\{\Psi(x; t)\}$ and $\mathrm{Im}\{\Psi(x; t)\}$ at several times are drawn in Fig. 3.3. Observe that the function, which is initially real, evolves a non-trivial phase during free evolution. This is not only true for the example we have considered, but is in fact quite general: under free evolution, all packets that are initially real at $t = 0$ cease to be essentially real for $t \neq 0$. A simple way of seeing this consists of expanding the packet in plane waves:

$$\Psi(\boldsymbol{x}; t) = (2\pi\hbar)^{-3/2} \int_{\mathbb{R}^3} d^3p\ \hat{\psi}(\boldsymbol{p})\mathrm{e}^{\mathrm{i}(\boldsymbol{p}\cdot\boldsymbol{x} - Et)/\hbar} \ , \quad E \equiv \frac{\boldsymbol{p}^2}{2M} \ . \tag{3.120}$$

Taking the complex conjugate and requiring that $\Psi^*(\boldsymbol{x}; t) = \mathrm{e}^{\mathrm{i}\beta(t)}\Psi(\boldsymbol{x}; t)$ for all $\boldsymbol{x}$ results in $\hat{\psi}(\boldsymbol{p})$ satisfying

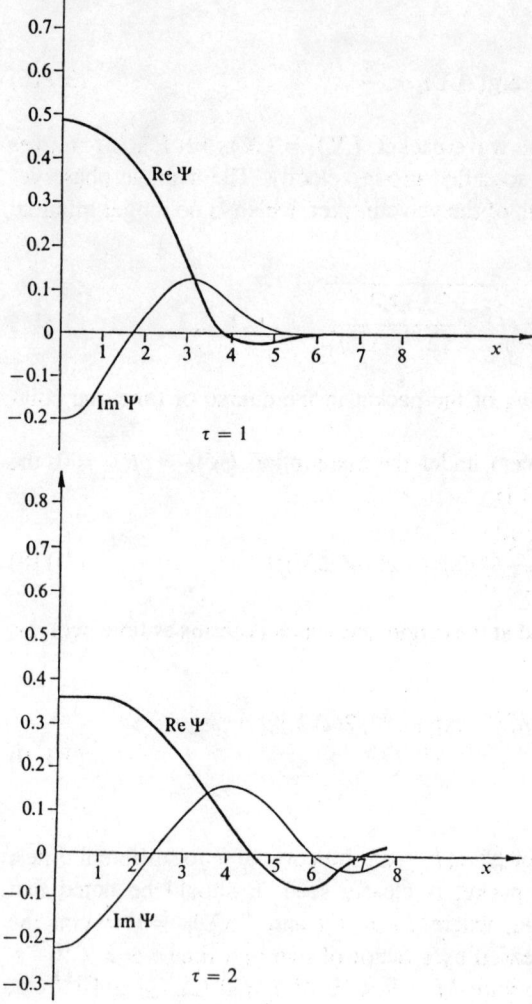

Fig. 3.3. Plots of $\text{Re}\{\Psi(x; t)\}$ and $\text{Im}\{\Psi(x; t)\}$ for different values of the parameter $\tau = \hbar t/2M(\Delta X)_0^2$ with $(\Delta X)_0 = 1$

$$\hat{\psi}^*(-\boldsymbol{p}) = \mathrm{e}^{\mathrm{i}\beta(t)}\mathrm{e}^{-2\mathrm{i}Et/\hbar}\hat{\psi}(\boldsymbol{p}) \ . \tag{3.121}$$

Since initially $\Psi(\boldsymbol{x}; 0)$ is real, $\hat{\psi}^*(-\boldsymbol{p}) = \hat{\psi}(-\boldsymbol{p})$; then (3.121) requires

$$\beta(t) = \frac{2Et}{\hbar} \quad (\bmod\ 2\pi) \tag{3.122}$$

for all $\boldsymbol{p}$ in the support of $\hat{\psi}$. The $|\boldsymbol{p}|$ solutions of (3.122) for $t \neq 0$ form a discrete set of points, while the support of $\hat{\psi}$ cannot have zero measure if $0 < \|\psi\| < \infty$, demonstrating the proposition.

Finally, we could have defined the *advanced propagator* G_-, responsible for evolution into the past, in the same way as we have introduced the retarded propagators, just by substitution of $\theta(t_1 - t_2)$ for $\theta(t_2 - t_1)$ in (3.100). From its definition it follows that $G_-(x_2, t_2; x_1, t_1) = G_+^*(x_1, t_1; x_2, t_2)$, so that everything established for the retarded propagators can be translated immediately to advanced propagators. Similarly, with very little effort, the above can be extended to non-conservative evolutions.

3.9 The Classical Limit of the Schrödinger Equation

In the preceding sections, a relationship has been established between the quantum equations of motion for expectation values and observables and the corresponding classical equations. Here we will complete the analogy by showing first that the Hamilton-Jacobi equation follows from the Schrödinger equation in the limit $\hbar \to 0$, and, second, by summarizing some rigorous results which amplify Ehrenfest's theorem and permit recovery of a classical description in an elegant and unambiguous way. If for simplicity we take a particle of mass M in a potential $V(x)$, then solving Hamilton's classical equations is equivalent [GO 70] to finding the complete integral of the Hamilton-Jacobi equation

$$\frac{\partial S}{\partial t} + \frac{1}{2M}(\nabla S)^2 + V = 0 . \tag{3.123}$$

This is a first-order partial differential equation in $3 + 1$ variables, x and t, and thus the complete integral, called Hamilton's principal function, depends on four independent constants of integration. Since S occurs in (3.123) only through its first derivatives, one of the constants is purely additive and may be taken to be zero, so that Hamilton's principal function is usually written as $S(x, t; \alpha_1, \alpha_2, \alpha_3)$. The momenta are then given by

$$p_i = \frac{\partial S(x, t; \alpha_1, \alpha_2, \alpha_3)}{\partial x_i} . \tag{3.124}$$

These equations permit the calculation of the constants α_1, α_2, and α_3 at the initial time t_0 in terms of the initial values of x and p. On the other hand, it can be shown that the quantities

$$\frac{\partial S(x, t; \alpha_1, \alpha_2, \alpha_3)}{\partial \alpha_i} \equiv \beta_i \tag{3.125}$$

are constants of the motion. These are determined by using the previous equation at $t = t_0$ in terms of the initial momenta and coordinates. The equations (3.125) in turn permit the evaluation of $x_i = x_i(t; \alpha_j, \beta_j)$, solving the problem of the motion. Noticing that ∇S is normal to the surfaces of constant S, we see that (3.124) expresses the fact that the particle trajectories must always be perpendicular to these surfaces.

If, as is true in our case, the Hamiltonian does not depend explicitly on time, S can be separated into a sum of a term which depends only on x and another depending only on t; then (3.123) leads to

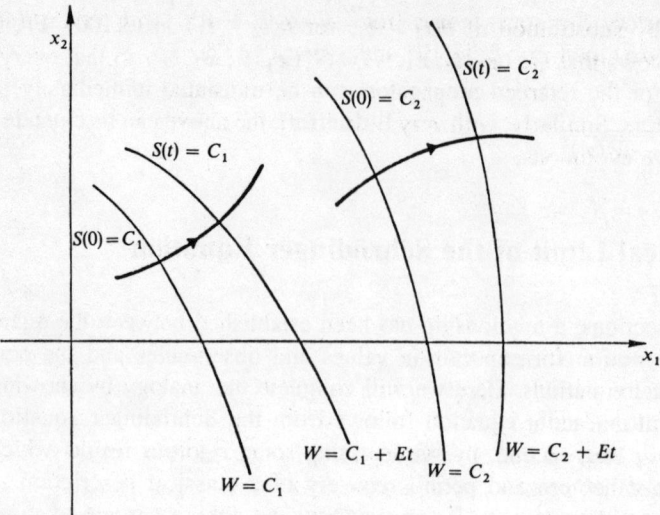

Fig. 3.4. Motion of the surfaces of constant S in a two-dimensional configuration space, and the trajectories of some particles, assuming $E > 0$

$$S(\boldsymbol{x}, t) = W(\boldsymbol{x}) - Et \,, \tag{3.126}$$

where $W(\boldsymbol{x})$, Hamilton's characteristic function, is the complete integral of the equation

$$\frac{1}{2M}(\nabla W)^2 + V = E \,. \tag{3.127}$$

The separation constant E in this case is equal to the total energy of the system. Hamilton's characteristic function depends, in addition to E, on three constants of integration, one of which is additive and can be arbitrarily set equal to zero, so that $S(\boldsymbol{x}, t; \alpha_1, \alpha_2, \alpha_3) = W(\boldsymbol{x}; \alpha_1, \alpha_2, \alpha_3) - Et$, where, for example, $\alpha_1 = E$. In configuration space, $\boldsymbol{x}$ in our case, the surfaces $W(\boldsymbol{x}; \boldsymbol{\alpha}) = C$ are fixed, while the surface $S(\boldsymbol{x}, t; \boldsymbol{\alpha}) = C$ moves in this space, coinciding at time t with the surface $W(\boldsymbol{x}; \boldsymbol{\alpha}) = C + Et$.

Figure 3.4 shows the situation for a two-dimensional case. It can be seen that the surfaces $S(\boldsymbol{x}, 0; \boldsymbol{\alpha}) = C_i$ coincide with $W(\boldsymbol{x}; \boldsymbol{\alpha}) = C_i$, and after some time t, the surfaces $S(\boldsymbol{x}, t; \boldsymbol{\alpha}) = C_i$ become the surfaces $W(\boldsymbol{x}; \boldsymbol{\alpha}) = C_i + Et$. According to (3.124), $\boldsymbol{p} = \nabla S$, and thus the trajectories of the particles are normal to the surfaces of constant S; hence these surfaces can be viewed as wavefronts propagating in configuration space. Since they deform in general as time elapses, the velocity of such waves, that is to say, the velocity with which the surfaces move, will not be uniform at all points. The velocity u, at a given point of the surface $S = \text{const}$, is given by the normal displacement dS by which the wave front advances in the direction of ∇S in the infinitesimal time interval dt, divided by dt; i.e., $u = dS/dt$. Taking into account that during the time interval dt the surface $S = $ const moves from W to $W + dW$ (with $dW = E\,dt$, and $dW = |\nabla W| dS$) then $u = E/|\nabla W|$ and, using (3.127),

$$u = \frac{E}{\sqrt{2M[E - V(\boldsymbol{x})]}} \; . \tag{3.128}$$

Since $2M(E - V) = p^2$, we have

$$u = \frac{E}{p} = \frac{E}{Mv} \; . \tag{3.129}$$

The phase velocity of a point on the surface $S = $ const is inversely proportional to the velocity of the particle whose motion describes the trajectory orthogonal to the surface S at that point.

Let us assume that $S_{\text{cl}}(\boldsymbol{x}, t)$ satisfies (3.123). Consider a fluid consisting of a collection of particles whose velocities, at a given instant t_0, are $(1/M)(\nabla S_{\text{cl}})(\boldsymbol{x}, t_0)$. These particles, under the action of the potential V, will move, and at each instant t their velocities are given by $(1/M)(\nabla S_{\text{cl}})(\boldsymbol{x}, t)$ if the particle is at the position $\boldsymbol{x}$. If ϱ_{cl} is the density of the fluid, we have the continuity equation

$$0 = \frac{\partial \varrho_{\text{cl}}}{\partial t} + \nabla \cdot \left(\varrho_{\text{cl}} \frac{\boldsymbol{p}}{M} \right) = \frac{\partial \varrho_{\text{cl}}}{\partial t} + \frac{1}{M} \nabla \cdot (\varrho_{\text{cl}} \nabla S_{\text{cl}}) \; . \tag{3.130}$$

After this digression into classical mechanics, we move on to our problem. The Schrödinger equation for a particle of mass M in a potential $V(\boldsymbol{x})$ is

$$i\hbar \frac{\partial \Psi(\boldsymbol{x}; t)}{\partial t} = \left[-\frac{\hbar^2}{2M} \Delta + V(\boldsymbol{x}) \right] \Psi(\boldsymbol{x}; t) \; . \tag{3.131}$$

Writing

$$\Psi(\boldsymbol{x}; t) = N \exp \left[\frac{i}{\hbar} S(\boldsymbol{x}, t) \right] \; , \tag{3.132}$$

where N is a normalization constant, (3.131) requires

$$\frac{\partial S}{\partial t} + \frac{1}{2M} (\nabla S)^2 + V - \frac{i\hbar}{2M} \Delta S = 0 \; , \tag{3.133}$$

which in the limit $\hbar \to 0$ coincides with the Hamilton-Jacobi equation. The last term is clearly the one responsible for quantum effects, $\hbar$ being the measure of these effects. Expanding S in a series of powers of $\hbar$ gives

$$S = S_0 - i\hbar S_1 + (-i\hbar)^2 S_2 + \dots \; . \tag{3.134}$$

Substituting into (3.133) and collecting terms of the same order in $\hbar$ leads to

$$\frac{\partial S_0}{\partial t} + \frac{1}{2M}(\nabla S_0)^2 + V = 0 \; ,$$

$$\frac{\partial S_1}{\partial t} + \frac{1}{M}(\nabla S_0) \cdot (\nabla S_1) + \frac{1}{2M}\Delta S_0 = 0 \; ,$$

$$\frac{\partial S_2}{\partial t} + \frac{1}{M}(\nabla S_0) \cdot (\nabla S_2) + \frac{1}{2M}(\nabla S_1)^2 + \frac{1}{2M}\Delta S_1 = 0 \tag{3.135}$$

$$\dots\dots\dots\dots\dots\dots\dots\dots\dots\dots\dots\dots \; .$$

Keeping only S_0 and S_1 in (3.134), the wave function would be

$$\Psi(\boldsymbol{x}; t) = N \, e^{S_1} e^{iS_0/\hbar} \; . \tag{3.136}$$

The first of equations (3.135) coincides with (3.123), and, since the equation is real, we can take its solution to be real (note that this is only possible in that region of space which is classically accessible; this is where we attempt to establish the analogy). Thus, the phase of this wave function is nothing more than Hamilton's principal function, to this order of approximation. The second of the equations coincides with the continuity equation (3.61) for the probability density, $|\Psi(\boldsymbol{x}; t)|^2 = |N|^2 \exp(2S_1)$, in which S_1 is assumed real (as is S_0). Now turning to the classical fluid mentioned previously, with $\varrho_{\mathrm{cl}} \equiv |N|^2 \exp(2S_1)$, and moving according to $S_{\mathrm{cl}}(\boldsymbol{x}; t) \equiv S_0(\boldsymbol{x}; t)$, the fluid's classical evolution will be approximately dictated by that of the quantum probability density.

Selecting a particular stationary state of energy E, for which $S_0 = -Et + W(\boldsymbol{x})$, the first equation of (3.135) shows that the local, reduced wavelength, $\bar{\lambda}(\boldsymbol{x}) \equiv \lambda(\boldsymbol{x})/2\pi$, is given by

$$\bar{\lambda}(x) = \frac{\hbar}{\sqrt{2M[E - V(\boldsymbol{x})]}} \; . \tag{3.137}$$

To conclude, we establish qualitatively the limits of validity of the approximation (3.136), and thus those of the classical-quantum analogy. Granted that the constant parts of $S_0, S_1, S_2, \dots$ can always be combined into the constant N, and if we limit our consideration to packets which are for practical purposes monochromatic (i.e. $\partial S_0/\partial t \simeq -E$, $\partial S_i/\partial t \simeq 0$ for $i \geq 1$), then an estimate of the possible validity of the series expansion (3.134) will be reflected in the conditions $|\nabla S_0| \gg \hbar|\nabla S_1| \gg \hbar^2|\nabla S_2| \gg \dots$. It is not possible to estimate these quantities rigorously by means of (3.135) in the general case, but the following order of magnitude relations are easily shown: $|\nabla S_0| \simeq \hbar/\bar{\lambda}$, $|\nabla S_1| \simeq |\nabla\bar{\lambda}|/\bar{\lambda}$, $|\nabla S_2| \simeq |\nabla\bar{\lambda}|^2/\hbar\bar{\lambda}$, so that the previous conditions can be summarized by the following:

$$|\nabla\bar{\lambda}| \ll 1 \; , \tag{3.138}$$

that is, the wavelength associated with the particle should change only slightly in one wavelength.

The rigorous procedure due to *Hepp* [HE 74] for obtaining the classical limit stands in sharp contrast to the above qualitative description. Even though the method is applicable to a system with an arbitrary number of degrees of freedom and with potential and magnetic interactions, we will restrict ourselves, for the sake of brevity, to displaying it for a one-dimensional case with a Hamiltonian

$$H = \frac{P^2}{2M} + V(X) \, . \tag{3.139}$$

The observables X and P are sensitive to a change in $\hbar$, since $[X, P] = i\hbar I$. In order to make the following arguments clearer, we need to express them in the form $X = X_\hbar \equiv \hbar^{1/2}\bar{X}$ and $P = P_\hbar \equiv \hbar^{1/2}\bar{P}$, where $[\bar{X}, \bar{P}] = iI$, and $\bar{X}, \bar{P}$ are fixed with respect to variations of $\hbar$.

We consider the unitary operator

$$W(\xi, \pi) = \exp\left[\frac{i}{\hbar}(\pi X_\hbar - \xi P_\hbar)\right] \, , \quad \xi, \pi \in \mathbf{R} \, . \tag{3.140}$$

This operator transforms an arbitrarily chosen state $|\phi\rangle$ into another one whose limit, as $\hbar \to 0$, represents a particle exactly localized at the position ξ and with momentum π. More definitively, we have (in the sense of strong convergence)

$$W^\dagger(\xi, \pi) \exp[i(\beta X_\hbar - \alpha P_\hbar)] W(\xi, \pi) \xrightarrow[\hbar \to 0]{} \exp[i(\beta\xi - \alpha\pi)] \, , \tag{3.141}$$

so that

$$\begin{aligned} W^\dagger(\xi, \pi) X_\hbar W(\xi, \pi) &\underset{\hbar \to 0}{\sim} \xi \, , \\ W^\dagger(\xi, \pi) P_\hbar W(\xi, \pi) &\underset{\hbar \to 0}{\sim} \pi \, . \end{aligned} \tag{3.142}$$

Operators of this type arise naturally in the study of coherent states (see Sect. 4.4).

It can be shown, under certain conditions on V (namely, regularity and some restriction upon its increase for large $|x|$) that the operators $X_\hbar(t)$ and $P_\hbar(t)$ – the position and momentum operators in the Heisenberg picture – evolving under H from initial time $t = 0$ are transformed under $W(x_{cl}(0), p_{cl}(0))$ in such a way that

$$\begin{aligned} W^\dagger(x_{cl}(0), p_{cl}(0)) X_\hbar(t) W(x_{cl}(0), p_{cl}(0)) &\underset{\hbar \to 0}{\sim} x_{cl}(t) + \hbar^{1/2}\bar{X}_{osc}(t) \, , \\ W^\dagger(x_{cl}(0), p_{cl}(0)) P_\hbar(t) W(x_{cl}(0), p_{cl}(0)) &\underset{\hbar \to 0}{\sim} p_{cl}(t) + \hbar^{1/2}\bar{P}_{osc}(t) \, . \end{aligned} \tag{3.143}$$

Here $(x_{cl}(t), p_{cl}(t))$ is the classical trajectory corresponding to the Hamiltonian associated with (3.139), with the initial conditions $(x_{cl}(0), p_{cl}(0))$. On the other hand, $\bar{X}_{osc}(t), \bar{P}_{osc}(t)$ are operators which satisfy the equations

$$\frac{d}{dt}\bar{X}_{osc}(t) = \frac{1}{M}\bar{P}_{osc}(t) \, , \quad \frac{d}{dt}\bar{P}_{osc}(t) = -V''(x_{cl}(t))\bar{X}_{osc}(t) \, , \tag{3.144}$$

with the initial conditions $\bar{X}_{osc}(0) = \bar{X}$, $\bar{P}_{osc}(0) = \bar{P}$.

The interpretation of these results is clear: given $|\phi\rangle$, and abbreviating $W(x_{cl}(0), p_{cl}(0))$ by W, we have

$$\langle W\phi|X_\hbar(t)|W\phi\rangle \underset{\hbar\to 0}{\sim} x_{\text{cl}}(t) + \hbar^{1/2}\langle\phi|\bar{X}_{\text{osc}}(t)|\phi\rangle \ ,$$

$$\langle W\phi|P_\hbar(t)|W\phi\rangle \underset{\hbar\to 0}{\sim} p_{\text{cl}}(t) + \hbar^{1/2}\langle\phi|\bar{P}_{\text{osc}}(t)|\phi\rangle \ . \tag{3.145}$$

Thus, $W|\phi\rangle$, initially localized on the trajectory in the limit $\hbar \to 0$, evolves under H in such a way that it follows the classical trajectory, except for corrections of order $\hbar^{1/2}$, which represent harmonic quantum fluctuations around the classical trajectory.

3.10 The Virial Theorem

In classical mechanics, the virial theorem affirms that for all systems of point-like particles with masses M_i, position vectors r_i and forces F_i, undergoing periodic motion, the relation

$$\overline{K} = -\frac{1}{2}\overline{\sum_i r_i \cdot F_i} \tag{3.146}$$

holds. Here K is the kinetic energy, and the bar over a quantity indicates its average over a period. [In reality, the theorem has a more general validity. It suffices that the positions and velocities remain bounded, if the bar is instead interpreted as a time average in the interval $(0, t)$, with $t \to \infty$.] This theorem has a quantum equivalent which we now develop.

Consider a conservative system, with Hamiltonian H, and let A, B be any two time-independent observables in the Schrödinger picture. An easy calculation permits us to see that in the Heisenberg picture

$$\frac{i}{\hbar}[H, A_{\text{H}}B_{\text{H}} + B_{\text{H}}A_{\text{H}}] = \dot{A}_{\text{H}}B_{\text{H}} + B_{\text{H}}\dot{A}_{\text{H}} + A_{\text{H}}\dot{B}_{\text{H}} + \dot{B}_{\text{H}}A_{\text{H}} \ , \tag{3.147}$$

where a dot indicates a derivative with respect to time. If $|\Psi\rangle_{\text{H}}$ is a bound state of the system, the expectation value of the l.h.s. is zero, and thus

$$\langle\dot{A}_{\text{H}}B_{\text{H}} + B_{\text{H}}\dot{A}_{\text{H}} + A_{\text{H}}\dot{B}_{\text{H}} + \dot{B}_{\text{H}}A_{\text{H}}\rangle_{\Psi_{\text{H}}} = 0 \ . \tag{3.148}$$

In the case of a particle with a Hamiltonian

$$H = \frac{P^2}{2M} + V(X) \ , \tag{3.149}$$

and taking $A = X_i$, $B = P_i$, the Heisenberg equation (2.205) tells us that

$$\dot{X}_{\text{H}} = \frac{1}{M}P_{\text{H}} \ , \quad \dot{P}_{\text{H}} = -(\nabla V)_{\text{H}} \ . \tag{3.150}$$

Substituting into (3.148), summing over $i = 1, 2, 3$ and recalling that the average values are independent of the picture used, we get

$$\boxed{\langle K \rangle_\Psi = \tfrac{1}{2} \langle \boldsymbol{X} \cdot (\nabla V) \rangle_\Psi}, \tag{3.151}$$

whose resemblance to (3.146) is evident. This result can be immediately generalized to any collection of particles with local interactions. If the potentials are not local, (3.151) is replaced by

$$\langle \boldsymbol{V} \cdot \boldsymbol{P} + \boldsymbol{P} \cdot \boldsymbol{V} \rangle_\Psi = -\langle \boldsymbol{X} \cdot \boldsymbol{F} + \boldsymbol{F} \cdot \boldsymbol{X} \rangle_\Psi, \quad \text{where} \tag{3.152}$$

$$V_{\text{H}} \equiv \frac{1}{i\hbar} [X_{\text{H}}, H], \quad F_{\text{H}} \equiv \frac{1}{i\hbar} [P_{\text{H}}, H] \tag{3.153}$$

are the velocity and force operators, respectively.

For the case of a particle with Hamiltonian (3.149), if V is a homogeneous function of order n in the variables x_1, x_2 and x_3,

$$V(\lambda x_1, \lambda x_2, \lambda x_3) = \lambda^n V(x_1, x_2, x_3), \tag{3.154}$$

then Euler's identity permits us to write

$$\boldsymbol{x} \cdot (\nabla V) = nV; \tag{3.155}$$

thus, for these potentials (3.151) becomes

$$\langle K \rangle_\Psi = \frac{n}{2} \langle V \rangle_\Psi, \tag{3.156}$$

an expression valid with $n = -1$ for a particle in a Coulomb potential, and with $n = 2$ for a harmonic oscillator.

Our demonstration of (3.148) and in particular (3.151) has been completely formal. There exists a very large class of potentials $V(x)$, $x \equiv (x_1, x_2, \ldots x_N)$, for which the virial theorem in the form (3.151) can be rigorously proved [WE 67], [AL 72] for a system of several particles. This class is composed of those V which satisfy:

1) $V \in Q_\alpha(\mathbf{R}^N)$ for some $\alpha > 0$, where the *Stummel* spaces Q_α are defined as

$$Q_\alpha(\mathbf{R}^N) \equiv \left\{ f : \sup_x \int_{|x-y| \leq 1} d^N y \, |f(y)|^2 |x-y|^{4-N-\alpha} < \infty \right\}; \tag{3.157}$$

2) There exists the derivative $V_r(x) \equiv \sum_{i=1}^N x_i \partial V / \partial x_i$, for all $x \neq 0$, a function $V_0(x) \in Q_\beta(\mathbf{R}^N)$ for some $\beta > 0$, and some $\varepsilon_0 > 0$, such that

$$\frac{1}{\varepsilon} |V((1+\varepsilon)x) - V(x)| \leq V_0(x) \tag{3.158}$$

whenever $0 < \varepsilon < \varepsilon_0$.

These conditions ensure that H is essentially self-adjoint on $C_0^\infty(\mathbf{R}^N)$, that (3.151) is valid for all normalizable eigenstates $|\Psi\rangle$ of H, and that $K|\Psi\rangle$, $V|\Psi\rangle$ and $V_r|\Psi\rangle$ are also normalizable.

Important subspaces of $Q_\alpha(\mathbf{R}^N)$ are: $L^p(\mathbf{R}^N) + L^\infty(\mathbf{R}^N)$, if $p \geq 2$ and $0 < \alpha < 4$, provided $p > 2N/(4 - \alpha)$; $L^2(\mathbf{R}^3) + L^\infty(\mathbf{R}^3)$, if $0 < \alpha \leq 1$. The potentials satisfying (3.157) admit singularities as strong as $|x|^{-1/2+\epsilon}$, if $N = 1$, and $|x|^{-3/2+\epsilon}$ for $N = 3$ ($\epsilon > 0$).

The virial theorem continues to be satisfied for quantum systems with several particles interacting via potentials from one, two, etc. bodies, so long as each one of these potentials satisfies the conditions (1) and (2) in the corresponding number of dimensions. This is essentially due to the inclusion $Q_\alpha(\mathbf{R}^N) \subset Q_\alpha(\mathbf{R}^{N+r})$, for all integers $r \geq 0$. In particular, this includes non-relativistic atomic systems with pure Coulomb interactions. It is convenient to note that when the dynamics is invariant under translations, there cannot be any bound states of the total Hamiltonian, but the results discussed here persist for the bound states of the Hamiltonian relative to the center of mass.

Potentials are frequently used which do not obey (3.157) – the harmonic oscillator, for example. Nevertheless, the virial theorem continues to hold for (attractive) harmonic potentials, because the wave functions of eigenstates have such well-behaved properties that all steps leading to (3.151) in the formal proofs remain valid.

The virial theorem has interesting applications for the spectral structure of Hamiltonians in addition to providing the useful relationship (3.156) between the expectation values for kinetic and potential energies [WE 67].

3.11 Path Integration

Feynman [FE 48] introduced a new formulation of quantum mechanics in 1948 which has been subsequently developed in the book by *Feynman* and *Hibbs* [FH 63]. The basic idea of the Feynman method is the following: in classical mechanics, the trajectory of a particle moving between the points (x_1, t_1) and (x_2, t_2) is determined (locally) by the principle of least action:

$$\delta S_{\text{cl}} \equiv \delta \int_{t_1}^{t_2} dt \, L(x(t), \dot{x}(t); t) = 0 \, , \qquad (3.159)$$

L being the Lagrangian of the system and S_{cl} the action. In quantum mechanics, the probability amplitude for the particle to pass from (x_1, t_1) to (x_2, t_2) receives contributions from all possible paths connecting the two points, whether or not they are classically realizable within the predetermined dynamics (we recall in this connection the Young double slit experiment). Feynman's fundamental hypothesis is that all of these paths which are physically indistinguishable (there being no detector which indicates the actual path taken by the particle) contribute coherently to the quantum amplitude with equal weight but with phases $\exp(iS_{\text{cl}}/\hbar)$, S_{cl} being the classical action over this path. At the macroscopic level, $|S_{\text{cl}}| \gg \hbar$, and thus contributions from separate paths will interfere destructively in general, there remaining only the contribution from that path which makes S_{cl} stationary, that is, from the true path followed by the classical motion (3.159).

Feynman brilliantly deduced the fundamental results of quantum mechanics starting from this intuitive idea. Here we will proceed backwards: relying on the quantum theory developed thus far, we will obtain Feynman's central formula for calculation of propagators. Consider a system with a Hamiltonian $H(t)$; the probability amplitude that if the particle is at $\boldsymbol{x}'$ at time t', it will be at $\boldsymbol{x}$ at $t > t'$ is given by

$$G(\boldsymbol{x}, t; \boldsymbol{x}', t') = \langle \boldsymbol{x}|U(t, t')|\boldsymbol{x}' \rangle , \tag{3.160}$$

where $U(t, t')$ is the evolution operator defined by (2.127). Now consider the time interval $[t', t]$ divided into $N + 1$ equal parts, and let the end times for these parts be $t_n = n\varepsilon + t'$, with $\varepsilon \equiv (t - t')/(N + 1)$, $t_0 \equiv t'$ and $t_{N+1} \equiv t$. Repeated application of (2.129) permits us to write

$$U(t, t') = U(t, t_N)U(t_N, t_{N-1}) \dots U(t_2, t_1)U(t_1, t') , \tag{3.161}$$

from which it immediately results that

$$
\begin{aligned}
\langle \boldsymbol{x}|U(t, t')|\boldsymbol{x}' \rangle = \int_{\mathbf{R}^{3N}} d^3x_1 \dots d^3x_N \, \langle \boldsymbol{x}|U(t, t_N)|\boldsymbol{x}_N \rangle \\
\times \langle \boldsymbol{x}_N|U(t_N, t_{N-1})|\boldsymbol{x}_{N-1} \rangle \dots \\
\dots \langle \boldsymbol{x}_2|U(t_2, t_1)|\boldsymbol{x}_1 \rangle \langle \boldsymbol{x}_1|U(t_1, t')|\boldsymbol{x}' \rangle .
\end{aligned}
\tag{3.162}
$$

Defining $\boldsymbol{x}_{N+1} \equiv \boldsymbol{x}$ and $\boldsymbol{x}_0 \equiv \boldsymbol{x}'$, the above equality in the limit $\varepsilon \downarrow 0$ (or equivalently $N \rightarrow \infty$) can be written as

$$G(\boldsymbol{x}, t; \boldsymbol{x}', t') = \lim_{\varepsilon \downarrow 0} \int_{\mathbf{R}^{3N}} d^3x_1 \dots d^3x_N \prod_{j=1}^{N+1} G(\boldsymbol{x}_j, t_j; \boldsymbol{x}_{j-1}, t_{j-1}) . \tag{3.163}$$

From the integral equation (2.127) which defines $U(t_j, t_{j-1})$, we can easily deduce that to order ε

$$G(\boldsymbol{x}_j, t_j; \boldsymbol{x}_{j-1}, t_{j-1}) \simeq \delta(\boldsymbol{x}_j - \boldsymbol{x}_{j-1}) - \frac{\mathrm{i}\varepsilon}{\hbar} \langle \boldsymbol{x}_j|H\left(\frac{t_j + t_{j-1}}{2}\right)|\boldsymbol{x}_{j-1} \rangle . \tag{3.164}$$

If we assume that the Hamiltonian for the problem has the form

$$H = \frac{1}{2M}\boldsymbol{P}^2 + V(\boldsymbol{X}, t) , \tag{3.165}$$

then

$$
\begin{aligned}
\langle \boldsymbol{x}_j|H|\boldsymbol{x}_{j-1} \rangle = \frac{1}{2M} \int_{\mathbf{R}^6} d^3p_j d^3p_{j-1} \langle \boldsymbol{x}_j|\boldsymbol{p}_j \rangle \langle \boldsymbol{p}_j|\boldsymbol{P}^2|\boldsymbol{p}_{j-1} \rangle \langle \boldsymbol{p}_{j-1}|\boldsymbol{x}_{j-1} \rangle \\
+ \delta(\boldsymbol{x}_j - \boldsymbol{x}_{j-1})V\left(\frac{\boldsymbol{x}_j + \boldsymbol{x}_{j-1}}{2}, t\right) .
\end{aligned}
\tag{3.166}
$$

Taking into account the expression (3.30), that $\langle p_j | P^2 | p_{j-1} \rangle = p_j^2 \delta(p_j - p_{j-1})$, and the integral representation of the Dirac delta-function, we obtain

$$\langle x_j | H(t) | x_{j-1} \rangle = \frac{1}{(2\pi\hbar)^3} \int_{\mathbf{R}^3} d^3 p_j \exp[i p_j \cdot (x_j - x_{j-1})/\hbar]$$
$$\times \left[\frac{1}{2M} p_j^2 + V\left(\frac{x_j + x_{j-1}}{2}, t\right) \right] , \qquad (3.167)$$

where the quantity in square brackets is nothing else than the classical Hamiltonian $H((x_j + x_{j-1})/2, p_j, t)$. Substituting this result in (3.164),

$$G(x_j, t_j; x_{j-1}, t_{j-1})$$
$$\simeq \int_{\mathbf{R}^3} \frac{d^3 p_j}{(2\pi\hbar)^3} \left[1 - \frac{i\varepsilon}{\hbar} H\left(\frac{x_j + x_{j-1}}{2}, p_j, \frac{t_j + t_{j-1}}{2}\right) \right]$$
$$\times \exp[i p_j \cdot (x_j - x_{j-1})/\hbar] , \qquad (3.168)$$

which, to the order considered, can be written as

$$G(x_j, t_j; x_{j-1}, t_{j-1}) \simeq \int_{\mathbf{R}^3} \frac{d^3 p_j}{(2\pi\hbar)^3} \exp\left(\frac{i\varepsilon}{\hbar} \left[p_j \cdot \left(\frac{x_j - x_{j-1}}{\varepsilon}\right)\right.\right.$$
$$\left.\left. - H\left(\frac{x_j + x_{j-1}}{2}, p_j, \frac{t_1 + t_{j-1}}{2}\right)\right]\right) . \qquad (3.169)$$

Substituting this into (3.163) gives

$$G(x; t; x', t') = \lim_{\varepsilon \downarrow 0} \int_{\mathbf{R}^{3(2N+1)}} d^3 x_1 \ldots d^3 x_N \frac{d^3 p_1}{(2\pi\hbar)^3} \cdots \frac{d^3 p_{N+1}}{(2\pi\hbar)^3}$$
$$\times \exp\left(\frac{i\varepsilon}{\hbar} \sum_{j=1}^{N+1} \left[p_j \cdot \left(\frac{x_j - x_{j-1}}{\varepsilon}\right) - H\left(\frac{x_j + x_{j-1}}{2}, p_j, \frac{t_j + t_{j-1}}{2}\right)\right]\right) . $$
$$(3.170)$$

With H given by (3.165), the integration over the p_j variables can be carried out explicitly, since

$$\int_{\mathbf{R}^3} \frac{d^3 p_j}{(2\pi\hbar)^3} \exp\left\{i \left[p_j \cdot (x_j - x_{j-1}) - \varepsilon p_j^2 / 2M \right] / \hbar \right\}$$
$$= \left(\frac{M}{2\pi i \hbar \varepsilon}\right)^{3/2} \exp[i M (x_j - x_{j-1})^2 / 2\hbar\varepsilon] ; \qquad (3.171)$$

the result is

$$G(x, t; x', t') = \lim_{\varepsilon \downarrow 0} \frac{1}{A} \int_{\mathbf{R}^{3N}} \frac{d^3 x_1}{A} \cdots \frac{d^3 x_N}{A}$$
$$\times \exp\left(\frac{i\varepsilon}{\hbar} \sum_{j=1}^{N+1} \left[\frac{M}{2} \left(\frac{x_j - x_{j-1}}{\varepsilon}\right)^2 - V\left(\frac{x_j + x_{j+1}}{2}, \frac{t_j + t_{j-1}}{2}\right)\right]\right) , \quad (3.172)$$

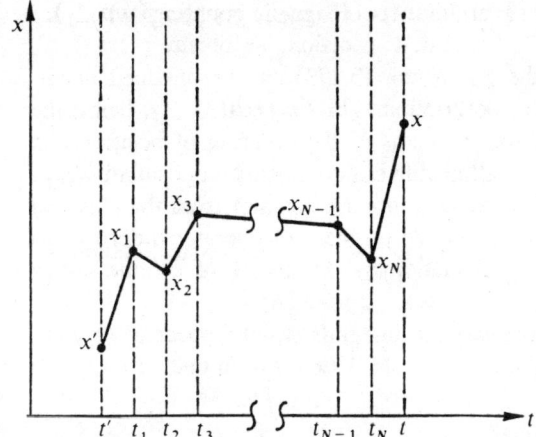

Fig. 3.5. A possible path from (x', t') to (x, t)

where $A \equiv (2\pi i\hbar\varepsilon/M)^{3/2}$. Given that $(x_j - x_{j-1})/\varepsilon$ is the average velocity between t_{j-1} and t_j, the quantity in the square brackets inside the summation in (3.172) is the Lagrangian of the system, and this quantity, multiplied by ε, is, to within quantities of order ε^3, the action $S_{cl}(x_j, t_j; x_{j-1}, t_{j-1})$ defined over the rectilinear trajectory connecting (x_{j-1}, t_{j-1}) and (x_j, t_j). The nomenclature "path integral" used by Feynman stems from the expression (3.172). In Fig. 3.5 is shown one possible polygonal trajectory connecting (x', t') and (x, t). Its contribution to $G(x, t; x', t')$ is proportional to the exponential factor which appears in (3.172), and the amplitude G is just the sum of all the contributions due to these polygonal paths, in the limit $\varepsilon \downarrow 0$.

It is convenient to rewrite (3.172) in the form

$$
G(x, t; x', t') = \lim_{\varepsilon\downarrow 0} \frac{1}{A} \int_{\mathbf{R}^{3N}} \frac{d^3 x_1}{A} \cdots
$$
$$
\frac{d^3 x_N}{A} \exp\left[\frac{i}{\hbar} \sum_{j=1}^{N+1} S_{cl}^{(R)}(x_j, t_j; x_{j-1}, t_{j-1}) \right] ,
$$

(3.173)

where $S_{cl}^{(R)}(x_j, t_j; x_{j-1}, t_{j-1})$ is the classical action between (x_{j-1}, t_{j-1}) and (x_j, t_j), calculated along the straight line trajectory which connects them. The formula (3.173), which has been formally justified here only for Hamiltonians of the type (3.165), is the starting point for Feynman's theory and yields the corresponding quantum propagator for all dynamical systems with a classical description. [Nevertheless, in its derivation, use has been made of (3.171); if the kinetic energy had, for example, a position dependent mass, the constant A in (3.173) would not be constant. It would then be necessary to substitute a new effective action for S_{cl} taking this fact into account in calculating the propagator. Hence, (3.170) is actually an expression which is valid for situations more general than those described by (3.165), or even for electromagnetic couplings which are in fact also covered by (3.173).]

In principle, it appears that the propagator (3.173) can be obtained unambiguously starting from a classical description. This is because S_{cl}, being the classical action, does not present the problems of the ordering of position and momentum observables to which we alluded in our comments on postulate VI, in connection with the construction of the Hamiltonian operator from the classical Hamiltonian. Nevertheless, this is not so; the problem of operator ordering has been replaced in the path integral calculation by the choice of a finite lattice approximation and its transition to the limit $\varepsilon \downarrow 0$ [DO 76].

In general, it is not easy to evaluate the integrals which appear in (3.173); consider, for example, a case of the free particle, $V(\boldsymbol{x}) \equiv 0$. In that case (3.172) gives

$$G^{(0)}(\boldsymbol{x}, t; \boldsymbol{x}', t') = \lim_{\varepsilon \downarrow 0} \left(\frac{\alpha}{\varepsilon} \right)^{3(N+1)/2} \int_{\mathbf{R}^{3N}} d^3 x_1 \ldots d^3 x_N$$

$$\times \exp \left[-\frac{\pi \alpha}{\varepsilon} \sum_{j=1}^{N+1} (\boldsymbol{x}_j - \boldsymbol{x}_{j-1})^2 \right] , \tag{3.174}$$

with $\alpha \equiv M/2\pi i\hbar$. If we do the integrals over $\boldsymbol{x}_1, \boldsymbol{x}_2, \ldots, \boldsymbol{x}_N$ successively, we always encounter integrals of the type

$$n^{-3/2} \left(\frac{\alpha}{\varepsilon} \right)^3 \int_{\mathbf{R}^3} d^3 x_n \exp \left\{ -\pi \alpha [(\boldsymbol{x}_0 - \boldsymbol{x}_n)^2 + n(\boldsymbol{x}_n - \boldsymbol{x}_{n+1})^2] / n\varepsilon \right\}$$

$$= \left[\frac{\alpha}{(n+1)\varepsilon} \right]^{3/2} \exp[-\pi \alpha (\boldsymbol{x}_0 - \boldsymbol{x}_{n+1})^2 / (n+1)\varepsilon] , \tag{3.175}$$

and thus we easily obtain

$$G^{(0)}(\boldsymbol{x}, t; \boldsymbol{x}', t') = \lim_{\varepsilon \downarrow 0} \left[\frac{\alpha}{(N+1)\varepsilon} \right]^{3/2} \exp[-\pi \alpha (\boldsymbol{x}_0 - \boldsymbol{x}_{N+1})^2 / (N+1)\varepsilon] , \tag{3.176}$$

that is,

$$G^{(0)}(\boldsymbol{x}, t; \boldsymbol{x}', t') = e^{-i3\pi/4} \left[\frac{M}{2\pi \hbar (t - t')} \right]^{3/2} \exp \left[\frac{iM}{2\hbar} \frac{(\boldsymbol{x} - \boldsymbol{x}')^2}{t - t'} \right] , \tag{3.177}$$

which coincides with the result given in (3.113), obtained by completely different methods (recall that $t > t'$).

It is interesting to note that the second exponential appearing in (3.177) is just $\exp[(i/\hbar)S_{cl}(\boldsymbol{x}, t; \boldsymbol{x}', t')]$. *Feynman* has demonstrated [FH 65] that for Lagrangians that are at most quadratic in $(x, \dot{x})$, the propagator is of the form [MP 79], [SC 81a]

$$G(x, t; x', t') = F(t, t') \exp[(i/\hbar)S_{cl}(x, t; x', t')] ,$$

$$F(t, t') = \sqrt{\frac{i}{2\pi \hbar} \frac{\partial^2 S_{cl}}{\partial x \, \partial x'}} , \tag{3.178}$$

where now S_{cl} should be calculated over that classical trajectory which the particle actually follows. For conservative systems, $F(t, t') = F(t - t', 0) \equiv f(t - t')$.

In the simple case of a harmonic oscillator of mass M and angular frequency ω, the classical action S_{cl} taken over a real trajectory is

$$S_{cl}(x, t; x', t') = \frac{M\omega}{2\sin(\omega\tau)}[(x^2 + x'^2)\cos(\omega\tau) - 2xx'] \tag{3.179}$$

for $0 < \tau \equiv t - t' < \pi/\omega$. To calculate the factor $f(\tau)$ we can either use (3.178) or, alternatively, argue as follows: a direct application of the Hamiltonian H for this system on the wave function

$$\psi_0(x) = \left(\frac{M\omega}{\pi\hbar}\right)^{1/4}\exp\left[-\frac{M\omega}{2\hbar}x^2\right] \tag{3.180}$$

yields $H|\psi_0\rangle = (\hbar\omega/2)|\psi_0\rangle$, i.e., $|\psi_0\rangle$ is an eigenvector of H. Thus, $\exp(-itH/\hbar)|\psi_0\rangle = \exp(-i\omega t/2)|\psi_0\rangle$. It suffices, then, in agreement with (3.99), to require that

$$\int_{\mathbf{R}} dx' \, G(x, \tau; x', 0)\psi_0(x') = \exp(-i\omega\tau/2)\psi_0(x) \tag{3.181}$$

in order to determine $f(\tau)$. The result leads immediately to the so-called Mehler formula for the propagator of a harmonic oscillator:

$$G(x, \tau; x', 0) = \sqrt{\frac{M\omega}{2\pi i\hbar \sin(\omega\tau)}}$$
$$\times \exp\left(i\frac{M\omega}{2\hbar\sin(\omega\tau)}[(x^2 + x'^2)\cos(\omega\tau) - 2xx']\right), \tag{3.182}$$
$$0 < \tau < \pi/\omega.$$

To extend this formula to arbitrary values of τ it is enough to use the group law (3.162) for the evolution operator. Taking the special case of (3.182) with $\tau = \pi/2\omega$, we get

$$G(x, \pi/2\omega; x', 0) = \sqrt{\frac{M\omega}{2\pi i\hbar}}\exp\left[-i\frac{M\omega}{\hbar}xx'\right]. \tag{3.183}$$

The group law and the invariance under time translations imply

$$G(x, \pi/\omega; x', 0) = \int dx'' \, G(x, \pi/\omega; x'', \pi/2\omega)G(x'', \pi/2\omega; x', 0)$$
$$= -i\delta(x + x'), \tag{3.184}$$

reflecting the fact that the classical particle will be on the opposite side after a half-period. By iterating the group law, we obtain for any $\tau = n\pi/\omega$ the expression

$$G(x, n\pi/\omega; x', 0) = (-i)^n\delta(x - (-1)^n x'), \tag{3.185}$$

and hence the propagator for every $\tau \neq n\pi/\omega$ is

$$G(x,\tau;x',0) = \exp\left[-\tfrac{i\pi}{2}(1/2 + [\omega\tau/\pi]_0)\right] \sqrt{\frac{M\omega}{2\pi\hbar|\sin(\omega\tau)|}}$$

$$\times \exp\left(i\frac{M\omega}{2\hbar\sin(\omega\tau)}[(x^2 + x'^2)\cos(\omega\tau) - 2xx']\right), \qquad (3.186)$$

where $[r]_0$ denotes the largest integer smaller than r. The continuity in τ of $U(\tau,0)$ guarantees that (3.186) goes over to (3.185) when $\tau \to n\pi/\omega$. Extension (3.186) of Mehler's formula is due to Souriau.

It has also been possible to obtain explicit formulae for the propagators for other important systems, such as the hydrogen atom [DK 79], [PS 84]. Numerical calculation of energy levels, correlations, etc., using this formulation via Monte Carlo techniques similar to those used in field theories on a lattice, can be found in [CF 81].

In the Feynman formulation, propagators occupy a central role while the Schrödinger equation, the commutation rules, etc., emerge as consequences of his general scheme. We will here illustrate this by formally deriving the Schrödinger equation. The function $G(x_1,t_1;x_0,t_0)$, $t_1 > t_0$, represents the probability amplitude for finding the particle in (x_1,t_1), knowing that it was at (x_0,t_0); and it defines, following Feynman, the wave function $\Psi(x_1,t_1)$ of the state which at t_0 represents the particle as localized at x_0 [see (3.160) for some understanding of just how natural this assumption is in the context of the usual formalism]. With (3.173) as Feynman's expression for the propagator, one gets

$$\Psi(x_2;t_2) = G(x_2,t_2;x_0,t_0) = \int_{\mathbf{R}^3} d^3x_1\, G(x_2,t_2;x_1,t_1)G(x_1,t_1;x_0,t_0)$$

$$= \int_{\mathbf{R}^3} d^3x_1\, G(x_2,t_2;x_1,t_1)\Psi(x_1;t_1) . \qquad (3.187)$$

This is identical to (3.99), and it gives the time evolution for any wave function, ignoring its previous history.

If we let $t_1 = t$ and $t_2 = t + \varepsilon$ in (3.187), then

$$\Psi(x;t+\varepsilon) = \int_{\mathbf{R}^3} d^3y\, G(x,t+\varepsilon;y,t)\Psi(y;t) . \qquad (3.188)$$

Taking as our classical Lagrangian

$$L = \frac{M}{2}\dot{x}^2 - V(x) , \qquad (3.189)$$

then (3.169) permits us to write, in the limit $\varepsilon \downarrow 0$,

$$G(x,t+\varepsilon;y,t) \simeq \int_{\mathbf{R}^3} \frac{d^3p}{(2\pi\hbar)^3}$$

$$\times \exp\left(\frac{i\varepsilon}{\hbar}\left[p\cdot\frac{x-y}{\varepsilon} - \frac{p^2}{2M} - V\left(\frac{x+y}{2}\right)\right]\right), \qquad (3.190)$$

from which

$$G(\boldsymbol{x}, t + \varepsilon; \boldsymbol{y}, t) \simeq \left(\frac{M}{2\pi i\hbar\varepsilon}\right)^{3/2}$$

$$\times \exp\left(\frac{i\varepsilon}{\hbar}\left[\frac{M}{2}\left(\frac{\boldsymbol{x}-\boldsymbol{y}}{\varepsilon}\right)^2 - V\left(\frac{\boldsymbol{x}+\boldsymbol{y}}{2}\right)\right]\right) , \qquad (3.191)$$

and thus

$$\Psi(\boldsymbol{x}; t + \varepsilon) \simeq \left(\frac{M}{2\pi i\hbar\varepsilon}\right)^{3/2} \int_{\mathbf{R}^3} d^3y$$

$$\times \exp\left(\frac{i\varepsilon}{\hbar}\left[\frac{M}{2}\left(\frac{\boldsymbol{x}-\boldsymbol{y}}{\varepsilon}\right)^2 - V\left(\frac{\boldsymbol{x}+\boldsymbol{y}}{2}\right)\right]\right) \Psi(\boldsymbol{y}; t) . \quad (3.192)$$

A change of variables $\boldsymbol{y} = \boldsymbol{x} + \boldsymbol{\eta}$ then gives

$$\Psi(\boldsymbol{x}; t + \varepsilon) \simeq \left(\frac{M}{2\pi i\hbar\varepsilon}\right)^{3/2} \int_{\mathbf{R}^3} d^3\eta$$

$$\times \exp\left[i\frac{M\boldsymbol{\eta}^2}{2\hbar\varepsilon}\right] \exp\left[-i\frac{\varepsilon}{\hbar}V(\boldsymbol{x}+\boldsymbol{\eta}/2)\right] \Psi(\boldsymbol{x}+\boldsymbol{\eta}; t) . \qquad (3.193)$$

The rapidly oscillating character of the integrand makes plausible the fact that only small values of $|\boldsymbol{\eta}|$ contribute to the integral. Thus we should expect that only the lowest orders in an expansion in powers of η_j in the last two factors in (3.193) will be important. If we use the formal "trick" of interpreting M as $M + i0$ to define the formal integrations, we have

$$\left(\frac{M}{2\pi i\hbar\varepsilon}\right)^{3/2} \int_{\mathbf{R}^3} d^3\eta \exp\left[i\frac{M\boldsymbol{\eta}^2}{2\hbar\varepsilon}\right] \eta_{i_1}\eta_{i_2}\ldots\eta_{i_{2N+1}} = 0$$

$$\left(\frac{M}{2\pi i\hbar\varepsilon}\right)^{3/2} \int_{\mathbf{R}^3} d^3\eta \exp\left[i\frac{M\boldsymbol{\eta}^2}{2\hbar\varepsilon}\right] \eta_{i_1}\eta_{i_2}\ldots\eta_{i_{2N-1}}\eta_{i_{2N}} \qquad (3.194)$$

$$= \left(\frac{i\hbar\varepsilon}{M}\right)^N \left[\delta_{i_1 i_2}\ldots\delta_{i_{2N-1}i_{2N}} + \ldots\right] ,$$

where the last square bracket contains the $(2N-1)!!$ different terms which result from all permutations of the lower indices. The expression (3.193) can be written as

$$\Psi(\boldsymbol{x}; t) + \varepsilon\frac{\partial\Psi(\boldsymbol{x}; t)}{\partial t} = \Psi(\boldsymbol{x}; t) - \frac{i\varepsilon}{\hbar}V(\boldsymbol{x})\Psi(\boldsymbol{x}; t)$$

$$+ \frac{i\hbar\varepsilon}{2M}\delta_{ij}\frac{\partial^2\Psi(\boldsymbol{x}; t)}{\partial x_i\partial x_j} + \mathrm{O}(\varepsilon^2) , \qquad (3.195)$$

and thus the evolution equation for the wave function ends up being

$$i\hbar \frac{\partial \Psi(\boldsymbol{x}; t)}{\partial t} = \left[-\frac{\hbar^2}{2M} \Delta + V(\boldsymbol{x}) \right] \Psi(\boldsymbol{x}; t) , \tag{3.196}$$

which is the usual Schrödinger equation.

The path integrations developed in an intuitive form by Feynman had already been studied by *Wiener* [WI 23] from a mathematical viewpoint, in connection with Brownian motion, and they currently constitute a highly relevant tool in questions dealing with statistical mechanics and quantum field theory. Feynman's basic formula (3.173) can be symbolically written as

$$G(\boldsymbol{x}, t; \boldsymbol{x}'t') = \int [\mathcal{D}\omega] \exp \left[\frac{i}{\hbar} S_{cl}(\omega) \right] , \tag{3.197}$$

where ω is an arbitrary classical path which connects $(\boldsymbol{x}', t')$ with $(\boldsymbol{x}, t)$, $S_{cl}(\omega)$ is the classical action over this path, and where we denote by $[\mathcal{D}\omega]$ the "measure" over the set of all such paths represented by the limit $A^{-1} \prod_{i=1}^{N}(A^{-1}d^3 x_i)$, $N \to \infty$. Unfortunately, $[\mathcal{D}\omega]$ does not exist as an adequate measure, and thus, aside from the enormous heuristic value of (3.197), its utility for rigorous analysis is very limited.

The fundamental observation that the change $t \to -it$ [i.e., the consideration of $\exp(-tH/\hbar)$, with $t > 0$, instead of $\exp(-itH/\hbar)$] allows us to make sense of path integrations is due to Kac [KA 50a], [KA 59]. The reason for this hinges on the fact that with this change, the calculation of $\langle \boldsymbol{x}| \exp(-tH/\hbar)|\boldsymbol{x}'\rangle$, carried out in a form analogous to that followed in (3.160), leads to the expression

$$\langle \boldsymbol{x}| \exp(-tH/\hbar)|\boldsymbol{x}'\rangle$$
$$= \int [\mathcal{D}\omega] \exp \left(-\frac{1}{\hbar} \int_0^t d\tau \left[\frac{M}{2} \dot{\omega}^2(\tau) + V(\omega(\tau)) \right] \right) . \tag{3.198}$$

Even though $[\mathcal{D}\omega]$ does not properly exist, it can be shown that the symbolic product $[\mathcal{D}\omega] \exp \left[-\hbar^{-1} \int_0^t d\tau(M/2)\dot{\omega}^2(\tau) \right]$ has the desired properties for a good measure, accounting for this the fact that "almost all" paths which contribute to the integral are not differentiable at any point, and the exponential goes to zero in the limit, which counterbalances the "infinity" in the Lebesgue measure mentioned previously. The measure obtained in this manner coincides with Wiener's original measure on Brownian paths with endpoints $\boldsymbol{x}, \boldsymbol{x}'$ and time lapse $t > 0$. Calling it $d\mu_{\boldsymbol{x},\boldsymbol{x}',t}(\omega)$, we arrive at the important Feynman-Kac formula:

$$\langle \boldsymbol{x}| \exp(-tH/\hbar)|\boldsymbol{x}'\rangle = \int d\mu_{\boldsymbol{x},\boldsymbol{x}',t}(\omega) \exp \left[-\frac{1}{\hbar} \int_0^t d\tau V(\omega(\tau)) \right] , \tag{3.199}$$

or equivalently

$$(\exp(-tH/\hbar)\psi)(\boldsymbol{x}) = \int d\mu_{\boldsymbol{x}}(\omega) \exp \left[-\frac{1}{\hbar} \int_0^t d\tau V(\omega(\tau)) \right] \psi(\omega(\tau)) , \tag{3.200}$$

where $d\mu_{\boldsymbol{x}}(\omega)$ is the Wiener probability measure for paths starting at $\boldsymbol{x}$.

Both (3.199) and (3.200) are valid under very general conditions [SI 79]. It suffices, for example, that in an N-dimensional problem the negative and positive parts of the potential, $V_\pm(x) \equiv \pm\max[\pm V(x), 0]$ satisfy: $V_+ \in L^1_{loc}(\mathbf{R}^N)$, $V_- \in L^p(\mathbf{R}^N) + L^\infty(\mathbf{R}^N)$ with $p = 1$ if $N = 1$, $p > 1$ when $N = 2$, and $p = N/2$ for $N \geq 3$.

The Feynman-Kac formula enables us to obtain several profound results in quantum dynamics. As an example of a surprisingly beautiful and simple application of it, let us prove, that, under the conditions we just have stated, the ground energy level $E_0 \equiv \inf \sigma(H)$, if it belongs to the point spectrum, is necessarily non-degenerate, and the corresponding normalizable wave function $\psi_0(x)$ can always be taken to be $\psi_0(x) > 0$ for all $x \in \mathbf{R}^N$. In effect, if $R \equiv \exp(-tH/\hbar)$, $t > 0$, it is clear that $R|\psi_0\rangle = r_0|\psi_0\rangle$, $r_0 \equiv \exp(-tE_0/\hbar)$, and $\|R\| = r_0$. As H is real, no generality is lost by assuming that $\psi_0(.)$ is real. Expression (3.199) shows that the integral kernel is non-negative, and thus that R preserves positivity: if $\psi(x) \geq 0$, for all x, then $(R\psi)(x) \geq 0$, almost everywhere. Further, under the assumed conditions, we can prove that R strictly conserves positivity: $\psi(.) \geq 0$, $\|\psi\| \neq 0 \Rightarrow (R\psi)(x) > 0$ almost everywhere. The following chain of inequalities follows immediately:

$$r_0\|\psi_0\|^2 = \langle \psi_0, R\psi_0 \rangle \leq \langle |\psi_0|, R|\psi_0| \rangle \leq \|R\|\|\psi_0\|^2 = r_0\|\psi_0\|^2 , \qquad (3.201)$$

and hence $\langle \psi_0, R\psi_0 \rangle = \langle |\psi_0|, R|\psi_0| \rangle$. If we write $\psi_0 = \psi_{0+} + \psi_{0-}$, where $\psi_{0\pm}(x) \equiv \pm\max[\pm\psi_0(x), 0]$, we have $\langle \psi_{0+}, R\psi_{0-} \rangle = 0$. The property of strict positivity requires that $\|\psi_{0+}\| \cdot \|\psi_{0-}\| = 0$. Then $\psi_0(x)$ may always be assumed to be ≥ 0 almost everywhere. But $\psi_0(x) = r_0^{-1}(R\psi_0)(x) > 0$ for all x, i.e., ψ_0 is strictly positive almost everywhere. Finally, if E_0 were degenerate, there would exist in addition to ψ_0 another eigenstate ψ_0', also with $\psi_0'(x) > 0$ for identical reasons, and which would be orthogonal to ψ_0. This is clearly impossible, proving our initial statement.

In passing, it should be observed that the ground state energy E_0 can be calculated, in principle, using the formula

$$E_0 = -\lim_{t\to\infty} \frac{\hbar}{t} \ln\langle \psi, \exp(-tH/\hbar)\psi \rangle \qquad (3.202)$$

for any $\psi \neq 0$, such that $\psi(x) \geq 0$ everywhere. Similarly,

$$|\psi_0\rangle = \lim_{t\to\infty} \frac{1}{\langle \psi | \exp(-2tH/\hbar)|\psi\rangle^{1/2}} \exp(-tH/\hbar)|\psi\rangle . \qquad (3.203)$$

We conclude by noting that the expression (3.173) for the propagator is justified under quite general conditions [NE 64], [FA 67a], [SI 71]. One way, among others, of giving meaning to the limit measure consists of adding a small positive imaginary component to the mass $M(M \to M + i\eta, \eta > 0)$, which we take to zero after the evaluation of the r.h.s. of (3.173).

4. One-Dimensional Problems

4.1 Introduction

In the previous chapters, we have laid down the physical and mathematical foundations of QM, as well as some of the general consequences. We continue by analyzing some simple problems, which will aid us in becoming familiar with some of the techniques used in QM. In addition to its educational value, there is currently real interest in this subject due to the modern advances in microfabrication of semiconductors, which permit the design and control of essentially one-dimensional potentials [SA 84], [KH 84].

We consider a one-dimensional stationary Schrödinger equation, for a particle of mass M with no internal degrees of freedom and under the influence of a potential $V(x)$:

$$\frac{d^2\psi(x)}{dx^2} + [\varepsilon - U(x)]\psi(x) = 0 \; ,$$

$$\varepsilon \equiv \frac{2M}{\hbar^2}E, \quad U(x) \equiv \frac{2M}{\hbar^2}V(x) \; ,$$

(4.1)

where E is the energy of the particle. For the eigenvalue problem posed by (4.1) to be well defined, it is necessary that $V(x)$ satisfy certain mathematical conditions; in the last section of this chapter we present a summary of sufficient conditions on $V(x)$.

Under these conditions, one can prove that, of the solutions of (4.1), only those ψ's that are (polynomially) bounded and such that ψ and ψ' are absolutely continuous are admissible as generalized eigenfunctions of the self-adjoint energy operator, and appear, therefore, in its spectral decomposition.

However, in practice some problems arise in which $V(x)$ does not satisfy the conditions that we stipulate. For example, there may be regions where $V(x)$ is infinite (barriers of infinite potential), or it may even happen that $V(x)$ includes in its expression some Dirac delta function. In these cases the wave function $\psi(x)$ is still continuous, but its derivative is not at points of infinite jumps of the potential. So, for example, if $V(x) = \infty$ for $x \leq a$, it suffices to consider the problem as a limiting case of another for which $V(x) = V_0$ in the zone $x \leq a$ (without changing anything else) and, at the end, taking the limit $V_0 \to \infty$. Bearing in mind the boundedness of $\psi(x)$ in the region $x \leq a$, it is necessary, as we will see, that in this region $\psi(x) = 0$ in the limit $V_0 \to \infty$. In the case of $U(x) = g\,\delta(x - y) + U_1(x)$, where U_1 is regular in a neighborhood of $x = a$, the formal integration of (4.1) leads to

$$\left[\frac{d\psi}{dx}\right]_{a+\eta} - \left[\frac{d\psi}{dx}\right]_{a-\eta} = \psi(a) \int_{a-\eta}^{a+\eta} dx\, U(x) + \mathrm{o}(\eta) , \qquad (4.2)$$

and thus, the discontinuity of the derivative at the point $x = a$ is $g\psi(a)$.

To discuss the general properties of the energy spectrum in (4.1), we assume that $U(x)$ is a piecewise continuous function over $\mathbf{R}$, that is, continuous except for at most a discrete set of points at which the potential has finite but discontinuous jumps. Furthermore, we will use the following notation:

$$U_0 \equiv \inf U(x) , \quad U_\pm \equiv \lim_{x \to \pm\infty} U(x) . \qquad (4.3)$$

The limits $U_\pm$, assuming they exist, may be finite or infinite (we will later find a very important case where this is not satisfied). When one of them is finite, we will assume that $U(x)$ tends towards it faster than $1/x$; more specifically, that for $x \to \pm\infty$ (depending on the case), $U(x) = U_\pm + B|x|^{-s_\pm}$ with B a constant and $s_\pm > 1$. When one of the limits $U_\pm$ is infinite and positive, there are no restrictions on how fast it increases; but if it is negative, in order not to have to impose at the corresponding endpoint artificial boundary conditions having no physical justification, we impose the condition $U(x) \geq -ax^2 - b$, where $a, b > 0$, for all x. Under all these assumptions it can be shown that $H \equiv -d^2/dx^2 + U(x)$ is essentially self-adjoint on $C_0^\infty(\mathbf{R})$.

The general results to be presented with only qualitative justification concerning the spectral properties of H can be rigorously proved and extended to more general cases, as we discuss in the last section. Finally, we will discuss some specific examples of bound states as well as scattering states.

4.2 The Spectrum of H

To make things concrete, we assume that $U_- \leq U_+$ (in the opposite case the results obtained here apply with trivial modifications). We begin by analyzing the asymptotic behavior of the solutions of (4.1) in particular cases of interest.

1) $\varepsilon > U_+$, with U_+ finite. We introduce a new unknown function, $h(x)$, by the relation

$$\psi(x) = \exp\left[\pm i k_+ x + \int_a^x dx\, h(x)\right] , \qquad (4.4)$$

where $k_+ \equiv |\varepsilon - U_+|^{1/2}$ and a is an arbitrary constant. For sufficiently large positive x values, $h(x)$ satisfies the equation

$$h'(x) \pm 2i k_+ h(x) + h^2(x) - \frac{B}{x^{s_+}} = 0 , \qquad (4.5)$$

from which asymptotically $h(x) \sim \mathrm{const} \times x^{-s_+}$; since we have assumed $s_+ > 1$, the integral in (4.4) has a constant as its asymptotic limit. Thus in this case there exist pairs of independent solutions behaving asymptotically as

$$e^{\pm ik_+ x}, \quad x \to +\infty, \quad \varepsilon > U_+ \text{ finite},$$
$$e^{\pm ik_- x}, \quad x \to -\infty, \quad \varepsilon > U_- \text{ finite}, \tag{4.6}$$

where $k_\pm \equiv |\varepsilon - U_\pm|^{1/2}$. Differentiating (4.4), one similarly concludes that the derivatives of these solutions behave like the derivatives of their asymptotic forms.

2) $\varepsilon < U_+ < \infty$. In this case, reasoning analogous to the above allows us to show that $\psi(x)$ tends asymptotically to a linear combination of

$$e^{\pm k_+ x}, \quad x \to +\infty, \quad \varepsilon < U_+ < \infty. \tag{4.7}$$

When $x \to -\infty$ and $\varepsilon > U_-$, U_- finite, the asymptotic behavior is again a combination of those given in (4.6). In the opposite case, of

$$e^{\pm k_- x}, \quad x \to -\infty, \quad \varepsilon < U_- < \infty. \tag{4.8}$$

(Similar behavior holds for the derivatives.)

3) If $U(x) \to \infty$ as $x \to +\infty$ or $-\infty$, one can infer from the asymptotic study of (4.1) that this equation has two linearly independent solutions, say ψ_1 and ψ_2, whose asymptotic behaviors are (as $x \to +\infty$ or $x \to -\infty$, depending on the case)

$$\psi_{1,2}(x) \sim \frac{1}{\sqrt{k(x)}} \exp\left[\pm \int_{x_0}^x dx' \, k(x')\right], \quad k(x) \equiv |U(x) - \varepsilon|^{1/2}. \tag{4.9}$$

(This behavior is guaranteed so long as $U(x)$ is twice differentiable in a neighborhood of the endpoint in question, and in addition $U'(x) = O(|U|^\alpha)$, where $0 < \alpha < 3/2$, in such a neighborhood.) Similarly

$$\psi'_{1,2}(x) \sim \pm\sqrt{k(x)} \exp\left[\pm \int_{x_0}^x dx' \, k(x')\right]. \tag{4.10}$$

4) If $U(x) \to -\infty$ as $x \to +\infty$ or $-\infty$, then, under the same conditions on U and U' as in (3) and using the same notation,

$$\psi_{1,2}(x) \sim \frac{1}{\sqrt{k(x)}} \exp\left[\pm i \int_{x_0}^x dx' \, k(x')\right],$$
$$\psi'_{1,2}(x) \sim \pm i\sqrt{k(x)} \exp\left[\pm i \int_{x_0}^x dx' \, k(x')\right]. \tag{4.11}$$

Given all this information, one can readily demonstrate that the spectrum of H, $\sigma(H)$, is of the typical form shown in Fig. 4.1. The point spectrum $\sigma_p(H)$ lies in $(U_0, U_-]$ and can be empty. The interior of the continuous spectrum $\sigma_c(H)$ lies in the interval (U_-, ∞), nondegenerate in the open interval $\sigma_{c1} \equiv (U_-, U_+)$ and doubly degenerate in $\sigma_{c2} \equiv (U_+, \infty)$. In fact:

1) $\varepsilon \leq U_0$. Let $\varepsilon < U_0$. In this case, $\varepsilon - U(x) < 0$, and therefore, $\psi''/\psi > 0$ for all values of x. Taking the decreasing solution as $x \to -\infty$ and extending

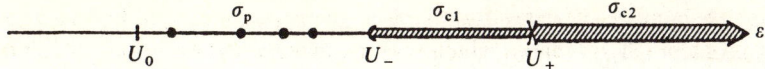

Fig. 4.1. An example of the spectrum for $H = -(d^2/dx^2) + U(x)$

it by means of (4.1), since $\psi(x)$ must always be convex towards the x-axis, it cannot tend to zero as $x \to +\infty$. It must thus grow at least exponentially in this region and therefore it is unacceptable as a generalized eigenfunction; hence, such values of ε do not belong to the spectrum. Classically also, the particle cannot have such energies. A simpler argument is that if $\varepsilon < U_0$, $H - \varepsilon > 0$, and therefore $\langle H \rangle > \varepsilon$, so that the spectrum of H is to the right of ε. Finally, $U_0 \in \sigma_p(H)$ implies that $0 \in \sigma_p(-d^2/dx^2)$, which is impossible.

2) $U_0 < \varepsilon < U_-$. The solution should decrease at least exponentially as $x \to -\infty$. Extending this solution to the region $x \to +\infty$, we find, in general, a linear combination of increasing and decreasing exponential functions. On varying the value of ε in the interval (U_0, U_-), the coefficients of this linear combination will change, and it may happen that there exist values of ε for which the coefficient of the increasing exponential is zero. The set of these ε values belong to the point spectrum σ_p and is a non-degenerate discrete set. Note that the wave function is non-zero in the classically inaccessible regions $\{x : U(x) > \varepsilon\}$, so that there exists a penetration probability.

3) $U_- < \varepsilon < U_+$. In this case there exist for all ε unique solutions which decrease, at least exponentially, as $x \to +\infty$. Therefore the spectrum is continuous and non-degenerate, and covers all of σ_{c1}.

4) $\varepsilon > U_+$. All solutions of (4.1) are admissible as generalized eigenfunctions, and for each value of ε, there are two linearly independent solutions; therefore, σ_{c2} is covered by the spectrum which is now continuous and doubly degenerate.

It is evident that, U being real, the arbitrary phase of the wave function in the nondegenerate case can always be selected in such a way that the wave function is real: it suffices to observe that if $\psi(x)$ is an admissible solution of (4.1), then so is $\psi^*(x)$, since ε as well as U is real.

The values of ε belonging to σ_p correspond to bound states; they are non-degenerate and, as we have just seen, the corresponding wave functions can be taken to be real. It has been noted that this set can be empty, which happens in particular in potentials such that $U(x) \geq \min(U_+, U_-)$ for all x. In other cases, for example the harmonic oscillator, there can be an infinite number of bound states.

The lower bound for the value $\inf \sigma(p^2/2M + V)$ given by V_0 is usually not very good. When $V \in L^1(\mathbf{R})$, the inequality

$$\inf \sigma(p^2/2M + V) \geq -M\|V\|_1^2/2\hbar^2 \tag{4.12}$$

produces on occasion a better estimate [FA 78].

Finding the number of states bound by a given potential is a particularly interesting problem [SI 76], about which one can make a few assertions of some practical utility.

In the first place, it is intuitively clear and rigorously provable [NA 68] that if $U(x) \to \infty$ as $|x| \to \infty$, then the energy spectrum is discrete and the number of bound states N is thus infinite. Suppose that $U_+ > U_-$, where U_- is finite. By the hypotheses made in Sect. 4.1, U_0 is finite, and for some $\varepsilon > 0$, $|x|^{1+\varepsilon}(U_- - U(x)) \to 0$ as $x \to -\infty$. It can be shown [SW 68], [SC 81] that N is always finite when $U(x) \geq U_- - 1/4x^2$ for all x values less than or equal to some a_-, while if there exists an $\alpha_- > 1/4$ such that $U(x) \leq U_- - \alpha_-/x^2$ for $x \leq a_-$, then N is infinite. (An analogous conclusion is reached when the roles of U_+ and U_- are interchanged.) When both limits U_+ and U_- are equal and finite, N is finite so long as $U(x) \geq U_\pm - 1/4x^2$ if $|x| \geq a$, and infinite if $U(x) \leq U_\pm - \alpha/x^2$, some $\alpha > 1/4$, in a neighborhood of one (or both) of the endpoints $\pm\infty$.

For finite N, it is interesting to know some bounds on the value of N. There is an essential difference between one and two dimensions, on the one hand, and dimensions greater than 2 on the other. For the one- and two-dimensional cases, any attractive potential [that is, $V(x) \leq 0$], which tends to zero at large distances and is not identically zero, has bound states, independent of its depth and range. Furthermore, a beautiful result of *Simon* [SI 76a] guarantees that, for one dimension, any $V(x)$ satisfying

$$\int_{\mathbf{R}} dx \, (1 + x^2)|V(x)| < \infty \, , \quad V \not\equiv 0 \, , \tag{4.13}$$

possesses some bound state of negative energy if and only if it encloses a non-positive area

$$\int_{\mathbf{R}} dx \, V(x) \leq 0 \, . \tag{4.14}$$

The easiness with which one-dimensional potential wells (attractive potentials) have bound states prevents the existence of upper bounds for N of homogeneous type, say $N \leq c(V)$, with $c(\lambda V) = |\lambda|^\alpha c(V)$, since evidently no function of this type is compatible with the result $N \geq 1$ for all potential wells.

If $U(x)$ is symmetric [$U(x) = U(-x)$], attractive (in the sense that $U(x) \leq U_\pm$), and monotone in the region $x > 0$, one can show [CA 67], [CO 65] that

$$N < 1 + \frac{2}{\pi} \int_{-\infty}^{+\infty} dx \, |U(x) - U_\pm|^{1/2} \, , \tag{4.15}$$

a bound which can be saturated by a square well potential with appropriate parameters. There also exist in this case lower bounds for N which are not so simply expressed as the previous one [CA 67]. The bound (4.15) suggests that N increases as the square root of the depth of the potential when this latter tends to infinity. One can rigorously prove [CH 68] under the above conditions, plus the requirement that $|U(x)|x^2$ be integrable at infinity (and in particular $U_\pm = 0$), that the increase of N is a universal function of the potential given by

$$N \sim \frac{1}{\pi} \int_{-\infty}^{+\infty} dx \; |U(x)|^{1/2} \tag{4.16}$$

whenever this integral is finite. This expression is known as the semiclassical estimate, since letting the potential depth increase indefinitely is equivalent to the limit $\hbar \downarrow 0$. It reflects the old idea, based in the Bohr-Sommerfeld method, that there is a bound state for each cell of area $h(= 2\pi\hbar)$ in the phase space region occupied by the classical motion, that is to say, $\{(x,p) : p^2/2M + V(x) \le 0\}$. Hence, being θ the step function,

$$N \underset{\hbar \to 0}{\sim} \frac{1}{2\pi\hbar} \int dx \; dp \; \theta[-(p^2/2M + V(x))] , \tag{4.17}$$

which coincides with (4.16).

When $U(x)$ is attractive and symmetric, and has k regions of monotony on the real axis, then [CO 65]

$$N < k - 1 + \frac{2}{\pi} \int_{-\infty}^{+\infty} dx \; |U(x) - U_\pm|^{1/2} . \tag{4.18}$$

There can be cases in which $U(x)$ does not satisfy the conditions stipulated for (4.18). In such cases one can often estimate N using (4.18) for another potential $U_0(x)$ to which (4.18) is applicable, according to the following: If $U_0(x) \le U(x)$ for all x, and $\min(U_-, U_+) = \min(U_{0-}, U_{0+})$, then one can prove, using the Min-Max principle (Sect. C.9), that the numbers of bound states $N(U)$ and $N(U_0)$ for U and U_0, with energies lower than $\min(U_-, U_+)$, satisfy $N(U) \le N(U_0)$, and if $\varepsilon_0(U) < \varepsilon_1(U) < \dots, \varepsilon_0(U_0) < \varepsilon_1(U_0) < \dots$ are the energies of these states under U and U_0, respectively, then $\varepsilon_n(U) \ge \varepsilon_n(U_0)$ for $n = 0, 1, 2, \dots$.

Another bound, interesting because of its generality, is the following [DT 79]:

$$N \le 1 + \int_{-\infty}^{+\infty} dx \; |xU_-(x)| , \tag{4.19}$$

where $U_-(x) \equiv \min(U(x), 0)$. Its validity is guaranteed so long as $(1 + |x|)U(x)$ is integrable.

An estimate of the number of bound states, which customarily gives very satisfactory results for the most common potentials in physics, is

$$N \simeq \frac{1}{2} + \frac{1}{\pi} \int dx \; \sqrt{\min(U_+, U_-) - U(x)} , \tag{4.20}$$

where the integral extends over the region in which the argument of the square root is positive. This equation is an immediate consequence of the WKB method, which is studied later in the book.

An interesting property of the bound states is related to the number of its nodes or zeros. One can demonstrate [DS 64] *Sturm's theorem*: If $\psi_0, \psi_1, \dots,$

$\psi_n, \ldots$ are the wave functions of the bound states with energies $\varepsilon_0 < \varepsilon_1 < \ldots < \varepsilon_n < \ldots$, then ψ_n has n nodes: between two consecutive nodes of ψ_n, there is a node of ψ_{n-1}, and moreover ψ_{n+r} has at least one zero for all $r \geq 1$. The state ψ_0 is called the *ground state*, and is the state with minimum energy; $\psi_1, \psi_2, \ldots$ are called *excited states*.

The fact that the energies increase with the number of nodes can be intuitively understood with the following reasoning: If $\psi(x_0) = 0$, then $\psi''(x_0) = 0$, and therefore, $\psi'(x)$ has an extremum at $x = x_0$. This makes possible a large contribution to the kinetic energy and, therefore, to the energy of the state.

4.3 Square Wells

We continue the consideration of bound states in some examples of potential wells.

Example 1. Consider the potential well in Fig. 4.2. In this case, the point spectrum is in the interval $(U_0, 0]$. For each eigenvalue ε, the wave function must be bounded, and since the potential is symmetric, such wave function will always be either symmetric or antisymmetric. This is because if $\psi(x)$ satisfies (4.1), so does $\psi(-x)$, and ε being non-degenerate, this forces $\psi(-x) = \lambda \psi(x)$, whence $\lambda = \pm 1$. Therefore, such a wave function is of the form

Region I : $\psi(x) = A_1 e^{\beta x}$

Region II : $\psi(x) = A_2(e^{i\alpha x} \pm e^{-i\alpha x})$ $\qquad\qquad$ (4.21)

Region III : $\psi(x) = \pm A_1 e^{-\beta x}$,   where

$$\alpha \equiv (\varepsilon - U_0)^{1/2} > 0 , \quad \beta \equiv |\varepsilon|^{1/2} ,\qquad\qquad (4.22)$$

and where A_1 and A_2 are constants. It suffices now to impose the continuity of the wave function and of its first derivative at the point $x = -L$, since as a consequence of the symmetry, it will also be automatically imposed at the point $x = L$. One obtains in this fashion the equations

$$e^{-\beta L} A_1 - [e^{-i\alpha L} \pm e^{i\alpha L}] A_2 = 0 ,$$
$$\beta e^{-\beta L} A_1 - i\alpha [e^{-i\alpha L} \mp e^{i\alpha L}] A_2 = 0 .\qquad (4.23)$$

In requiring the compatibility of these two systems of equations, one finds the eigenvalues as the solutions of

$$\alpha \tan \alpha L = \beta \quad \text{(positive parity)}$$
$$\alpha \cot \alpha L = -\beta \quad \text{(negative parity)}\qquad (4.24)$$

according to the parity of the wave function. In these equations (4.24), we must exclude the case $\alpha = 0$, since from (4.21), it leads to functions which are constant in region II, and therefore constants in all of $\mathbf{R}$, thus being unnormalizable or

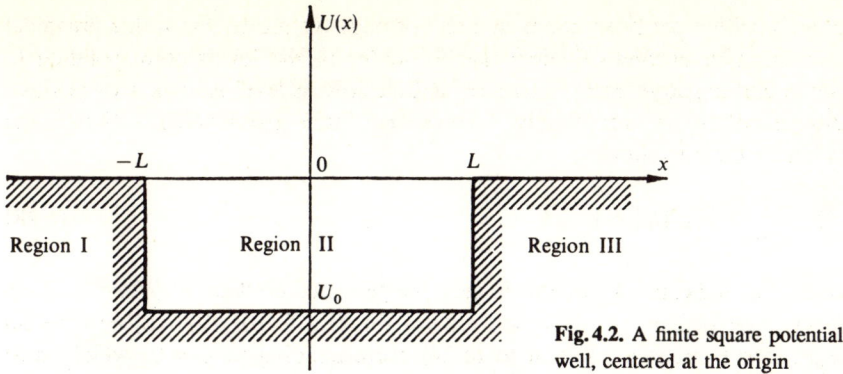

Fig. 4.2. A finite square potential well, centered at the origin

zero. (This is in agreement with the fact that $\sigma_p(H) \subset (U_0, 0]$.) The normalized wave functions in the three regions are given by

$$
\left.
\begin{aligned}
\psi(x) &= N_+(\cos \alpha L)e^{\beta(x+L)} \\
\psi(x) &= N_+ \cos \alpha x \\
\psi(x) &= N_+(\cos \alpha L)e^{-\beta(x-L)}
\end{aligned}
\right\} \quad \text{(positive parity)}
$$

$$
\left.
\begin{aligned}
\psi(x) &= -N_-(\sin \alpha L)e^{\beta(x+L)} \\
\psi(x) &= N_- \sin \alpha x \\
\psi(x) &= N_-(\sin \alpha L)e^{-\beta(x-L)}
\end{aligned}
\right\} \quad \text{(negative parity)} ,
$$

(4.25)

where, aside from phase factors,

$$
N_+ = N_- = \sqrt{\frac{\beta}{1 + \beta L}} . \tag{4.26}
$$

In particular, we must have $\beta > 0$, that is, $\varepsilon \in (U_0, 0)$, to achieve a bound solution.

Although explicit expressions are known for solutions of (4.24) in terms of integrals that require numerical evaluation [SI 78], it is most convenient to proceed directly to the numerical resolution. With the goal of obtaining an estimate of these eigenvalues, one can use the following graphical method: We first replace (4.24) by the equivalent equations (taking into account that $\alpha, \beta > 0$)

$$
\cos \alpha L = \pm \frac{\alpha}{\sqrt{|U_0|}} , \quad \tan \alpha L > 0 \quad \text{(positive parity)} ,
$$

$$
\sin \alpha L = \pm \frac{\alpha}{\sqrt{|U_0|}} , \quad \cot \alpha L < 0 \quad \text{(negative parity)} .
$$

(4.27)

Finding graphical solutions is now reduced to finding the intersections of the half-lines $y = \pm x/L|U_0|^{1/2}$, where $x \equiv \alpha L > 0$, with the curves $y = \cos x$ (positive parity) and $y = \sin x$ (negative parity), making sure that the x value found is in the correct quadrant. In Fig. 4.3, this method has been applied for the case $L|U_0|^{1/2} = 4$. Of the four intersections found, only the points marked 1, 3 and 4 in the figure are in the right quadrant, and one sees, therefore, that there are two positive parity solutions and one of negative parity. From this graphical construction, it becomes evident that the order of the levels corresponding to positive and negative parity alternate, and the lowest level is always of positive parity. Similarly it immediately follows that for a given value $L|U_0|^{1/2}$, the number of bound states is

$$N = \left[1 + \frac{2}{\pi}L|U_0|^{1/2}\right]_0 , \qquad (4.28)$$

where $[x]_0$ indicates again the largest integer smaller than x. When x is an integer, then $[x]_0 = x - 1$, which reflects the fact that if $(2/\pi)L|U_0|^{1/2}$ is an integer, there exists a solution to (4.24) corresponding to $\varepsilon = 0$, which must be eliminated as it gives rise to a wave function either identically zero or not normalizable (the case $\beta = 0$ previously rejected). [Note that this argument for excluding the eigenvalue $\varepsilon = 0$ is applicable to any integrable one-dimensional potential having compact support.]

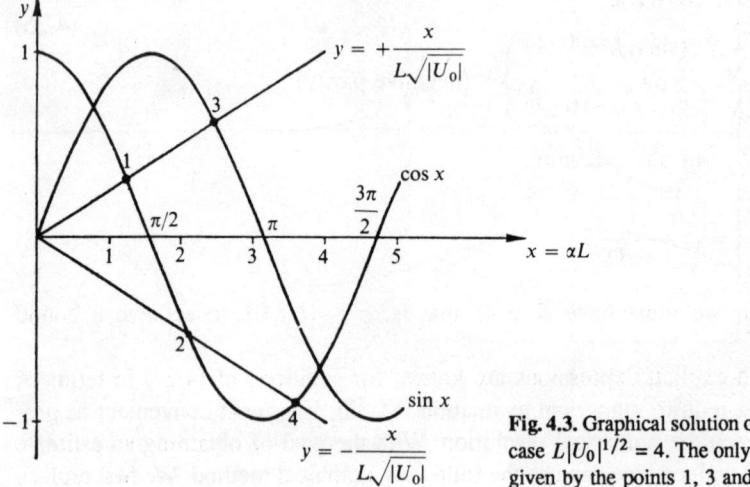

Fig. 4.3. Graphical solution of (4.27) for the case $L|U_0|^{1/2} = 4$. The only valid roots are given by the points 1, 3 and 4

It should be noted that the upper bound on N given by (4.15) leads to

$$N < 1 + \frac{4}{\pi}L|U_0|^{1/2} , \qquad (4.29)$$

and can thus be saturated in potential wells which are not too deep. However, for square wells with a large number of bound states, this bound is approximately twice the value of N. Also, it is worthwhile to observe that, according to (4.28), an arbitrary one-dimensional square well possesses at least one bound state; consequently, any non-constant potential $U(x)$ with $U_- = U_+$ and attractive $(U(x) \leq U_{\pm})$, in which a conveniently displaced square well can always be inscribed, also has at least one bound state, as was already stated in Sect. 4.2.

For the square well in our example, if $|U_0|^{1/2}L \ll 1$, the only bound state has the energy $\varepsilon_0 \simeq -U_0^2 L^2$.

Figure 4.4 shows the wave functions ψ_0, ψ_1, and ψ_2 for the bound states of a square well with parameters $L|U_0|^{1/2} = 4$. It is observed that in the regions I and III, classically forbidden for these energies, the wave functions decrease exponentially, while in region II, permitted for the classical particle, the behavior is oscillatory and the number of nodes of ψ_n is n.

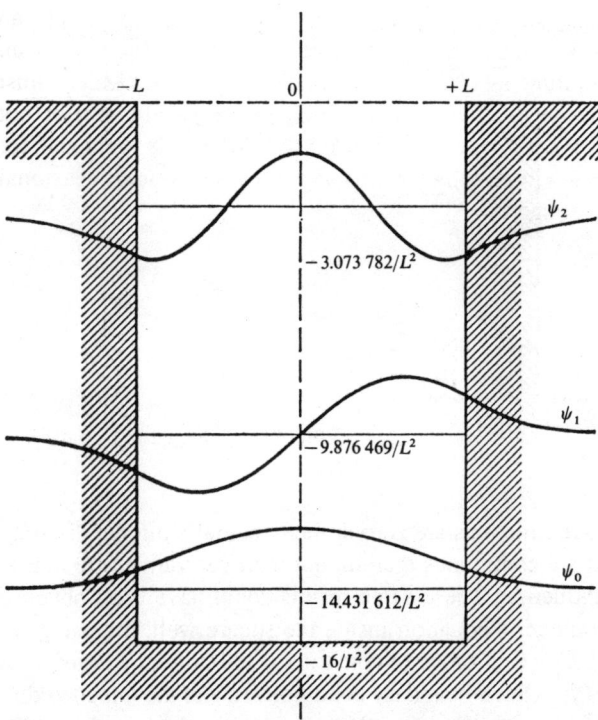

Fig. 4.4. Wave functions and potential for a square potential well with $L|U_0|^{1/2} = 4$

Example 2. If in the preceding square well we let U_0 be the zero of energy, and raise the walls to infinity, what results is an infinite square potential well represented in Fig. 4.5. With this change of the zero of energy, the parameters α and β in (4.22) become

$$\alpha = \varepsilon^{1/2}, \quad \beta = |\varepsilon + U_0|^{1/2}, \tag{4.30}$$

where ε is now referred to the new origin, and therefore, $\varepsilon > 0$. Letting $U_0 \to -\infty$, that is, raising the walls without limit, the eigenfunctions (4.25) tend to zero in the classically forbidden regions. On the other hand, the energy spectrum is obtained from (4.24) by letting $\beta \to \infty$, so that, for states of positive parity, $\tan \alpha L = \infty$, that is, $\alpha L = (r+1/2)\pi$, and for those of negative parity, $\tan \alpha L = 0$, that is, $\alpha L = r\pi$ (r an integer). Consequently, the normalizable wave functions and the energies for this infinite square well are

$$\psi_{n-1}(x) = \frac{1}{\sqrt{L}} \sin \frac{n\pi}{2L}(x+L) , \quad |x| \le L$$

$$\psi_{n-1}(x) = 0 , \quad |x| \ge L \qquad\qquad (4.31)$$

$$\varepsilon_{n-1} = \left(\frac{\pi}{2L}\right)^2 n^2 , \quad n = 1, 2, \dots ,$$

where $n = 1, 3, \dots$ corresponds to even, and $n = 2, 4, \dots$ corresponds to odd parity.

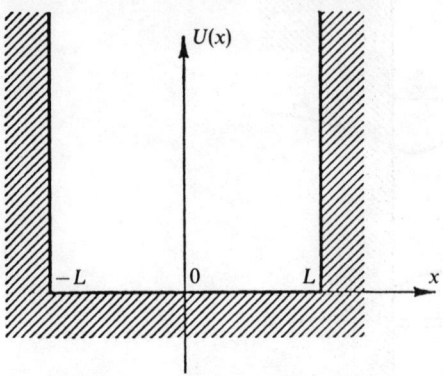

Fig. 4.5. Infinite square potential well

Observe that these wave functions are zero at the potential's infinite discontinuity points $x = \pm L$, and are continuous therein, but their derivatives are not, as foreshadowed in the introduction. These same results could have been obtained by directly solving Schrödinger's equation inside the square well, imposing the boundary conditions $\psi(\pm L) = 0$ in order to construct a self-adjoint extension of the energy operator $-d^2/dx^2$. Such boundary conditions reflect the impenetrability of the infinite walls. Hence, whenever we find ourselves with an impenetrable potential barrier, we require the wave function to be zero at it.

An identical conclusion is arrived at by taking the kinetic energy operator as the Friedrichs extension of the operator $-d^2/dx^2$ initially defined on C^∞ functions in $(-L, +L)$, with support in this open interval. This extension is the natural one associated with the energy operator as a quadratic form [SI 71].

The spectrum we have obtained is discrete. There is no continuous spectrum, since all the solutions of (4.1) inside the square well are continuous and hence normalizable if not identically zero. These results persist even if the potential energy $U(x)$ is not zero inside the square well: it suffices that $U(x)$ be integrable in the interior for the spectrum of H, with the boundary conditions indicated above, to be discrete and bounded from below [NA 68].

Finally, note that in the infinite square well the energy of the ground state is not zero, but $E_0 = \pi^2 \hbar^2 / 8ML^2$, whereas classically it is zero. That is to say, there exists a non-zero minimum energy, called the *zero-point energy*. This is an immediate consequence of the uncertainty principle; since Δx is necessarily $\lesssim L$, one has that $\Delta P \gtrsim \hbar/L$ and therefore, $E \gtrsim \hbar^2/2ML^2$.

Example 3. Finally, we consider a semi-infinite square well represented in Fig. 4.6. Bearing in mind what was said earlier, the wave functions are identically zero for $x \geq 0$, and for $x < 0$, the problem is analogous to that of Example 1. Since the wave functions must be continuous at $x = 0$, only the negative parity solutions of that case must be considered, and thus the eigenvalues are the solutions of $\alpha \cot \alpha L = -\beta$, while the normalizable eigenfunctions are given, in the regions I, II and III, respectively, by

$$\psi(x) = N(\sin \alpha L)e^{\beta(x+L)}, \quad \psi(x) = -N \sin \alpha x, \quad \psi(x) = 0$$

$$|N| = \sqrt{\frac{2\beta}{1+\beta L}} \; . \tag{4.32}$$

The energy eigenvalues are those of the odd states in Example 1. Therefore, the number of bound states in this case is

$$N = \left[\frac{1}{2} + \frac{L|U_0|^{1/2}}{\pi} \right]_0 , \tag{4.33}$$

where $[.]_0$ has the same meaning as in (4.28).

In contrast to Example 1, it can now occur that there are no bound states; it suffices that $|U_0|^{1/2}L \leq \pi/2$.

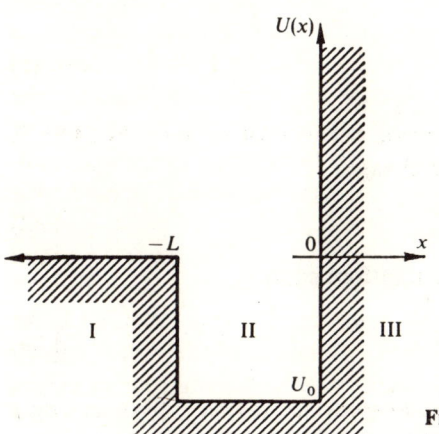

Fig. 4.6. A square potential well with an infinite barrier at $x \geq 0$

4.4 The Harmonic Oscillator

Without a doubt, of all the one-dimensional potentials, the harmonic oscillator is the most important, because its spectral characteristics have wide applications, both in first and in second quantization. The Hamiltonian is

$$
H = \frac{P^2}{2M} + \frac{k}{2}X^2 \ , \tag{4.34}
$$

where k is called the spring constant. In accordance with the general discussion given in Sect. 4.2, the spectrum has only a discrete part and is formed by an infinite set of values in the interval $(0, \infty)$. The eigenfunctions have well-defined parities.

For convenience of notation, we introduce the quantities

$$
\omega \equiv \sqrt{k/M} \ , \quad \alpha \equiv \sqrt{M\omega/\hbar} \ , \quad \xi \equiv \alpha x \ , \quad \lambda \equiv 2E/\hbar\omega \ , \tag{4.35}
$$

and the Hamiltonian operator becomes

$$
H = \frac{\hbar\omega}{2} \left[-\frac{d^2}{d\xi^2} + \xi^2 \right] \ . \tag{4.36}
$$

Due to the importance of the harmonic oscillator, we will present two ways of solving this problem.

1) **Analytical Treatment.** We will look for the bound states ψ of the Schrödinger equation

$$
-\frac{d^2\psi(\xi)}{d\xi^2} + \xi^2\psi(\xi) = \lambda\psi(\xi) \ . \tag{4.37}
$$

Since the dominant part of the asymptotic behavior of ψ, given by (4.9), is $\exp(-\xi^2/2)$, it is convenient to make the change

$$
\psi(\xi) = e^{-\xi^2/2}u(\xi) \ , \tag{4.38}
$$

so that the equation to be studied is now transformed to

$$
\frac{d^2u(\xi)}{d\xi^2} - 2\xi\frac{du(\xi)}{d\xi} + (\lambda - 1)u(\xi) = 0 \ . \tag{4.39}
$$

Given the definite parity of u, it suffices to study the region $\xi \geq 0$, and by a change of variable $y = \xi^2$, (4.39) converts into the following confluent hypergeometric equation (Sect. A.10):

$$
y\frac{d^2v(y)}{dy^2} + \left(\frac{1}{2} - y \right) \frac{dv(y)}{dy} - \frac{1}{4}(1 - \lambda)v(y) = 0 \ , \tag{4.40}
$$

where $v(y) \equiv u(\xi)$. Comparing this equation with (A.136), we see that its most general solution is

$$v(y) = AM\left(\frac{1-\lambda}{4}, \frac{1}{2}, y\right) + By^{1/2}M\left(\frac{3-\lambda}{4}, \frac{3}{2}, y\right) , \tag{4.41}$$

where A and B are constants of integration. Except in the cases for which

$$\Gamma\left(\frac{1}{4} - \frac{\lambda}{4}\right) = \infty \text{ and } B = 0 , \quad \Gamma\left(\frac{3}{4} - \frac{\lambda}{4}\right) = \infty \text{ and } A = 0 , \tag{4.42}$$

we conclude from (A.144) that in general the function $v(y)$ has an asymptotic behavior of the type $\exp(y)[= \exp(\xi^2)]$ and $\psi(x)$ would not be normalizable. The above conditions imply that $\lambda = 2n+1$, and therefore the energy eigenvalues are

$$\boxed{E_n = \hbar\omega\left(n + \tfrac{1}{2}\right)} , \quad n = 0, 1, 2, \ldots . \tag{4.43}$$

Replacing these λ's back in (4.39), one next sees that this equation is the one satisfied by the Hermite polynomials, so that

$$\boxed{\psi_n(x) = N_n H_n(\alpha x)e^{-\alpha^2 x^2/2}} , \quad n = 0, 1, 2, \ldots , \tag{4.44}$$

is the eigenfunction corresponding to E_n, where N_n is a normalization constant. The function $\psi_{2n}(x)$ is even and $\psi_{2n+1}(x)$ is odd. Equation (A.54) shows that in order for the wave functions to be normalized, N_n has to be, within a phase factor,

$$N_n = \sqrt{\frac{\alpha}{\sqrt{\pi}2^n n!}} . \tag{4.45}$$

Figure 4.7 displays the lowest bound states of the harmonic oscillator.

The energy spectrum given by (4.43) is very much like the one postulated by Planck (Sect. 1.2), namely $E_n = n\hbar\omega$, which can also be obtained using the Sommerfeld-Wilson-Ishiwara quantization rules. In both cases the levels are equidistant, with separation $\hbar\omega$, but the origins are different, since (4.43) leads to a zero-point energy $\hbar\omega/2$, as qualitatively required by the uncertainty principle.

For small quantum numbers ($n \sim 1$), the difference between the classical and quantum behavior of the oscillator are noticeable. To convince ourselves of this, we will consider an assembly of classical harmonic oscillators, of equal mass and frequency, all with the same energy E, and with equiprobable distribution of phases. The classical probability density $\mathcal{P}_{cl}(x)$ of finding a harmonic oscillator a distance x from the origin is then

$$\mathcal{P}_{cl}(x) = \frac{1}{\pi\sqrt{A^2 - x^2}} , \quad A = (2E/M\omega^2)^{1/2} , \tag{4.46}$$

where $|x| \leq A$, A being the amplitude of their oscillations. Equation (4.46) simply expresses that $\mathcal{P}_{cl}(x)$ is inversely proportional to the velocity of the oscillator at the point x. On the other hand, the quantum mechanical probability density $\mathcal{P}_{QM}(x)$ for the state ψ_n, of energy E_n, is $|\psi_n(x)|^2$, and a glance at Fig. 4.7

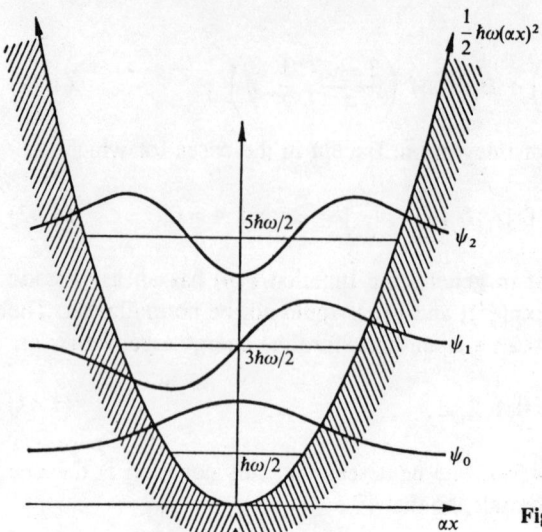

Fig. 4.7. Wave functions and potential energy for a harmonic oscillator

suffices to convince ourselves that, if $n \sim 1$, this probability density is very different from $\mathcal{P}_{cl}$, for the same energy. However, for $n \gg 1$, the correspondence principle implies that these differences should disappear. Using an asymptotic expansion for $H_n(\alpha x)$ for large n values, one obtains

$$\mathcal{P}_{\mathrm{QM}}(x) \underset{n \gg 1}{\sim} \frac{2\alpha}{\pi \sqrt{2n - \alpha^2 x^2}} \cos^2 \left[\left(2n + \frac{1}{2} \right) \frac{\alpha x}{\sqrt{2n}} - n \frac{\pi}{2} \right] . \tag{4.47}$$

Since the oscillations of the r.h.s. are more and more closely spaced as n increases, we can substitute the average value of $\cos^2$, i.e., $1/2$, thus recovering (4.46), for $E = E_n$. Figure 4.8 illustrates this for $n = 50$.

An elementary application of the virial theorem (3.156) indicates that for the harmonic oscillator in its state ψ_n

$$\langle T \rangle_n = \langle V \rangle_n = \left(n + \tfrac{1}{2} \right) \hbar\omega/2 \text{ , and hence,} \tag{4.48}$$

$$\langle P^2 \rangle_n = \left(n + \tfrac{1}{2} \right) M\hbar\omega = M E_n$$
$$\langle X^2 \rangle_n = \left(n + \tfrac{1}{2} \right) \hbar/M\omega = E_n/M\omega^2 \text{ ,} \tag{4.49}$$

which coincides with the classical expressions. This result, which is expected for $n \gg 1$ due to the correspondence principle, is accidentally satisfied in the harmonic oscillator for all n. Given that $\langle P \rangle_n = \langle X \rangle_n = 0$ (keep in mind the real character of ψ_n and its definite parity) (4.49) results in

$$\Delta_n X \cdot \Delta_n P = \left(n + \tfrac{1}{2} \right) \hbar \text{ .} \tag{4.50}$$

The ground state is, therefore, a minimal wave packet [as we already knew, since $\psi_0(x)$ is gaussian].

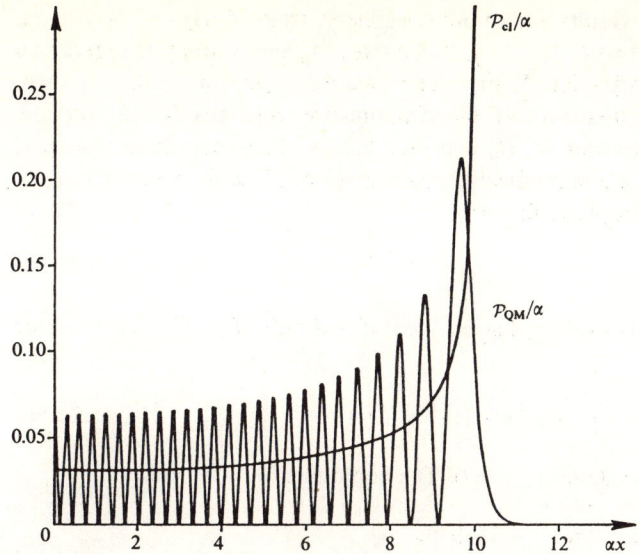

Fig. 4.8. Comparison of the classical and quantum mechanical probability densities for the $n = 50$ state

Finally, the wave functions $\hat{\psi}_n(p)$ of the bound states in the momentum representation can be obtained by Fourier transforming $\psi_n(x)$. However, we do not have to do this explicitly; it is enough to note that these $\hat{\psi}_n(p)$ functions should have the same structure as $\psi_n(x)$, given the invariance of the Hamiltonian (4.34) and the canonical commutation rules under the transformation $X \to P/M\omega$ and $P \to -M\omega X$. This argument assures us that

$$\hat{\psi}_n(p) = \frac{c_n}{\sqrt{M\omega}}\psi_n(p/M\omega) , \tag{4.51}$$

where $|c_n| = 1$, since $\hat{\psi}_n(p)$ must also have unit norm. Analyzing the dominant term in each member at large momenta, one obtains $c_n = (-\mathrm{i})^n$.

2) Operator Treatment. We define the operators

$$a \equiv \frac{1}{\sqrt{2}}\left(\xi + \frac{d}{d\xi}\right) , \quad a^\dagger = \frac{1}{\sqrt{2}}\left(\xi - \frac{d}{d\xi}\right) , \quad N \equiv a^\dagger a . \tag{4.52}$$

It is evident that N is formally a self-adjoint and positive definite operator. One can immediately obtain the commutation relations

$$\boxed{[a, a^\dagger] = I , \quad [N, a] = -a , \quad [N, a^\dagger] = a^\dagger} . \tag{4.53}$$

In terms of these operators, the Hamiltonian (4.36) can be written as

$$H = \hbar\omega\left[a^\dagger a + \tfrac{1}{2}\right] = \hbar\omega\left[N + \tfrac{1}{2}\right] . \tag{4.54}$$

From the previous results and, more concretely, from (4.43) and (4.54), the spectrum of N is $\sigma(N) = \{0, 1, 2, \ldots\}$. It is interesting and instructive to rederive this result, which justifies for N the name "number operator". The argument is based on a formal treatment of the commutation relations (4.53), and the knowledge that the spectrum of H, and thus that of N, is non-degenerate and discrete. We will denote a normalizable eigenvector of N, with eigenvalue μ, by $|\mu\rangle$, unique except for a phase factor:

$$N|\mu\rangle = \mu|\mu\rangle \ . \tag{4.55}$$

Since $N = a^\dagger a$, it is evident that $a|\mu\rangle = 0$ if and only if $\mu = 0$. On the other hand, $a^\dagger|\mu\rangle \neq 0$:

$$\|a^\dagger|\mu\rangle\|^2 = \langle\mu|aa^\dagger|\mu\rangle = \langle\mu|a^\dagger a + I|\mu\rangle = \mu + 1 \ , \tag{4.56}$$

and since N is positive definite, $\mu \geq 0$. The commutation rules imply

$$\begin{aligned} Na|\mu\rangle &= (aN - a)|\mu\rangle = (\mu - 1)a|\mu\rangle \ , \\ Na^\dagger|\mu\rangle &= (a^\dagger N + a^\dagger)|\mu\rangle = (\mu + 1)a^\dagger|\mu\rangle \ . \end{aligned} \tag{4.57}$$

One sees immediately that if $\mu \neq 0$, then $a|\mu\rangle$ is an eigenvector of N with eigenvalue $\mu - 1$, while $a^\dagger|\mu\rangle$ is always an eigenvector of N with eigenvalue $\mu + 1$. Therefore, the operator a lowers N by one unit and $a^\dagger$ raises it by the same quantity. Starting from the eigenvalue μ, we can form a set of eigenvalues $\ldots, \mu - 2, \mu - 1, \mu, \mu + 1, \mu + 2, \ldots$, by iterative application of a and $a^\dagger$. Since $N \geq 0$, and $a|\nu\rangle = 0$ if and only if $\nu = 0$, μ is forced to be an non-negative integer, and the previous series is $0, 1, 2, \ldots$. Hence, $\sigma(N) = \{0, 1, 2, \ldots\}$ and (4.54) leads to (4.43).

Had we not assumed it, the non-degeneracy of the spectrum of N would be guaranteed by irreducibility of the system of operators a and $a^\dagger$.

From the above we see that the normalized eigenvectors are, up to a phase,

$$\boxed{|n\rangle = \frac{1}{\sqrt{n!}}(a^\dagger)^n|0\rangle} \tag{4.58}$$

and that they satisfy $\langle n|n'\rangle = \delta_{nn'}$. Furthermore,

$$\boxed{a|n\rangle = \sqrt{n}|n-1\rangle \ , \quad a^\dagger|n\rangle = \sqrt{n+1}|n+1\rangle \ , \quad N|n\rangle = n|n\rangle} \ . \tag{4.59}$$

Here these operators have no direct physical significance, but the problem of finding the eigenvalues of H can be seen in another way which gives an immediate physical interpretation. Since the eigenvalues of H are equidistant, one can then consider that $H - \hbar\omega/2$ is the Hamiltonian of a system of indistinguishable particles, all of them with energy $\hbar\omega$. Then $|n\rangle$ represents the state with n particles, and the operators a, $a^\dagger$, and N play the role of *annihilation*, *creation*, and *number operators*, respectively. Their central importance in the study of quantum mechanical systems with many identical particles stems from this interpretation.

In the position representation, the wave functions $\psi_n(x)$ are $\psi_n(x) = \langle x | n \rangle$. From $a|0\rangle = 0$, one has

$$\left(\xi + \frac{d}{d\xi} \right) \psi_0(x) = 0 \ , \tag{4.60}$$

whose normalized solution is

$$\psi_0(x) = \alpha^{1/2} \pi^{-1/4} e^{-\xi^2/2} \ . \tag{4.61}$$

From (4.58), one obtains

$$\psi_n(x) = \frac{1}{\sqrt{n!}} \frac{1}{2^{n/2}} \left(\xi - \frac{d}{d\xi} \right)^n \psi_0(x) \ . \tag{4.62}$$

Using the identity

$$\xi - \frac{d}{d\xi} = -e^{\xi^2/2} \frac{d}{d\xi} e^{-\xi^2/2} \tag{4.63}$$

and (A.53), we immediately obtain

$$\psi_n(x) = \sqrt{\frac{\alpha}{\sqrt{\pi} 2^n n!}} H_n(\alpha x) e^{-\alpha^2 x^2/2} \ , \tag{4.64}$$

which are the solutions previously found.

From (4.52), one sees finally that the position and momentum operators are related to a and $a^\dagger$ as follows:

$$\boxed{\begin{aligned} a &= \frac{1}{\sqrt{2}}(\alpha X + iP/\hbar\alpha) \ , & a^\dagger &= \frac{1}{\sqrt{2}}(\alpha X - iP/\hbar\alpha) \ , \\[2mm] X &= \frac{1}{\sqrt{2}\alpha}(a^\dagger + a) \ , & P &= \frac{i\alpha\hbar}{\sqrt{2}}(a^\dagger - a) \ . \end{aligned}} \tag{4.65}$$

To end this brief exposition of the harmonic oscillator, we will discuss its time evolution in the Heisenberg picture and present a family of states, called coherent states, which play a very important role in quantum optics.

1) Heisenberg Picture. Consider the harmonic oscillator problem in the Heisenberg picture. Since all the operators that we will use are assumed to be in this picture, we omit the subscript H that denotes it. From the expression for H, one obtains [keeping in mind that the commutation rules (4.53) are maintained under a change of picture]

$$[a(t), H] = \hbar\omega a(t) \ , \quad [a^\dagger(t), H] = -\hbar\omega a^\dagger(t) \ . \tag{4.66}$$

Thus, the evolution equations (2.205) for the creation and annihilation operators are

$$i\frac{da(t)}{dt} = \omega a(t) , \quad i\frac{da^\dagger(t)}{dt} = -\omega a^\dagger(t) , \tag{4.67}$$

and, therefore,

$$a(t) = a(0)e^{-i\omega t} , \quad a^\dagger(t) = a^\dagger(0)e^{i\omega t} . \tag{4.68}$$

From (4.65), one thus obtains

$$X(t) = X(0)\cos \omega t + \frac{P(0)}{M\omega}\sin \omega t , \tag{4.69}$$

$$P(t) = P(0)\cos \omega t - M\omega X(0)\sin \omega t ,$$

identical to the classical expressions. These are as expected, since we saw (Sect. 3.7) that the quadratic character of the potential guaranteed that the mean values follow exactly the classical trajectories (3.96).

2) Coherent States. We have just derived the fact that, for a wave packet obeying the dynamics of the harmonic oscillator, the mean values of the position and momentum operators follow the classical equations of motion. In general, however, the form of the wave packet changes with time; but there exists a special class of non-stationary states that do not change form in the course of time. These states are of great interest in the quantum study of optical coherence and some of their properties will be studied here.

The eigenstates of the operator a are called *coherent states*:

$$a|z\rangle = z|z\rangle , \tag{4.70}$$

where z is a complex number. To find these states we express $|z\rangle$ as a linear combination of $|n\rangle$ states:

$$|z\rangle = \sum_{n=0}^{\infty} \frac{c_n}{\sqrt{n!}}|n\rangle . \tag{4.71}$$

Substituting this expression into (4.70) and bearing in mind (4.59), one deduces that $c_n = zc_{n-1}$, and thus, $c_n = z^n c_0$. Therefore, the states $|z\rangle$, normalized and up to a phase factor, are

$$|z\rangle = \exp\left(-\frac{1}{2}|z|^2\right) \sum_{n=0}^{\infty} \frac{z^n}{\sqrt{n!}}|n\rangle , \tag{4.72}$$

which permits us to assert that $\sigma_p(a) = \mathbb{C}$ and that the probabilities of finding the states $|n\rangle$ in $|z\rangle$ follow Poisson's law. An equivalent form of writing (4.72) is, evidently,

$$|z\rangle = \exp\left(-\frac{1}{2}|z|^2\right) \exp(za^\dagger)|0\rangle . \tag{4.73}$$

Rewriting the quantity $za^\dagger = z(a^\dagger - a) + za$ in (4.73) and using the Campbell-Hausdorff special relation (2.178), as well as (4.65), one obtains

$$|\underline{z}\rangle = \exp\left[-\frac{1}{2}(|z|^2 - z^2)\right] \exp\left[-\frac{i\sqrt{2}zP}{\alpha\hbar}\right] |0\rangle .$$ (4.74)

Equation (2.178), for a- and $a^\dagger$-type operators, is guaranteed when acting on finite linear combinations of $|n\rangle$, since these vectors are analytic for such operators, and the power series expansions of the exponentials which contain these operators are then convergent.

The operator that appears in (4.74) is formally $V_{z\sqrt{2}/\alpha}$, defined in (2.174). Since z is complex, this operator is not unitary if $\mathrm{Im}\, z \neq 0$. But since the wave function $|0\rangle$ is gaussian, the power series expansion of V converges when acting on this state, and produces a translation of $\psi_0(x)$ by the amount $z\sqrt{2}/\alpha$ with the result

$$\psi_z(x) \equiv \langle x|\underline{z}\rangle = \sqrt{\frac{\alpha}{\sqrt{\pi}}} \exp\left[-\frac{1}{2}(|z|^2 - z^2)\right]$$
$$\times \exp\left[-\frac{1}{2}(\alpha x - 2^{1/2}z)^2\right] ,$$ (4.75)

showing that $|\underline{z}\rangle$ also represents a minimal wave packet.

To find the uncertainties of the position and momentum operators for the state $|\underline{z}\rangle$, one can proceed as follows: From (4.70) we see that $\langle\underline{z}|a|\underline{z}\rangle = z$, as well as $\langle\underline{z}|a^\dagger|\underline{z}\rangle = z^*$, $\langle\underline{z}|a^2|\underline{z}\rangle = z^2$, $\langle\underline{z}|a^{\dagger 2}|\underline{z}\rangle = z^{*2}$, and $\langle\underline{z}|a^\dagger a|\underline{z}\rangle = |z|^2$; the commutation rules add $\langle\underline{z}|aa^\dagger|\underline{z}\rangle = 1 + |z|^2$. Using (4.65), we obtain

$$\langle\underline{z}|X|\underline{z}\rangle = \frac{1}{\sqrt{2}\alpha}(z^* + z) , \quad \langle\underline{z}|P|\underline{z}\rangle = \frac{i\alpha\hbar}{\sqrt{2}}(z^* - z) ,$$

$$\langle\underline{z}|X^2|\underline{z}\rangle - \langle\underline{z}|X|\underline{z}\rangle^2 = \frac{1}{2\alpha^2} , \quad \langle\underline{z}|P^2|\underline{z}\rangle - \langle\underline{z}|P|\underline{z}\rangle^2 = \frac{\alpha^2\hbar^2}{2} ,$$ (4.76)

and thus $\Delta_z X = 1/\sqrt{2}\alpha$ and $\Delta_z P = \alpha\hbar/\sqrt{2}$, which confirm that the wave packet $\langle x|\underline{z}\rangle$ is minimal.

If a state $|\underline{z}, 0\rangle \equiv |\underline{z}\rangle$ is prepared at the instant $t = 0$, its time evolution is given by

$$|\underline{z}, t\rangle = e^{-iHt/\hbar}|\underline{z}\rangle = e^{-|z|^2/2} \sum_{n=0}^{\infty} \frac{z^n}{\sqrt{n!}} e^{-i\omega t(n+1/2)}|n\rangle ,$$ (4.77)

so that

$$|\underline{z}, t\rangle = e^{-i\omega t/2}|\underline{z e^{-i\omega t}}\rangle ,$$ (4.78)

which permits us to assert that the coherence of these states is maintained under time evolution. It is easily shown that the most general Hamiltonian $H(t)$ under which any evolving coherent state maintains itself as such is

$$H(t) = H + f_1(t)X + f_2(t)P + g(t) ,$$

where H is the Hamiltonian of the harmonic oscillator (4.34) and f_1, f_2, and g are arbitrary real functions of time.

From expressions (4.76) and (4.78)

$$\langle \underline{z}, t | X | \underline{z}, t \rangle = \langle \underline{z} | X | \underline{z} \rangle \cos \omega t + \frac{\langle \underline{z} | P | \underline{z} \rangle}{M\omega} \sin \omega t ,$$
$$\langle \underline{z}, t | P | \underline{z}, t \rangle = \langle \underline{z} | P | \underline{z} \rangle \cos \omega t - M\omega \langle \underline{z} | X | \underline{z} \rangle \sin \omega t , \tag{4.79}$$

and thus the mean values of the position and momentum operators in the state $|\underline{z}, t\rangle$ oscillate in agreement with classical laws. The wave function corresponding to the state $|\underline{z}, t\rangle$ is

$$\Psi_z(x; t) = \sqrt{\frac{\alpha}{\sqrt{\pi}}} e^{-i\omega t/2} \exp\left[-\frac{1}{2}(|z|^2 - z^2 e^{-2i\omega t}) \right]$$

$$\times \exp\left[-\frac{1}{2}(\alpha x - 2^{1/2} z e^{-i\omega t})^2 \right] . \tag{4.80}$$

From this, and using (4.76) and (4.79), a simple calculation allows us to write

$$|\Psi_z(x; t)|^2 = \frac{\alpha}{\sqrt{\pi}} \exp\left(-\alpha^2 [x - \langle \underline{z}, t | X | \underline{z}, t \rangle]^2 \right) , \tag{4.81}$$

that is to say, the probability density of the wave packet remains invariant in time relative to its center $\langle \underline{z}, t | X | \underline{z}, t \rangle$, which oscillates according to the laws of classical physics.

It is interesting to show that coherent states are not only minimal with respect to the conjugate position and momentum variables, but also with respect to the energy and time. Indeed, the previous computations imply

$$\langle \underline{z} | H | \underline{z} \rangle = \left(|z|^2 + \tfrac{1}{2} \right) \hbar\omega , \quad \langle \underline{z} | H^2 | \underline{z} \rangle = \left(|z|^4 + 2|z|^2 + \tfrac{1}{4} \right) \hbar^2 \omega^2 , \tag{4.82}$$

and therefore, the uncertainty in the energy is $\Delta_z H = \hbar\omega |z|$. On the other hand, the width of the wave packet is $\Delta_z X = 1/\sqrt{2}\alpha$, and its center, in its oscillatory motion, has a maximum velocity $v = (\sqrt{2}\alpha\hbar/M)|z|$, as can be deduced from (4.79). If this maximum velocity is reached at the instant t_0, it is clear that the characteristic time $\tau_{|\underline{z}, t_0\rangle}(X)$ is $\hbar/2\Delta_z H$, so that $|\underline{z}, t_0\rangle$ is minimal with respect to the pair energy-time. Therefore, since $\tau_{\Psi(t)}(A) = \tau_{\Psi(0)}(A_H(t))$, then $|\underline{z}\rangle$ is also minimal in energy-time, as claimed.

Finally, we will collect some mathematical properties of coherent states:

(a) Two coherent states $|\underline{z}_1\rangle$ and $|\underline{z}_2\rangle$ always overlap, since from (4.72)

$$\langle \underline{z}_2 | \underline{z}_1 \rangle = \exp\left[-\tfrac{1}{2} |\underline{z}_2 - \underline{z}_1|^2 + i \, \text{Im} \, (z_2^* z_1) \right] \neq 0 . \tag{4.83}$$

(b) The coherent states $|\underline{z}\rangle$ are continuous functions of z. In fact, it is easy to show, starting from (4.83), that $\| \, |\underline{z}_1\rangle - |\underline{z}_2\rangle \|$ tends to zero as $z_2 \to z_1$.

(c) In the linear space of entire analytic functions $f(z)$, such that

$$f(z) = \sum_{n=0}^{\infty} \frac{c_n}{\sqrt{n!}} z^n \qquad (4.84)$$

with $\sum_{n=0}^{\infty} |c_n|^2 < \infty$, one can introduce a scalar product

$$(f, g) \equiv \frac{1}{\pi} \int_{\mathbf{R}^2} d^2z \, f^*(z) g(z) e^{-|z|^2} , \qquad (4.85)$$

where $d^2z \equiv dx \, dy$, giving a Hilbert-space structure to this space: the Bargmann-Segal space. Note that if $|\psi\rangle$ is a state of the harmonic oscillator and we define $\psi(z^*) \equiv \langle \underline{z}|\psi\rangle$, then

$$\psi(z) = e^{-|z|^2/2} f_\psi(z) , \quad f_\psi(z) \equiv \sum_{n=0}^{\infty} \frac{z^n}{\sqrt{n!}} \langle n|\psi\rangle , \qquad (4.86)$$

where $f_\psi(z)$ belongs to the Bargmann-Segal space. This correspondence $|\psi\rangle \rightarrow f_\psi(z)$ establishes an isomorphism between the Hilbert space of states of the harmonic oscillator and the Bargmann-Segal space. In particular,

$$|n\rangle \rightarrow \frac{z^n}{\sqrt{n!}} . \qquad (4.87)$$

(d) Under this isomorphism, the annihilation and creation operators a and $a^\dagger$ act as follows:

$$a : f(z) \rightarrow \frac{df(z)}{dz} , \quad a^\dagger : f(z) \rightarrow z f(z) . \qquad (4.88)$$

(e) Bearing in mind this isomorphism, it is clear that $\{|\underline{z}\rangle\}$ is a basis, although it is not orthogonal. This basis is overcomplete and from the analyticity of $f(z)$ it follows that any set $\{|\underline{z}_\alpha\rangle\}$ with a finite accumulation point is also a basis. One can prove, in addition, that every finite set $\{|\underline{z}_\alpha\rangle\}$ is linearly independent, and that, for example,

$$\int_{\mathbf{R}^2} d^2z \, z^p |\underline{z}\rangle = 0 \qquad (4.89)$$

for $p = 1, 2, \dots$. One also has a closure relation, with pair-wise non-orthogonal projections

$$\frac{1}{\pi} \int_{\mathbf{R}^2} d^2z |\underline{z}\rangle\langle \underline{z}| = I , \qquad (4.90)$$

as seen by taking arbitrary matrix elements

$$\langle \psi|\varphi\rangle = \frac{1}{\pi} \int_{\mathbf{R}^2} d^2z \, e^{-|z|^2} f_\psi(z)^* f_\varphi(z) , \qquad (4.91)$$

and keeping in mind that this equality is satisfied by the previous isomorphism.

For more details and a more complete bibliography on the formalism of coherent states, see [KS 68].

These operational techniques, as well as the notion of coherent states, have been extended to other potentials [IH 51], [NS 79], [NI 80], and [GN 80].

4.5 Transmission and Reflection Coefficients

In the previous sections, we centered our attention on $\sigma_p(H)$, obtaining information on the bound energy levels and wave functions for some important physical systems. In general, however, $\sigma_c(H) \neq \emptyset$, and we will spend most of the rest of this chapter studying this part of the spectrum. Under quite general mathematical conditions (which are discussed in detail later) the wave functions belonging to the continuous spectrum correspond to physical situations in which asymptotically free particles impinge on the potential and, after interacting, move off, with free asymptotic motion.

Let $U_\pm$ be finite, and for concreteness we assume, as in Sect. 4.2, that $U_- \leq U_+$. It is known from the general discussion there, that for an energy $\varepsilon > U_+$, $\varepsilon \in \sigma_c(H)$ and it is doubly degenerate. There are two linearly independent solutions to (4.1), with asymptotic behavior

$$
\begin{aligned}
\psi^{(-)}(x) &\sim \begin{cases} e^{ik_- x} + \varrho_-(E)e^{-ik_- x}, & x \to -\infty, \\ \sigma_-(E)e^{ik_+ x}, & x \to +\infty, \end{cases} \\
\psi^{(+)}(x) &\sim \begin{cases} e^{-ik_+ x} + \varrho_+(E)e^{ik_+ x}, & x \to +\infty, \\ \sigma_+(E)e^{-ik_- x}, & x \to -\infty, \end{cases}
\end{aligned}
\tag{4.92}
$$

where $k_\pm \equiv |\varepsilon - U_\pm|^{1/2}$. The wave function $\psi^{(-)}(x)$ represents a particle that is incident on the potential from $-\infty$ with wave function $\exp(ik_- x)$. The interaction with the potential produces a reflected wave $\sim \varrho_-(E)\exp(-ik_- x)$, which escapes to $-\infty$, and a transmitted wave, $\sim \sigma_-(E)\exp(ik_+ x)$, which moves off to $+\infty$. The wave function $\psi^{(+)}(x)$ can be interpreted in a similar fashion.

In dealing with stationary waves such as those above, it may be surprising that one can speak of non-trivial time evolution. The relation between the stationary method and collision theory viewed as a time evolution is discussed in Chap. 8 for the three-dimensional case, which is physically more important. But intuitively we can understand the evolution process by constructing wave packets with such stationary waves. We initially prepare a wave packet with wave functions $\exp(ik_- x)$ peaked around k_-, in a left-hand region that is very distant from the potential. This wave packet advances freely (Fig. 4.9) with group velocity $\hbar k_-/M$ until it reaches the potential. Then, it undergoes an interaction, and finally two wave packets emerge. One is a superposition of waves of type $\varrho_-(E)\exp(-ik_- x)$, the other of type $\sigma_-(E)\exp(ik_+ x)$; these move with group velocity $-\hbar k_-/M$ and $\hbar k_+/M$, respectively, in the asymptotic regions.

Since the Wronskian $W(y_1, y_2) \equiv y_1 y_2' - y_1' y_2$ of two solutions of (4.1) is constant, and ε and $U(x)$ being real, y_1^* and y_2^* are also solutions, it then follows immediately, using the asymptotic relations (4.92), that

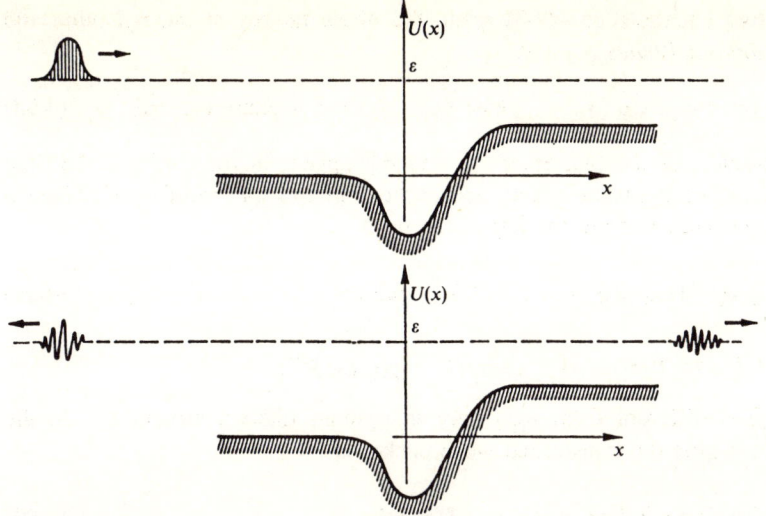

Fig. 4.9. Reflection and transmission of a wave packet

$$\frac{i}{2}W(\psi^{(-)}, \psi^{(-)*}) = k_-[1 - |\varrho_-|^2] = k_+|\sigma_-|^2 ,$$

$$\frac{i}{2}W(\psi^{(+)}, \psi^{(+)*}) = -k_-|\sigma_+|^2 = -k_+[1 - |\varrho_+|^2] ,$$

$$\frac{i}{2}W(\psi^{(-)}, \psi^{(+)}) = k_-\sigma_+ = k_+\sigma_- ,$$

$$\frac{i}{2}W(\psi^{(-)}, \psi^{(+)*}) = -k_-\varrho_-\sigma_+^* = k_+\varrho_+^*\sigma_- .$$

(4.93)

The first two relations describe the *conservation of flux*: the incident flux minus the reflected flux equals the transmitted flux. If, for solutions $\psi^{(\mp)}(x)$, we define the *transmission coefficient* as

$$T_\mp(E) = \frac{\text{transmitted flux}}{\text{incident flux}}(\rightleftharpoons) = \frac{k_\pm}{k_\mp}|\sigma_\mp|^2 ,$$

(4.94)

the third equation (4.93) implies the *reciprocity* relation

$$T_+(E) = T_-(E) \equiv T(E) .$$

(4.95)

In the same manner, we define the *reflection coefficient* as

$$R_\mp(E) = \frac{\text{reflected flux}(\rightleftharpoons)}{\text{incident flux}(\rightleftharpoons)} = |\varrho_\mp(E)|^2 ,$$

(4.96)

which [along with (4.93)] results in

$$R_+(E) = R_-(E) \equiv R(E) = 1 - T(E) .$$

(4.97)

The last two equations in (4.93) relate the phase factors of the *reflection* and *transmission amplitudes* $\sigma_\pm$ and $\varrho_\pm$:

$$\text{phase}\,(\sigma_-) = \text{phase}\,(\sigma_+)\,, \quad \text{phase}\,(\varrho_-/\sigma_-) = \pi - \text{phase}\,(\varrho_+/\sigma_+)\,. \qquad (4.98)$$

The importance of the phases of $\varrho_\pm$ and $\sigma_\pm$ appear in the study of the time delays that a wave packet suffers upon reflection and transmission. Consider a wave packet incident from the left of the form

$$\int_{\mathbf{R}} dk'_-\, f(k'_-)\exp[i(k'_- x - E't/\hbar)]\,, \quad \text{where} \qquad (4.99)$$

$$k'_- = [\varepsilon' - U_-]^{1/2}\,, \quad f(k'_-) = |f(k'_-)|\exp[i\alpha(E')]\,,$$

and $f(k'_-)$ is different from zero only in a small interval around k_-. In the asymptotic region the transmitted wave packet is

$$\int_{\mathbf{R}} dk'_-\, f(k'_-)\sigma_-(E')\exp[i(k'_+ x - E't/\hbar)]\,, \qquad (4.100)$$

and the reflected wave packet is

$$\int_{\mathbf{R}} dk'_-\, f(k'_-)\varrho_-(E')\exp[i(-k'_- x - E't/\hbar)]\,. \qquad (4.101)$$

Writing

$$\sigma_-(E') = |\sigma_-(E')|e^{2i\delta_-(E')}\,, \quad \varrho_-(E') = |\varrho_-(E')|e^{2i\phi_-(E')}\,, \qquad (4.102)$$

it can be easily shown from arguments analogous to those in Sect. 1.12 that the centers of the respective wave packets move according to

$$x_{\text{inc}} = v_- t - \hbar v_- \left[\frac{d\alpha(E')}{dE'}\right]_{E'=E}\,,$$

$$x_{\text{trans}} = v_+ t - \hbar v_+ \left[\frac{d\alpha(E')}{dE'} + 2\frac{d\delta_-(E')}{dE'}\right]_{E'=E}\,, \qquad (4.103)$$

$$x_{\text{refl}} = -v_- t + \hbar v_- \left[\frac{d\alpha(E')}{dE'} + 2\frac{d\varphi_-(E')}{dE'}\right]_{E'=E}\,,$$

where $v_\pm \equiv \hbar k_\pm/M$ and $E \equiv V_- + Mv_-^2/2$. The transmitted wave packet suffers a delay with respect to the incident one: its center passes a point b which is far to the right of the interaction region at a time τ_{trans}^- later than that at which the center of the incident wave packet passes. This time delay is

$$\tau_{\text{trans}}^-(E, b) = 2\hbar \left[\frac{d\delta_-(E')}{dE'}\right]_{E'=E} + b\left(\frac{1}{v_+} - \frac{1}{v_-}\right)\,. \qquad (4.104)$$

The second term is there to account for any possible difference of velocities v_- and v_+, which could cause the incident and the transmitted wave packets to advance in different ways through the free motion region to the right of the barrier, even in the absence of the barrier interaction. Defining

$$2\delta_-^{\text{trans}}(E, b) \equiv 2\delta_-(E) + (k_+ - k_-)b \,, \tag{4.105}$$

one sees that $2\delta_-^{\text{trans}}(E, b)$ is just the phase difference at the point b of the transmitted wave relative to the incident wave. Hence,

$$\tau_{\text{trans}}^-(E, b) = 2\hbar \frac{d}{dE} \delta_-^{\text{trans}}(E, b) \,. \tag{4.106}$$

Only if $U_- = U_+$ is δ^{trans} independent of b, and one can then speak of a unique time delay in transmission which is a function of energy only. Relation (4.106) is the one-dimensional analog of the *Eisenbud-Wigner* [WI 55] relation for scattering in three dimensions. (For a precise definition of these time delays see [JS 72a], [AJ 77] and [RS 79].)

Similarly, the reflected wave packet suffers a delay τ_{refl}^- with respect to the incident wave, which, for a point a far to the left of the barrier is given by

$$\tau_{\text{refl}}^-(E, a) = 2\hbar \left[\frac{d\varphi_-(E')}{dE'} \right]_{E'=E} - \frac{2a}{v_-} \,. \tag{4.107}$$

We must keep in mind that the phase $\varphi_-(E)$ changes to $[\varphi_-(E) - kd]$ under a translation of the coordinate origin by an amount d, as can be seen from (4.92). Thus, expression (4.107), when applied to this new phase shift, leads to the same numerical result as before, since $(a - d)$ is now the abscissa of the observation point for the reflected wave packet with respect to the new coordinate origin. To be able to speak of a reflection delay, it is necessary to compare the time τ_{refl}^- given by (4.107) with the time which the classical particle with velocity v_- would take in the round trip from the point a to another arbitrarily selected point a_0 in the region where the potential is large. The difference between these two times is evidently

$$\tau_{\text{refl}}^-(E, a_0) = 2\hbar \left[\frac{d\varphi_-(E')}{dE'} \right]_{E'=E} - \frac{2a_0}{v_-} \,, \tag{4.108}$$

where we again stress that a_0 is a point in the region where the potential is important.

It is clear that the time defined by (4.108) represents some intrinsic property of the potential, so long as there is no appreciable change when a_0 varies in that region. If D is the range of the potential, then $|\tau_{\text{refl}}^-(E, a_0)| \gg D/v_-$ needs to be satisfied for this to be true.

Completely analogous considerations apply for wave packets incident from $+\infty$ to the region of interaction. Relations (4.98) permit us to show that, when $U_- = U_+$, the time delays for transmission are independent of the sense of direction of the incident wave, and therefore we can simply write τ_{trans}.

Obviously, expressions (4.103) are valid only if $\varphi_-(E')$ and $\delta_-(E')$ do not vary much in the region of integration, i.e.,

$$\Delta k \ \hbar v_- \left| \frac{d\varphi_-(E)}{dE} \right| \ll 1 \ , \quad \Delta k \ \hbar v_+ \left| \frac{d\delta_-(E)}{dE} \right| \ll 1 \ . \tag{4.109}$$

Also, since the extent of the wave packet under consideration is $\Delta x \sim 1/\Delta k$, under these conditions the time delays satisfy

$$|\tau| \ll \frac{\Delta x}{v} \ , \tag{4.110}$$

that is, the retardation is small compared with the time that the center of the incident wave packet takes in crossing a distance equal to its own extent. This indicates the practical difficulties in measuring these time delays in the situation considered here [conditions (4.109)]. However, when the phases change very rapidly with energy, typical of resonant phenomena, it is possible to measure the associated time delay (4.106) indirectly. This delay can now be considerable for wave packets which clearly violate (4.109). We will discuss this in Sect. 4.7.

One could ask why one speaks of retardation times and not of advanced times. The reason for this is causality in local interactions, which requires that no appreciable wave amplitude gets out of the interaction region before the relevant part of the incident wave packet arrives at this region. Causality is intuitively guaranteed since $V(x)$ is a local interaction which significantly affects only the non-negligible wave functions in the region where $V(x)$ acts. If D is the size of the scattering region, the arrival time of the incident wave packet has an uncertainty of D/v, so that any advanced times much greater than D/v would violate causality. Therefore,

$$\tau \gtrsim -D/v \ . \tag{4.111}$$

In the particularly interesting case $U_+ = U_-$, one has $k_- = k_+ \equiv k$, and the relations (4.93) lead to

$$\sigma_+ = \sigma_- = \sigma \ , \quad |\sigma|^2 + |\varrho_\pm|^2 = 1 \ , \quad \varrho_-\sigma^* + \varrho_+^*\sigma = 0 \ . \tag{4.112}$$

If we represent by $|\pm k \ \text{in}\rangle$ the state incident from the left (right) with the wave function $\exp(\pm ikx)$, and by $|\pm k \ \text{out}\rangle$ the corresponding states emerging after collision with the potential, it follows from (4.92) that

$$|k \ \text{out}\rangle = \varrho_-|-k \ \text{in}\rangle + \sigma_-|k \ \text{in}\rangle, \ |-k \ \text{out}\rangle = \sigma_+|-k \ \text{in}\rangle + \varrho_+|k \ \text{in}\rangle. \tag{4.113}$$

As a consequence of (4.112), the matrix

$$S \equiv \begin{pmatrix} \sigma_- & \varrho_+ \\ \varrho_- & \sigma_+ \end{pmatrix} \tag{4.114}$$

is *unitary*. This matrix, which contains all the information about the collision, is known as the *S matrix* or the *scattering matrix*.

The first of the relations (4.112), which is a consequence of the real character of the potential $V(x)$, simply expresses the formal equality

$$\langle -k \text{ in}| - k' \text{out}\rangle = \langle k' \text{in}|k \text{ out}\rangle \tag{4.115}$$

which follows, as we will soon see, from the invariance of the scattering problem under time reversal.

Since S is unitary, its eigenvalues are of the form $\exp(2i\delta_0)$ and $\exp(2i\delta_1)$, where δ_0 and δ_1 are real, and are the *phase shifts* in the scattering process. In the particular case in which $V(x)$ is symmetric with respect to the origin [at which the phases of the waves $\exp(\pm ikx)$ are taken to be equal], that is to say, $V(x) = V(-x)$, it is obvious that $\varrho_+ = \varrho_- \equiv \varrho$, and in this case the incident eigenstates of the S matrix,

$$S|k,j\rangle = e^{2i\delta_j}|k,j\rangle , \quad j = 0, 1 \tag{4.116}$$

are simply the even and odd combinations of $|\pm k \text{ in}\rangle$:

$$\langle x|k,0\rangle = A\cos kx , \quad \langle x|k,1\rangle = B\sin kx , \tag{4.117}$$

where A and B are constants of equal modulus. The solutions of (4.1), arising from the evolution of $|k,0\rangle$ and $|k,1\rangle$ behave asymptotically as $A\cos(kx+\delta_0)$ and $B\sin(kx+\delta_1)$ respectively, when $x \to +\infty$, which justifies calling δ_0 and δ_1 phase shifts. Similarly, for a symmetric potential, the reflection and transmission amplitudes, as functions of the phase shifts δ_0 and δ_1, are given by

$$\sigma = \cos(\delta_0 - \delta_1)e^{i(\delta_0+\delta_1)} , \quad \varrho = \sin(\delta_0 - \delta_1)e^{i(\delta_0+\delta_1+\pi/2)} , \tag{4.118}$$

as can be obtained directly from the calculation of the eigenvalues of the S matrix, which turn out to be

$$e^{2i\delta_0} = \sigma + \varrho , \quad e^{2i\delta_1} = \sigma - \varrho . \tag{4.119}$$

Formally, one can define a total diffusion coefficient I for the potential

$$I \equiv \frac{\dot{j}_{\text{diffusion}}}{\dot{j}_{\text{incident}}} = |\varrho|^2 + |\sigma - 1|^2 , \tag{4.120}$$

where j is the probability current density.

From (4.119) and (4.120) one obtains

$$I = 2\left(\sin^2\delta_0 + \sin^2\delta_1\right) , \tag{4.121}$$

so that evidently $I \leq 4$ (a restriction due to the unitarity of S), and also

$$I = -2\text{Re}(\sigma - 1) , \tag{4.122}$$

an expression known as the *optical theorem*, by analogy to similar results in three-dimensional scattering.

Finally if $U_- < \varepsilon < U_+$, according to the general discussion of Sect. 4.2, $\varepsilon \in \sigma_c(H)$ and is simple. In this case, the only admissible solution, $\psi^{(-)}(x)$, has the asymptotic behavior

$$\psi^{(-)}(x) \sim \begin{cases} e^{ik_-x} + \varrho_-(E)e^{-ik_-x} , & x \to -\infty \\ \sigma_-(E)e^{-k_+x} , & x \to +\infty \end{cases} \tag{4.123}$$

and, calculating the Wronskian $W[\psi^{(-)}, \psi^{(-)*}]$, we find immediately that $|\varrho_-(E)|^2 = 1$, so that we have total reflection.

The formal nature of the preceding developments is conspicuous. However, the conclusions obtained can be made rigorous under certain conditions on the potential. In particular, if the potential

$$V \in L_\delta^1 \equiv \left\{ f : \int_{\mathbf{R}} dx\, (1+x^2)^{\delta/2} |V(x)| < \infty \right\} ,$$

for some $\delta > 1$ [which is the case, for example, if $V \in L_{\text{loc}}^1$ and $V(x) = O(|x|^{-(2+\varepsilon)})$, as $|x| \to \infty$, some $\varepsilon > 0$], one can demonstrate [FA 67], [DT 79], [SC 81] the validity of these results for scattering states. It is now convenient to make a change in notation: instead of expressing the amplitudes $\varrho_\pm$ and $\sigma_\pm$ as functions of E, we will treat them as functions of $k = |\varepsilon|^{1/2}$, and define its extension to $k < 0$ [hence outside the physical context of (4.92)], by letting $\varrho_\pm(-k) = \varrho_\pm^*(k)$ and $\sigma_\pm(-k) = \sigma_\pm^*(k)$, compatible with the real character of the time-independent Schrödinger equation. Then, if $V \in L_\delta^1$, one proves that:

1) The matrix $S(k)$ is continuous if k is real and non-zero (continuity at $k = 0$ is also the case if $V \in L_2^1$).

2) Relations (4.112) are valid for those k's in 1). In particular, $S(k)$ is unitary.

3) The amplitude $\sigma(k)$ admits an analytic extension to the half plane $\text{Im } k > 0$, giving rise to a meromorphic function with a finite number of (simple) poles $k = i\kappa_n$, $n = 1, 2, \ldots$ and $\kappa_n > 0$. The energies $(i\kappa_n)^2$ are precisely the ε_n energies of the bound states of H (which are also simple).

4) $\sigma(k) = 1 + O(1/|k|)$ as $|k| \to \infty$, $\text{Im } k \geq 0$.

$\varrho_\pm(k) = O(1/|k|)$ as $|k| \to \infty$, k real . $\qquad$ (4.124)

5) $|\sigma(k)| > 0$ if $k \neq 0$, $\text{Im } k \geq 0$.

$|k| \leq \text{const} \times |\sigma(k)|$ when $k \to 0$.

For $V \in L_2^1$, we have the following generic properties :

As $k \to 0$, $\sigma(k) \sim \alpha k$, $\alpha \neq 0$, if $\text{Im } k \geq 0$.

As $k \to 0$, $\varrho_\pm(k) \sim -1 + \alpha_\pm k$, for k real.

In general, $\varrho_\pm(k)$ cannot be extended off the real axis. If the potential decreases very rapidly at large distances, the extension is possible to a neighborhood of this axis. In particular, for a V of compact support, σ as well as $\varrho_\pm$ are meromorphic throughout the complex plane.

The asymptotic behavior (4.124) simply expresses that $R(E) \to 0$ and $T(E) \to 1$, when $E \to \infty$, as we would expect, since at high energies the particle is hardly affected by the potential. On the other hand, the connection between the poles of σ and the bound state energies of H stems from the following argument: For all $k \neq 0$, with $\operatorname{Im} k \geq 0$, the Schrödinger equation (4.1) has two solutions $f_1(x, k)$ and $f_2(x, k)$, with asymptotic behavior

$$f_1(x, k) \sim e^{ikx}, \quad x \to \infty, \quad \text{and} \quad f_2(x, k) \sim e^{-ikx}, \quad x \to -\infty. \quad (4.125)$$

If, in addition, $k > 0$, then f_1 and f_2 are proportional to $\psi^{(-)}$ and $\psi^{(+)}$ given in (4.92), and

$$
\begin{aligned}
f_1(x, k) &= \frac{\varrho_-(k)}{\sigma(k)} f_2(x, k) + \frac{1}{\sigma(k)} f_2(x, -k), \\
f_2(x, k) &= \frac{\varrho_+(k)}{\sigma(k)} f_1(x, k) + \frac{1}{\sigma(k)} f_1(x, -k),
\end{aligned}
\qquad (4.126)
$$

whence

$$W(f_1(\cdot, k), f_2(\cdot, k)) = 2ik/\sigma(k),$$

also valid for $k < 0$ and which holds as well for the analytic extension to $\operatorname{Im} k > 0$. In order that $\varepsilon = k^2 < 0$ be the energy of a bound state, it is necessary that $k = i\kappa$, $\kappa > 0$. For such k's, $f_1(x, k)$ and $f_2(x, k)$ are increasing (decreasing) exponentially when $x \to \mp\infty$ ($\pm\infty$), respectively; therefore, while its Wronskian is non-zero, no solution of (4.1) with $\varepsilon = -\kappa^2$ is square integrable, that is, $\varepsilon \notin \sigma_p(H)$. On the other hand, if this Wronskian is zero, and thus $i\kappa$ is a pole of σ, then f_1 and f_2 are proportional, and any of them represents a normalizable eigenfunction of H, i.e., $\varepsilon \in \sigma_p(H)$.

The generic behavior referred to in 5) always obtains when

$$W(f_1(\cdot, 0), f_2(\cdot, 0)) \neq 0.$$

In the opposite case,

$$|\sigma(k)| \geq \text{const} > 0, \quad |\varrho_\pm(k)| \leq \text{const} < 1$$

for all real k.

For a rigorous study of potentials with different asymptotic limits U_- and U_+ see [DS 78].

4.6 Delta Function Potentials

Even with the simplest potentials, it is very difficult to carry out analytical calculations to completion. For this reason, we begin with the presentation of a potential that, although it does not satisfy the conditions we have required for the largest part in this chapter, does lend itself to complete calculations. This potential, described by delta functions, leads to a good energy operator in the sense of quadratic forms (Appendix C), and we will formally apply the methods developed previously to resolve scattering problems.

The potential selected is

$$U(x) = B[\delta(x + a) + \delta(x - a)] , \quad a \geq 0 , \tag{4.127}$$

which is evidently symmetric. If the wave is incident from the left, the solution $\psi^{(-)}$ is

$$
\begin{aligned}
\psi^{(-)}(x) &= e^{ikx} + \varrho e^{-ikx}, & x &\leq -a \\
\psi^{(-)}(x) &= A_1 e^{ikx} + A_2 e^{-ikx}, & -a &\leq x \leq a \\
\psi^{(-)}(x) &= \sigma e^{ikx}, & x &\geq a
\end{aligned} \tag{4.128}
$$

The continuity condition for the wave function at the points $x = \pm a$, leads to

$$A_1 = \frac{\sigma e^{2i\mu} - e^{-2i\mu} - \varrho}{2i \sin 2\mu} , \quad A_2 = \frac{\varrho e^{2i\mu} + 1 - \sigma}{2i \sin 2\mu} , \quad \mu \equiv ka . \tag{4.129}$$

Relation (4.2), which gives the discontinuity of the first derivative at the points $x = \pm a$, allows us to calculate σ and ϱ, and obtain, for the phase shifts,

$$
\begin{aligned}
e^{2i\delta_0} &= \sigma + \varrho = \frac{\phi_0(-\mu)}{\phi_0(\mu)} , \quad e^{2i\delta_1} = \sigma - \varrho = \frac{\phi_1(-\mu)}{\phi_1(\mu)} \\
\phi_0(\mu) &\equiv \frac{1}{\mu}[\mu + i\beta e^{i\mu} \cos \mu] , \quad \phi_1(\mu) \equiv \frac{1}{\mu}[\mu + \beta e^{i\mu} \sin \mu]
\end{aligned} \tag{4.130}
$$

where $\beta \equiv Ba$. The functions ϕ_0 and ϕ_1 are called the *Jost functions*: note that the Jost functions tend to one in the non-interaction limit $B \to 0$.

From (4.130),

$$\delta_0 = \arctan \left[-\frac{2\beta \cos^2 \mu}{2\mu - \beta \sin 2\mu} \right] , \quad \delta_1 = \arctan \left[-\frac{2\beta \sin^2 \mu}{2\mu + \beta \sin 2\mu} \right] . \tag{4.131}$$

Using (4.94) and (4.118), the transmission coefficient is

$$T = \cos^2(\delta_0 - \delta_1) , \tag{4.132}$$

giving explicitly

$$T = \frac{\mu^4}{\mu^4 + [\mu\beta \cos 2\mu + (\beta^2/2) \sin 2\mu]^2} . \tag{4.133}$$

Finally, selecting the origin as the reference point inside the potential zone [that is, $a_0 = 0$ in (4.108)], and given that, from (4.118), the phases ϱ and σ differ by a constant, the time delays are such that $\tau_{\text{trans}} = \tau_{\text{refl}} \equiv \tau$, where

$$\tau = \frac{Ma^2}{\hbar} \frac{1}{\mu} \frac{d}{d\mu}(\delta_0 + \delta_1) . \tag{4.134}$$

The functions $\mu\phi_0(\mu)$ and $\phi_1(\mu)$, given by (4.130), evidently have analytic extensions to the whole μ-plane. It is clear that $\phi_i(-\mu) = \phi_i^*(\mu^*)$, that is, ϕ_i are hermitian analytic. As already stated in Sect. 4.5, their zeros, which can be poles of σ, have in general a physical interpretation. It is thus of some interest to locate them. We denote by $\mu_i = \alpha_j + i\gamma_j$, where α_j and γ_j are real, a possible root of the equation $\phi_j(\mu) = 0$. Separating it into its real and imaginary parts, we obtain

$$1 + e^{2\gamma_0} \cos 2\alpha_0 = -2\gamma_0/\beta , \quad e^{-2\gamma_0} \sin 2\alpha_0 = 2\alpha_0/\beta ,$$
$$- 1 + e^{-2\gamma_1} \cos 2\alpha_1 = 2\gamma_1/\beta , \quad e^{-2\gamma_1} \sin 2\alpha_1 = -2\alpha_1/\beta . \tag{4.135}$$

From these equations (or from hermitian analyticity noted above) one sees immediately that if $\alpha_j + i\gamma_j$ is a solution, then so is $-\alpha_j + i\gamma_j$, i.e., the roots are symmetric with respect to the imaginary axis. The pure imaginary roots ($\alpha_j = 0$) are determined, according to (4.135), by

$$-\frac{2\gamma_0}{\beta} = 1 + e^{-2\gamma_0} , \quad \frac{2\gamma_1}{\beta} = -1 + e^{-2\gamma_1} . \tag{4.136}$$

A simple numerical study of (4.136) allows us to conclude that for the first equation there are no roots if $\beta > \beta_0$; only one double root, with $\gamma_0 < 0$, if $\beta = \beta_0$; two simple roots, with $\gamma_0 < 0$, if $0 < \beta < \beta_0$; and only one simple root, with $\gamma_0 > 0$, if $\beta < 0$, where $\beta_0 \simeq 0.278465$. The second equation has only one root, with $\gamma_1 < 0$, if $-1 < \beta < 0$ and with $\gamma_1 \geq 0$ if $\beta \leq -1$. To calculate the roots which are not pure imaginary it is convenient to transform (4.135) and write them in the form

$$\frac{1}{\beta}e^{-\beta} = \frac{\sin 2\alpha_0}{2\alpha_0}e^{2\alpha_0 \cot 2\alpha_0}, \quad \gamma_0 = \frac{1}{2} \ln \left[\beta \frac{\sin 2\alpha_0}{2\alpha_0} \right] ,$$
$$-\frac{1}{\beta}e^{-\beta} = \frac{\sin 2\alpha_1}{2\alpha_1}e^{2\alpha_1 \cot 2\alpha_1}, \quad \gamma_1 = \frac{1}{2} \ln \left[-\beta \frac{\sin 2\alpha_1}{2\alpha_1} \right] . \tag{4.137}$$

The previous equations have an infinite number of non-real roots, all with $\gamma < 0$; that is, the complex roots are all found in the lower half-plane. Their position in the complex μ-plane varies with β, and in Fig. 4.10 the behavior is given for the two roots with smaller $|\alpha|$ in the half-plane $\alpha \geq 0$.

We will now see the physical significance of the zeros of $\phi_0(\mu)$, with the understanding that the same discussion can be applied to the roots of $\phi_1(\mu) = 0$ with similar conclusions. To proceed to this discussion we consider three different cases:

1) $\mu_0 = 0 + i\gamma_0, \gamma_0 > 0$

If $\phi_0(\mu_0) = 0$, then from (4.130) and (4.136) it follows that σ and ϱ have a pole at μ_0. Since (4.128), with the matching conditions at $x = \pm a$, is a solution to the Schrödinger equation for k real as well as for k complex, taking $\mu = \mu_0$, one checks that there exists a solution such that

$$\psi(x) \sim e^{\gamma_0 x/a}, \quad x \leq -a$$
$$\psi(x) \sim e^{-\gamma_0 x/a}, \quad x \geq a . \tag{4.138}$$

This solution, therefore, represents a bound state with energy $\varepsilon = -(\gamma_0/a)^2$. Repeating this process for ϕ_1, and recalling the results from the analysis of

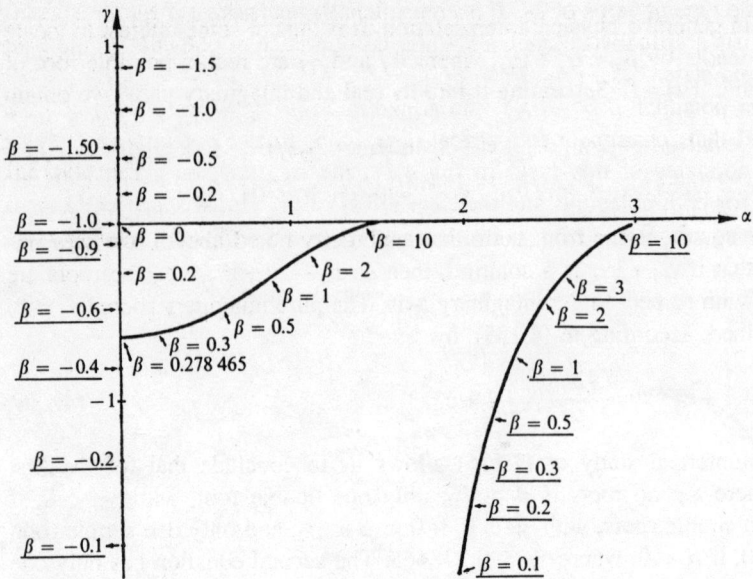

Fig. 4.10. Trajectories of the zeros of $\phi_j(\mu)$ described in the complex μ-plane as β varies. The underlined values correspond to a zero of $\phi_1(\mu)$ and the others to a zero of $\phi_0(\mu)$

(4.136), one concludes that the potential considered here has a unique bound state, which is even if $-1 \le \beta < 0$, and two bound states (one even, the other odd) if $\beta < -1$. An analogous conclusion can clearly be reached by looking directly for normalizable solutions to the Schrödinger equation, which again leads to (4.136) with $\gamma_i \equiv a|\varepsilon_i|^{1/2}$.

2) $\mu_0 = 0 + i\gamma_0$, $\gamma_0 < 0$

If $|\gamma_0| \ll 1$ and $0 \le \mu \lesssim |\gamma_0|$, then a Taylor expansion for ϕ_i around the point $\mu_0 = i\gamma_0$ gives

$$\phi_0(\mu) = -i\gamma_0^{-1}(1 + \beta + 2\gamma_0)(\mu - i\gamma_0) + O((\mu - i\gamma_0)^2) ,$$
$$\phi_1(\mu) = \gamma_0^{-1}(\beta + 2\gamma_0) + O(\mu - i\gamma_0)$$

(4.139)

and using relations (4.130) one gets

$$e^{2i\delta_0} \simeq -\frac{\mu + i\gamma_0}{\mu - i\gamma_0} , \quad e^{2i\delta_1} \simeq 1 .$$

(4.140)

Therefore, δ_0 varies substantially in this zone, while δ_1 remains practically equal to zero. Hence, one obtains for the transmission coefficient:

$$T(\mu) \simeq \frac{\mu^2}{\mu^2 + \gamma_0^2} .$$

(4.141)

Zeros of this type are said to represent *virtual states*, and their typical effect, shared also by loosely bound states, is to make the transmission coefficient increase rapidly at low energies from its zero value at $\mu = 0$. Similar results hold for the same type of zeros of ϕ_1. It is convenient to realize that if for an intensity β there exists a virtual state, then a slightly more attractive potential would have a new bound state.

For the potential $\beta = -0.98$, the equation $\phi_1(\mu) = 0$ has a zero at $\mu_1 = 0 - i\,0.0201$ that corresponds to a virtual state, while for the potential $\beta = -0.50$, there are no states of this type. In Fig. 4.11, the transmission coefficients are compared for both potentials and the clear effect of the virtual state can be seen.

The zeros $\mu_0 = 0 + i\gamma_0$ with γ_0 negative and large lack a physical interpretation.

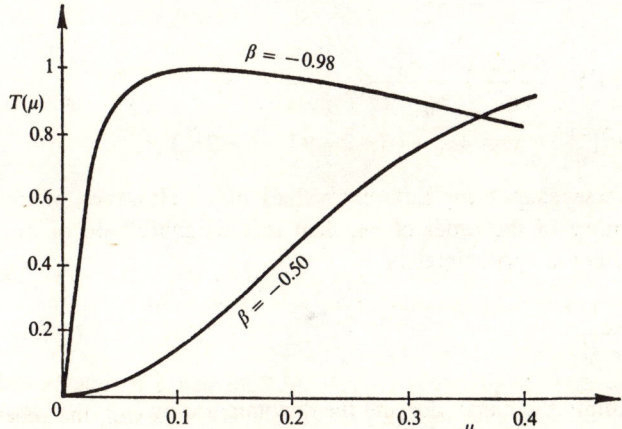

Fig. 4.11. Comparison of the transmission coefficients in the presence ($\beta = -0.98$) and absence ($\beta = -0.50$) of an odd virtual state

3) $\mu_0 = \alpha_0 + i\gamma_0$, where $\alpha_0 > 0$ and $\gamma_0 < 0$

Consider r ow a root of this class with $|\gamma_0| \ll 1$. For values of μ close to μ_0, one can calculate $\phi_0(\mu)$ and $\phi_0(-\mu) = \phi_0^*(\mu^*)$ by a Taylor expansion around the point μ_0. This way one obtains

$$\phi_0(\mu) = \frac{-i}{\mu_0}[2\alpha_0 + i(1 + \beta + 2\gamma_0)](\mu - \mu_0) + O((\mu - \mu_0)^2),$$

$$\phi_0(-\mu) = \frac{i}{\mu_0^*}[2\alpha_0 - i(1 + \beta + 2\gamma_0)](\mu - \mu_0^*) + O((\mu - \mu_0^*)^2),$$

(4.142)

and analogously,

$$\phi_1(\mu) = \frac{1}{\mu_0}[2\alpha_0 + i(\beta + 2\gamma_0)] + O(\mu - \mu_0),$$

$$\phi_1(-\mu) = \frac{1}{\mu_0^*}[2\alpha_0 - i(\beta + 2\gamma_0)] + O(\mu - \mu_0^*).$$

(4.143)

If μ is real and close to α_0, and $x \equiv \mu - \alpha_0$, the phase shifts satisfy

$$
\begin{aligned}
e^{2i\delta_0} &\simeq \frac{-2\alpha_0 + i(1 + \beta + 2\gamma_0)}{2\alpha_0 + i(1 + \beta + 2\gamma_0)} \left(\frac{\mu_0}{\mu_0^*} \right) \frac{x + i\gamma_0}{x - i\gamma_0}, \\
e^{2i\delta_1} &\simeq \frac{2\alpha_0 - i(\beta + 2\gamma_0)}{2\alpha_0 + i(\beta + 2\gamma_0)} \left(\frac{\mu_0}{\mu_0^*} \right),
\end{aligned}
\tag{4.144}
$$

and therefore, while δ_1 remains practically constant in a neighborhood of $\mu = \alpha_0$, δ_0 varies considerably, increasing by an amount of the order of π in crossing this region in the direction of increasing energies. These rapid increases by π for the phase shifts characterize the *resonances* or *resonant states*, which are thus associated with the zeros of the type being analyzed now. For the transmission coefficient, one obtains

$$
\begin{aligned}
T(\alpha_0 + x) &\simeq T(\alpha_0) \left(1 - \frac{2\alpha_0 x}{r\gamma_0} \right)^2 \frac{\gamma_0^2}{\gamma_0^2 + x^2}, \\
T(\alpha_0) &\equiv [1 + 4\alpha_0^2/r^2]^{-1}, \quad r \equiv 4\alpha_0^2 + (\beta + 2\gamma_0)(1 + \beta + 2\gamma_0),
\end{aligned}
\tag{4.145}
$$

which can present diverse shapes for different values of x. However, when $2\alpha_0 \ll r$ and x is at most of the order of γ_0, then in a neighborhood of α_0, the transmission coefficient is approximately

$$
T(\mu) \simeq \frac{\gamma_0^2}{(\mu - \alpha_0)^2 + \gamma_0^2}.
\tag{4.146}
$$

If in (4.146) we multiply the numerator and the denominator by $(\mu + \alpha_0)^2$ and take into account that $(\mu + \alpha_0)^2 \gamma_0^2 \simeq 4\alpha_0^2 \gamma_0^2$, then we obtain for the transmission coefficient as a function of energy

$$
\begin{aligned}
T(E) &\simeq \frac{(\Gamma_R/2)^2}{(E - E_R)^2 + (\Gamma_R/2)^2}, \\
E_R &= \frac{\hbar^2}{2Ma^2} \alpha_0^2, \quad \Gamma_R = 4\sqrt{E_R} \sqrt{\frac{\hbar^2}{2Ma^2} \gamma_0^2},
\end{aligned}
\tag{4.147}
$$

where E_R is the *resonant energy* and Γ_R is the *resonance width*. This relation is called the *Breit-Wigner formula*.

If β is large enough and negative, one deduces from the analysis of (4.137) that the zeros of ϕ_i corresponding to resonances are, to a good approximation, given by the asymptotic expansions

$$
\begin{aligned}
\alpha_n &\sim \frac{n\pi}{2} \left[1 + \frac{1}{|\beta|} + \frac{1}{|\beta|^2} + \frac{3 - n^2\pi^2}{3|\beta|^3} + \dots \right], \\
\gamma_n &\sim -\frac{n^2\pi^2}{4|\beta|^2} + \dots, \quad n = 1, 2, 3, \dots,
\end{aligned}
\tag{4.148}
$$

where odd indices correspond to the zeros of ϕ_0 and even indices to those of ϕ_1. Note that in the limit $\beta \to -\infty$, these α_n values reduce to the energy levels of the infinite square well potential of width $2a$. This is not surprising since in this

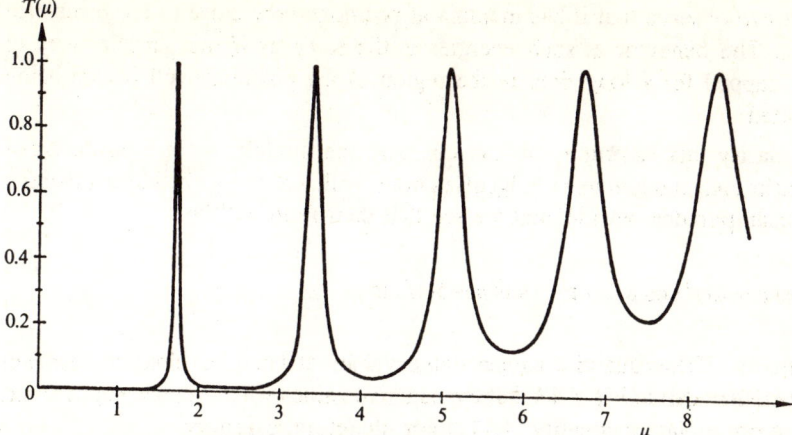

Fig. 4.12. The transmission coefficient for $\beta = -11$

limit the resonances become infinitely narrow, which means that the particle is trapped inside the potential, which behaves then just like a pair of impenetrable walls.

In Fig. 4.12, the transmission coefficient is represented as a function of μ for $\beta = -11$. It is clearly seen that for a given potential, the widths of the resonances increase with energy and become progressively less pronounced above the background.

Let us examine what happens to the time delays. As previously stated, δ_0 (δ_1) changes rapidly near a resonance corresponding to a zero of ϕ_0 (ϕ_1), and therefore it is expected that a study of these time delays will permit us to characterize the positions of the resonances. In Fig. 4.13, the time delay is plotted against μ,

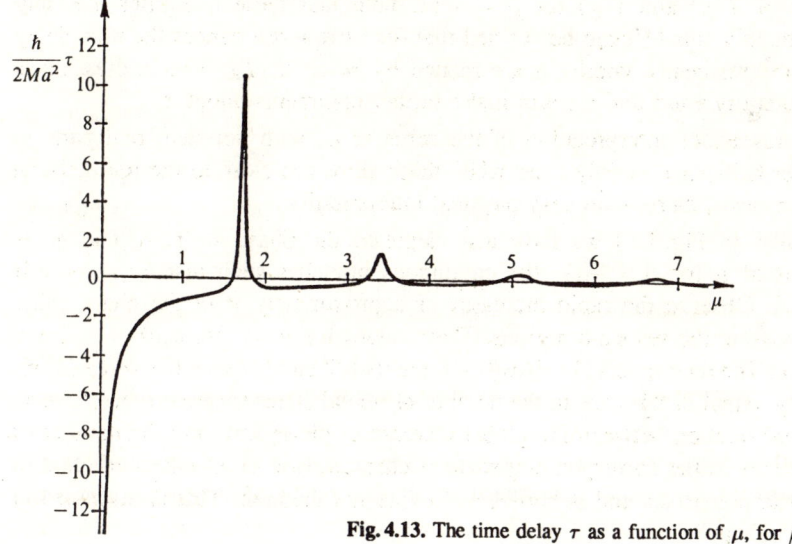

Fig. 4.13. The time delay τ as a function of μ, for $\beta = -11$

and one can observe that it has maxima at positions very close to the resonance energies. The behavior at such energies is the same as if the particle were to remain trapped for a long time in the region of the potential well before being transmitted.

To clarify this assertion, we can compare the moduli of the amplitudes of symmetric and antisymmetric eigenfunctions with energy ε, inside a potential with the amplitudes outside, and we see that their ratio will be

$$|\cos(\mu + \delta_0)/\cos\mu| \quad \text{or} \quad |\sin(\mu + \delta_1)/\sin\mu|$$

respectively. These quantities reach large values at energies close to resonant energies. Note also in Fig. 4.13 that the negative values which represent advanced times do not violate inequality (4.111) nor, therefore, causality.

Table 4.1. Data for the first five resonances of the potential with $\beta = -11$

Zeros of $\phi_0(\mu)$	Zeros of $\phi_1(\mu)$	Values of μ corresponding to the maxima of T	Values of μ corresponding to the maxima of τ
1.721 8-i 0.025 6		1.722 6	1.721 7
	3.416 0-i 0.087 2	3.419 8	3.415 1
5.077 9-i 9.162 1		5.085 5	5.076 3
	6.715 1-i 0.236 8	6.725 9	6.713 5
8.335 1-i 0.306 8		8.348 1	8.333 8

Table 4.1 gives the first roots of $\phi_j(\mu) = 0$, as well as the positions of the maxima of $T(\mu)$ and $\tau(\mu)$ for $\beta = -11$; note that these quantities are only approximately equal. It can be verified that for narrow resonances the time delay τ_R and the resonance width Γ_R are related by $\tau_R \simeq 2\hbar/\Gamma_R$. This expression is quite generally valid and we will make further comments about it.

The resonance interpretation of the zeros of ϕ_i with non-zero real parts in the lower half-plane is only clear when these zeros are close to the real axis. In the other cases, there is no easy physical interpretation.

Finally, in Fig. 4.14 we have also depicted the phase shifts δ_0 and δ_1 as functions of μ for $\beta = -11$ [the continuous branches such that $\delta_i(\mu) \to 0$ if $\mu \to \infty$]. Observe the rapid increases of approximately π in the phase shifts when crossing the resonant regions. Their values for $\mu = 0$ are $\delta_0(0) = \pi/2$ and $\delta_1(0) = \pi$. The increments $(1/\pi)(\delta_0(0) - \delta_0(\infty)) + 1/2$ and $(1/\pi)(\delta_1(0) - \delta_1(\infty))$ are, evidently, equal in this case to the number of bound states for each type of wave. This close relation between the global increase of phase shift and the number of bound states in the corresponding wave is characteristic of an extended class of symmetric potentials, and is known as Levinson's theorem. This is discussed in detail in Chap. 8.

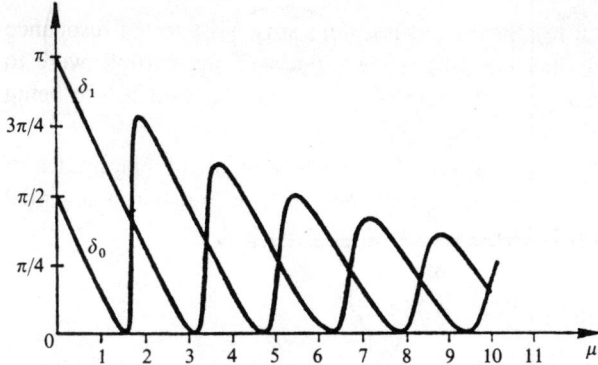

Fig. 4.14. Phase shifts δ_0 and δ_1 for $\beta = -11$

Frequently, one considers the poles of the transmission amplitude, $\sigma(\varepsilon)$, in the complex ε-plane. Keeping in mind that if $\mu = |\mu| \exp(\mathrm{i}\alpha)$, then $\varepsilon = a^2 |\mu|^2 \exp(2\mathrm{i}\alpha)$, it is necessary to use two Riemann sheets joined along the real non-negative axis to describe $\sigma(\varepsilon)$; one passes from one to another by crossing from below anywhere along the cut half-axis $[0, \infty)$. The first sheet corresponds to Im $\mu > 0$, and is called the physical sheet, while the second, corresponding to Im $\mu < 0$, is called the unphysical sheet. In the complex ε-plane, the transmission amplitude has poles located at points corresponding to zeros of the functions $\phi_j(\mu)$. [Notice that $\phi_j(\mu) = 0$ and $\phi_j(-\mu) = 0$ do not have common roots.] In Fig. 4.15, the pole structure of $\sigma(\varepsilon)$ is represented in the complex ε-plane.

The important information which the analyticity and the pole structure of $\sigma(\varepsilon)$ yield about the physics of the problem, and which we show here explicitly in an example, is of very general validity. As we have already said in Sect. 4.5, whenever $U(x)$ tends to zero as $|x| \to \infty$ faster than $1/x^2$, $\sigma(\varepsilon)$ will be analytic in the half-plane Im $k > 0$, and there exists a one-to-one correspondence between the poles of σ in this region and the bound states of negative energy. To be able to talk about analyticity in the lower half-plane Im $k < 0$, or in part of it, it is necessary that $U(x)$ tend to zero much faster, e.g., exponentially. In such a case, the previous discussion on the virtual and resonant states carries over to this situation.

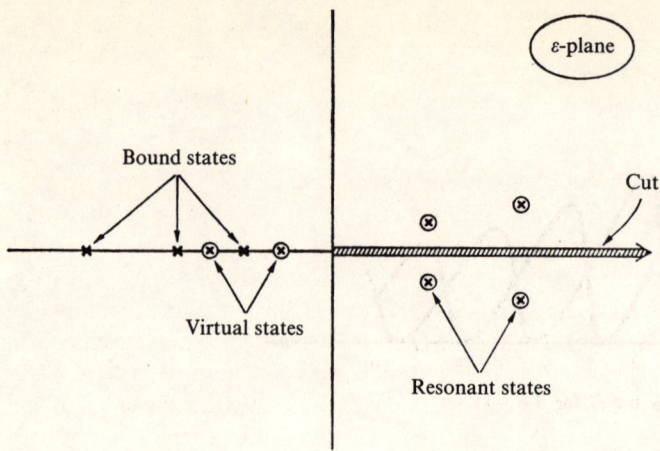

Fig. 4.15. The pole structure of $\sigma(\varepsilon)$ in the complex ε-plane. The points marked with $\times$ correspond to the first Riemann sheet, and those with $\otimes$ correspond to the second sheet

4.7 Square Potentials

Continuing with examples, we will consider a few more simple potentials.

Example 1. In the case of Fig. 4.6, we have $U_- = 0$ and $U_+ = \infty$, and therefore, $\sigma_c(H)$ covers the interval $(0, \infty)$ and the spectrum is simple. Keeping in mind that $\psi(x) = 0$ for $x \geq 0$, the wave function, for energy $\varepsilon > 0$, is

$$
\begin{aligned}
\psi(x) &= 2A \sin k'x , & -L \leq x < 0 , \\
\psi(x) &= \mathrm{e}^{\mathrm{i}kx} + \varrho_-(E)\mathrm{e}^{-\mathrm{i}kx}, & x \leq -L ,
\end{aligned}
\tag{4.149}
$$

where $k' = (\varepsilon - U_0)^{1/2}$ and $k = \varepsilon^{1/2}$. Imposing the continuity of the wave function and of its first derivative at the point $x = -L$,

$$
\varrho_-(E) = \mathrm{e}^{2\mathrm{i}[\varphi(E)-kL]}, \quad A = -\frac{\mathrm{e}^{-\mathrm{i}kL}}{2 \sin k'L}[1 + \mathrm{e}^{2\mathrm{i}\varphi(E)}] ,
\tag{4.150}
$$

where

$$
\mathrm{e}^{2\mathrm{i}\varphi(E)} = \frac{\tan k'L - \mathrm{i}k'/k}{\tan k'L + \mathrm{i}k'/k} .
\tag{4.151}
$$

Note that $|\varrho_-(E)| = 1$, and therefore there is always total reflection.
From (4.150) and (4.151), one deduces that

$$
\begin{aligned}
|A|^2 &= \varepsilon \left(\varepsilon - U_0 \cos^2 L\sqrt{\varepsilon - U_0}\right)^{-1} , \\
\varphi(E) &= -\tfrac{\pi}{2} + \arctan\left[\sqrt{\varepsilon/(\varepsilon - U_0)} \tan L\sqrt{\varepsilon - U_0}\right] ,
\end{aligned}
\tag{4.152}
$$

and, taking the origin as the point of reference, the time delay is

$$\tau_{\text{refl}}^-(E) = \frac{4M}{\hbar} \left\{ \frac{L\varepsilon\sqrt{\varepsilon - U_0} - (1/2)U_0 \sin 2L\sqrt{\varepsilon - U_0}}{2\sqrt{\varepsilon}\sqrt{\varepsilon - U_0}[\varepsilon - U_0 \cos^2 L\sqrt{\varepsilon - U_0}]} - \frac{L}{2\sqrt{\varepsilon}} \right\}, \quad (4.153)$$

which has a zero limit when $U_0 \to 0$. The typical behavior of $|A|^2$ and $\tau_{\text{refl}}^-(E)$ is represented in Fig. 4.16. It should be noted that $|A|^2$ reaches its maximum value, $|A|^2 = 1$, at $L\sqrt{\varepsilon - U_0} = (n + 1/2)\pi$, where n takes on all integer values which make $\varepsilon \geq 0$; the resonant energies are approximately $\varepsilon_n = (n+1/2)^2\pi^2/L^2 + U_0 \geq 0$. It is seen in Fig. 4.16 that $\tau_{\text{refl}}^-(E)$ has a maximum for energies close to these. Everything occurs as if for a resonant energy the particle remains trapped in the region of the square well for a long time before reflection. By (4.152), one deduces that the peaks of $|A|^2$ become less pronounced over the background and wider as ε increases.

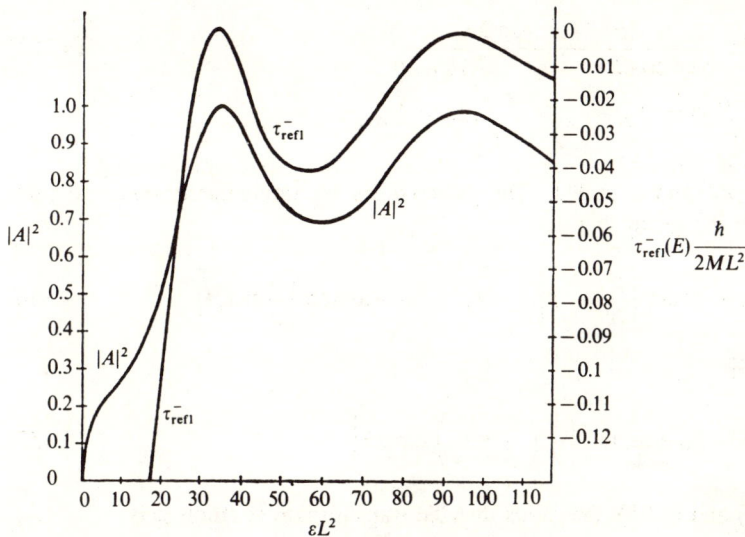

Fig. 4.16. Behavior of $|A|^2$ and $\tau_{\text{refl}}^-(E)$ for $L^2|U_0| = 27$. The first two maxima of $|A|^2$ correspond to $\varepsilon L^2 = 34.685$ and $\varepsilon L^2 = 93.902$. The corresponding maxima in τ_{refl}^- are situated at $\varepsilon L^2 = 33.659$ and $\varepsilon L^2 = 92.892$, and the values of τ_{refl}^- at these points are $\tau^- = 0.000562(2ML^2/\hbar)$ and $\tau^- = 0.0000627(2ML^2/\hbar)$

If we extend expressions (4.149) to the region $U_0 \leq \varepsilon \leq 0$ and introduce $k = i\beta$, with $\beta > 0$, the requirement for no increasing wave for $x \leq -L$ (condition for a bound state), is that $\varrho_-(E) = \infty$, that is, $\tan k'L + k'/\beta = 0$, which is the equation previously obtained for the energies of bound states. In this particular example the study of $\varrho_-(E)$ in the complex ε-plane provides information on the physical properties of the system considered, much as $\sigma(E)$ did in the previous example.

Example 2. For the symmetric square potential in Fig. 4.2, one has $U_- = U_+ = 0$, and therefore $\sigma_c(H)$ fills the interval $(0, \infty)$ and is doubly degenerate. If the incident wave is coming from the left, the solution $\psi^{(-)}(x)$ is given by

$$
\begin{aligned}
\psi^{(-)}(x) &= e^{ikx} + \varrho e^{-ikx}, & -\infty < x \le -L\,; \\
\psi^{(-)}(x) &= A e^{ik'x} + B e^{-ik'x}, & -L \le x \le L\,; \\
\psi^{(-)}(x) &= \sigma e^{ikx}, & L \le x < \infty\,,
\end{aligned}
\tag{4.154}
$$

where $k \equiv \sqrt{\varepsilon}$ and $k' \equiv \sqrt{\varepsilon - U_0}$. The quantities A, B, ϱ and σ are determined from the required continuity of the wave function and its first derivative at $x = \pm L$. A simple calculation leads to

$$
\begin{aligned}
\sigma &= e^{-2i\alpha} \frac{2\alpha\beta}{2\alpha\beta \cos 2\beta - i(\alpha^2 + \beta^2) \sin 2\beta}\,, \\
\varrho &= e^{-2i\alpha} \frac{i(\beta^2 - \alpha^2) \sin 2\beta}{2\alpha\beta \cos 2\beta - i(\alpha^2 + \beta^2) \sin 2\beta}\,, \\
A &= \frac{\beta + \alpha}{2\beta} e^{i(\alpha - \beta)} \sigma\,, \quad B = \frac{\beta - \alpha}{2\beta} e^{i(\alpha + \beta)} \sigma\,,
\end{aligned}
\tag{4.155}
$$

where $\alpha \equiv kL$ and $\beta \equiv k'L$. The phase shifts are immediately obtained from (4.119) with the result that

$$
\delta_0 = -\alpha + \arctan\left[\frac{\beta}{\alpha} \tan \beta\right]\,, \quad \delta_1 = -\alpha + \arctan\left[\frac{\alpha}{\beta} \tan \beta\right]\,,
\tag{4.156}
$$

and therefore

$$
\delta_0 + \delta_1 = -2\alpha + \arctan\left[\frac{1}{2}\left(\frac{\alpha}{\beta} + \frac{\beta}{\alpha}\right) \tan 2\beta\right]\,.
\tag{4.157}
$$

From (4.94) and (4.95), one finds that the transmission coefficient is

$$
T(E) = \left[1 + \frac{U_0^2}{4\varepsilon(\varepsilon - U_0)} \sin^2 2L\sqrt{\varepsilon - U_0}\right]^{-1}\,.
\tag{4.158}
$$

In Fig. 4.17, we give the typical behavior of the transmission coefficient. From (4.158), one sees that $T(E)$ has maxima, with $T(E) = 1$, for energies such that $2L\sqrt{\varepsilon - U_0} = n\pi$, where n is a positive integer. These energies are the resonance energies, and for them the transmission is total.

As in the example of the delta-like potentials, the transmission amplitude $\sigma(\varepsilon)$ is an analytic function of ε in the plane cut along $[0, \infty)$. In the first Riemann sheet the function $\sigma(\varepsilon)$ is analytic with poles at those points satisfying

$$
\cot 2k'L = \frac{i}{2}\left(\frac{k}{k'} + \frac{k'}{k}\right)\,.
\tag{4.159}
$$

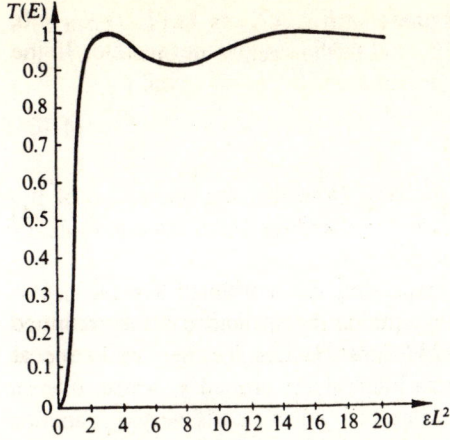

Fig. 4.17. Transmission coefficient $T(E)$ as a function of energy, for the potential $L^2|U_0| = 27/4$

For this equation to be satisfied on this sheet it is necessary that $k = i\beta_0$ with $\beta_0 > 0$, and so, using simple trigonometric relations, the poles satisfy one of the following relations:

$$k' \tan k'L = \beta_0 , \quad k' \cot k'L = -\beta_0 , \tag{4.160}$$

which are equivalent to (4.24). Therefore, once again we find that the poles of $\sigma(\varepsilon)$ in the first Riemann sheet lie on the negative real axis and correspond to the energies of the bound states.

If ε_r is a resonant energy, then for energies ε, close to ε_r, one can write

$$\left(\frac{k}{k'} + \frac{k'}{k}\right) \tan 2k'L \simeq \frac{4}{\gamma_r}(\varepsilon - \varepsilon_r) , \tag{4.161}$$

where γ_r, which has the dimensions of ε, is defined by

$$\frac{4}{\gamma_r} \equiv \left[\frac{d}{d\varepsilon}\left(\frac{k}{k'} + \frac{k'}{k}\right)\tan 2k'L\right]_{\varepsilon=\varepsilon_r} = \frac{L(2\varepsilon_r - U_0)}{(\varepsilon_r - U_0)\sqrt{\varepsilon_r}} . \tag{4.162}$$

The quantity $\Gamma_r \equiv (\hbar^2/2M)\gamma_r > 0$, with the dimensions of energy, is the width of the resonance. Keeping in mind (4.155), one finds that for $\varepsilon \simeq \varepsilon_r$, the transmission amplitude can be written as

$$\sigma(E) \simeq \pm e^{-2ikL} \frac{i\gamma_r/2}{\varepsilon - \varepsilon_r + i\gamma_r/2} , \tag{4.163}$$

where the plus sign corresponds to the case in which $2k'L = 2n\pi$ for this resonance, and the minus sign applies otherwise. One sees, therefore, that $\sigma(E)$ behaves as if it had a pole at $\varepsilon_r - i\gamma_r/2$, which is necessarily in the unphysical sheet. As in the earlier example, the resonances appear as complex poles of $\sigma(E)$ in the unphysical sheet. In any case, the width γ_r is in general very large and (4.163) is of very restricted applicability, except when $L|U_0|^{1/2} \gg 1$.

On the other hand, if we define the phase shift $\delta_{res}(E)$ as $\delta_{res}(E) = \delta_-(E) - \delta_{bg}(E)$ (bg = background), where $\delta_{bg}(E) = -kL$, then near a resonance we have

$$\delta_{res}(E) \simeq \frac{1}{2} \arctan \left[\frac{2}{\gamma_r}(\varepsilon - \varepsilon_r) \right] , \tag{4.164}$$

as can be deduced immediately from (4.163). Note that the derivative of this quantity reaches its maximum value $1/\gamma_r$ at $\varepsilon = \varepsilon_r$ and that, furthermore, $\delta_{res}(\varepsilon_r) = n\pi/2$ and $\delta_{res}(\varepsilon_r \pm \gamma_r/2) = n\pi/2 \pm \pi/8$.

We finish this example by briefly discussing the motion of a wave packet constructed with momenta approximately equal to the resonance momentum. Let the resonance energy be E_r, and $k_r \equiv (2ME_r)^{1/2}/\hbar$. Let $f(k)$ be a real, smooth function of the variable k, zero outside an interval Δk around k_r whose interior has no other resonance. Assume furthermore that f is a slowly varying function over the resonance peak. Consider now the incident wave packet

$$\Psi_{inc}(x; t) = \int_{\Delta k} dk \, f(k) \exp[i(kx - Et/\hbar)] ; \tag{4.165}$$

the transmitted wave packet is

$$\Psi_{trans}(x; t) = \int_{\Delta k} dk \, f(k)\sigma(E) \exp[i(kx - Et/\hbar)] . \tag{4.166}$$

Using (4.163), we get

$$\Psi_{trans}(x; t) = \pm \int_{\Delta k} dk \, f(k)\frac{i\Gamma_r/2}{E - E_r + i\Gamma_r/2}$$
$$\times \exp[i(k(x - 2L) - Et/\hbar)] . \tag{4.167}$$

In spite of the approximation used for $\sigma(E)$, calculating the integral in (4.167) is quite complicated, even for simple functions $f(k)$. However, for the case of a practically constant amplitude $f(k)$ over $\Delta E \gg \Gamma_r$ (and thus for wave packets which are highly localized in position) it is not difficult to obtain an estimate of the expectation value $\langle x \rangle_{\Psi_{trans}} \simeq 2L + \hbar k_r t/M$. Observe that Γ_r does not appear. In other words, the resonance time delay of $2\hbar/\Gamma_r$ at $k = k_r$, important if $\Delta E \ll \Gamma_r$, does not affect the wave packet considered here. This conclusion, which indicates the difficulty of direct measurement of time delays, does not diminish the essential role which these play in resonance phenomena in general. An indirect manifestation can be seen in a rigorous and detailed analysis of (4.167). *Newton* [NE 82] studies (4.167) for an $f(k)$ that is gaussian in energy. This is the generic expression (if $L = 0$) for the transmitted wave function for any potential in an energy region that includes only one resonance and has no background phase shifts. He proves that if $\Delta E \gtrsim \Gamma_r$, then $|\Psi_{trans}(x; t)|^2$ has a behavior with time at each fixed $x > 0$ which is, to a very good approximation, exponentially decreasing,

$$|\Psi_{\text{trans}}(x;t)|^2 \propto \exp(-t\Gamma_{\text{r}}/\hbar) \, ,\tag{4.168}$$

when the condition

$$\left(\frac{E_{\text{r}}}{\Delta E}\right)^2 (k_{\text{r}}x)^{-2/3} \gtrsim \frac{t-x/v_{\text{r}}}{\hbar/\Gamma_{\text{r}}} \gg 4\left(\frac{\Gamma_{\text{r}}}{\Delta E}\right) + \frac{1}{2}\left(\frac{\Gamma_{\text{r}}}{\Delta E}\right)^2 \tag{4.169}$$

is met. For times shorter or larger, the exponential law fails. At a resonance, therefore, we can imagine that the particle is captured in the potential well for a time $\sim \hbar/\Gamma_{\text{r}}$, and then escapes, following the typical exponential law for radioactive phenomena.

The analysis which has been carried out becomes more precise for sharper resonances. In the potential considered here, and with the data used for the graphs, the resulting resonances are not particularly narrow; we could, for example, substitute for the $U(x)$ used in Fig. 4.2 that given in Fig. 4.18. The barriers B_1 and B_2 make the escape of the particle more difficult once it is inside the well. The resonance energies would then be close to the bound state energies of the corresponding infinite square well, i.e., with both barriers infinitely high, and their widths would decrease as these barriers increase. In our case, the resonance condition $2k'L = n\pi$ is precisely the bound state condition in the infinite square well potential.

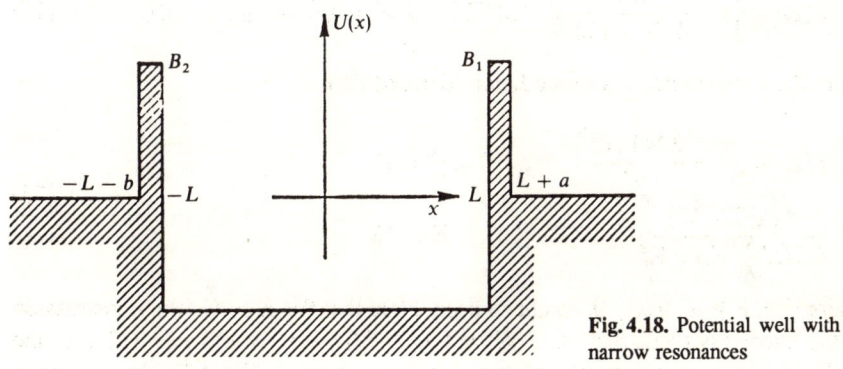

Fig. 4.18. Potential well with narrow resonances

Example 3. We now examine the rectangular barrier in Fig. 4.19. This potential does not have bound states. The continuous spectrum extends from 0 to ∞ and is doubly generate. If the incident wave comes from the left, the solution $\psi^{(-)}(x)$ is

$$\begin{aligned}
\psi^{(-)}(x) &= e^{ikx} + \varrho(E)e^{-ikx}, & x &\leq -L\,; \\
\psi^{(-)}(x) &= Ae^{k'x} + Be^{-k'x}, & -L &\leq x \leq L\,; \\
\psi^{(-)}(x) &= \sigma(E)e^{ikx}, & x &\geq L\,,
\end{aligned}\tag{4.170}$$

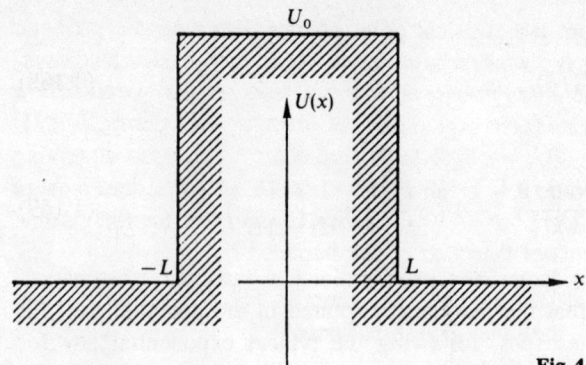

Fig. 4.19. Rectangular potential barrier

where it is supposed that $0 < \varepsilon < U_0$, $k = \varepsilon^{1/2}$ and $k' = |\varepsilon - U_0|^{1/2}$. Requiring the continuity of the wave function and of its first derivative at the points $x = \pm L$, one obtains for the transmission coefficient

$$T(E) = \left[1 + \frac{U_0^2}{4\varepsilon(U_0 - \varepsilon)} \sinh^2 2k'L\right]^{-1}, \tag{4.171}$$

$$k' = \sqrt{U_0 - \varepsilon}, \ 0 \le \varepsilon \le U_0 .$$

Extending this result to values of $\varepsilon > U_0$,

$$T(E) = \left[1 + \frac{U_0^2}{4\varepsilon(\varepsilon - U_0)} \sin^2 2k''L\right]^{-1}, \ k'' = \sqrt{\varepsilon - U_0}, \ \varepsilon \ge U_0 . \tag{4.172}$$

From these relations it is immediately derived that

$$T(E) < \left[1 + \frac{2MV_0L^2}{\hbar^2}\right]^{-1}, \quad 0 < E < V_0$$

$$\frac{4E(E - V_0)}{4E(E - V_0) + V_0^2} \le T(E) \le 1, \quad E \ge V_0 . \tag{4.173}$$

where $\sinh x \ge x$, $\forall x \ge 0$, has been used. Note that for $E > V_0$, the transmission of the barrier is total, that is, $T(E) = 1$ if $2k''L = n\pi$, where $n = 1, 2, \ldots$, and its behavior is completely analogous to that given in Fig. 4.17. For energies $E < V_0$, the transmission is always less than unity and rapidly decreases as $V_0 - E$ increases. When $(U_0 - \varepsilon)^{1/2}L \gg 1$, then

$$T(E) \simeq \frac{16E(V_0 - E)}{V_0^2} \exp\left[-\frac{4L}{\hbar}\sqrt{2M(V_0 - E)}\right], \tag{4.174}$$

i.e., the transmission coefficient decreases exponentially as $(V_0 - E)^{1/2}$ increases.

Classically, for $E < V_0$, there is no transmission. Quantum mechanically, due to the finiteness of $\hbar$, it is possible for the particle to cross the barrier. This effect is called *tunneling*, and is of great importance. It may seem paradoxical that the particle as seen from the classical point of view has a negative kinetic energy inside the barrier. It is the uncertainty principle (which makes incompatible the position and kinetic energy observables) that resolves the question: For the particle to have an appreciable probability of crossing the barrier, (4.171) tells us that we must have $\sqrt{U_0 - \varepsilon}\, L \lesssim 1$. On the other hand, upon observing its passage through the barrier, it is given a momentum $\Delta p \sim \hbar/L$ and thus an energy of the order of $\hbar^2/2ML^2 \gtrsim V_0 - E$, which is sufficient for the particle to appear with an energy greater than that of the barrier.

Example 4. Finally, we consider the potential given in Fig. 4.20. If an incident beam comes from $x = -\infty$, then the wave function $\psi^{(-)}(x)$ is

$$\varepsilon > U_0 : \quad \psi^{(-)}(x) = \begin{cases} e^{ikx} + \varrho_-(E)e^{-ikx}, & x \leq 0 , \\ \sigma_-(E)e^{ik'x}, & x \geq 0 ; \end{cases}$$

$$\varepsilon < U_0 : \quad \psi^{(-)}(x) = \begin{cases} e^{ikx} + \varrho_-(E)e^{-ikx}, & x \leq 0 , \\ \sigma_-(E)e^{-k''x}, & x \geq 0 , \end{cases} \tag{4.175}$$

where $k = \varepsilon^{1/2}$, $k' = (\varepsilon - U_0)^{1/2}$ and $k'' = (U_0 - \varepsilon)^{1/2}$. Matching them at the origin, we find

$$\varepsilon > U_0 : \quad \varrho_-(E) = \frac{k - k'}{k + k'} , \qquad \sigma_-(E) = \frac{2k}{k + k'} ;$$

$$\varepsilon < U_0 : \quad \varrho_-(E) = \frac{k - ik''}{k + ik''} , \qquad \sigma_-(E) = \frac{2k}{k + ik''} , \tag{4.176}$$

and the transmission coefficient for $\varepsilon > U_0$ is

$$T(E) = \frac{4kk'}{(k + k')^2} , \tag{4.177}$$

which is zero for $\varepsilon = U_0$ and increases monotonically until it reaches the value 1 as $\varepsilon \to \infty$. Evidently, $T(E) = 0$ if $E < V_0$, since then there is total reflection.

4.8 Periodic Potentials

For the potentials considered up to now, the continuous spectrum in general fills the right half-line, and the discrete spectrum is situated to its left. There exist a class of potentials, however, which are physically very interesting, but do not share these characteristics. These are the periodic potentials:

$$U(x + a) = U(x) , \tag{4.178}$$

whose shape is repeated indefinitely, with period a. Such potentials are produced in three dimensions by any perfect lattice structure, and thus they are of great importance for solid-state physics. Here we will limit ourselves to one dimension, which is sufficient to reveal the principal features. We assume that $U(x) \in L^2(0, a)$.

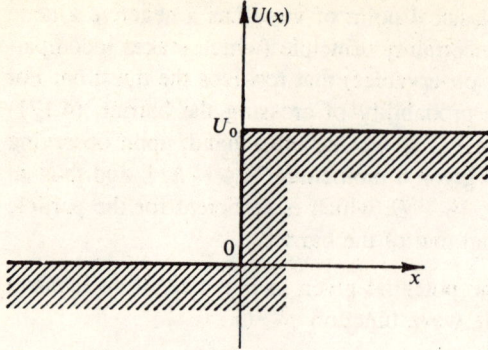

Fig. 4.20. Potential barrier of finite height and infinite extent

The first fact to emphasize is that there are no bound states: $\sigma_p(H) = \emptyset$. The proof is elementary: We suppose that $\varepsilon \in \sigma_p(H)$, and let M_ε be the corresponding eigensubspace. The differential equation (4.1) is of second order; hence $d \equiv \dim(M_\varepsilon) \leq 2$. Given the periodicity (4.178), it is clear that if $\psi(x) \in M_\varepsilon$, then so is $\psi(x - a) \in M_\varepsilon$. Therefore M_ε is stable under the unitary operator V_a (2.174) for a translation by a. Diagonalizing this operator in this subspace, we can find an orthonormal basis $\psi_1, \ldots, \psi_d$ in M_ε such that

$$V_a \psi_j = e^{i\theta_j} \psi_j , \quad \theta_j \text{ real} , \tag{4.179}$$

that is,

$$\psi_j(x - a) = e^{i\theta_j} \psi_j(x) . \tag{4.180}$$

This last relation is clearly incompatible with the square integrability of ψ_j. Hence, $M_\varepsilon = \{0\}$.

To determine the continuous spectrum $\sigma_c(H)$, we will base our reasoning on the following criterion [SI 82]: Except at most for energies in a set of zero spectral measure (which will be ignored), $\varepsilon \in \sigma_c(H)$ if and only if some non-trivial solution of (4.1) is polynomially bounded. Let us fix some initial conditions $\psi(0)$ and $\psi'(0)$, and let $\psi(x)$ be the solution of (4.1) with these boundary conditions. The values $\psi(a)$ and $\psi'(a)$ are linear combinations of the initial conditions

$$\begin{pmatrix} \psi(a) \\ \psi'(a) \end{pmatrix} = B(\varepsilon) \begin{pmatrix} \psi(0) \\ \psi'(0) \end{pmatrix} . \tag{4.181}$$

The matrix $B(\varepsilon)$ generally depends on the choice of starting point, which here is $x = 0$. The translational invariance (4.178) implies that

$$\begin{pmatrix} \psi(na) \\ \psi'(na) \end{pmatrix} = (B(\varepsilon))^n \begin{pmatrix} \psi(0) \\ \psi'(0) \end{pmatrix} , \tag{4.182}$$

and therefore the boundedness or unboundedness of $\psi(x)$ is linked to the size of the eigenvalues of the $B(\varepsilon)$ matrix. Let us take two pairs of initial conditions which diagonalize $B(\varepsilon)$:

$$B(\varepsilon) \begin{pmatrix} \psi_i(0) \\ \psi_i'(0) \end{pmatrix} = \lambda_i(\varepsilon) \begin{pmatrix} \psi_i(0) \\ \psi_i'(0) \end{pmatrix} . \tag{4.183}$$

[In principle such pairs $\psi_i(0), \psi_i'(0), i = 1, 2$, do not always exist. Only if the roots of the secular equation for $B(\varepsilon)$ are distinct is it possible to guarantee that such a diagonalization exists. Since $\det B(\varepsilon) = 1$ [which is simply an expression of the constancy of the Wronskian for the solutions determined by $\psi(0) = 1, \psi'(0) = 0$ and $\psi(0) = 0, \psi'(0) = 1$], the roots of $B(\varepsilon)$ are always distinct when $\text{Tr}\{B(\varepsilon)\} \neq \pm 2$. It can be demonstrated [HO 71] that $\text{Tr}\{B(\varepsilon)\}$ is an entire analytic function of ε, of order $1/2$, so that the zeros of $\text{Tr}\{B(\varepsilon)\} - (\pm 2)$ form a discrete set. As far as $\sigma_c(H)$ is concerned, these values of ε can be neglected.]

Relation (4.182), together with (4.183), leads to

$$\psi_i(x + na) = (\lambda_i(\varepsilon))^n \psi_i(x) . \tag{4.184}$$

Therefore, if $|\lambda_i(\varepsilon)| \neq 1$, then $\psi_i(x)$ increases exponentially in some direction. Consequently, such ε do not belong to the continuous spectrum. Only if $|\lambda_i(\varepsilon)| = 1$ is $\psi_i(x)$ (polynomially) bounded, and $\varepsilon \in \sigma_c(H)$. Hence, in this case, writing $\lambda_1(\varepsilon) = \exp(ika)$ and $\lambda_2(\varepsilon) = \exp(-ika)$, k real, we have that there exist two linearly independent solutions ψ_1 and ψ_2, such that

$$\psi_1(x - a) = e^{-ika} \psi_1(x) , \quad \psi_2(x - a) = e^{ika} \psi_2(x) . \tag{4.185}$$

If $\varphi_1(x) \equiv e^{-ikx} \psi_1(x)$ and $\varphi_2(x) \equiv e^{ikx} \psi_2(x)$, there results

$$\varphi_i(x - a) = \varphi_i(x) ,$$
$$\psi_1(x) = e^{ikx} \varphi_1(x) , \tag{4.186}$$
$$\psi_2(x) = e^{-ikx} \varphi_2(x) .$$

To physicists, this expression is known as *Bloch's theorem* (for mathematicians it is *Floquet's theorem*), according to which there are two independent solutions for (4.1) of the plane wave type modulated by a periodic function with period a.

Since the secular equation for $B(\varepsilon)$ is

$$\lambda^2 - [\text{Tr}\{B(\varepsilon)\}]\lambda + 1 = 0 , \tag{4.187}$$

and $\text{Tr}\, B\{(\varepsilon)\}$ is real for ε real [it suffices to recall that the $B(\varepsilon)$ matrix has real coefficients in this case], then $\varepsilon \in \sigma_c(H)$ if and only if $[\text{Tr}\{B(\varepsilon)\}]^2 \leq 4$. Since $B(\varepsilon)$ is an analytic function of ε, the strict inequality is satisfied over the set constructed by the union of open disjoint intervals, whose endpoints satisfy $\text{Tr}\, B\{(\varepsilon)\} = \pm 2$. More concretely, Birkhoff's oscillation theorem proves the following [HO 71]: Let $\varepsilon_0, \varepsilon_1, \ldots$ be the roots of $\text{Tr}\{B(\varepsilon)\} = 2$, ordered in an increasing sense, and each counted as many times as its multiplicity indicates. Analogously, let $\varepsilon_0', \varepsilon_1', \ldots$ be the roots of $\text{Tr}\{B(\varepsilon)\} = -2$. Then

$$\varepsilon_0 < \varepsilon_0' \leq \varepsilon_1' < \varepsilon_1 \leq \varepsilon_2 < \varepsilon_2' \leq \varepsilon_3' < \ldots , \quad \text{and} \tag{4.188}$$

$$\sigma_c(H) = [\varepsilon_0, \varepsilon_0'] \cup [\varepsilon_1', \varepsilon_1] \cup [\varepsilon_2, \varepsilon_2'] \cup \ldots . \tag{4.189}$$

Consequently, the spectrum has a *band structure*, that is, intervals of allowed energies (stability zones), separated by intervals of forbidden energies (instability zones) – see Fig. 4.21.

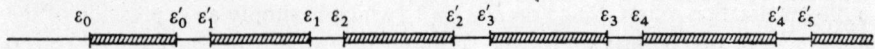

Fig. 4.21. Band structure in $\sigma(H)$

This structure always results if $U(x) \neq$ const, although it may occur in some cases that many of the forbidden zones disappear. The widths of these forbidden zones always tend to zero as $\varepsilon \to +\infty$, in which limit the spectrum naturally resembles that of the kinetic energy. For a rigorous study of Schrödinger's equation with periodic potentials, in one or several dimensions, see [RS 78].

We will illustrate these ideas with an elementary example, corresponding to the potential in Fig. 4.22, of period $(a + b)$:

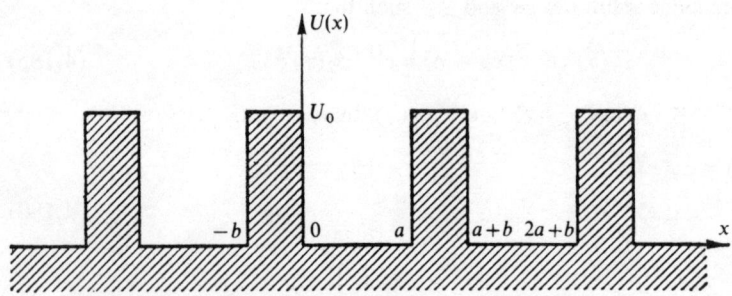

Fig. 4.22. Example of a periodic potential: the Kronig-Penney model

We begin by determining $\operatorname{Tr} B\{(\varepsilon)\}$. Since $U(x) \geq 0$, it suffices to consider $\varepsilon > 0$. For the initial conditions $\psi(0) = 1$ and $\psi'(0) = 0$, in the interval $(0, a)$ the wave function is $\cos \alpha x$, with $\alpha \equiv \sqrt{\varepsilon}$. In the region $(a, a + b)$, it is a linear combination of $\sin \beta(x - a)$ and $\cos \beta(x - a)$, with $\beta \equiv \sqrt{\varepsilon - U_0}$. Matching as usual at $x = a$, in the second region this function becomes

$$\psi(x) = -\frac{\alpha}{\beta} \sin \alpha a \sin \beta(x - a) + \cos \alpha a \cos \beta(x - a) \, , \tag{4.190}$$

and therefore, from (4.181):

$$B_{11} = -\frac{\alpha}{\beta} \sin \alpha a \sin \beta b + \cos \alpha a \cos \beta b \, . \tag{4.191}$$

Similarly, taking the function $(1/\alpha) \sin \alpha x$ in the interval $(0, a)$, one obtains

$$B_{22} = \cos \alpha a \cos \beta b - \frac{\beta}{\alpha} \sin \alpha a \sin \beta b \, . \tag{4.192}$$

Hence

$$\mathrm{Tr}\{B(\varepsilon)\} = 2\cos\alpha a\cos\beta b - \left[\frac{\alpha}{\beta} + \frac{\beta}{\alpha}\right]\sin\alpha a\sin\beta b\ . \tag{4.193}$$

The allowed values for ε should satisfy $[\mathrm{Tr}\{B(\varepsilon)\}]^2 \le 4$, and for such energies (except, at most, for an irrelevant discrete set), there are two linearly independent solutions to (4.1) of Bloch type (4.186), where $\exp[\pm ik(a+b)]$ are the roots of the secular equation for $B(\varepsilon)$

$$\lambda^2 - [\mathrm{Tr}\{B(\varepsilon)\}]\lambda + 1 = 0\ . \tag{4.194}$$

Therefore, the *momentum* $\hbar k$ (a designation which we will immediately justify) satisfies

$$2\cos k(a+b) = \mathrm{Tr}\{B(\varepsilon)\}\ , \tag{4.195}$$

an equation which implicitly determines the dispersion relation $\varepsilon = \varepsilon(k)$. Note that this function is even, and that it suffices to consider the values of k in the interval $[-\pi/(a+b),\ \pi/(a+b)]$, called the *Brillouin zone*.

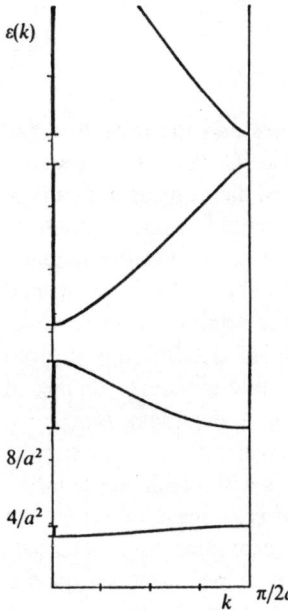

Fig. 4.23. The functions $\varepsilon(k)$ for the first few energy bands, with $a = b$ and $U_0 = 10/a^2$

The energy $\varepsilon(k)$ is represented in Fig. 4.23 for different energy bands for the potential of Fig. 4.22 with some appropriate parameters. Without going into details, which can be found in any solid state physics book, we want to make some observations concerning the physics which emerges from an inspection of Fig. 4.23. In the first place, if we form a narrow wave packet with Bloch functions from a fixed energy band, the usual argument shows that such a wave packet moves through the periodic structure with a group velocity $v_g = (dE/dk)/\hbar$. This velocity is zero at the band endpoint energies. On the other hand, the particle acquires an effective mass $M^*(k)$, due to its interaction with the lattice. In fact, if such particle is acted upon by an external force F, produced, for example, by an electric field, the change in its energy and the acceleration of the center of the wave packet are given by

$$\frac{dE}{dt} = F v_g , \qquad \frac{dv_g}{dt} = \frac{1}{\hbar} \frac{d^2 E}{dk^2} \frac{dk}{dt} , \qquad (4.196)$$

and hence,

$$\hbar \frac{dk}{dt} = F , \qquad \frac{dv_g}{dt} = \frac{F}{M^*(k)} , \qquad M^*(k) \equiv \left[\frac{1}{\hbar^2} \frac{d^2 E}{dk^2} \right]^{-1} . \qquad (4.197)$$

Therefore, $\hbar k$ can be interpreted as the momentum of the wave packet; the effective mass $M^*(k)$ changes sign as the energy changes across the energy band, so that the electron can simulate a positive charge when its energy is near some endpoint of each band.

4.9 Inverse Spectral Problem

Given a potential $V(x)$, the energies of its bound states and the reflection and transmission amplitudes are, in principle, determinable. Quite often, however, while one can experimentally determine these quantities, direct measurements of the potential are not possible. This had resulted in substantial effort, dating back to the 1940s, to reconstruct V from these data. It is evident that the function $V(.)$ is not determined by its discrete spectrum alone; this latter is in general finite. We need at least an infinite collection of data to determine V. For a while, the community of physicists envisaged the possibility of determining V from the collision amplitudes at all incident energies (the "fine-structure" details of V require very small wavelengths and therefore very high energies). *Bargmann* [BA 49a] dispelled the notion of being able to do this rigorously by giving the first example of two distinct three-dimensional potentials with identical scattering amplitudes for s-waves (the three- dimensional context is an inessential piece of information for his arguments). Furthermore, in his example, both potentials have a unique bound state, with the same energy. An immediate consequence of this unexpected fact is that we need something more than just the scattering amplitudes and the energies of the bound states to determine V.

The complete solution to this important inverse spectral problem was discovered by *Marchenko* [MA 50], [MA 55], *Gel'fand-Levitan* [GL 51] and *Jost-Kohn* [JK 52]. Excellent expositions of these and other results can be found in *Faddeev* [FA 59], *Chadan-Sabatier* [CS 77] and [DT 79]. The interest in these reconstruction techniques for the potential has reached unsuspected heights after the discovery by *Gardner, Green, Kruskal* and *Miura* [GG 77] of its direct and enormous importance for the explicit solution of the Cauchy problem for certain non-linear equations like that of Kortweg–de Vries (KdV). Today, it has become an indispensible tool in this field.

It is beyond the scope of this book to discuss in detail the general method for the reconstruction of V. We limit ourselves to stating the relevant facts and to arguing some of its more elementary aspects.

Let us assume that $U(x) [\equiv 2MV(x)/\hbar^2]$ is a potential with N bound states with energies $-\kappa_1^2 < -\kappa_2^2 < -\kappa_3^2 < \ldots < -\kappa_N^2 < 0$. To find another potential which possesses in addition to these levels an additional level $-\kappa^2 < -\kappa_1^2$, we begin by determining the solutions $f_{1,2}(x, i\kappa)$, $\kappa > 0$, for the Schrödinger equation with $U(x)$ and the asymptotic behavior (4.125). We then construct a linear combination

$$u_\alpha(x) \equiv f_1(x, i\kappa) + \alpha f_2(x, i\kappa) , \quad \alpha > 0 , \tag{4.198}$$

which grows exponentially as $|x| \to \infty$. The functions $f_1(. , i\kappa)$, $f_2(. , i\kappa)$ and as a consequence u_α are strictly positive. If for example f_1 had a zero, say at x_0, the function $f_1(x, i\kappa)\theta(x - x_0)$, which is square integrable, would be an eigenfunction of the Hamiltonian with potential

$$U_{x_0}(x) = \begin{cases} U(x), & \text{if } x > x_0 , \\ \infty, & \text{if } x < x_0 , \end{cases} \tag{4.199}$$

with eigenvalue $-\kappa^2$. But it is evident that $U_{x_0}(x) \geq U(x)$, $\forall x$, and thus, according to the Min-Max principle, the potential U should support some bound state of energy $\leq -\kappa^2$, contrary to our assumption. Consequently, $u_\alpha^{-1}(.) \in L^2(\mathbb{R})$.

Let us see now that any potential of the family

$$U_\alpha(x) \equiv U(x) - 2\frac{d^2}{dx^2} \ln u_\alpha(x) \tag{4.200}$$

has, in addition to the bound levels of U, a new level of energy $-\kappa^2$. To this end, we observe that if A is the differential operator $A \equiv u_\alpha(x)(d/dx)u_\alpha^{-1}(x)$, one has

$$-\frac{d^2}{dx^2} + U(x) + \kappa^2 = A^\dagger A , \quad -\frac{d^2}{dx^2} + U_\alpha(x) + \kappa^2 = AA^\dagger . \tag{4.201}$$

Formally, it is easy to see that $A^\dagger A$ and $AA^\dagger$ have the same spectrum, except at most for the point 0: If $f \neq 0$ is such that $(A^\dagger A)f = \lambda f$, and therefore $Af \neq 0$ if $\lambda \neq 0$, then $(AA^\dagger)Af = \lambda Af$. But $0 \notin \sigma(A^\dagger A)$, since $-\kappa^2 < \inf \sigma(-d^2/dx^2 + U)$. Therefore, $\sigma(A^\dagger A) \subset \sigma(AA^\dagger)$. On the other hand, it is obvious that $(AA^\dagger)u_\alpha^{-1} = 0$, that is,

$$\left(-\frac{d^2}{dx^2} + U_\alpha\right) u_\alpha^{-1} = -\kappa^2 u_\alpha^{-1} \tag{4.202}$$

as we claimed.

We illustrate this addition technique of new levels with a typical example. Let $U(x) = 0$. The previous method will allow us to generate a potential with a unique bound state. We now have

$$u_\alpha(x) = e^{-\kappa x} + \alpha e^{\kappa x} , \tag{4.203}$$

and therefore,

$$U_\alpha(x) = -2\kappa^2 \cosh^{-2}[\kappa(x - a)] , \quad \alpha \equiv e^{-2\kappa a} . \tag{4.204}$$

Note that the freedom of choosing $\alpha > 0$ reflects the invariance of the spectrum under spatial translations. Furthermore it is remarkable that this potential (which is a soliton of the KdV equation) is transparent to collisions: its transmission amplitude is $\sigma_\alpha(k) = (k+i\kappa)/(k-i\kappa)$ and the transmission is total for any energy. Although this can be shown directly, it is a consequence of the following results.

The construction method for U_α is rigorous if $U \in L_1^1(\mathbb{R})$, in which case the new potential $U_\alpha \in L_1^1(\mathbb{R})$. For details see [DT 79], where it is also shown that the reflection and transmission amplitudes for U_α are obtained from those for U by means of these relations:

$$\varrho_{\alpha,+}(k) = -\frac{k + i\kappa}{k - i\kappa} \varrho_+(k) , \quad \sigma_\alpha(k) = \frac{k + i\kappa}{k - i\kappa} \sigma(k) . \tag{4.205}$$

To reiterate the process of addition of levels, it is important to note that, explicitly,

$$f_{\alpha,j}(x, k) = \frac{(-1)^{j-1}}{ik - \kappa} A f_j(x, k) , \quad j = 1, 2 . \tag{4.206}$$

It is clear that by successive application of this technique, one can add levels at will, each being more deeply bound. It is just as easy to eliminate the deepest level of the potential U, with wave function ψ_0, by taking the potential $U(x) - 2(d^2/dx^2) \ln \psi_0$ as immediately deduced from above. Finally, compact formulae are known which add or subtract an arbitrary number of the deepest bound levels. All the potentials obtained this way have the same transmission coefficients.

The previous algorithm requires a fixed starting potential whose discrete spectrum we can alter with the given techniques, but it does not allow us to modify its transmission coefficients. In this sense, the method is not useful for constructing a potential with predetermined transmission coefficients.

The solution of the general inverse problem for a unique reconstruction of potentials requires the following information:

a) The energies of the bound states: $-\kappa_1^2 < \ldots < -\kappa_n^2 < 0$.
b) The so-called normalization constants: $c_1, \ldots, c_N > 0$, as many as there are bound states.
c) The reflection amplitude $\varrho_-(k)$, $\forall k \in \mathbb{R}$.

The neccessity of (b) can be understood in terms of the previously considered method for the addition of levels; there the new potential depended not only on $-\kappa^2$, but also on the parameter $\alpha > 0$, which in a certain sense expresses the freedom in localizing the potential (recall the example presented). For N levels, it is expected that the same discrete spectrum can be attained with potentials of very diverse localizations, from N potential wells with practically disjoint supports, each one responsible for a level, to a unique structure which sustains all of them. On the other hand, invariance of the transmission amplitude under spatial translations makes impossible its substitution for the reflection in (c).

With the information just mentioned, the reconstruction of the potential U is obtained in the following fashion: we define the function

$$\Omega(x) \equiv 2\sum_{j=1}^{N} c_j e^{-2\kappa_j x} + \frac{1}{\pi}\int_{\mathbf{R}} dk\, \varrho_-(k)e^{2ikx}\,, \qquad \kappa_j > 0\,, \tag{4.207}$$

and consider Marchenko's integral equation

$$B(x,y) + \Omega(x+y) + \int_0^{\infty} d\xi\, \Omega(x+y+\xi)B(x,\xi) = 0\,, \qquad y > 0\,. \tag{4.208}$$

If there exists a $U \in L_2^1(\mathbf{R})$, with bound levels a), reflection coefficient c), and for which the function $f_1(x,k)$ satisfies

$$c_j = \left(\int_{\mathbf{R}} dx\, f_1^2(x, i\kappa_j)\right)^{-1}\,, \qquad j = 1,\ldots,N\,, \tag{4.209}$$

then Marchenko's equation has a unique solution $B(x,y)$, absolutely continuous in x and y, and furthermore

$$U(x) = -\frac{d}{dx}B(x,+0)\,. \tag{4.210}$$

A more difficult technical problem is that of the characterization of those scattering matrices $S(k)$ that arise from potentials in $L_2^1(\mathbf{R})$ [CS 77], [DT 79].

4.10 Mathematical Conditions

In this last section we collect some mathematical information for one-dimensional Hamiltonians in $L^2(I)$, $I \equiv (a,b)$, associated with the differential operator

$$\tau \equiv -\frac{d^2}{dx^2} + U(x)\,. \tag{4.211}$$

Given their simplicity, there are abundant results about them, which we select mostly from [DS 64], [NA 68], [WE 67a] and [SC 81]. The differential operator τ (4.211) can be studied in three different circumstances according to whether I is finite, semi-infinite, or all of $\mathbf{R}$. In the following we assume that $U \in L_{\text{loc}}^1(I)$, namely $U \in L^1(\alpha,\beta), \forall \alpha,\beta$ such that $a < \alpha < \beta < b$, even though this is not strictly necessary (see Sect. C.8), and we have already analyzed in Sect. 4.6 situations where this condition does not hold.

Let τ_1 be the differential operator (4.211) whose domain $D(\tau_1)$ contains the functions $\psi(x)$ defined in I, such that ψ, ψ' are absolutely continuous in any compact subinterval of (a, b), $\psi, \tau_1(\psi) \in L^2(I)$. Similarly, τ_0 indicates the closure of $\tau_1 \upharpoonright \{\psi \in D(\tau_1) : \text{supp } \psi \text{ is compact in } I\}$. Then τ_0 is closed symmetric, $\tau_0^\dagger = \tau_1$, and the defect indices of τ_0 are equal [note that (4.211) is real] and not larger than two (because τ is second order):

$$d(\tau_0) \equiv d_+(\tau_0) = d_-(\tau_0) \leq 2 . \tag{4.212}$$

The Hamiltonian operator H must be a self-adjoint extension of τ_0. It will therefore be a symmetric restriction of τ_1 with a domain obtained from $D(\tau_1)$ by imposing some *boundary conditions* necessary whenever $d(\tau_0) \neq 0$. This explains why it is important to compute this defect index:

1) If τ is *regular* (I finite, $U \in L^2(I)$), $d(\tau_0) = 2$.
2) If a is a *regular endpoint* (a finite, $U \in L^1(a, \beta), \beta < b$), either

$$\left.\begin{array}{l} d(\tau_0) = 1 \text{ (``limit point'' case)} \\ d(\tau_0) = 2 \text{ (``limit circle'' case)} \end{array}\right\} \text{(Weyl \textit{alternative})}$$

(The same applies if b is a regular endpoint.)
3) If $c \in I$, and τ_0^-, τ_0^+ are the τ_0 associated with τ in (a, c), (c, b) respectively:

$$d(\tau_0) = d(\tau_0^-) + d(\tau_0^+) - 2 . \tag{4.213}$$

In the first case, an important collection of self-adjoint extensions of τ_0 is obtained by imposing "decoupled" boundary conditions

$$\left.\begin{array}{ll} \cos \alpha\, \psi(a) + \sin \alpha\, \psi'(a) & = 0 \\ \cos \beta\, \psi(b) + \sin \beta\, \psi'(b) & = 0 \end{array}\right\} (\alpha, \beta \text{ arbitrary real}) \tag{4.214}$$

to restrict $D(\tau_1)$. In particular the choice $\alpha = \beta = 0$ provides the Friedrichs extension of τ_0 which is semi-bounded.

In the "limit point" case of the regular endpoint a, every self-adjoint extension of τ_0 is determined by

$$\cos \alpha\, \psi(a) + \sin \alpha\, \psi'(a) = 0 \quad (\alpha \text{ arbitrary real}) . \tag{4.215}$$

Whatever the self-adjoint extension of τ_0 chosen, the *essential spectrum* is independent of the extension and it will be called $\sigma_{\text{ess}}(\tau_0)$. The same applies to the absolutely continuous spectrum, and it will be denoted by $\sigma_{\text{a.c.}}(\tau_0)$, and the a.c. parts of these extensions are equivalent (isometrically isomorphic). Furthermore, with the notation of the third case:

$$\sigma_{\text{ess}}(\tau_0) = \sigma_{\text{ess}}(\tau_0^-) \cup \sigma_{\text{ess}}(\tau_0^+) , \quad \sigma_{\text{a.c.}}(\tau_0) = \sigma_{\text{a.c.}}(\tau_0^-) \cup \sigma_{\text{a.c.}}(\tau_0^+) \tag{4.216}$$

where $H_{\text{a.c.}} \simeq (H^-)_{\text{a.c.}} \oplus (H^+)_{\text{a.c.}}$, calling H, H^-, H^+ self-adjoint extensions of τ_0, τ_0^-, and τ_0^+, respectively.

The relations (4.213) and (4.216) allow us to obtain information about the case $I = \mathbf{R}$ for example, from results in $\mathbf{R}_+$.

We next formulate some results about the spectral properties of self-adjoint extensions of τ_0 in situations of interest for our purposes:

1) Regular Case

Every self-adjoint extension H of τ_0 is semi-bounded, and $\sigma(H) = \sigma_{\text{disc}}(H)$. If the boundary conditions defining H are decoupled (4.214), then $\sigma(H)$ is simple, and if $\varepsilon_0 < \varepsilon_1 < \ldots$ are the points in $\sigma(H)$ and $\psi_0, \psi_1 \ldots$ the corresponding eigenfunctions, they satisfy Sturm's theorem.

2) Semi-infinite Case, Regular Endpoint

For concreteness, we fix $I = \mathbf{R}_+$. Let H be a self-adjoint extension of τ_0. We will distinguish several behaviors of $U(x)$ when $x \to \infty$:

2.1) $U = a + U_1$, with $U_1 \in L^1_\delta(\mathbf{R}_+)$, some $\delta \geq 0$.

Then:

$d(\tau_0) = 1$, H is semi-bounded, $\sigma(H)$ is simple, and $\sigma_{\text{ess}}(H) = \sigma_{\text{a.c.}}(H)$
$= [a, \infty)$.

2.2) $U \in C^2(x_0, \infty)$; U', U'' do not change sign in (x_0, ∞) and
$U' = O(|U|^\alpha)$, $0 < \alpha < 3/2$, $x \to \infty$.

Then:

2.2.1) $U(x) \to \infty$, $x \to \infty \Rightarrow d(\tau_0) = 1$, H is semi-bounded,
$\sigma(H)$ is simple, and $\sigma_{\text{ess}}(H) = \emptyset$.

2.2.2) $U(x) \to -\infty$, $x \to \infty$, $|U|^{-1/2}$ not summable at ∞
$\Rightarrow d(\tau_0) = 1$, $\sigma(H)$ is simple and $\sigma(H) = \sigma_c(H) = \mathbf{R}$.

[In the case 2.2), if $U(x) \to -\infty$, $x \to \infty$, with $|U|^{-1/2}$ summable at ∞, it can be shown that $d(\tau_0) = 2$ and $\sigma_{\text{ess}}(H) = \emptyset$. Such a situation will not be considered here.]

In the spectral decomposition of H [except for the case 2.2.1)] there appears a continuous part. The generalized eigenfunctions $\psi_\lambda(x)$, $\lambda \in \sigma(H)$, satisfy the differential equation $(\tau - \lambda)\psi(x) = 0$, as well as the boundary condition (4.215) associated with H, and they are therefore completely determined up to normalization. Moreover, $\psi_\lambda(x)$ as a function of x is polynomially bounded for almost all λ (with respect to the spectral measure μ of H). The expansion of a function $f(.) \in L^2(\mathbf{R}_+)$ in terms of $\psi_\lambda(.)$ is of the following form: Defining

$$\tilde{f}(\lambda) \equiv \lim_{J_n \to \mathbf{R}_+} \int_{J_n} dx \, \psi_\lambda^*(x) f(x) \tag{4.217}$$

where J_n is an increasing sequence of compact sets, such limit exists in the strong sense with respect to $L^2(\sigma(H), d\mu)$ and

$$f(x) = \lim_{A \to \infty} \int_{[-A,A] \cap \sigma(H)} d\sigma(\lambda)\, \tilde{f}(\lambda)\psi_\lambda(x) \tag{4.218}$$

where the limit is now in the strong sense in $L^2(\mathbf{R}_+)$ and σ is a measure equivalent to μ.

In the case 2.2.1), (4.218) reduces to an ordinary sum over the eigenfunctions $\psi_{\lambda_j}(x)$ of H. In case 2.1) and $\delta > 1$ it can be proved that $\sigma_{\text{s.c.}}(H) = \emptyset$, $\psi_\lambda \in L^\infty(\mathbf{R}_+)$ for almost all λ in $[a, \infty)$, and we can normalize the ψ_λ so that

$$f(x) = \sum_{\lambda_j \in \sigma_{\text{p}}(H)} \langle \psi_{\lambda_j} | f \rangle \psi_{\lambda_j}(x) + \lim_{A \to \infty} \int_a^A d\lambda\, \tilde{f}(\lambda)\psi_\lambda(x)\,. \tag{4.219}$$

Those ψ_{λ_j} with

$$\lambda_j \in \sigma_{\text{p}}(H) \quad \text{and} \quad \lambda_j < \lambda^* \equiv \inf\{\lambda : \lambda \in \sigma_{\text{ess}}(H)\}$$

($= a$ in 2.1) and $\equiv \infty$ for 2.2.1)) satisfy Sturm's theorem and there is a finite number of them in the case 2.1) if

$$\underline{\lim}\, x^2(U(x) - \lambda^*) > -\tfrac{1}{4}\,, \quad x \to \infty \tag{4.220}$$

and an infinite number if

$$\overline{\lim}\, x^2(U(x) - \lambda^*) < -\tfrac{1}{4}\,, \quad x \to \infty\,. \tag{4.221}$$

3) *Semi-infinite Case, Non-regular Endpoint*

We keep the previous notation and assume that the behavior of $U(x)$ as $x \to \infty$ is of the form 2.1) or 2.2.1) or 2.2.2). With respect to the absence of regularity at the origin, we limit our consideration to the case where

$$U(x) \geq -\frac{\alpha}{x^2}\,, \quad \alpha < \frac{1}{4}\,, \quad x \to 0 \tag{4.222}$$

distinguishing only the possibilities

$$\underline{\lim}_{x \to 0} x^2 U(x) \geq \tfrac{3}{4} \tag{4.223}$$

$$\overline{\lim}_{x \to 0} |x^2 U(x)| < \tfrac{3}{4}\,. \tag{4.224}$$

Then, in the alternative (4.223), $d(\tau_0) = 0$, namely, $H = \tau_0$; under (4.224), $d(\tau_0) = 1$ and it is necessary to impose a boundary condition to determine H. This boundary condition is associated with the finite endpoint. In the first case, for all $\psi \in D(H)$ the $\lim_{x \to 0} \psi(x) \equiv \psi(0) = 0$ exists, and in the case (4.224) we choose the boundary condition $\psi(0) = 0$, which is the one convenient for our purposes (see Chap. 6).

It can be shown that

2.1) + (4.222) $\Rightarrow H$ semi-bounded, $\sigma(H)$ simple,
$$\sigma_{\text{ess}}(H) = \sigma_{\text{a.c.}}(H) = [a, \infty).$$

2.2.1) + (4.222) $\Rightarrow H$ semi-bounded, $\sigma(H)$ simple, $\sigma_{\text{ess}}(H) = \emptyset$.

2.2.2) + (4.222) $\Rightarrow \sigma(H)$ simple, $\sigma(H) = \sigma_{\text{c}}(H) = \mathbb{R}$.

The spectral decomposition of H is analogous to the one established in the case 2). The functions $\psi_\lambda(x)$ (polynomially bounded almost everywhere with respect to μ as functions of x) must satisfy the corresponding boundary condition at $x = 0$ if (4.224) is satisfied, and as explained before we choose $\psi_\lambda(0) = 0$. In the case (4.223) it is not necessary to impose a boundary condition; however, since the essential spectrum of τ_0 in the interval $(0, a), a < \infty$, is empty, ψ_λ must belong to $L^2(0, a)$ and it is therefore the solution to $(\tau - \lambda)\psi = 0$ satisfying $\psi_\lambda(0) = 0$.

Finally Sturm's theorem is still valid as well as the conditions (4.220), (4.221) with respect to the number of eigenvalues to the left of $\sigma_{\text{ess}}(H)$ in the case of behavior of the type 2.1 as $x \to \infty$.

4) Cases $I = \mathbb{R}$

Now at each endpoint $x \to \pm\infty$ we can have the behavior 2.1), 2.2.1) or 2.2.2). The relations (4.213), (4.216) allow us to extract the corresponding conclusions. Thus for example:

$2.1)_\infty + 2.1)_{-\infty} \Rightarrow d(\tau_0) = 0, \ H$ is semi-bounded ,
$$\sigma_{\text{ess}}(H) = \sigma_{\text{a.c.}}(H) = [\min(a_-, a_+), \infty) ,$$
$$\sigma_{\text{p}}(H) \text{ is simple },$$
$$(\min(a_-, a_+), \ \max(a_-, a_+)) \text{ has multiplicity 1 },$$
$$(\max(a_-, a_+), \infty) \text{ has multiplicity 2 };$$

$2.1)_\infty + 2.2.1)_{-\infty} \Rightarrow d(\tau_0) = 0, \ H$ is semi-bounded, $\sigma(H)$ is simple ,
$$\sigma_{\text{ess}}(H) = \sigma_{\text{a.c.}}(H) = [a_+, \infty) ;$$

etc., indicating by $a_\pm$ the values of $U(x)$ as $x \to \pm\infty$.

In the spectral decomposition we will have to take two solutions $\psi_{\lambda,1}, \psi_{\lambda,2}$ of $(\tau - \lambda)\psi = 0$ for those λ with multiplicity 2, and in (4.218) σ will become a 2×2 spectral matrix. When the multiplicity of λ is 1, ψ_λ will be an L^2 function in that endpoint $\pm\infty$, such that a neighborhood of λ does not contain any point of the essential spectrum of τ_0 restricted to $\mathbb{R}_+$ respectively. Thus, if $a_- < a_+$ in the case $2.1)_\infty + 2.1)_{-\infty}$ ψ_λ should be in $L^2(\mathbb{R}_+)$ if $\lambda \in (a_-, a_+)$. Finally, Sturm's theorem holds as well as the criteria (4.220) and (4.221) in the sense that in the case $2.1)_\infty + 2.1)_{-\infty}$ the finiteness of the number of eigenvalues is ensured when (4.220) is satisfied in whichever endpoint where $U(x) \to \lambda^*$, while it is enough for (4.221) to be satisfied in one of the endpoints to imply that the number of $\lambda_j \in \sigma_{\text{p}}(H)$, with $\lambda_j < \lambda^*$ is infinite. When one endpoint is of type 2.1) and the other of type 2.2.1) the criteria is applied in the endpoint 2.1).

The generalized eigenfunctions $\psi_{\lambda,i}, \psi_\lambda$ are μ-a.e. polynomially bounded as functions of x (bounded if $\delta > 1$).

To finish with this short summary, if $U \in L^1(\mathbf{R})$ and therefore $\sigma_{\text{a.c.}}(H) = [0, \infty)$, the results of Sect. C.8 imply that $H_{\text{a.c.}} \simeq -d^2/dx \equiv H_0$, and the isomorphism is realized by any of the isometric Möller operators

$$\Omega_\pm(H, H_0) \equiv \lim_{t \to \mp\infty} e^{itH} e^{-itH_0} \tag{4.225}$$

satisfying $\Omega_\pm^\dagger \Omega_\pm = I$. If $U \in L_\delta^1(\mathbf{R})$, some $\delta > 1$, then

$$\Omega_\pm \mathcal{H} = \mathcal{H}_{\text{a.c.}}(H), \quad \mathcal{H}_{\text{s.c.}}(H) = \{0\}, \quad S \equiv \Omega_-^\dagger \Omega_+ \text{ unitary} . \tag{4.226}$$

The existence of $\Omega_\pm$ and the unitarity of S are basic ingredients in the development of a rigorous scattering theory in one- dimensional collisions along the lines presented in Chap. 8. Taking as $\psi_{\lambda,1}, \psi_{\lambda,2}$ for $\lambda > 0$ those $\psi^{(\mp)}$ of the form (4.92) which will be represented as $\sqrt{2\pi}\langle x|\Omega_+|k\rangle$, with $k^2 \equiv \lambda$, $k \gtrless 0$, and $\sqrt{2\pi}\langle x|k\rangle = \exp(ikx)$, it is easy to see that these generalized eigenfunctions form an orthonormal basis with respect to dk for the continuous spectrum. Its use simplifies considerably the action of Ω_+:

$$\Omega_+ : \int_{\mathbf{R}} dk\, \hat{\psi}(k)\langle x|k\rangle \to \int_{\mathbf{R}} dk\, \hat{\psi}(k)\langle x|\Omega_+|k\rangle \tag{4.227}$$

and similarly for Ω_- with the appropriate basis $\psi^{(\pm)*}$ which we denote by $\sqrt{2\pi}\langle x|\Omega_-|k\rangle, k \gtrless 0$.

The relation between the two bases is

$$\left. \begin{array}{l} \Omega_+|k\rangle = \sigma \Omega_-|k\rangle + \varrho_- \Omega_-|-k\rangle \\ \Omega_+|-k\rangle = \varrho_+ \Omega_-|k\rangle + \sigma \Omega_-|-k\rangle \end{array} \right\} (k > 0) , \tag{4.228}$$

from which, applying $\Omega_-^\dagger$ and using (4.226):

$$\left. \begin{array}{l} S|k\rangle = \sigma|k\rangle + \varrho_-|-k\rangle \\ S|-k\rangle = \varrho_+|k\rangle + \sigma|-k\rangle \end{array} \right\} (k > 0) \tag{4.229}$$

where σ ($\equiv \sigma_- = \sigma_+$), ϱ_-, ϱ_+ are the amplitudes introduced in Sect. 4.5 and which are identified with the S matrix elements on the λ energy shell, as indicated in (4.114).

For a rigorous formulation of one dimensional multichannel scattering when $U(x)$ has different asymptotic behaviors as $x \to \pm\infty$, see [DS 78].

5. Angular Momentum

5.1 Introduction

Angular momentum plays a central role in quantum mechanics. A careful understanding of its properties and of methods for dealing with it can be very useful
not only in providing a better understanding of a given physical problem, but
also in simplifying its mathematical treatment.

Many of the results of this chapter are immediate consequences of general
theorems in the theory of Lie groups and their representations. Nevertheless,
we intend to provide here direct demonstrations, even though not all of these
will be completely rigorous, relying on the general theory only when absolutely
necessary. Because of the importance of angular momentum in general, we have
collected in Appendix B a series of relevant formulae, which are either proved
in this chapter or whose proof we often outline in that appendix. Further details
can be found in the references [WI 31], [RO 57], [ED 57], [FR 59] and [BL 81],
among others.

5.2 The Definition of Angular Momentum

In a physical system consisting of N particles with Cartesian coordinates $r_1, r_2,$
$\ldots, r_N$ and conjugate momenta $p_1, p_2, \ldots, p_N$, respectively, the dynamical variable called orbital angular momentum L is defined as

$$L = \sum_{i=1}^{N} r_i \times p_i \ . \tag{5.1}$$

According to postulate VI, the operator corresponding to the orbital angular
momentum observable is given by the same expression, where now r_i and p_i
are the position and momentum operators (which we will henceforth denote with
the same symbols in order to simplify the notation). In particular, in the position
representation, the orbital angular momentum operator is

$$\boxed{L = -\mathrm{i}\hbar \sum_{j=1}^{N} r_j \times \nabla_j} \ , \tag{5.2}$$

and the components of L are essentially self-adjoint in $S(\mathbf{R}^{3N})$ (the space of rapidly decreasing C^{∞} functions). From (5.1), taking into account the commutation rules for the position and momentum operators, it is easily obtained that

$$\boxed{[L_i, L_j] = i\hbar\varepsilon_{ijk}L_k}\ , \tag{5.3}$$

where ε_{ijk} is the totally antisymmetric Levi-Cività tensor, with $\varepsilon_{123} = +1$. Just as in classical mechanics, it is to be expected that L be related to the generators of rotations. In order to analyze this question in quantum mechanics, we consider the simple case of a single particle for which x_1, x_2, and x_3 form a complete set of compatible observables, and which a certain experimental apparatus prepares in a state described by the wave function $\psi(\boldsymbol{x})$. If we now rotate this apparatus about an axis given by the unit vector $\hat{\boldsymbol{n}}$ through the angle $\theta\,(0 \leq \theta \leq \pi)$, the rotated apparatus will prepare the particle in a state expressible by a new wave function $\psi'(\boldsymbol{x})$. Under this rotation, a coordinate point $\boldsymbol{x}$ acquires the coordinates $\boldsymbol{x}'$, which as functions of the previous ones are

$$\begin{aligned}
x_i' &= \mathcal{R}_{ij}(\hat{\boldsymbol{n}}, \theta)x_j \\
\mathcal{R}_{ij}(\hat{\boldsymbol{n}}, \theta) &\equiv \cos\theta\,\delta_{ij} + (1 - \cos\theta)n_i n_j - \sin\theta\,\varepsilon_{ijk}n_k\ .
\end{aligned} \tag{5.4}$$

Thus, if the particle was in the state $|\boldsymbol{x}\rangle$, representing its localization at the point $\boldsymbol{x}$, after a rotation $\mathcal{R}$ its state is $\propto |\mathcal{R}\boldsymbol{x}\rangle$. Since we assume that space is isotropic, the transition probabilities between any two states and between those states after the rotation are equal. A theorem of Wigner's, which will be proven in Chap. 7, enables us to assert that rotations are implementable by unitary or antiunitary operators; we discard the latter possibility on the basis of continuity arguments presented there or, in this case, by taking into account the fact that each rotation is the square of some other rotation. Thus,

$$|x\rangle \xrightarrow{\mathcal{R}(\hat{\boldsymbol{n}}, \theta)} R(\hat{\boldsymbol{n}}, \theta)|x\rangle \propto |\mathcal{R}(\hat{\boldsymbol{n}}, \theta)\boldsymbol{x}\rangle\ , \tag{5.5}$$

$R(\hat{\boldsymbol{n}}, \theta)$ being a unitary operator. Therefore,

$$R|\boldsymbol{x}\rangle = \lambda(\mathcal{R}\boldsymbol{x})|\mathcal{R}\boldsymbol{x}\rangle\ , \tag{5.6}$$

where λ is complex and of unit modulus.

If the translation by a vector $\boldsymbol{a}$ is denoted by T_a, it is clear that $T_a T_b = T_{a+b}$, $T_{\mathcal{R}a}\mathcal{R}T_{-a} = \mathcal{R}$. These operations of translation and rotation carried out on a physical system immersed in a homogeneous and isotropic space can be mathematically implemented, according to Wigner's theorem and continuity arguments, by unitary operators V_a, R, which represent, apart from a phase, the action of the Euclidean group. The reason for this phase lies in the fact that although the results coincide for pure physical states, that is, on unit rays, there is no reason for them to coincide on the representative state vectors which we have chosen in that ray. But an important theorem by *Bargmann* [BA 52a] allows us to affirm that it is possible to redefine the global phases for these operators in such a way that

$$V_a V_b = V_{a+b} , \quad V_{\mathcal{R}a} R V_{-a} = R . \tag{5.7}$$

With this choice, application of the second equality in (5.7) to $|x\rangle$, and using (5.6) we immediately find that $\lambda(\mathcal{R}, x)$ does not depend on x so that, by redefining the phase of R [which does not modify (5.7)], we can always write

$$R|x\rangle = |\mathcal{R}x\rangle . \tag{5.8}$$

Finally, with the choice of phases for $\{|p\rangle\}$ which led to (3.30), we deduce from (5.8) that

$$R|p\rangle = |\mathcal{R}p\rangle . \tag{5.9}$$

Since R is unitary,

$$\boxed{\begin{aligned} (R\psi)(x) &= \langle x|R|\psi\rangle = \langle \mathcal{R}^{-1}x|\psi\rangle = \psi(\mathcal{R}^{-1}x) , \\ (R\psi)\hat{}(p) &= \langle p|R|\psi\rangle = \langle \mathcal{R}^{-1}p|\psi\rangle = \hat{\psi}(\mathcal{R}^{-1}p) , \end{aligned}} \tag{5.10}$$

which shows that the wave function for a particle without internal degrees of freedom transforms under rotation as a scalar function.

The expression (5.10) allows us to find the explicit unitary operator $R(\hat{n}, \theta)$. In fact, for an infinitesimal rotation through an angle $\delta\theta$ around the axis $\hat{e}_3 \equiv (0, 0, 1)$ the first of the relations (5.10) can be written as

$$\psi(x_1 + x_2\delta\theta, x_2 - x_1\delta\theta, x_3) = (R(\hat{e}_3, \delta\theta)\psi)(x) , \tag{5.11}$$

and using a Taylor expansion of the l.h.s. gives, to first order,

$$R(\hat{e}_3, \delta\theta) \simeq I + \delta\theta \left(x_2\frac{\partial}{\partial x_1} - x_1\frac{\partial}{\partial x_2} \right) = I - \frac{i}{\hbar}\delta\theta\hat{e}_3 \cdot \boldsymbol{L} . \tag{5.12}$$

The same argument, applied to an infinitesimal rotation through the angle $\delta\theta$ about $\hat{n}$, results in

$$R(\hat{n}, \delta\theta) \simeq I - \frac{i}{\hbar}\delta\theta\hat{n} \cdot \boldsymbol{L} . \tag{5.13}$$

On the other hand, $R(\hat{n}, \theta + \delta\theta) = R(\hat{n}, \delta\theta)R(\hat{n}, \theta)$; then using (5.13) and a Taylor expansion of $R(\hat{n}, \theta + \delta\theta)$, we obtain a differential equation for $R(\hat{n}, \theta)$

$$\frac{dR(\hat{n}, \theta)}{d\theta} = -\frac{i}{\hbar}(\hat{n} \cdot L)R(\hat{n}, \theta) . \tag{5.14}$$

Adding the boundary condition $R(\hat{n}, 0) = I$, an integration immediately yields

$$\boxed{R(\hat{n}, \theta) = \exp\left[-\frac{i}{\hbar}\theta\hat{n} \cdot \boldsymbol{L} \right]} . \tag{5.15}$$

Both (5.14) and (5.15) demonstrate that $\boldsymbol{L}$ is the generator of rotations.

In this derivation, we have supposed that the particle is totally specified by its position or momentum. We know, however, that in general these attributes do not suffice to constitute a complete set of compatible observables, as the particle may possess internal degrees of freedom, which we have labelled spin. The particle having spin implies that it is necessary to complete the set of observables x by a new observable A compatible with, but independent of, them, with a discrete and finite spectrum $\sigma(A) = \{a_i, 1 \leq i \leq n\}$. An orthonormal basis will thus be of the form $\{|x, a_i\rangle\}$. In the case of the electron, for example, we have seen that $n = 2$, and for such cases we say that the particle has spin 1/2. In general, $n = 2s + 1$, with s being a non-negative integer or half-integer (properly a half-odd-integer), and we say that the particle has spin s. In agreement with experiment, we will assume that A may be chosen such that the displacements V_β have no effect on the spin indices, in other words, that $V_\beta|x, a_i\rangle$ is an eigenstate of A with the same eigenvalue a_i. This permits us to choose the local basis $\{|x, a_i\rangle\}$ such that

$$V_\beta|x, a_i\rangle = |x + \beta, a_i\rangle , \tag{5.16}$$

a consistent choice due to (5.7). The generators of this unitary representation of translations are precisely the self-adjoint operators p which we call momentum. The expression (5.16) assures us that there is an orthonormal basis $\{|p, a_i\rangle\}$ associated with the complete set of compatible observables $\{p, A\}$ such that

$$V_\beta|p, a_i\rangle = \exp\left[-\frac{i}{\hbar}p \cdot \beta\right] |p, a_i\rangle , \tag{5.17}$$

and, in addition,

$$\langle x, a_j|p, a_i\rangle = (2\pi\hbar)^{-3/2} \exp\left[\frac{i}{\hbar}p \cdot x\right] \delta_{ji} . \tag{5.18}$$

Even though translations do not affect the observable A, rotations in general do. Recall a typical Stern-Gerlach apparatus (Fig. 2.4): If the first magnet which prepares the state $|a_{+1}\rangle$ has the same orientation as the second magnet ($\hat{n}_A = \hat{n}_B$), this one will not produce a new doubling of the beam ($|a_{+1}\rangle = |b_{+1}\rangle$). But if we rotate $\hat{n}_A$ such that $\hat{n}_A \neq \hat{n}_B$, the state $|a_{+1}\rangle$ thus prepared will appear as a linear combination of $|b_{+1}\rangle$ and $|b_{-1}\rangle$. In other words, the rotation has affected the spin indices relative to the direction $\hat{n}_B$. Thus

$$R|x, a_i\rangle = \sum_j D(\mathcal{R})_{ji}|\mathcal{R}x, a_j\rangle , \tag{5.19}$$

where the possible dependence of the matrix D on x can be eliminated just as in the case of spinless particles. Using (5.18) and the unitarity of R,

$$R|p, a_i\rangle = \sum_j D(\mathcal{R}_{ji})|\mathcal{R}p, a_j\rangle , \tag{5.20}$$

$D(\mathcal{R})$ must be a unitary matrix for each rotation $\mathcal{R}$, because R is.

If $\mathcal{R}_1, \mathcal{R}_2, \mathcal{R}_3$ are rotations such that $\mathcal{R}_3 = \mathcal{R}_1 \mathcal{R}_2$, the unitary operators R_i which implement them preserve this group law, apart from a phase; that is, $R_3 = \omega(\mathcal{R}_1, \mathcal{R}_2) R_1 R_2$, with $|\omega| = 1$. The phase $\omega(\mathcal{R}_1, \mathcal{R}_2)$ can be locally chosen to be 1 since SO(3) is semisimple. But since the rotation group is not simply connected, this trivially local choice has no reason to be globally valid. Thus, using (5.19) or (5.20),

$$D(\mathcal{R}_1, \mathcal{R}_2) = \omega(\mathcal{R}_1, \mathcal{R}_2) D(\mathcal{R}_1) D(\mathcal{R}_2) , \tag{5.21}$$

so that $D(\mathcal{R})$ produces a unitary projective representation of SO(3). In general, this representation $D(\mathcal{R})$, and hence that of the Euclidean group acting on the states, is reducible; that is, there exist sets of internal states which remain uncoupled under rotations. When this does not occur, we say that the particle is *elementary* under the Euclidean group.

Summarizing, we understand the term elementary particle, from a mathematical perspective, to mean a physical system for which the position observables x and the observable A, with a finite and discrete spectrum, form a complete set of compatible observables, and the set of its states is irreducible under the Euclidean group. Thus, for example, a physical system of two free pions whose states can be arbitrarily prepared, is not elementary in the above sense. Nevertheless, if we restrict ourselves to the subset of states formed by that pair of pions whose relative motion is completely specified and is, for example, invariant under rotations with respect to the center of mass of the two particles, this particular system is elementary. This shows that the mathematical concept of elementarity can be very far removed from that which we would intuitively understand it to be: In the above example the second case is imbedded in a set of analogous systems which form a continuum from which it is not separable except in the ideal limit.

Consider another example: Among the possible states of a proton-neutron (p-n) system is the deuteron. The states of the deuteron form an elementary system with spin 1, and are separated in internal energy from the rest of the p-n states by a finite non-zero amount. As a consequence, at sufficiently low energies we may consider the deuteron as a physically elementary system, and only above certain energies will it be convenient to consider it as part of the general p-n system. These nucleons will of course now appear as constituents of the deuteron, and the observables which formed a complete set of compatible observables for the deuteron will no longer be sufficient. It is a matter of experience and of the nature of the phenomena of interest to decide whether it is convenient to consider a physical system as elementary.

The multipliers $\omega(\mathcal{R}_1, \mathcal{R}_2)$ for particles or other elementary systems for which $D(\mathcal{R})$ is irreducible may be chosen in such a way that ω is always ± 1 [BA 52a]. This is due to the fact that the universal covering group for SO(3) is the simply connected group SU(2), which covers the rotation group twice: given a matrix $A \in$ SU(2) and $x \in \mathbf{R}^3$, it is easy to see that

$$Ax \cdot \sigma A^\dagger = (\mathcal{R}(A)x) \cdot \sigma , \tag{5.22}$$

where $\mathcal{R}(A) \in \mathrm{SO}(3)$, and the σ are the Pauli matrices (2.112). The correspondence $A \to \mathcal{R}(A)$ constitutes a representation of SU(2), and every rotation is an image of some element A. Given the rotation $\mathcal{R}(\hat{n}, \theta)$, $0 \le \theta \le \pi$, both the matrices $A(\hat{n}, \theta) \equiv \exp(-\mathrm{i}\theta\hat{n} \cdot \sigma/2)$ and $A(\hat{n}, \theta+2\pi) [= -A(\hat{n}, \theta)]$ lead to $\mathcal{R}(\hat{n}, \theta)$. Notice in particular that the matrices $\hbar\sigma/2$ satisfy the commutation relations (5.3), and are the generators of SU(2). For this group, the parameter θ should lie between 0 and 2π, since $A(\hat{n}, 2\pi + \alpha) = A(-\hat{n}, 2\pi - \alpha)$. We glean from (5.22) that $\mathcal{R}(A) = \mathcal{R}(-A)$, so that the above correspondence produces a double covering of SO(3) by SU(2). The group SU(2) is simply connected, and constitutes the universal covering group of the rotation group. Given an irreducible, projective, unitary and continuous representation $D(\mathcal{R})$ of SO(3), it can be shown [BA 52a] that there exists an irreducible, unitary and continuous representation $D'(A)$ of SU(2) such that $D(\mathcal{R}) = D'(A)$ or $= D'(-A)$. Since the element $-I \in$ SU(2) is central and D' is irreducible, $D'(-I)$ should be a multiple of the identity; but $(-I)^2 = I$, so that $D'(-I) = \pm I$, justifying $\omega = \pm 1$ when D is irreducible.

To determine the general form of the representation $D(\mathcal{R})$, we note that, given any rotation $\mathcal{R}(\hat{n}, \theta)$, there exists a rotation through an angle sufficiently small, and therefore close to the identity, which generates it through repeated multiplication:

$$\mathcal{R}(\hat{n}, \theta) = \left[\mathcal{R}\left(\hat{n}, \frac{\theta}{N} \right) \right]^N .$$

In a neighborhood of the identity, the multipliers may be chosen to be 1, and with this choice D becomes a unitary representation of the local group, which we can represent in infinitesimal terms:

$$D\left(\mathcal{R}\left(\hat{n}, \frac{\theta}{N} \right) \right) \simeq I - \frac{\mathrm{i}}{\hbar}\frac{\theta}{N}\hat{n} \cdot S , \tag{5.23}$$

S being the generators, which play a role analogous to that of L in (5.13). Thus, proceeding to a finite rotation:

$$D(\mathcal{R}(\hat{n}, \theta)) = \pm \exp\left[-\frac{\mathrm{i}}{\hbar}\theta\hat{n} \cdot S \right] , \tag{5.24}$$

where the sign $\pm$ is due to the projective character of the association $\mathcal{R} \to D(\mathcal{R})$ and is globally unavoidable. The operators S, which are $n \times n$ matrices acting on the spin indices, must be self-adjoint since D is unitary.

Since both S and L are generators of representations of the same group, they must satisfy the same commutation rules. One direct method for finding these rules is from the use of the identity

$$e^{\lambda A}e^{\lambda B}e^{-\lambda A}e^{-\lambda B} = \exp\{\lambda^2[A, B] + O(\lambda^3)\} , \tag{5.25}$$

which is easily verified, either by series expansion or by using the Campbell-Hausdorff formula (2.190), and observing that if $\lambda A = -\mathrm{i}\theta S_1/\hbar$, $\lambda B = -\mathrm{i}\theta S_2/\hbar$, with θ very small, then the l.h.s. of (5.25) represents a rotation through the angle θ^2 about the 3-axis. Thus, we can write

$$\boxed{[S_i, S_j] = i\hbar\varepsilon_{ijk}S_k}\,, \tag{5.26}$$

which is identical to (5.3).

In the next section we show that S_j has a simple spectrum formed by the points $-s\hbar, -(s-1)\hbar, \ldots, (s-1)\hbar, s\hbar$, with $s = (n-1)/2$.

Given that our Hilbert space $\mathcal{H}$ is realizable as $L^2(\mathbb{R}^3) \otimes \mathbb{C}^n$ through the basis $|x, a_i\rangle$, that is, each state is represented by n complex square-integrable wave functions, and the matrices $D(\mathcal{R})$ act on $\mathbb{C}^n$, the simplicity of the spectrum of any operator $\hat{n} \cdot S$ permits us to take the operator $A = I \otimes (\hat{n} \cdot S)$ as the observable A which completes the set of compatible observables. We write for brevity $A = \hat{n} \cdot S$, with the understanding that this last operator acts on $\mathcal{H}$ through the factor $\mathbb{C}^n$.

The action of a rotation on the wave function (relative to the complete set of compatible observables $\{x, S_3\}$) for a state of the system

$$\boxed{\psi_m(x) \equiv \langle x, m\hbar|\psi\rangle}\,, \quad m = -s, \ldots, +s\,, \tag{5.27}$$

thus can be written as

$$\boxed{\begin{aligned} (R\psi)_m(x) &= \pm \sum_{m'} D_{m\,m'}(\mathcal{R})\psi_{m'}(\mathcal{R}^{-1}x) \\ &= \pm \left(\exp\left[-\frac{i}{\hbar}\theta\hat{n} \cdot (L+S)\right]\psi\right)_m (x)\,, \end{aligned}} \tag{5.28}$$

an expression in which L acts on x and S acts on the spin components. (Strictly speaking, this should read $L \otimes I + I \otimes S$ instead of $L + S$.)

Since L and S act on separate independent variables, they commute with each other, so that if $J \equiv L + S$, the components of J also satisfy

$$\boxed{[J_i, J_j] = i\hbar\varepsilon_{ijk}J_k}\,. \tag{5.29}$$

This operator J, which from (5.28) can be seen to be the generator of all the effects of a rotation, is called the *total angular momentum* of the system, and is an observable composed of the *orbital angular momentum* L and the *spin angular momentum* S.

For systems composed of N elementary particles, the effect of a rotation is obtained by applying it to each particle; thus, if $S_1, \ldots, S_N$ are the corresponding spin operators, the system wave function $\psi_{m_1,\ldots,m_N}(x_1, \ldots, x_N) \equiv \langle x_1, m_1\hbar; \ldots; x_N, m_N\hbar|\psi\rangle$ changes under $\mathcal{R}$ according to

$$
\begin{aligned}
&(R\psi)_{m_1,\dots,m_N}(\boldsymbol{x}_1,\dots,\boldsymbol{x}_N) \\[6pt]
&= \pm \sum_{m_1'\dots m_N'} D^{(1)}_{m_1 m_1'}(\mathcal{R}) \cdots D^{(N)}_{m_N m_N'}(\mathcal{R}) \psi_{m_1',\dots,m_N'}(\mathcal{R}^{-1}\boldsymbol{x}_1,\dots,\mathcal{R}^{-1}\boldsymbol{x}_N) \\[6pt]
&= \pm \left(\exp\left[-\frac{\mathrm{i}}{\hbar}\theta\hat{n}\cdot\sum_j (L_j + S_j) \right] \psi \right)_{m_1,\dots,m_N} \quad (\boldsymbol{x}_1,\dots,\boldsymbol{x}_N)
\end{aligned}
$$

$$(5.30)$$

where each S_j acts only on the corresponding m_j, $m_j = -s_j,\dots,s_j$, and each L_j on the $\boldsymbol{x}_j$. The total angular momentum of the system is $J = L + S$, with $L = \sum L_i$, and $S = \sum S_i$. Each J_i commutes with the other partial angular momenta, and thus the three components of J satisfy (5.29). This is true also of L and S.

In Chap. 1 we discussed the experimental evidence for the spin 1/2 of the electron. The Stern-Gerlach technique is not actually applicable in determining the spin of a free electron, due to its electric charge and to the uncertainty principle. It can be shown that the deflection of the electron trajectory produced by the Lorentz force would completely mask that which is associated with the magnetic moment. For an exposition of this question, see *Mott* and *Massey* [MM 65]. The determination of the spin of the various particles is achieved through an analysis of the processes of production and/or decay, where angular correlations, polarization, etc., are measured. There also exist experimental techniques (double collisions, etc.) for the preparation of polarized beams of a variety of particles.

5.3 Eigenvalues of Angular Momentum Operators

Let J be the total angular momentum operator for a physical system. It is convenient to introduce other operators which are intimately related to J. First, we define the operator for the square of the angular momentum as

$$J^2 = J_1^2 + J_2^2 + J_3^2 . \tag{5.31}$$

It is a self-adjoint operator which satisfies the commutation rules

$$[J^2, J_i] = 0 , \tag{5.32}$$

an immediate consequence of (5.29).

Given three self-adjoint operators J, with commutation relations (5.29), defined in a domain which is dense and stable under these operators, and such that J^2 is essentially self-adjoint in that common domain, it can be shown [NE 59] that they are generators of a continuous unitary representation of SU(2). The famous Peter-Weyl theorem affirms that these representations are the direct sum of finite irreducible ones, as a result of the compact nature of SU(2). We are thus led to a study of irreducible representations of (5.29) in a Hilbert space of finite dimension; the general solution can be obtained as a direct sum of those found in this way.

Consider, then, the problem of determining three self-adjoint operators J, in a Hilbert space $\mathcal{H}$ of finite dimension, satisfying (5.29) in an irreducible way. The expression (5.32) tells us that J^2 commutes with J, and, the representation being irreducible, J^2 must be a multiple of the identity, say $J(J+1)\hbar^2 I$, where we can without loss of generality take $J \geq 0$ since J^2 is a positive definite operator. Since the components of J do not generally commute, they cannot all be simultaneously diagonalized. Let us take any one of them, say J_3, and let $|JM\rangle$ be the normalized eigenvector of that operator, where the index J reminds us of the constant value of J^2, and $M\hbar$ is the eigenvalue of J_3:

$$\boxed{J^2|JM\rangle = \hbar^2 J(J+1)|JM\rangle, \quad J_3|JM\rangle = \hbar M|JM\rangle} \ . \tag{5.33}$$

The quantities J and M are real, and we say of a state such as $|JM\rangle$ that, in units of $\hbar$, it has an angular momentum J and the component of J in the direction of quantization (the third component in this example) is M.

To determine J and the spectrum of J_3 we introduce the operators

$$\boxed{J_\pm = J_1 \pm iJ_2 , \quad J_\pm^\dagger = J_\mp} \ . \tag{5.34}$$

With the help of the commutation relations (5.29) it is easy to prove that

$$[J_3, J_\pm] = \pm\hbar J_\pm , \quad [J_+, J_-] = 2\hbar J_3 \ . \tag{5.35}$$

Furthermore, since

$$J^2 = \tfrac{1}{2}(J_+J_- + J_-J_+) + J_3^2 \ , \tag{5.36}$$

the commutation rules (5.35) allow us to write

$$J_-J_+ = J^2 - J_3(J_3 + \hbar) , \quad J_+J_- = J^2 - J_3(J_3 - \hbar) \ . \tag{5.37}$$

Using these equalities,

$$
\begin{aligned}
J_-J_+|JM\rangle &= \hbar^2[J(J+1) - M(M+1)]|JM\rangle \\
&= \hbar^2(J-M)(J+M+1)|JM\rangle \\
J_+J_-|JM\rangle &= \hbar^2[J(J+1) - M(M-1)]|JM\rangle \\
&= \hbar^2(J+M)(J-M+1)|JM\rangle ,
\end{aligned}
\tag{5.38}
$$

and thus the squared norms of the vectors $J_\pm|JM\rangle$ are

$$\|J_\pm|JM\rangle\|^2 = \hbar^2(J \mp M)(J \pm M + 1) \ . \tag{5.39}$$

Therefore, the second member of (5.39) should be greater than or equal to zero, and consequently

$$-J \leq M \leq +J . \tag{5.40}$$

In the same way,

$$\begin{aligned}
J_+|JM\rangle &= 0 \quad \text{if and only if} \quad M = J \\
J_-|JM\rangle &= 0 \quad \text{if and only if} \quad M = -J .
\end{aligned} \tag{5.41}$$

Starting with a vector $|JM\rangle$ and using the commutation relations (5.32) and (5.35), we get

$$\begin{aligned}
\boldsymbol{J}^2(J_\pm|JM\rangle) &= J_\pm \boldsymbol{J}^2|JM\rangle = \hbar^2 J(J+1)(J_\pm|JM\rangle) \\
J_3(J_\pm|JM\rangle) &= (J_\pm J_3 \pm \hbar J_\pm)|JM\rangle = \hbar(M \pm 1)(J_\pm|JM\rangle) .
\end{aligned} \tag{5.42}$$

This indicates that when $J_+(J_-)$ is applied to the state $|JM\rangle$ with $M \neq J$ ($M \neq -J$), the result is a vector, which properly normalized we shall designate by $|JM \pm 1\rangle$; it has the same angular momentum J (this is evident since $\boldsymbol{J}^2$ is a multiple of the identity), but its third component has increased (decreased) by one unit. (Note the similarity of this argument with the one used in the operator treatment of the harmonic oscillator.) Choosing the relative phase between $J_\pm|JM\rangle$ and $|JM \pm 1\rangle$ to be zero, and taking into account that $|JM\rangle$ was chosen to be normalized, it results from (5.39) that

$$J_\pm|JM\rangle = \hbar[(J \mp M)(J \pm M + 1)]^{1/2}|JM \pm 1\rangle . \tag{5.43}$$

This choice of relative phases guarantees that the matrix elements of $J_\pm$ are all non-negative; this is called the Condon-Shortley phase convention and is practically the only one used.

Note that if we start with the state $|JM\rangle$ and repeatedly apply the operator J_+ (J_-), we obtain the states $|JM+1\rangle$, $|JM+2\rangle$, ... ($|JM-1\rangle$, $|JM-2\rangle$, ...), and, according to (5.41), the only way in which this chain can terminate and not violate the condition (5.40) is for the state $|JJ\rangle$ ($|J,-J\rangle$) to form part of this succession. This implies, together with the condition $J \geq 0$, that J must be a non-negative integer or half-integer, and that M differs from J by an integer.

On the other hand, the sequence of vectors obtained in this way starting from the initial $|JM\rangle$ state must be a basis for the Hilbert space of this representation, because the subspace generated by these elements is, by construction, invariant under $\boldsymbol{J}$. Since the representation we seek is irreducible, the subspace in question must be trivial.

In summary, we can say that the possible values of J are $J = 0, 1/2, 1, 3/2, \ldots$ and for each J the possible values of M are $M = -J, -J+1, \ldots, J-1, J$. These are non-degenerate when the $\boldsymbol{J}$ system is irreducible.

We denote by $\mathcal{H}^{(J)}$ the space spanned by the vectors $|JM\rangle$ (where $M = -J, -J+1, \ldots, J-1, J$), where an irreducible representation of $\boldsymbol{J}$ is defined. Since an arbitrary element of SU(2) has the form $\exp[-\mathrm{i}(\theta/2)\hat{n} \cdot \boldsymbol{\sigma}]$, the representation of $\hbar\boldsymbol{\sigma}/2$ given by $\boldsymbol{J}$ automatically leads to a representation D^J of SU(2). This representation is also irreducible, since a subspace of $\mathcal{H}^{(J)}$ which is stable under the action of SU(2) is also stable under the generators $\boldsymbol{J}$. Note that

$$J_k|\psi\rangle = \lim_{\delta\theta \to 0} \frac{i\hbar}{\delta\theta} [D'(A(\hat{e}_k, \delta\theta)) - \Pi]|\psi\rangle \; . \tag{5.44}$$

Conversely, if a continuous, unitary and irreducible representation D' of SU(2) is given in $\mathcal{H}^{(J)}$, the operators J obtained from (5.44) provide an irreducible representation of the commutation rules for angular momentum. The irreducibility is due to the fact that if J were reducible, then since D' is obtained by exponentiation of its generators, it too would be reducible, contrary to the assumption.

In the language of group theory, our exposition thus far can be summarized as follows: The set of all rotations constitutes a connected Lie group whose generators J_1, J_2, and J_3 form a Lie algebra determined by the commutation relations (5.29), and which is isomorphic to that for the group SU(2). The continuous, unitary and irreducible representations of this latter group are completely characterized by specifying a non-negative integer or half-integer J, and we denote them by $\mathcal{D}^{(J)}$. The basis states of the irreducible representation $\mathcal{D}^{(J)}$ are the $|JM\rangle$, where the weight M takes on the values $M = -J, -J+1, \ldots, J-1, J$. SU(2) being simply connected and the covering group for the group of rotations, the representations of SO(3), both the single-valued and the double-valued ones, always follow from the (single-valued) representations $\mathcal{D}^{(J)}$ of SU(2). We will use the same symbol $\mathcal{D}^{(J)}$ to designate these projective representations of SO(3), and when we speak of the representative of a rotation, it will be implicitly understood that such a representative is fully determined except at most in its overall sign.

Note that in the space $\mathcal{H}^{(J)}$, containing the irreducible representation $\mathcal{D}^{(J)}$, the following $(2J+1) \times (2J+1)$ hermitian matrices correspond to the operators J and J^2:

$$\begin{aligned}
\langle JM'|J_1|JM\rangle &= \frac{\hbar}{2}[\delta_{M',M+1} + \delta_{M',M-1}]\sqrt{J(J+1) - MM'} \; , \\
\langle JM'|J_2|JM\rangle &= \frac{\hbar}{2i}[\delta_{M',M+1} - \delta_{M',M-1}]\sqrt{J(J+1) - MM'} \; , \\
\langle JM'|J_3|JM\rangle &= \hbar\delta_{M'M}M \; , \\
\langle JM'|J^2|JM\rangle &= \hbar^2\delta_{M'M}J(J+1) \; ,
\end{aligned} \tag{5.45}$$

using the *standard basis* $\{|JM\rangle\}$, chosen according to the Condon-Shortley convention. This can be deduced immediately from (5.33), (5.34), and (5.43). Observe that while all elements of J_1, J_3 and J^2 are real in this basis, those of J_2 are all pure imaginary.

In particular, for the case $J = 1/2$ we find, as expected,

$$J = \frac{\hbar}{2}\sigma$$

$$\sigma_1 = \begin{pmatrix} 0 & 1 \\ 1 & 0 \end{pmatrix}, \quad \sigma_2 = \begin{pmatrix} 0 & -i \\ i & 0 \end{pmatrix}, \quad \sigma_3 = \begin{pmatrix} 1 & 0 \\ 0 & -1 \end{pmatrix}, \tag{5.46}$$

where the σ are the Pauli spin matrices. Their most important properties, all simply derivable, are given in Appendix B.

Until now we have limited ourselves to the study of irreducible representations of J, necessarily of finite dimension. Given a physical system with an associated Hilbert space $\mathcal{H}$, the representation of rotations in it is not, in general, irreducible; it can be decomposed into a direct sum of irreducible ones. As a consequence, J^2 and J_3 do not suffice to specify a state, and it is necessary to add new observables to form a complete set of compatible observables. Of course, any self-adjoint operator which can distinguish among these irreducible representations suffices (if we understand distinguishing to mean leaving invariant each subspace of an irreducible representation and having distinct values in each one of them). This ensures that there exist complete sets of compatible observables containing J^2 and J_3 and for which the added observables commute with J. Denoting the vectors of an orthonormal basis associated with one such set of compatible observables by $|JM\alpha\rangle$, these vectors generate an $\mathcal{H}^{(J)}$ as previously introduced for each fixed α, in which the representation of rotations is of the $\mathcal{D}^{(J)}$ type. The formulae (5.45) continue to be valid upon substitution of $|JM\alpha\rangle$ for $|JM\rangle$, $\langle JM'\alpha'|$ for $\langle JM'|$ and introducing a $\delta_{\alpha'\alpha}$ in the right-hand sides. Of course, the index α is not necessarily discrete, and the basis vectors $|JM\alpha\rangle$ can on occasion be orthonormal in a generalized sense. Thus, for example, for a free, spinless particle, for which $J = L$, the complete set of compatible observables can be $\{L^2, L_3, p^2/2M\}$.

5.4 Orbital Angular Momentum

According to the contents of Sect. 5.2, the orbital angular momentum operator for a single particle in the position representation is given by

$$L = -i\hbar r \times \nabla . \tag{5.47}$$

In spherical coordinates, its components and its square have the form

$$L_1 = i\hbar \left[\sin\phi \frac{\partial}{\partial\theta} + \cot\theta \cos\phi \frac{\partial}{\partial\phi} \right] ,$$

$$L_2 = i\hbar \left[-\cos\phi \frac{\partial}{\partial\theta} + \cot\theta \sin\phi \frac{\partial}{\partial\phi} \right] ,$$

$$L_3 = -i\hbar \frac{\partial}{\partial\phi} , \tag{5.48}$$

$$L^2 = -\hbar^2 \left[\frac{1}{\sin\theta} \frac{\partial}{\partial\theta} \left(\sin\theta \frac{\partial}{\partial\theta} \right) + \frac{1}{\sin^2\theta} \frac{\partial^2}{\partial\phi^2} \right] ,$$

and thus, the equations equivalent to (5.33) are, omitting the variable r which plays no role,

$$\left[\frac{1}{\sin\theta}\frac{\partial}{\partial\theta}\left(\sin\theta\frac{\partial}{\partial\theta}\right) + \frac{1}{\sin^2\theta}\frac{\partial^2}{\partial\phi^2} + L(L+1) \right] F_{LM}(\theta,\phi) = 0 \,,$$

$$\left[i\frac{\partial}{\partial\phi} + M \right] F_{LM}(\theta,\phi) = 0 \,. \tag{5.49}$$

The second of equations (5.49) implies that the dependence of $F_{LM}(\theta,\phi)$ on ϕ is of the form $\exp(iM\phi)$, and since the points (r,θ,ϕ) and $(r,\theta,\phi+2\pi)$ coincide, it appears natural to conclude that M should be an integer. That is effectively correct, since the operator L^2, acting on the Hilbert space of square integrable functions defined on the unit sphere, is just the so-called Laplace-Beltrami operator, self-adjoint and with infinitely differentiable eigenfunctions on this surface. Therefore, using the results of Sect. 5.3, the possible values of L are seen to be $0, 1, 2, \ldots$; that is, the orbital angular momentum is always an integer. Given L, the possible values of M are $M = -L, -L+1, \ldots, L-1, L$. M is called the *magnetic quantum number*. Finally, from the results of Appendix A we have that, apart from a phase,

$$F_{LM}(\theta,\phi) = Y_L^M(\theta,\phi) \,, \tag{5.50}$$

that is, the normalized eigenfunctions of L^2 and L_3 on the unit sphere are just the spherical harmonics. These satisfy the Condon-Shortley convention.

5.5 Angular Momentum Uncertainty Relations

We first analyze the uncertainty relation for an angular variable and its conjugate angular momentum. More precisely, we will consider the azimuthal angle ϕ; if we define ϕ as a continuous variable over the interval $[-\pi, \pi]$, then the z-component of the orbital angular momentum can be represented by the differential operator

$$L_z = -i\hbar\frac{d}{d\phi} \,, \tag{5.51}$$

and the commutator of these two conjugate variables is

$$\boxed{[\phi, L_z] = i\hbar} \,. \tag{5.52}$$

It is necessary to exercise extreme care in the use of this relation, otherwise it is easy to end up with inconsistencies. Recall our discussion of the operator $(-id/dx)$ in Sect. 2.3, and the discussion of commutation relations in Sect. 3.12. In particular it is necessary to observe that

$$\int_{-\pi}^{\pi} d\phi\, \psi_2^*(\phi)[L_z\psi_1(\phi)] = \int_{-\pi}^{\pi} d\phi\, [L_z\psi_2(\phi)]^*\psi_1(\phi)$$
$$- i\hbar[\psi_2^*(\pi)\psi_1(\pi) - \psi_2^*(-\pi)\psi_1(-\pi)] \,. \tag{5.53}$$

Thus, only if $\psi_1(\phi)$ and $\psi_2(\phi)$ are such that the second (boundary) term on the right vanishes, will the operator L_z be symmetric; for arbitrary functions, its "adjoint" is

$$"L_z^\dagger = L_z + i\hbar[\delta(\phi - \pi) - \delta(\phi + \pi)] ." \tag{5.54}$$

The periodic nature of the angular position requires that the boundary conditions for determining the appropriate self-adjoint extension of L_z be periodic: $\psi(\pi) = \psi(-\pi)$. Since the angular variable ϕ defined above is not periodic, even when $\psi_1(\phi)$ and $\psi_2(\phi)$ are, the relation

$$\langle L_z\psi_2(\phi), \phi\psi_1(\phi)\rangle = \langle\psi_2(\phi), L_z\phi\psi_1(\phi)\rangle \tag{5.55}$$

will not be always assured, and for a state with an arbitrary, periodic and normalized wave function $\psi(\phi)$, it will not be possible to conclude in general that

$$\Delta\phi \cdot \Delta L_z \geq \frac{\hbar}{2} . \tag{5.56}$$

It is clear that a general relation of this type is inadmissible, since one can always find a state for which $\Delta L_z < \hbar/4\pi$, and thus $\Delta\phi$ should be greater than 2π; however, indeterminacies in ϕ larger than 2π have no physical meaning. Evidently, for a sufficiently differentiable state $\psi(\phi)$ for which $\psi(\pm\pi) = 0$, there arises no problem in demonstrating (5.56), so that the inequality should be satisfied.

We proceed to find the inequality which generalizes (5.56) for an arbitrary state with a normalized, periodic wave function $\psi(\phi)$ which is sufficiently differentiable. To this end, we introduce the operators $\alpha \equiv \phi - \langle\phi\rangle_\psi$ and $\beta \equiv L_z - \langle L_z\rangle_\psi$. The Schwarz inequality allows us to write

$$\Delta\phi \cdot \Delta L_z = \|\alpha\psi\|\,\|\beta\psi\| \geq |\langle\alpha\psi, \beta\psi\rangle| = |\langle\psi, \alpha\beta\psi\rangle| , \tag{5.57}$$

where we have used the fact that $\alpha = \alpha^\dagger$ in the last equality. But

$$\alpha\beta = \frac{1}{2}[\alpha, \beta] + \frac{1}{2}(\alpha\beta + \beta\alpha) = \frac{i\hbar}{2} + \frac{1}{2}(\alpha\beta + \beta\alpha) , \tag{5.58}$$

and

$$\begin{aligned}
\langle\psi, [\alpha\beta + \beta\alpha]\psi\rangle &= \langle\psi, \alpha\beta\psi\rangle + \langle\beta^\dagger\psi, \alpha\psi\rangle \\
&= \langle\psi, \alpha\beta\psi\rangle + \langle\alpha\beta^\dagger\psi, \psi\rangle \\
&= 2\mathrm{Re}\{\langle\psi, \alpha\beta\psi\rangle\} - 2\pi i\hbar|\psi(\pi)|^2
\end{aligned} \tag{5.59}$$

since $\psi(\pi) = \psi(-\pi)$. Using these three relations, we obtain

$$\Delta\phi \cdot \Delta L_z \geq \frac{\hbar}{2}\left([1 - 2\pi|\psi(\pi)|^2]^2 + \frac{4}{\hbar^2}[\mathrm{Re}\{\langle\psi, \alpha\beta\psi\rangle\}]^2\right)^{1/2} \tag{5.60}$$

and, *a fortiori*,

$$\Delta\phi \cdot \Delta L_z \geq \frac{\hbar}{2}|1 - 2\pi|\psi(\pi)|^2| \, . \tag{5.61}$$

For the equality in (5.61) to hold, it is necessary and sufficient that the Schwarz inequality used in (5.57) also becomes an equality, and that $\text{Re}\{\langle\psi, \alpha\beta\psi\rangle\} = 0$, which is equivalent to demanding that the periodic function $\psi(\phi)$ obey the equation $\beta\psi(\phi) = -i\mu\alpha\psi(\phi)$, where μ is a real parameter [recall (2.70)]. A particular solution of this equation is the family of periodic, normalized wave functions, $\psi_\lambda(\phi)$, which in the interval $[-\pi, +\pi]$ are given by

$$\psi_\lambda(\phi) = N_\lambda \exp\left[\frac{\lambda}{2\pi^2}\phi^2\right] \, , \quad N_\lambda^2 \equiv \left[\int_{-\pi}^{\pi} d\phi \, \exp\left[\frac{\lambda}{\pi^2}\phi^2\right]\right]^{-1} \, , \tag{5.62}$$

λ being an arbitrary real parameter. In this case (5.61) can be applied as an equality; thus,

$$\Delta\phi \cdot \Delta L_z = \frac{\hbar}{2}\left|1 - \frac{2\pi e^\lambda}{\int_{-\pi}^{\pi} d\phi \, \exp(\lambda\phi^2/\pi^2)}\right| \tag{5.63}$$

or equivalently

$$\frac{1}{\hbar}\Delta\phi \cdot \Delta L_z = \frac{1}{2}\left|1 - \frac{1}{\int_0^1 dx \, e^{-\lambda(1-x^2)}}\right| \equiv F(\lambda) \, . \tag{5.64}$$

This function is represented graphically in Fig. 5.1. For $\lambda \leq 1.13012$, we have $(1/\hbar)\Delta\phi \cdot \Delta L_z \leq 1/2$.

Another way of avoiding the difficulties due to the non-periodicity of ϕ is to use continuous periodic variables to localize the azimuthal position [LO 63], the simplest of these being to choose $\cos\phi$ and $\sin\phi$ instead of ϕ. Using (2.179) we have

$$[\sin\phi, L_z] = i\hbar\cos\phi \, , \quad [\cos\phi, L_z] = -i\hbar\sin\phi \, , \tag{5.65}$$

from which the uncertainty relations

$$(\Delta L_z) \cdot (\Delta\sin\phi) \geq \frac{\hbar}{2}|\langle\cos\phi\rangle| \, ,$$

$$(\Delta L_z) \cdot (\Delta\cos\phi) \geq \frac{\hbar}{2}|\langle\sin\phi\rangle| \tag{5.66}$$

can be deduced. As in [CN 68], both of these can be combined to produce a result symmetric in $\sin\phi$ and $\cos\phi$:

$$(\Delta L_z) \cdot \left[\frac{(\Delta\sin\phi)^2 + (\Delta\cos\phi)^2}{\langle\sin\phi\rangle^2 + \langle\cos\phi\rangle^2}\right]^{1/2} \geq \frac{\hbar}{2} \, . \tag{5.67}$$

If, on the other hand, the state is such that the probability density $\mathcal{P}(\phi) = |\psi(\phi)|^2$ is a function symmetric with respect to the point ϕ_0 and quite localized around it in the interval $[-\pi, +\pi]$, then

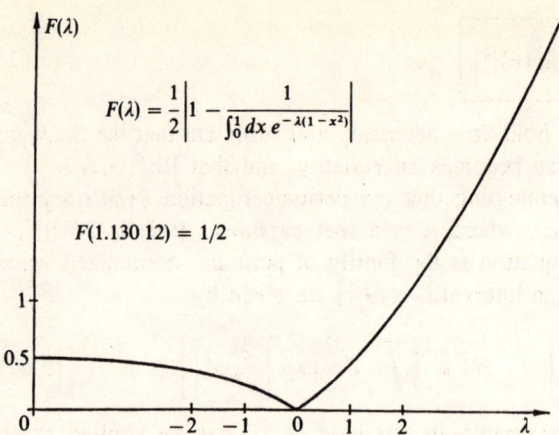

$$F(\lambda) = \frac{1}{2}\left| 1 - \frac{1}{\int_0^1 dx\, e^{-\lambda(1-x^2)}} \right|$$

$$F(1.130\,12) = 1/2$$

Fig. 5.1. Representation of $(\Delta\phi \cdot \Delta L_z)/\hbar$ as a function of the real parameter λ

$$\langle \phi \rangle = \phi_0 , \quad \langle \phi^2 \rangle = \phi_0^2 + \int_{-\pi}^{\pi} d\phi\, \mathcal{P}(\phi)(\phi - \phi_0)^2 , \tag{5.68}$$

and thus,

$$(\Delta\phi)^2 = \int_{-\pi}^{\pi} d\phi\, \mathcal{P}(\phi)(\phi - \phi_0)^2 . \tag{5.69}$$

It is easy to show that up to terms of second order in $(\phi - \phi_0)$ we have

$$\langle \cos\phi \rangle \simeq \left[1 - \tfrac{1}{2}(\Delta\phi)^2\right]\cos\phi_0 , \quad \langle \sin\phi \rangle \simeq \left[1 - \tfrac{1}{2}(\Delta\phi)^2\right]\sin\phi_0 ,$$
$$\langle \cos^2\phi \rangle \simeq [1 - (\Delta\phi)^2]\cos^2\phi_0 + (\Delta\phi)^2 \sin^2\phi_0 , \tag{5.70}$$
$$\langle \sin^2\phi \rangle \simeq [1 - (\Delta\phi)^2]\sin^2\phi_0 + (\Delta\phi)^2 \cos^2\phi_0 ,$$

and hence

$$(\Delta\cos\phi)^2 \simeq (\Delta\phi)^2 \sin^2\phi_0 , \quad (\Delta\sin\phi)^2 \simeq (\Delta\phi)^2 \cos^2\phi_0 ; \tag{5.71}$$

then, in this approximation, the inequalities (5.66) and (5.67) become

$$\Delta\phi \cdot \Delta L_z \geq \frac{\hbar}{2} . \tag{5.72}$$

Further details concerning these uncertainty relations may be found in [CN 68].

Finally, some comments on the commutation relation (5.52) are called for. In Chap. 3 we alluded to the fact that such a commutation relation must be supplemented with additional information if a solution is to be specified. There, we imposed Weyl's rule for X and P in order to be able to deduce the position and momentum observables, and in this form the solution was essentially unique. It is thus clear that ϕ and L_z cannot satisfy Weyl's relation, since the spectrum of ϕ in $L^2(0, 2\pi)$ is $[0, 2\pi]$ and that for $\overline{L}_z \equiv L_z/\hbar$ is $\{0, \pm 1, \pm 2, \dots\}$. One is bounded and the other is not, and the spectrum of the latter is discrete. It is of some interest to know the form of the Weyl relation which the unitary operators generated by ϕ and L_z would lead to, and it is not difficult to construct it.

We define

$$\overline{U}_\alpha \equiv e^{-i\alpha\phi}, \quad \overline{V}_\beta \equiv e^{-i\beta\overline{L}_z}, \quad \alpha \text{ and } \beta \text{ real} \tag{5.73}$$

and apply $\overline{U}_{-\alpha}\overline{V}_{-\beta}\overline{U}_\alpha\overline{V}_\beta$ to the eigenfunction $\exp(in\phi)$ of $\overline{L}_z$ with eigenvalue n. It is clear that

$$\overline{U}_\alpha\overline{V}_\beta e^{in\phi} = e^{-in\beta}e^{-i\alpha\phi}e^{in\phi} . \tag{5.74}$$

It would be an error to now act formally with $\overline{V}_{-\beta}$ since $\exp[i(n - \alpha)\phi]$ does not belong to the domain of L_z. This difficulty can be resolved by a Fourier expansion of this function in the interval $(0, 2\pi)$:

$$e^{i(n-\alpha)\phi} = \frac{1}{2\pi i}\sum_m \frac{1 - e^{-2\pi i\alpha}}{m + \alpha}e^{i(m+n)\phi}, \quad \phi \in (0, 2\pi) , \tag{5.75}$$

and now it suffices to apply $\overline{V}_{-\beta}$ term by term. As a consequence,

$$\overline{U}_{-\alpha}\overline{V}_\beta\overline{U}_\alpha\overline{V}_\beta e^{in\phi} = e^{i\alpha\phi}\left[\frac{1}{2\pi i}\sum_m \frac{1 - e^{-2\pi i\alpha}}{m + \alpha}e^{im(\phi+\beta)}\right]e^{in\phi} . \tag{5.76}$$

The function in brackets, defined for $\phi \in (0, 2\pi)$ and extended to $\mathbf{R}$ periodically, can be obtained from (5.75) with $n = 0$ and also extended periodically, by means of the displacement $-\beta$. This allows us to write, given the validity of (5.76) for all $\exp(in\phi)$ and thus for all $\psi(\phi) \in L^2(0, 2\pi)$, the equality

$$\overline{U}_{-\alpha}\overline{V}_{-\beta}\overline{U}_\alpha\overline{V}_\beta = e^{-i\alpha\beta^*}[\chi_{[0,2\pi-\beta^*]}(\phi) + e^{2\pi i\alpha}\chi_{[2\pi-\beta^*,2\pi]}(\phi)] , \tag{5.77}$$

where $\beta^* \equiv \beta \pmod{2\pi}$, $0 \le \beta^* < 2\pi$, and $\chi_{[a,b]}(\phi)$ is the spectral projection of ϕ associated to (or the characteristic function of) the interval $[a, b]$. The expression (5.77) is the Weyl form for ϕ, L_z, in $L^2(0, 2\pi)$.

Finally, attempts have been made to introduce the notion of coherent states, so useful in quantum optics, into the operator algebra of angular momentum. Thus, concepts such as *Bloch states* and *intelligent states* have been introduced. Due to the commutation relations (5.29), it is clear from (2.64) that

$$\Delta J_x \cdot \Delta J_y \ge \frac{\hbar}{2}|\langle J_z\rangle| \tag{5.78}$$

for any state in a space of bounded angular momentum. Limiting ourselves to a space $\mathcal{H}^{(J)}$, associated with a unitary, irreducible representation $\mathcal{D}^{(J)}$ of SU(2), we define *Bloch* states [BL 46] [AC 72], or coherent spin states, as

$$|z\rangle \equiv R(\hat{\mathbf{n}}, \theta)|J, -J\rangle = (1 + |z|^2)^{-J}\exp(zJ_+/\hbar)|J, -J\rangle , \tag{5.79}$$

where $\hat{\mathbf{n}} \equiv (\sin\phi, -\cos\phi, 0)$ and $z \equiv \tan(\theta/2)\exp(-i\phi)$. Note the analogy with expression (4.73). Even though these states do not satisfy the property $J_-|z\rangle \propto$

$|z\rangle$, they do share with the coherent states the saturation of the equality in the corresponding uncertainty relation [(5.78) for Bloch states, and $\Delta x \cdot \Delta p \geq \hbar/2$ for those of a harmonic oscillator]. Further, under evolution in a magnetic field with the Hamiltonian

$$H(t) = -\gamma \boldsymbol{J} \cdot \boldsymbol{B}(t) , \quad \boldsymbol{B}(t) = (B_1 \cos \omega t, \; B_1 \sin \omega t, \; B_0) , \tag{5.80}$$

the Bloch states remain so as time evolves.

In addition to these coherent spin states, there are others for which the equality in (5.78) holds. The label "intelligent states" has been given to all states for which this is true [AG 64] [AC 76].

5.6 Matrix Representations of the Rotation Operators

It is well known that three parameters must be specified to characterize a rotation. Until now we have used the angle θ and the two angles necessary to specify the direction of the unit vector $\hat{n}$. Nevertheless, in many theoretical developments, including the present one, it is better to use as parameters the three Euler angles which characterize a general rotation. Of the many definitions of Euler angles available in the literature, we will adopt the one used by *Rose* [RO 57]: let $Ox_1x_2x_3$ be the original right-handed coordinate system. We define three rotations as follows:

1) A positive rotation with angle α $(0 \leq \alpha < 2\pi)$ around the Ox_3 axis. Let $O\xi_1\xi_2\xi_3$ be the triad produced from $Ox_1x_2x_3$ under this rotation.
2) A positive rotation with angle β $(0 \leq \beta \leq \pi)$ around the $O\xi_2$ axis. Let the new triad be $O\eta_1\eta_2\eta_3$.
3) A positive rotation with angle γ $(0 \leq \gamma < 2\pi)$ around the $O\eta_3$ axis. Let $Ox_1'x_2'x_3'$ be the final triad.

Under a rotation, the coordinates of a point can change for two reasons: either because the rotation is "physically" applied to the point itself, with the coordinate axes held fixed (an active rotation), or because, the point being held fixed, the axes are "physically" rotated (passive rotation). The representative matrices for this rotation from these two perspectives are the inverses of each other. Note that in writing $\mathcal{R}\boldsymbol{x}$ we mean that rotation in the active sense.

The active rotation characterized by these Euler angles will be denoted by $\mathcal{R}(\alpha, \beta, \gamma)$. The explicit form of this matrix can be obtained taking into account the construction specified above:

$$\mathcal{R}^{-1}(\alpha, \beta, \gamma) = \mathcal{R}_{\text{passive}}(\alpha, \beta, \gamma)$$
$$= \begin{pmatrix} \cos\gamma & \sin\gamma & 0 \\ -\sin\gamma & \cos\gamma & 0 \\ 0 & 0 & 1 \end{pmatrix} \begin{pmatrix} \cos\beta & 0 & -\sin\beta \\ 0 & 1 & 0 \\ \sin\beta & 0 & \cos\beta \end{pmatrix} \begin{pmatrix} \cos\alpha & \sin\alpha & 0 \\ -\sin\alpha & \cos\alpha & 0 \\ 0 & 0 & 1 \end{pmatrix}$$
$$\tag{5.81}$$

and therefore,

$$\mathcal{R}(\alpha, \beta, \gamma)$$
$$= \begin{pmatrix} \cos\alpha \cos\beta \cos\gamma - \sin\alpha \sin\gamma & -\cos\alpha \cos\beta \sin\gamma - \sin\alpha \cos\gamma & \cos\alpha \sin\beta \\ \sin\alpha \cos\beta \cos\gamma + \cos\alpha \sin\gamma & -\sin\alpha \cos\beta \sin\gamma + \cos\alpha \cos\gamma & \sin\alpha \sin\beta \\ -\sin\beta \cos\gamma & \sin\beta \sin\gamma & \cos\beta \end{pmatrix}$$

$$(5.82)$$

Given the Euler angles α, β, and γ, one can obtain $\hat{n}$ and θ by means of the formulae

$$\cos\theta = \frac{1}{2}[\text{Tr}\{\mathcal{R}\} - 1] , \qquad n_k = \frac{-1}{2\sin\theta}\varepsilon_{kij}\mathcal{R}_{ij} , \qquad (5.83)$$

which are deduced from (5.4).

If we denote the component of $\boldsymbol{J}$ along the $O\eta_i$ ($O\xi_i$) axis by J_{η_i} (J_{ξ_i}), then the operator in each irreducible sub-representation corresponding to the rotation $\mathcal{R}(\alpha, \beta, \gamma)$ is, apart from a $\pm$ sign,

$$R(\alpha, \beta, \gamma) = e^{-i\gamma J_{\eta_3}/\hbar}e^{-i\beta J_{\xi_2}/\hbar}e^{-i\alpha J_3/\hbar} . \qquad (5.84)$$

From the form in which we have carried out the Euler rotations and the transformation laws for operators, there results

$$e^{-i\beta J_{\xi_2}/\hbar} = e^{-i\alpha J_3/\hbar}e^{-i\beta J_2/\hbar}e^{i\alpha J_3/\hbar}$$

$$e^{-i\gamma J_{\eta_3}/\hbar} = e^{-i\beta J_{\xi_2}/\hbar}e^{-i\gamma J_{\xi_3}/\hbar}e^{i\beta J_{\xi_2}/\hbar} \qquad (5.85)$$

$$= e^{-i\alpha J_3/\hbar}e^{-i\beta J_2/\hbar}e^{-i\gamma J_3/\hbar}e^{i\beta J_2/\hbar}e^{i\alpha J_3/\hbar} .$$

Substituting these expressions into (5.84), we see that the rotation operator we want can be written as

$$\boxed{R(\alpha, \beta, \gamma) = e^{-i\alpha J_3/\hbar}e^{-i\beta J_2/\hbar}e^{-i\gamma J_3/\hbar}} . \qquad (5.86)$$

This demonstrates that the same total rotation is obtained if the rotations are carried out in inverse order around the original axes. Note that (5.86) is also a simple consequence of the relation (5.81).

Here we have arrived at the necessary expression for discussing in a relatively simple form how the vector $|JM\rangle \in \mathcal{H}^{(J)}$ transforms under the group of rotations. Since we have already proven that $\mathcal{H}^{(J)}$ is stable under this group, we may write

$$\boxed{\begin{aligned} R(\alpha, \beta, \gamma)|JM\rangle &= \sum_{M'}\langle JM'|R(\alpha, \beta, \gamma)|JM\rangle|JM'\rangle \\ &\equiv \sum_{M'} D^J_{M'M}(\alpha, \beta, \gamma)|JM'\rangle , \end{aligned}} \qquad (5.87)$$

where $D^J_{M'M}(\alpha, \beta, \gamma)$ stands for the matrix elements of the rotation operator $R(\alpha, \beta, \gamma)$ in the basis formed by the vectors $|JM\rangle$. Keeping in mind (5.86), it immediately follows that

$$D^J_{M'M}(\alpha,\beta,\gamma) = e^{-i\alpha M'}\langle JM'|e^{-i\beta J_2/\hbar}|JM\rangle e^{-i\gamma M}$$
$$\equiv e^{-i\alpha M'} d^J_{M'M}(\beta) e^{-i\gamma M} . \tag{5.88}$$

Therefore, the dependence on the Euler angles α and γ is factorizable in a trivial way. The problem is thus reduced to the calculation of $d^J_{M'M}(\beta)$. We shall do this only for the case $J = 1/2$. According to the definition above and taking into account (5.46),

$$d^{1/2}_{M'M} = \langle \tfrac{1}{2} M' | e^{-i\beta\sigma_2/2} | \tfrac{1}{2} M\rangle . \tag{5.89}$$

From the immediately obvious statements $(\sigma_2)^{2n} = I$ and $(\sigma_2)^{2n+1} = \sigma_2$, it follows that

$$e^{-i\beta\sigma_2/2} = I\cos(\beta/2) - i\sigma_2\sin(\beta/2) ; \tag{5.90}$$

thus,

$$d^{1/2}_{1/2,1/2}(\beta) = d^{1/2}_{-1/2,-1/2}(\beta) = \cos(\beta/2) ,$$
$$d^{1/2}_{1/2,-1/2}(\beta) = -d^{1/2}_{-1/2,1/2}(\beta) = -\sin(\beta/2) . \tag{5.91}$$

The transformation laws for the states are therefore

$$R(\alpha,\beta,\gamma)|1/2\ 1/2\rangle = \cos(\beta/2)e^{-i(\alpha+\gamma)/2}|1/2\ 1/2\rangle$$
$$+ \sin(\beta/2)e^{i(\alpha-\gamma)/2}|1/2\ -1/2\rangle ,$$
$$R(\alpha,\beta,\gamma)|1/2\ -1/2\rangle = -\sin(\beta/2)e^{-i(\alpha-\gamma)/2}|1/2\ 1/2\rangle$$
$$+ \cos(\beta/2)e^{i(\alpha+\gamma)/2}|1/2\ -1/2\rangle . \tag{5.92}$$

The general expression for the $d^J_{M'M}(\beta)$ has been given by *Wigner* [WI 31] and can be seen in (B. 19). In Appendix B appear a number of interesting properties of the functions $d^J_{M'M}(\beta)$ and $D^J_{M'M}(\alpha,\beta,\gamma)$, as well as explicit expressions for the latter in the special cases $J = 1/2$ and 1.

We proceed to a discussion of when to expect a single-valued or a double-valued correspondence $\mathcal{R}(\alpha,\beta,\gamma) \rightarrow R(\alpha,\beta,\gamma)$ for each irreducible sub-representation. We shall see that this correspondence is single-valued only if J is an integer, and is double-valued for a half-integer J. To prove this, we consider the elements $A(\hat{n},\pi)$ and $A(-\hat{n},\pi)$ of SU(2) leading to the same rotation $\mathcal{R}(\pm\hat{n},\pi)$. The operators corresponding to these SU(2) matrices are

$$R(\pm\hat{n},\pi) = \exp\left[\mp\frac{i}{\hbar}\pi\hat{n}\cdot J\right] . \tag{5.93}$$

Since $\hat{n}\cdot J$ is a self-adjoint operator (equivalent to J_3), it is possible, starting from $\{|JM\rangle\}$, to find a new basis $\{|\overline{JM}\rangle\}$ such that $\hat{n}\cdot J$ will be diagonal in it; this is equivalent to taking the direction of quantization along the $\hat{n}$ axis. In this basis, the matrix elements of (5.93) are

$$D^J_{M'M}(\pm\hat{n},\pi) = \delta_{M'M}e^{\mp i\pi M} , \tag{5.94}$$

from which we see that the matrices representative of $A(\hat{n}, \pi)$ and $A(-\hat{n}, \pi)$ which lead to the same rotation are equal if and only if J is an integer. If J is a half-integer, these matrices differ in sign. The same thing happens if we consider the pair $A(\hat{n}, \theta)$ and $A(-\hat{n}, 2\pi - \theta)$ leading to the same rotation $\mathcal{R}(\hat{n}, \theta)$. We have thus proved the correspondence

$$
\boxed{
\begin{aligned}
\mathcal{R}(\hat{n}, \theta) &\rightarrow D^J_{M'M}(\hat{n}, \theta) & \text{if } J = \text{integer} , \\
\mathcal{R}(\hat{n}, \theta) &\rightarrow \pm D^J_{M'M}(\hat{n}, \theta) & \text{if } J = \text{half-integer} .
\end{aligned}
}
\tag{5.95}
$$

Let $R_{2\pi} \equiv \exp(-i2\pi\hat{n} \cdot \boldsymbol{J}/\hbar)$ be the representative operator for $A(\hat{n}, 2\pi)$ associated with a rotation of angle 2π around any axis $\hat{n}$. The preceding formulae show easily that $R_{2\pi}$, in each space $\mathcal{H}^{(J)}$, is a multiple of unity, $(-1)^{2J}$, and thus independent of $\hat{n}$. Under the action of this trivial rotation, any observable A is invariant, so that

$$
[A, R_{2\pi}] = 0 ,
\tag{5.96}
$$

i.e., $R_{2\pi}$ commutes with all observables, and thus permits us to establish a super-selection rule according to the discussion in Sect. 2.15. States which correspond to a linear superposition of vectors with different values for the operator $R_{2\pi}$, that is, with different $(-1)^{2J}$, are not realizable as pure states. Such a superse-lection operator gives rise to the univalence superselection rule; and it coincides with the parity of the number of fermions (as will be explained in Chap. 13).

With modern (spin-1/2) neutron interferometry, "double-slit" experiments have been performed [WC 75], [RZ 75] in which one of the sub-beams (say beam 1) traverses a local magnetic field which produces exactly the same effect in the corresponding spin amplitude ψ_1 as a rotation $R_{2\pi}$, without affecting the amplitude ψ_2 of the other sub-beam. Exactly the interference pattern has been observed which one would expect from the change of sign $\psi_1 \rightarrow -\psi_1$ produced by such a rotation. This is generally interpreted as a direct experimental demonstration of the double-valued character of $\mathcal{D}^{(1/2)}$. However, the "local" character of that 2π-rotation should be noted (if it were global, the effect would not be observable, since then it would also change $\psi_2 \rightarrow -\psi_2$). See [GR 83] for more details.

5.7 Addition of Angular Momenta

Let $\boldsymbol{J}_1$ and $\boldsymbol{J}_2$ be angular momentum operators of two kinematically independent subsystems 1 and 2, i.e., such that $[\boldsymbol{J}_1, \boldsymbol{J}_2] = 0$ (for example: the orbital angular momenta $\boldsymbol{L}_1$ and $\boldsymbol{L}_2$ of two different systems; the orbital angular momentum $\boldsymbol{L}$ and the spin $\boldsymbol{S}$ of the same system; ...). For the total system consisting of the union of 1 and 2, the operator $\boldsymbol{J} = \boldsymbol{J}_1 + \boldsymbol{J}_2$, which satisfies the commutation rules for angular momentum operators, is the total angular momentum operator for the system $1 + 2$.

The operators J_1^2, $(J_1)_z$, J_2^2, $(J_2)_z$ commute with one another and a complete set of compatible operators may be constructed by adding to them new observables A which commute with J_1 and J_2. One element of the associated orthonormal basis will be designated by $|J_1 M_1 J_2 M_2 \alpha\rangle$. These vectors satisfy

$$\left. \begin{array}{l} J_i^2 |J_1 M_1 J_2 M_2 \alpha\rangle = \hbar^2 J_i(J_i+1)|J_1 M_1 J_2 M_2 \alpha\rangle \\ (J_i)_z |J_1 M_1 J_2 M_2 \alpha\rangle = \hbar M_i |J_1 M_1 J_2 M_2 \alpha\rangle \end{array} \right\} (i = 1, 2) . \tag{5.97}$$

The values α correspond to the observables A and will in general be omitted, with the understanding that we are restricted to a (perhaps generalized) eigensubspace. For given values of J_1 and J_2, these vectors generate a $(2J_1 + 1) \times (2J_2 + 1)$-dimensional space, which we shall denote by $\mathcal{H}^{(J_1, J_2)}$ (the α being understood to be fixed). Inasmuch as J commutes with J_1^2 and J_2^2, vectors in $\mathcal{H}^{(J_1, J_2)}$ can be found which are simultaneous eigenvectors of J_1^2, J_2^2, J^2, and J_z. These normalized vectors shall be denoted by $|J_1 J_2 J M\rangle$, and by definition they satisfy

$$J_i^2 |J_1 J_2 J M\rangle = \hbar^2 J_i(J_i+1)|J_1 J_2 J M\rangle , \quad i = 1, 2 ;$$
$$J^2 |J_1 J_2 J M\rangle = \hbar^2 J(J+1)|J_1 J_2 J M\rangle ; \tag{5.98}$$
$$J_z |J_1 J_2 J M\rangle = \hbar M |J_1 J_2 J M\rangle .$$

The problem we want to address now, is to determine, given J_1 and J_2, the possible values of J and M and their degeneracies. Classically, the problem is elementary: J has any value between $|J_{1,\text{cl}} - J_{2,\text{cl}}|$ and $J_{2,\text{cl}} + J_{2,\text{cl}}$ and M fills the interval $[-J_{\text{cl}}, J_{\text{cl}}]$. Quantum mechanically, the existence of the commutation rules (5.29) discretizes the results, as we show in what follows.

To this end we must take into account that

$$J_z |J_1 M_1 J_2 M_2\rangle = \hbar(M_1 + M_2)|J_1 M_1 J_2 M_2\rangle . \tag{5.99}$$

For a given M, we indicate by $n(M)$ the number of states $|J_1 M_1 J_2 M_2\rangle$ such that $M_1 + M_2 = M$. To calculate $n(M)$ it is useful to construct the plot in Fig. 5.2 [ME 59], where the points with coordinates (M_1, M_2) have been indicated, and a set of broken lines connect the points with constant $M_1 + M_2$. By a simple inspection of Fig. 5.2, we easily conclude that

$$n(M) = \begin{cases} 0 & \text{if } |M| > J_1 + J_2 \\ J_1 + J_2 + 1 - |M| & \text{if } J_1 + J_2 \ge |M| \ge |J_1 - J_2| , \\ \min(2J_1 + 1, 2J_2 + 1) & \text{if } |J_1 - J_2| \ge |M| \ge 0 . \end{cases} \tag{5.100}$$

On the other hand, if $|J_1 J_2 J M\rangle \in \mathcal{H}^{(J_1, J_2)}$, then all the vectors $|J_1 J_2 J M'\rangle$ with $M' = -J, -J+1, \ldots, J-1, J$ will also belong to this space; it suffices to apply to the former the operators $J_\pm = (J_1)_\pm + (J_2)_\pm$ repeatedly, and to recall that the space is stable under the action of these operators. We then have that if $N(J)$ indicates the degeneracy of the total angular momentum J, we can write

$$n(M) = \sum_{J \ge |M|} N(J) , \tag{5.101}$$

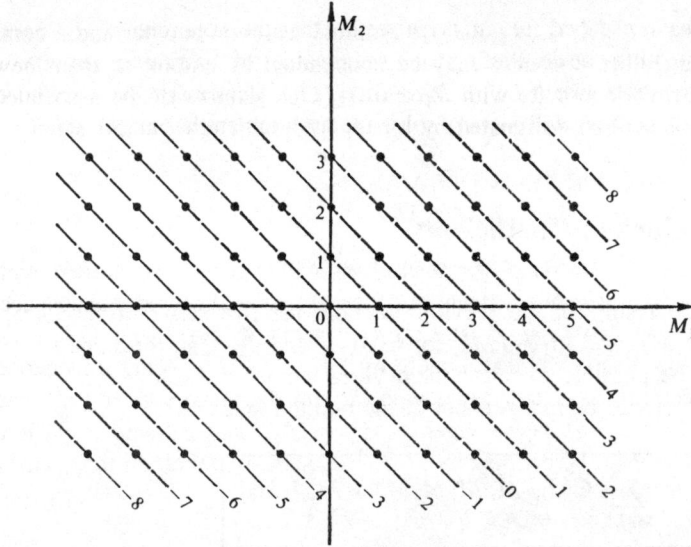

Fig. 5.2. Lattice for the calculation of $n(M)$ for $J_1 = 5$ and $J_2 = 3$. The points indicate the values of (M_1, M_2), and the dotted lines join values of (M_1, M_2) such that $M_1 + M_2 = M$

from which $N(J) = n(J) - n(J + 1)$ and using (5.100) we obtain

$$N(J) = 0 \quad \text{if } J > J_1 + J_2, \quad J < |J_1 - J_2|$$
$$N(J) = 1 \quad \text{if } |J_1 - J_2| \leq J \leq J_1 + J_2 . \tag{5.102}$$

Summarizing, for given values of J_1 and J_2, the possible values of the total angular momentum J are $|J_1 - J_2|$, $|J_1 - J_2| + 1, \ldots, J_1 + J_2 - 1$, $J_1 + J_2$. For each possible value of J there is one and only one series of eigenvectors $|J_1 J_2 J M\rangle$, where M takes on the values $M = -J, -J + 1, \ldots, J - 1, J$.

Using group theory language, this result can be presented in the following form: given the irreducible representations $\mathcal{D}^{(J_1)}$ and $\mathcal{D}^{(J_2)}$ of the rotation group with basis vectors $|J_1 M_1\rangle$ and $|J_2 M_2\rangle$, the vectors $|J_1 M_1 J_2 M_2\rangle \equiv |J_1 M_1\rangle \otimes |J_2 M_2\rangle$ give rise to a representation of the group, in general reducible, which can be decomposed into irreducible representations according to the *Clebsch-Gordan series*

$$\boxed{\mathcal{D}^{(J_1)} \otimes \mathcal{D}^{(J_2)} = \mathcal{D}^{(|J_1 - J_2|)} \oplus \mathcal{D}^{(|J_1 - J_2|+1)} \oplus \cdots \oplus \mathcal{D}^{(J_1+J_2-1)} \oplus \mathcal{D}^{(J_1+J_2)}} .$$

$$\tag{5.103}$$

The vectors $|J_1 J_2 J M\rangle$ are the basis vectors for the resulting irreducible representations.

Another important consequence of the previous result is that if $J_1^2, (J_1)_z, J_2^2$, $(J_2)_z$ and A form a complete set of compatible observables, where A commutes with J_1, J_2, then the observables J_1^2, J_2^2, J^2, J_z and A will also form a complete set.

It will not have escaped the attentive reader that the notations $|J_1 J_2 JM\alpha\rangle$ and $|J_1 M_1 J_2 M_2\alpha\rangle$ may be conflicting for special values of the quantum numbers. More detailed symbols like $|(J_1 J_2)JM\alpha\rangle$ for $|J_1 J_2 JM\alpha\rangle$ might have avoided it but anyhow the context will always solve the ambituities that might arise.

5.8 Clebsch-Gordan Coefficients

According to the results of the previous section, the two sets of orthonormal vectors $|J_1 M_1 J_2 M_2\rangle$ and $|J_1 J_2 JM\rangle$, for fixed J_1 and J_2, span the same space; thus, it should be possible to relate them via a unitary transformation, whose coefficients in fact will be independent of the omitted index α, i.e.

$$\boxed{|J_1 J_2 JM\rangle = \sum_{M_1 M_2} C(J_1 J_2 J; M_1 M_2 M)|J_1 M_1 J_2 M_2\rangle} . \tag{5.104}$$

It is evident that $C(J_1 J_2 J; M_1 M_2 M) = \langle J_1 M_1 J_2 M_2 | J_1 J_2 JM\rangle$. These scalar products are called the *Clebsch-Gordan coefficients* (C-G). Note that in reality the double sum which appears in (5.104) is redundant, since $M = M_1 + M_2$, and (5.104) reduces to

$$|J_1 J_2 JM\rangle = \sum_{M_1} C(J_1 J_2 J; M_1 M - M_1 M)|J_1 M_1 J_2 M - M_1\rangle . \tag{5.105}$$

For this reason, we will frequently use the symbol $C(J_1 J_2 J; M_1 M_2)$ instead of $C(J_1 J_2 J; M_1 M_2 M)$, and it is understood that the missing argument is $M = M_1 + M_2$. When the possibility of error arises, a comma will be inserted between M_1 and M_2.

We have explained how the phases of all the vectors $|J_1 M_1 J_2 M_2\rangle$ can be determined by fixing the phase of $|J_1 J_1 J_2 J_2\rangle$ and using the Condon-Shortley phase convention (that is, requiring that the matrix elements of $J_{1\pm}, J_{2\pm}$ in this basis be ≥ 0). In the same way, given the phase of $|J_1 J_2 JJ\rangle$, and using the same convention, the phases of the vectors $|J_1 J_2 JM\rangle$ can be determined. Nevertheless, in order to determine completely the Clebsch-Gordan coefficients, some criterion for fixing the phases of the vectors $|J_1 J_2 JJ\rangle$ with $J = |J_1 - J_2|, \ldots, J_1 + J_2$, relative to those of the vectors $|J_1 M_1 J_2 M_2\rangle$ must be given. Here we follow the generally used convention that the component of $|J_1 J_2 JJ\rangle$ along $|J_1 J_1 J_2 J - J_1\rangle$ be real and non-negative; that is,

$$C(J_1 J_2 J; J_1 J - J_1 J) \geq 0 . \tag{5.106}$$

We shall see that this condition guarantees that all the Clebsch-Gordan coefficients are real. In fact, applying $J_\pm$ to both sides of (5.104), and making use of (5.43), we obtain

$$\sqrt{(J \mp M)(J \pm M + 1)}|J_1 J_2 J M \pm 1\rangle$$
$$= \sum_{M_1 M_2} C(J_1 J_2 J; M_1 M_2 M)[\sqrt{(J_1 \mp M_1)(J_1 \pm M_1 + 1)}|J_1 M_1 \pm 1 J_2 M_2\rangle$$
$$+ \sqrt{(J_2 \mp M_2)(J_2 \pm M_2 + 1)}|J_1 M_1 J_2 M_2 \pm 1\rangle] . \tag{5.107}$$

Multiplying by $|J_1 M_1 J_2 M_2\rangle$ and taking into account the definition of the Clebsch-Gordan coefficients, there results

$$\sqrt{(J - M)(J + M + 1)}C(J_1 J_2 J; M_1 M_2 M + 1)$$
$$= \sqrt{(J_1 + M_1)(J_1 - M_1 + 1)}C(J_1 J_2 J; M_1 - 1 M_2 M)$$
$$+ \sqrt{(J_2 + M_2)(J_2 - M_2 + 1)}C(J_1 J_2 J; M_1 M_2 - 1 M) ;$$
$$\sqrt{(J + M)(J - M + 1)}C(J_1 J_2 J; M_1 M_2 M - 1) \tag{5.108}$$
$$= \sqrt{(J_1 - M_1)(J_1 + M_1 + 1)}C(J_1 J_2 J; M_1 + 1 M_2 M)$$
$$+ \sqrt{(J_2 - M_2)(J_2 + M_2 + 1)}C(J_1 J_2 J; M_1 M_2 + 1 M) .$$

If we take $M = J$ in the first equation, the l.h.s. becomes zero, and we obtain an equation which, through interated use, permits calculation of all the coefficients $C(J_1 J_2 J; M_1, J - M_1)$ in terms of $C(J_1 J_2 J; J_1, J - J_1)$. As all the coefficients in (5.108) are real, the condition (5.106) guarantees the reality of $C(J_1 J_2 J; M_1, J - M_1)$. Starting with these, and using the second equation in (5.108), the rest of the coefficients can be calculated; all of them are evidently real. The use of the above equations, together with the normalization conditions, totally determines the C-G coefficients.

To illustrate this procedure, we consider the case $J_1 = J_2 = 1/2$; the possible values of J are $J = 0, 1$. From the normalization condition and (5.106),

$$|1/2 \ 1/2 \ 1 \ 1\rangle = |1/2 \ 1/2 \ 1/2 \ 1/2\rangle , \tag{5.109}$$

that is, $C(1/2 \ 1/2 \ 1; \ 1/2 \ 1/2 \ 1) = +1$. Using the second equation in (5.108), with $J = 1, M = 1$, and $M_1 = -1/2, M_2 = +1/2$ and also with $M_1 = +1/2, M_2 = -1/2$, we obtain

$$C(1/2 \ 1/2 \ 1; \ 1/2 \ -1/2 \ 0) = C(1/2 \ 1/2 \ 1; -1/2 \ 1/2 \ 0) = 1/\sqrt{2}$$

from which

$$|1/2 \ 1/2 \ 1 \ 0\rangle = \frac{1}{\sqrt{2}}|1/2 \ 1/2 \ 1/2 \ -1/2\rangle + \frac{1}{\sqrt{2}}|1/2 \ -1/2 \ 1/2 \ 1/2\rangle. \tag{5.110}$$

From the same equation, with $J = 1$, $M = 0$, and $M_1 = M_2 = -1/2$, we obtain $C(1/2 \ 1/2 \ 1; -1/2 \ -1/2 \ -1) = 1$, that is

$$|1/2 \ 1/2 \ 1 \ -1\rangle = |1/2 \ -1/2 \ 1/2 \ -1/2\rangle . \tag{5.111}$$

Missing still is the determination of the state $|1/2 \ 1/2 \ 0 \ 0\rangle$. According to (5.106) $C(1/2 \ 1/2 \ 0; 1/2 \ -1/2) = A \geq 0$, and using the first of the equations (5.108), with $J = M = 0$ and $M_1 = M_2 = 1/2$, results in $C(1/2\ 1/2\ 0;\ -1/2\ 1/2) = -A$. The normalization condition for the state sought implies that $A = 1/\sqrt{2}$, and thus

$$|1/2\ 1/2\ 0\ 0\rangle = \frac{1}{\sqrt{2}}|1/2\ 1/2\ 1/2\ -1/2\rangle - \frac{1}{\sqrt{2}}|1/2\ -1/2\ 1/2\ 1/2\rangle\ ,(5.112)$$

which solves the problem.

A general explicit expression (B.41) for the C-G coefficients can be found following the above methods [WI 31]. With that as starting point, interesting properties of these coefficients can be deduced, some of which are given in Appendix B. The calculation of the C-G coefficients by means of (B.41) is somewhat involved, and because of the frequency with which they are used we have included some tables at the end of Appendix B. More complete sets of tables appear in [RB 59].

Finally, taking into account that

$$C(J_1 J_2 J; M_1 M_2 M) = \langle J_1 M_1 J_2 M_2 | J_1 J_2 J M \rangle$$

and also the reality of the coefficients, the inverse relation to (5.104) is

$$\boxed{|J_1 M_1 J_2 M_2\rangle = \sum_{JM} C(J_1 J_2 J; M_1 M_2 M)|J_1 J_2 J M\rangle}\ . \qquad (5.113)$$

The sum over M is evidently trivially realized, since $M = M_1 + M_2$, and the sum over J is from $|J_1 - J_2|$ to $J_1 + J_2$. Moreover, the orthogonality relations now appear in the form

$$\boxed{\begin{aligned} &\sum_{M_1} C(J_1 J_2 M; M_1, M - M_1)C(J_1 J_2 J'; M_1, M - M_1) = \delta_{JJ'} \\ &\sum_{J} C(J_1 J_2 J; M_1, M - M_1)C(J_1 J_2 J; M_1', M' - M_1') = \delta_{MM'}\delta_{M_1 M_1'}\ , \end{aligned}}$$

$$(5.114)$$

which just express the unitary (actually orthogonal) character of the transformation matrix between the bases.

It should be noted that relations like (5.114), (B.43) and other similar ones do not hold *stricto sensu*, the validity of the first in (5.114) requiring for instance that $J \in \{|J_1 - J_2|, |J_1 - J_2| + 1, \ldots, J_1 + J_2\}$ and $M \in \{-J, -J + 1, \ldots, J\}$. We keep however to the usual practice of tacitly taking for granted that such obvious additional conditions are met.

5.9 Irreducible Tensors Under Rotations

Frequently, observables and other operators appear in physics whose transformation properties under the action of rotations are specific and simple. Such operators are particular components of objects called tensors under SO(3), whose analysis we will sketch, given their practical importance.

An *irreducible tensor operator* of order J under the rotation group is defined as a linear combination of $(2J + 1)$ operators T_M^J, with $M = -J, -J + 1, \ldots, J - 1, J$, which transform according to the following law under the action of the group:

$$
\boxed{R(\alpha, \beta, \gamma) T_M^J R^{-1}(\alpha, \beta, \gamma) = \sum_{M'} D_{M'M}^J(\alpha, \beta, \gamma) T_{M'}^J} . \tag{5.115}
$$

In the first place, it is of interest to note that the above definition is equivalent (at least formally, and rigorously if T_M^J satisfies appropriate conditions) to requiring that the T_M^J satisfy the commutation relations

$$
\boxed{
\begin{aligned}
[J_\pm, T_M^J] &= \hbar \sqrt{(J \mp M)(J \pm M + 1)} T_{M \pm 1}^J \\
[J_3, T_M^J] &= \hbar M T_M^J ,
\end{aligned}
}
\tag{5.116}
$$

which characterize $\{T_M^J\}$ in particular as the standard basis for the tensor operator. In fact, for an infinitesimal rotation through angle $\delta\theta$ around the $\hat{n}$ axis, the transformation law (5.115) can be rewritten

$$
\begin{aligned}
\left[1 - \frac{i}{\hbar}\delta\theta\hat{n} \cdot \boldsymbol{J}\right] &T_M^J \left[1 + \frac{i}{\hbar}\delta\theta\hat{n} \cdot \boldsymbol{J}\right] \\
&= \sum_{M'} \langle JM'| \left[1 - \frac{i}{\hbar}\delta\theta\hat{n} \cdot \boldsymbol{J}\right] |JM\rangle T_{M'}^J .
\end{aligned}
\tag{5.117}
$$

Neglecting second- and higher-order infinitesimals, and remembering that $\hat{n}$ is an arbitrary vector, results in

$$
[\boldsymbol{J}, T_M^J] = \sum_{M'} \langle JM'|\boldsymbol{J}|JM\rangle T_{M'}^J , \tag{5.118}
$$

from which one immediately obtains the laws (5.116). Conversely, relations (5.116) imply that (5.115) is valid for any infinitesimal transformation and, by a composition of these, for any finite rotation.

Some properties of irreducible tensors, whose proof is relatively simple, are given in Appendix B. The calculation of the matrix elements of an irreducible tensor among states with well-defined angular momenta are exceptionally simple as a result of the *Wigner-Eckart theorem* [WI 31], [EC 30]:

$$\langle J_2 M_2 \alpha_2 | T_M^J | J_1 M_1 \alpha_1 \rangle$$
$$= \langle J_2 \alpha_2 \| T^J \| J_1 \alpha_a \rangle C(J_1 J J_2; M_1 M M_2) , \qquad (5.119)$$

that is to say, all the third component dependence is factorizable into a C-G coefficient; the quantity $\langle J_2 \alpha_2 \| T^J \| J_1 \alpha_1 \rangle$, called the *reduced matrix element*, does not depend on M_1, M_2 and M. Due to the importance of this theorem, we give here its proof: The $(2J + 1)(2J_1 + 1)$ vectors $T_M^J | J_1 M_1 \alpha_1 \rangle$ (not necessarily linearly independent) transform according to one of the representations of the rotation group contained in $\mathcal{D}^{(J)} \otimes \mathcal{D}^{(J_1)}$, and thus the matrix elements being considered are zero unless $M_2 = M_1 + M$, and J_2 appears in the series $|J - J_1|, |J - J_1| + 1, \dots J + J_1 - 1, J + J_1$. On the other hand, from the first of equations (5.116), together with (5.43), we obtain

$$\sqrt{(J_2 \pm M_2)(J_2 \mp M_2 + 1)} \langle J_2 M_2 \mp 1 \alpha_2 | T_M^J | J_1 M_1 \alpha_1 \rangle$$
$$- \sqrt{(J_1 \mp M_1)(J_1 \pm M_1 + 1)} \langle J_2 M_2 \alpha_2 | T_M^J | J_1 M_1 \pm 1 \alpha_1 \rangle$$
$$= \sqrt{(J \mp M)(J \pm M + 1)} \langle J_2 M_2 \alpha_2 | T_{M \pm 1}^J | J_1 M_1 \alpha_1 \rangle . \qquad (5.120)$$

With the changes $J \leftrightarrow J_2$ and $M \leftrightarrow M_2$, equations (5.108) can be rewritten as

$$\sqrt{(J_2 \pm M_2)(J_2 \mp M_2 + 1)} C(J_1 J J_2; M_1 M M_2 \mp 1)$$
$$- \sqrt{(J_1 \mp M_1)(J_1 \pm M_1 + 1)} C(J_1 J J_2; M_1 \pm 1 M M_2)$$
$$= \sqrt{(J \mp M)(J \pm M + 1)} C(J_1 J J_2 2; M_1 M \pm 1 M_2) . \qquad (5.121)$$

As mentioned above, the homogeneous system (5.121) can be used to determine the dependence of the C-G coefficients on M, M_1 and M_2 for fixed J, J_1 and J_2, starting with $C(J_1 J J_2; J_1, J_2 - J_1)$. Analogously, the system of equations (5.120) determines the third-component dependence of the matrix elements of T_M^J starting from $\langle J_2 J_2 \alpha_2 | T_{J_2 - J_1}^J | J_1 J_1 \alpha_1 \rangle$. The two systems of equations being formally analogous, the theorem is proved.

Note that the Wigner-Eckart theorem saves a large calculational effort, since to find the values of the $(2J_2 + 1)(2J + 1)(2J_1 + 1)$ matrix elements considered, it now suffices to use a table of C-G coefficients, calculating only one of the elements directly.

The simplest example of a tensor operator is that corresponding to $J = 0$, that is, an operator A invariant under rotations. Such operators are called *scalars*; an example is the kinetic energy. The tensor operators with $J = 1$ are labelled *vectors*, and these transform according to the defining representation of SO(3). These are customarily designated by boldface letters, as A, B, and the standard components are

$$A_{+1}^1 = -\frac{1}{\sqrt{2}}(A_1 + iA_2) , \quad A_0^1 = A_3 , \quad A_{-1}^1 = \frac{1}{\sqrt{2}}(A_1 - iA_2) . \qquad (5.122)$$

Thus, for example, the angular momenta, the position operator and the momentum operator are all vector operators. The commutation rules for J with the Euclidean components of a vector operator A are

$$[J_i, A_j] = i\hbar\varepsilon_{ijk}A_k ,\tag{5.123}$$

as can be derived from (5.118) and (5.122).

The product of two tensor operators $A^{J_1}_{M_1}$ and $B^{J_2}_{M_2}$ also has a specific transformation law under SO(3), namely a sub-representation of $\mathcal{D}^{(J_1)} \otimes \mathcal{D}^{(J_2)}$, as a consequence of (5.115). This will in general be reducible, and it can be decomposed into a direct sum of irreducible representations contained in the C-G series for $\mathcal{D}^{(J_1)} \otimes \mathcal{D}^{(J_2)}$. This decomposition is once again accomplished by the use of C-G coefficients.

Thus, given A, B, any component A_iB_j can be decomposed in the form

$$A_iB_j = \tfrac{1}{2}(A_iB_j + A_jB_i - \tfrac{2}{3}\delta_{ij}A \cdot B) + \tfrac{1}{2}(A_iB_j - A_jB_i) + \tfrac{1}{3}\delta_{ij}A \cdot B.\tag{5.124}$$

The first term on the right is a component of a second rank, symmetric traceless tensor, i.e., one with five (independent) components. The second term is an antisymmetric tensor of second rank having three components. The last term is a scalar operator, since $A \cdot B$ is a scalar under rotations, and δ_{ij} is invariant. Since $\mathcal{D}^{(1)} \otimes \mathcal{D}^{(1)} = \mathcal{D}^{(2)} \oplus \mathcal{D}^{(1)} \oplus \mathcal{D}^{(0)}$, it is clear that these terms correspond, respectively, to tensors with $J = 2, 1, 0$. The first gives the quadrupolar part of A_iB_j, the second gives the vector part (note that we can associate with it a component of $A \times B$ by using the tensor ε_{ijk}), and the third is the scalar part. It may occur that some of these components vanish (for example, if $A = B$, the vector part does not appear).

5.10 Helicity

For a particle with spin, $\{p, S_z\}$ is a complete set of compatible observables. But such a complete set requires the specification of an arbitrary direction of quantization not intrinsically tied to the particle. If instead of these observables we substitute S_z by

$$\boxed{\Lambda \equiv \frac{S \cdot p}{|p|}} ,\tag{5.125}$$

i.e., if we choose the direction of quantization along the momentum of the particle at each instant of time, we will have the complete set of compatible observables $\{p, \Lambda\}$, this one having an intrinsic meaning. This new observable Λ, the projection of the spin on the direction of motion, is called the *helicity* [JW 59]. The spectrum of Λ, as with any component of spin, is, in units of $\hbar$, $\{-s, -s + 1, \ldots, s\}$. We shall call a generic point of this spectrum "λ".

The associated orthonormal basis $\{|p, \lambda\rangle\}$ is extremely useful in a discussion of collisions, especially for relativistic energies.

To determine this basis relative to the vectors $|p, m\rangle$, where $m\hbar$ is the eigenvalue of S_z, we first observe an important property of Λ: helicity is a scalar operator under rotations. As a consequence, a given state $|p, \lambda\rangle$ can be obtained from the state $|p\hat{e}_3, \lambda\rangle$ by applying a rotation which takes $p\hat{e}_3$ into p. There are an infinite number of rotations which will do this, and their effects on the initial state can be distinguished only in the phase of the final state. We choose the rotation $\mathcal{R}(\phi, \theta, -\phi)$, where (θ, ϕ) are the polar angles of $\hat{p}$. Thus, we define, including its phase (but not the sign $\pm$ in the case of half-integer spins),

$$|p, \lambda\rangle \equiv R(\phi, \theta, -\phi)|p\hat{e}_3, \lambda\rangle . \tag{5.126}$$

When $p = p\hat{e}_3$, then $\Lambda = S_z$, and the state $|p\hat{e}_3, \lambda\rangle$ is part of the basis associated with $\{p, S_z\}$. We can then deduce from (5.27) and (5.28), as expressed in momentum space,

$$|p, \lambda\rangle = \sum_{m'} D^s_{m'\lambda}(\phi, \theta, -\phi)|p, m'\rangle \tag{5.127}$$

and thus we find the relation between the two bases.

Using (B.30) one immediately obtains

$$\langle p', \lambda'|p, \lambda\rangle = \delta(p' - p)\delta_{\lambda'\lambda} . \tag{5.128}$$

In practice, it is very common to use other different bases, such as the one associated with $\{p, J^2, J_z, \Lambda\}$, where p is the observable $|p|$. Denoting one of the states of this basis by $|pJM\lambda\rangle$, with normalization

$$\langle p'J'M'\lambda'|pJM\lambda\rangle = \frac{\delta(p' - p)}{p^2}\delta_{J'J}\delta_{M'M}\delta_{\lambda'\lambda} , \tag{5.129}$$

we can verify that with the definition

$$|pJM\lambda\rangle \equiv \sqrt{\frac{2J + 1}{2\pi}} \int d\Omega \, D^{J*}_{M\lambda}(\phi, \theta, -\phi)|p, \lambda\rangle , \tag{5.130}$$

these states satisfy the required properties:

a) Using the formula

$$\int d\Omega \, D^{J'*}_{M'\lambda'}(\phi, \theta, -\phi)D^J_{M\lambda}(\phi, \theta, -\phi) = \frac{4\pi}{2J + 1}\delta_{J'J}\delta_{M'M}\delta_{\lambda'\lambda} , \tag{5.131}$$

which is a special case of (B.33), and using (5.128), we can easily prove (5.129).

b) Taking into account the closure relation

$$\sum_{J,M} \frac{2J + 1}{4\pi}D^J_{M\lambda}(\phi, \theta, -\phi)D^{J*}_{M\lambda}(\phi', \theta', -\phi') = \delta(\Omega - \Omega') , \tag{5.132}$$

which follows immediately from (B.24), we can invert (5.130), with the result

$$|p, \lambda\rangle = \sum_{JM} \sqrt{\frac{2J+1}{4\pi}} D^J_{M\lambda}(\phi, \theta, -\phi) |pJM\lambda\rangle \,, \tag{5.133}$$

so that $\{|pJM\lambda\rangle\}$ is a complete basis.

c) It remains to be shown that $|pJM\lambda\rangle$ is an eigenstate of J^2, J_z with eigenvalues $J(J+1)\hbar^2$ and $M\hbar$. We begin by giving the transformation law for $|p, \lambda\rangle$ under rotations. Using $\mathcal{R}_p \equiv \mathcal{R}(\phi, \theta, -\phi)$, equation (5.126) becomes

$$R|p\lambda\rangle = D^s_{\lambda\lambda}(\mathcal{R}^{-1}_{\mathcal{R}p} R \mathcal{R}_p)|\mathcal{R}p, \lambda\rangle \,. \tag{5.134}$$

Noting that $\mathcal{R}^{-1}_{\mathcal{R}p} R \mathcal{R}_p$ is a rotation around $\hat{e}_3$, the associated matrix D^s is diagonal. Since for the same rotation around $\hat{e}_3$, $D^J_{\lambda\lambda'} = D^s_{\lambda\lambda'}$, given that the values of L are always integers so that s and J are both integers or both half-integers, and using (5.130), (5.134) and the unitarity of D^J, then we may write

$$
\begin{aligned}
R|pJM\lambda\rangle &= \sqrt{\frac{2J+1}{4\pi}} \int d\Omega_p \sum_{\lambda'} D^J_{\lambda'M}(\mathcal{R}^{-1}_p) D^J_{\lambda\lambda'}(\mathcal{R}^{-1}_{\mathcal{R}p} R \mathcal{R}_p)|\mathcal{R}p, \lambda\rangle \\
&= \sqrt{\frac{2J+1}{4\pi}} \int d\Omega_p \, D^J_{\lambda M}(\mathcal{R}^{-1}_{\mathcal{R}p} R)|\mathcal{R}p, \lambda\rangle \\
&= \sum_{M'} D^J_{M'M}(\mathcal{R}) \sqrt{\frac{2J+1}{4\pi}} \int d\Omega_p \, D^{J*}_{M'\lambda}(\mathcal{R}_{\mathcal{R}p})|\mathcal{R}p, \lambda\rangle \\
&= \sum_{M'} D^J_{M'M}(\mathcal{R})|pJM'\lambda\rangle \,,
\end{aligned} \tag{5.135}
$$

where we have used $d\Omega_p = d\Omega_{\mathcal{R}p}$. The previous relation shows that $|pJM\lambda\rangle$ transforms under rotations as do the states $|JM\rangle$ in (5.87), and thus they have angular momentum J and third component M.

Finally, for particles without spin, we derive from (5.130), specialized to the case $s = 0$, together with (B.32),

$$\boxed{\langle p'|pLM\rangle = \frac{\delta(p'-p)}{p^2} Y^M_L(\hat{p}')} \,, \tag{5.136}$$

a relation which permits us to go immediately from the momentum representation to the one associated with p, L^2, L_z, and vice versa. Using the expansion (A.42) of a plane wave in terms of spherical harmonics, (5.136) leads immediately to

$$\boxed{\langle x|pLM\rangle = \sqrt{\frac{2}{\pi\hbar^3}} i^L j_L(kr) Y^M_L(\hat{x})} \,, \tag{5.137}$$

useful in relating p, L^2, L_z with the position representation.

6. Two-Particle Systems: Central Potentials

6.1 Introduction

In this chapter, we will study some characteristics of the Schrödinger equation for an isolated system of two spinless particles of masses M_1 and M_2, which interact through a potential $V(r_1, r_2)$ dependent only on their positions. The Schrödinger equation for the stationary states of this system is

$$\left[-\frac{\hbar^2}{2M_1} \Delta_1 - \frac{\hbar^2}{2M_2} \Delta_2 + V(r_1, r_2) \right] \Psi(r_1, r_2) = E\Psi(r_1, r_2) . \tag{6.1}$$

It is of interest to introduce the center of mass position operator R and the relative position operator r defined by

$$R = \frac{M_1 r_1 + M_2 r_2}{M_1 + M_2} , \quad r = r_1 - r_2 . \tag{6.2}$$

The canonically conjugate variables are

$$P = p_1 + p_2 , \quad p = \frac{M_2 p_1 - M_1 p_2}{M_1 + M_2} , \tag{6.3}$$

i.e., the total and relative momenta, respectively.

It is easily shown that (6.1) can be rewritten in the form

$$\left[-\frac{\hbar^2}{2M} \Delta_R - \frac{\hbar^2}{2\mu} \Delta_r + \bar{V}(r, R) \right] \phi(r, R) = E\phi(r, R) , \tag{6.4}$$

where $\phi(r, R) \equiv \Psi(r_1, r_2), \bar{V}(r, R) \equiv V(r_1, r_2)$, M is the total mass and μ is the reduced mass of the system:

$$M = M_1 + M_2 , \quad \mu = \frac{M_1 M_2}{M_1 + M_2} . \tag{6.5}$$

For an isolated system, the problem being considered is invariant under translations (and rotations); therefore, the potential $\bar{V}$ cannot depend on R, making equation (6.4) separable. The equation which describes the motion of the center of mass is the Schrödinger equation for a free particle of mass M, and as a consequence, the general solution of (6.4) is a continuous linear superposition of functions of the type

$$\phi(r, R) = (2\pi\hbar)^{-3/2} e^{iP \cdot R/\hbar} \psi(r) . \tag{6.6}$$

In each one of these functions the center of mass has a well-defined momentum P.

The wave function $\psi(r)$ describes the relative motion of the two particles and must satisfy the equation

$$(H_{rel}\psi)(r) \equiv \left[-\frac{\hbar^2}{2\mu}\Delta + V(r)\right]\psi(r) = E_{rel}\psi(r) ,$$ (6.7)

where $V(r) \equiv \bar{V}(r, R)$ and $E_{rel} = E - P^2/2M$. In the discussion that follows, we write $H_{rel} \equiv H$, $E_{rel} \equiv E$ and $U \equiv 2\mu V/\hbar^2$.

The first question that arises concerning (6.7) addresses the self-adjointness of H. To guarantee this property, it is necessary to impose conditions on the potential, so that H can be essentially self-adjoint in some natural domain. Thus, one can demonstrate [FL 74] that if

$$V(r) = V_1(r) + V_2(r) ,$$
$$V_1 \in L^2_{loc}(\mathbb{R}^3) , \quad V_1(r) \geq -ar^2 - b ,$$ (6.8)
$$V_2 \in L^2(\mathbb{R}^3) ,$$

then H has $C_0^\infty(\mathbb{R}^3)$ as its natural domain, being essentially self-adjoint there. Note that each V can display singular behavior at some point, say the origin, where it may behave like $r^{-(3/2-\varepsilon)}, \varepsilon > 0$. The potential family (6.8) can be extended substantially while maintaining the essentially self-adjoint character of H in $C_0^\infty(\mathbb{R}^3)$. See, for example, *Simon* [SI 82] for a general study in any number of dimensions.

The potentials (6.8) contain the so-called Kato class formed by the potentials $V \in L^2 + L^\infty$. If the potential is Kato, it can be proved [KA 80] that, if $H_0 \equiv -\hbar^2\Delta/2\mu$, the domain $D(H) = D(H_0)$ consists of bounded, uniformly continuous, Hölder continuous functions and V is H_0-infinitesimal, namely: for all $\bar{a} > 0$, there exists a $\bar{b}$ such that

$$\|V\psi\| \leq \bar{a}\|H_0\psi\| + \bar{b}\|\psi\| , \quad \forall\psi \in D(H_0) .$$ (6.9)

Furthermore, H is semi-bounded, i.e. bounded from below.

It is clear that if V has a singularity at the origin of the type $r^{-\alpha}, \alpha \geq 3/2$, such a potential is not in the previous classes. However, from a physical viewpoint, the natural limit [SI 71] for singular behavior is r^{-2} and not $r^{-3/2}$, since, if $\alpha < 2$, a simple argument based on the uncertainty principle indicates that such a singularity does not destroy the semi-bounded character of the energy: let us suppose the particle to be localized in a neighborhood Δr around $r = 0$, so that its kinetic energy $\langle H_0 \rangle \sim \hbar^2/2\mu(\Delta r)^2$ and $\langle V \rangle \sim g/\langle\Delta r\rangle^\alpha$, $gr^{-\alpha}$ being the behavior of V at the origin. Therefore, $H_0 + V$ remains semi-bounded as $\Delta r \to 0$ if $\alpha < 2$, but not if $\alpha > 2$, for $g < 0$.

A natural class of potentials which covers the range $3/2 \leq \alpha < 2$, and which leads to self-adjoint operators H with the use of quadratic forms (see Appendix

C) has been widely studied by *Simon* [SI 71]. This class is formed by potentials $V \in R + L^\infty$, where R is the Rollnik family

$$R \equiv \left\{ V_1(\boldsymbol{x}) : \|V_1\|_R^2 \equiv \int d^3x \, d^3y \, \frac{|V_1(\boldsymbol{x})| \, |V_1(\boldsymbol{y})|}{|\boldsymbol{x} - \boldsymbol{y}|^2} < \infty \right\}. \tag{6.10}$$

These potentials V are such that $|V|^{1/2}$ is $H_0^{1/2}$-infinitesimal, and lead to semi-bounded operators H. A special subset of $R + L^\infty$ is $L^{3/2} + L^\infty$, and thus in particular the Kato class.

Acquiring knowledge of the spectral properties of H is an extremely important problem, but very delicate and difficult. Among the properties whose determination is important are: the essential spectrum, presence or absence of the continuous singular spectrum, and characteristics of the point spectrum (in particular, of its discrete part, with emphasis on the number of bound states and properties of their eigenfunctions).

For potentials which decrease to zero at large distances, one intuitively expects that the continuous spectrum should consist of all the positive energies, and if the potential increases indefinitely as $|\boldsymbol{x}| \to \infty$, the whole spectrum should be discrete.

The first of these intuitions is correct under very general conditions if, instead of the continuous spectrum, we refer to the essential spectrum. Concretely, whenever $V \in R + L^\infty$, the essential spectrum is always $\sigma_{\text{ess}}(H) = [0, \infty)$ [SI 71]. In particular, it suffices that $V = V_1 + V_2$, with $V_1 \in L^2, V_2 \in L_{\text{loc}}^2$ and $V_2(\boldsymbol{x}) \to 0$ as $|\boldsymbol{x}| \to \infty$. This is the case for the Coulomb potential, $V = \alpha/r$. In this kind of situation, which arises frequently in applications, the spectrum of H on the negative real axis consists of a discrete set of eigenvalues with finite multiplicity, with at most one accumulation point at the origin. The fact that $\sigma_{\text{ess}}(H) = [0, \infty)$ does not exclude the possibility of eigenvalues with positive energy, i.e., that $\sigma_{\text{p}}(H) \cap (0, \infty) \neq \emptyset$. There is the famous *von Neumann–Wigner* example [NW 29], which we will discuss shortly, where $V(\boldsymbol{r}) \to 0$ when $r \to \infty$, and which leads nevertheless to an H possessing a positive eigenvalue, contrary to naive physical intuition. Conditions on the potential which eliminate the possibility of eigenstates with positive eigenvalues will be discussed later.

With regard to the absence of an essential spectrum, our physical intuition proves to be correct for a large number of families of potentials. It suffices, for example, that $V = V_1 + V_2$, with $V_1 \geq 0$, $V_1(\boldsymbol{x}) \to \infty$ as $|\boldsymbol{x}| \to \infty$, $V_1 \in L_{\text{loc}}^1$ and $V_2 \in L^{3/2} + L^\infty$, to ensure that $\sigma_{\text{ess}}(H) = \emptyset$ [RS 78]. Furthermore, in this case H is semi-bounded and therefore the bound levels form a sequence which increases monotonically to ∞.

The rather "exotic" properties of states associated with the continuous singular spectrum make it desirable to know some conditions which ensure that $\sigma_{\text{s.c.}}(H) = \emptyset$. An important theorem by *Agmon-Kato-Kuroda* [RS 79] shows that $\sigma_{\text{s.c.}}(H) = \emptyset$ and $\sigma_{\text{a.c.}}(H) = [0, \infty)$, provided $(1 + |\boldsymbol{x}|^2)^{1/2+\varepsilon} V \in L^2 + L^\infty$ for some $\varepsilon > 0$. The Coulomb potential is not a member of this large family, but it can be proved that these results apply to very general kinds of potentials, essentially those which differ from Coulomb ones by potentials of shorter range.

Let us proceed now to the point spectrum. As mentioned before, positive eigenvalues may exist even if the potential tends to zero for large distances. Nevertheless it is known [SI 72] that H has no positive eigenvalues if $V \in L^2 + L_\varepsilon^\infty$, $V \in C^2$ except for a finite number of points, and $V = V_1 + V_2$, where $rV_1(r) \to 0$, $V_2(r) \to 0$, and $r \cdot \nabla V_2(r) \to 0$, as $r \to \infty$. Another interesting class of potentials for which also $\sigma_p(H) \cap (0, \infty) = \emptyset$ is *Ikebe*'s [IK 60], consisting of potentials V which are locally Hölder continuous apart from a finite number of singularities, and such that $V \in L^2$ and $V(r) = O(|r|^{-2-\varepsilon})$ as $r \to \infty$ for some $\varepsilon > 0$. These conditions are rather strong, and they guarantee, in addition, that $\sigma_{\text{a.c.}}(H) = [0, \infty)$, $\sigma_{\text{s.c.}}(H) = \emptyset$ and $\sigma_p(H) \cap (-\infty, 0)$ consist, at most, of a finite number of eigenvalues of finite multiplicity.

In the typical case of potentials for which H is semi-bounded and $\sigma_{\text{ess}}(H) = [0, \infty)$, it turns out to be very useful to know lower bounds on the energies and upper bounds on the number of bound states, multiplicities included. Defining $2\mu = \hbar = 1$, to simplify the expression that follows, it can be proven [GM 76], [TH 81] that

$$\inf \sigma(H_0 + V) \geq -c_p \|V_-\|_p^{2p/(2p-3)} , \quad p > 3/2 , \tag{6.11}$$

where $V_- \equiv \min(V, 0)$, and

$$c_p \equiv (4\pi)^{-2/(2p-3)} \left(1 - \frac{1}{p}\right)^2 \left[\Gamma\left(\frac{2p-3}{p-1}\right)\right]^{(2p-2)/(2p-3)} . \tag{6.12}$$

We will denote by $N(V)$ the number of bound levels with negative energy, counting each of them as many times as their multiplicities indicate, that is, $N(V) \equiv \text{rank } E_H((-\infty, 0))$, where $E_H(.)$ are the spectral projections of $H = H_0 + V$. For one-dimensional potentials, we have already stated that the finiteness of $N(V)$ depends on the behavior of V at large distances. The same thing occurs in three dimensions, as suggested again by the uncertainty principle: Suppose that V behaves as g/r^α when $r \to \infty$. If the particle is localized in a spherical shell $R \leq r \leq 2R$, and R is sufficiently large, its kinetic energy will be $\langle H_0 \rangle \sim \hbar^2/2\mu R^2$, and the potential energy $\langle V \rangle \sim g/R^\alpha$. Therefore, if $g < 0$ and $\alpha < 2$, the negative potential energy will dominate over the kinetic energy at large R, and thus the average value of H will be negative on a set of states of infinite linear dimension. This suggests that $N(V)$, under these circumstances, is infinite. This heuristic argument is easily made rigorous, using the Min-Max principle [SI 70]: If $V \in L^2 + L_\varepsilon^\infty$, and $V(r) < -gr^{-\alpha}$, $g > 0$, $\alpha < 2$, $\forall r \geq R_0$, then $N(V) = \infty$ (thus, for example, in the attractive Coulomb potential case); but if $V(r) > -gr^{-2}$, $\forall r \geq R_0$, then

(i) $g < \frac{1}{4}\hbar^2/2\mu \Rightarrow N(V) < \infty$, and

(ii) $g > \frac{1}{4}\hbar^2/2\mu \Rightarrow N(V) = \infty$.

Among the large variety of upper bounds to $N(V)$, we will discuss a few of special relevance. First, for its historical interest, is the so-called *Birman-Schwinger* bound [BI 61], [SC 61]:

$$N(V) \leq (4\pi)^{-2} \|U_-\|_R^2 , \tag{6.13}$$

also valid as a bound to rank $E_H((-\infty, 0])$. This bound is complicated to calculate, and does not give the correct asymptotic behavior when the potential strength increases indefinitely. Semi-classical arguments similar to the ones presented in Chap. 4 suggest that

$$N(\lambda V) \underset{\lambda \to \infty}{\sim} \frac{1}{(2\pi)^3} \int d^3x \, d^3k \, \theta(-k^2 - \lambda U(\boldsymbol{x}))$$

$$= \frac{1}{6\pi^2} \lambda^{3/2} \|U_-\|_{3/2}^{3/2} . \tag{6.14}$$

This expression can be proved rigorously for potentials $V \in L^{3/2}$. It is thus important to obtain bounds on $N(V)$ compatible with this asymptotic behavior. The closest to (6.14) is the elegant *Cwikel-Lieb-Rosenbljum* bound [CW 77], [LI 76], [RO 72]. With $N_\alpha(V) \equiv \operatorname{rank} E_H((-\infty, \alpha])$, $\alpha \leq 0$, one has

$$N_\alpha(V) \leq L_3 \|(U - \bar\alpha)_-\|_{3/2}^{3/2} , \quad \bar\alpha \equiv 2\mu\alpha/\hbar^2 ; \tag{6.15}$$

where $0.1156 \geq L_3 \geq 4/3^{3/2}\pi^2$. Another simple bound for $N(V)$ with the correct asymptotic behavior is that of *Martin* [MA 75]:

$$N(V) \leq (2\pi)^{-1} \|U_-\|_1^{1/2} \|U_-\|_2 . \tag{6.16}$$

For more details on bounds consult [RS 78].

Because of the complexities of calculating the eigenfunctions (except for spherically symmetric potentials), it becomes especially important to be able to derive some of their general properties. We have already mentioned (Sect. 3.11) that if V is such that $V_+ \in L_{loc}^1$ and $V_- \in L^{3/2} + L^\infty$, and the ground state energy $E_0 \equiv \inf \sigma(H)$ belongs to $\sigma_p(H)$, then this level is not degenerate and the associated wave function is, up to a phase, strictly positive at all points. On the other hand, referring to the local properties and behavior at infinity, one can demonstrate the following [SI 82]: If $H\psi = E\psi$, $\psi \in L^2$, then

(i) $V_+ \in L_{loc}^p, V_- \in L^p + L^\infty$, for some $p > 3/2 \Rightarrow \psi$ is continuous almost everywhere and tends to 0 as $|\boldsymbol{x}| \to \infty$.

(ii) If in addition to (i), $V_+ \to \infty$ as $|\boldsymbol{x}| \to \infty$, then for all $a > 0$, there exists some c such that $|\psi(\boldsymbol{x})| \leq c \exp(-a|\boldsymbol{x}|)$, for all $\boldsymbol{x}$.

(iii) $V \in L^p + L_\varepsilon^\infty$, for some $p > 3/2$ and $E < 0 \Rightarrow$ given any $a < |E|^{1/2}$, there exists a constant c such that $|\psi(\boldsymbol{x})| \leq c \exp(-a|\boldsymbol{x}|)$, for all $\boldsymbol{x}$.

Properties (ii) and (iii), while very complicated to prove, are not surprising in view of analogous results for one-dimensional problems.

6.2 The Radial Equation

Given the rotational invariance of an isolated system, the potential $V(\boldsymbol{r})$ is spherically symmetric, and we will represent it by $V(r)$. Introducing polar coordinates, equation (6.7) becomes

$$\left[\frac{p_r^2}{2\mu} + \frac{L^2}{2\mu r^2} + V(r) \right] \psi(\boldsymbol{r}) = E\psi(\boldsymbol{r}) , \tag{6.17}$$

where L is the (relative) orbital angular momentum (5.48), and p_r is the so-called radial momentum operator

$$p_r \equiv -i\hbar \left[\frac{\partial}{\partial r} + \frac{1}{r} \right] = -i\hbar \frac{1}{r} \frac{\partial}{\partial r} r . \tag{6.18}$$

This operator has as its natural domain $C_0^\infty(\mathbf{R}^3 - \{0\})$, and in this domain it is symmetric. However, it lacks self-adjoint extensions, so that it can never be considered as an observable: The equation $p_r^\dagger \psi = \pm(i\hbar/r_0)\psi$, providing the deficiency indices, is solved by looking for functions $\psi \in L^2$, such that the distribution $(-i\hbar/r)(\partial(r\psi)/\partial r)$ in $\mathbf{R}^3 - \{0\}$ is in L^2 and satisfies the differential equation

$$-i\frac{1}{r}\frac{\partial}{\partial r} r\psi = \pm\frac{i}{r_0}\psi , \tag{6.19}$$

whose solutions are of the form $(1/r)\exp(\pm r/r_0)F(\theta, \phi)$, where $F(\theta, \phi)$ is arbitrary. Therefore, the deficiency indices of p_r are $(0, \aleph_0)$.

It is evident that

$$[H, L] = 0 , \tag{6.20}$$

and therefore, the search for stationary states is reduced to those which, in addition to having well-defined energies, are eigenstates of L^2 and L_z. In accordance with this, the most general solution of (6.17) is a linear combination of solutions of the form

$$\psi_{lm}(\boldsymbol{r}) = R_l(r)Y_l^m(\theta, \phi) , \tag{6.21}$$

where $R_l(r)$, the *radial wave function*, is determined by the radial equation

$$\boxed{\left[\frac{p_r^2}{2\mu} + \frac{l(l+1)\hbar^2}{2\mu r^2} + V(r) \right] R_l(r) = ER_l(r)} . \tag{6.22}$$

The first term of H represents the *radial kinetic energy*, and the second is the *rotational kinetic energy*, which acts as a repulsive potential, added to $V(r)$ (the *centrifugal barrier*).

The decomposition of the wave function ψ in terms of eigenfunctions ψ_{lm} of L^2 and L_3 is the so-called *partial wave decomposition*. When $l = 0, 1, 2, 3, \ldots$

we speak of $s, p, d, f, \ldots$-waves following the spectroscopic tradition. Obviously the s-waves have spherical symmetry.

From the discussion of p_r, one could question also whether $p_r^2/2\mu$ is an observable. However, it is a real operator, and therefore it has self-adjoint extensions. It is precisely its Friedrichs extension (Appendix C) which is adopted as the observable, so that it coincides with the kinetic energy operator $p^2/2\mu$ for s-waves, that is, those with $l = 0$.

To simplify the notation, we introduce the variables $\varepsilon \equiv 2\mu E/\hbar^2$ and $U(r) \equiv 2\mu V(r)/\hbar^2$, which we continue to call the energy and the potential energy, respectively. With this in mind, equation (6.22) is written in the following manner:

$$\frac{d^2 R_l}{dr^2} + \frac{2}{r}\frac{dR_l}{dr} + \left[\varepsilon - U(r) - \frac{l(l+1)}{r^2}\right] R_l = 0 . \tag{6.23}$$

Given the form of p_r, it proves convenient to make the following substitution:

$$\boxed{R_l(r) \equiv \frac{1}{r} u_l(r)} , \tag{6.24}$$

whence $u_l(r)$, the *reduced radial function*, satisfies the *reduced radial equation* for the partial wave l:

$$\boxed{H_l u_L \equiv -u_l'' + \left[\frac{l(l+1)}{r^2} + U(r)\right] u_l = \varepsilon u_l} . \tag{6.25}$$

The normalization condition for ψ_{lm} is expressed as

$$\int_0^\infty dr\, |u_l(r)|^2 = 1 . \tag{6.26}$$

The projection of H onto the $\mathcal{H}^{(l)}$-spaces of angular momentum l has reduced the original eigenvalue problem to a collection of analogous problems for H_l, $l = 0, 1, 2, \ldots$. The first question that arises is: Under what conditions are these H_l self-adjoint in $L^2(0, \infty)$? Second, under what circumstances can we guarantee that $H = \oplus_0^\infty \tilde{H}_l$? Here $\tilde{H}_l \simeq H_l \otimes I_l$, I_l being the identity operator on the angular variables in $\mathcal{H}^{(l)}$.

It can be shown, using the decomposition theorem of a differential operator and Levinson's theorem [NA 68], that if $U(r) \in L^2_{\text{loc}}(0, \infty)$, $U(r) \geq -ar^2 - b$, $r \geq R_0$, and $r^2 U(r) \geq -[(l+1/2)^2 - 1]$ in a neighborhood of the origin, then H_l is essentially self-adjoint in $C_0^\infty(0, \infty)$. However, even for potentials which are well-behaved at the origin, the Hamiltonian $H_{l=0}$ for the s-wave sector is not essentially self-adjoint, having deficiency indices $(1, 1)$, so that for these waves it is necessary to impose a boundary condition at the origin. These conditions are always chosen to be of the type $u_0(0) = 0$, that is, the type associated with the Friedrichs extension. The reason is not arbitrary: if, for example, V is of

Kato class, the wave functions of the domain of H are continuous, and in order for $H = \oplus_0^\infty \tilde{H}_l$, it is necessary that the functions $u_l(r)/r$ be continuous, which forces $u_l(0) = 0$ for all $u_l \in D(H_l)$.

A simple argument leading to this condition, called the *regularity condition*, for the functions u_l is as follows: Suppose that $U(r) \sim \lambda r^{-s}$, $s < 2$, as $r \to 0$. Thus, in a neighborhood of the origin, the dominant part of $U(r) + l(l+1)/r^2$ is the centrifugal term, and in such a region,

$$u_l''(r) \simeq \frac{l(l+1)}{r^2} u_l(r) , \qquad (6.27)$$

and hence (6.25) has two linearly independent solutions, with asymptotic behavior, for $r \to 0$, of the form

$$u_l(r) \sim r^{l+1} , \quad u_l(r) \sim r^{-l} . \qquad (6.28)$$

If $l \neq 0$, the second solution violates the integrability condition (6.26), leaving only the first solution as a possible solution to the eigenvalue problem. This solution obviously satisfies the regularity condition. If $l = 0$, this argument is no longer applicable; but $u_0(r) \sim 1$ leads to $R_0(r) \sim r^{-1}$, which, following application of $-\Delta$, produces a $\delta(r)$, which is impossible to cancel with any other terms of the Schrödinger equation. Again, only the regular solution remains.

The operational equality $H = \oplus_0^\infty \tilde{H}_l$, in which $H_{l=0}$ has been extended with the boundary condition $u_0(0) = 0$, can be proved under very general conditions for $U(r)$; it suffices, for example, to restrict oneself to potentials in the class (6.8). By means of quadratic forms, one can extend this operational equality to central potentials $U = U_1 + U_2 + U_3$, that, as functions in $\mathbf{R}^3$, satisfy

   (i) $U_j \in L^1_{loc}(\mathbf{R}^3 - \{0\})$, $j = 1, 2, 3$;
   (ii) $U_1(\cdot) \geq 0$;
   (iii) $U_2 \in L^p(\mathbf{R}^3) + L^\infty(\mathbf{R}^3)$,   for some $p > 3/2$;
   (iv) $|U_3(r)| \leq \alpha/4r^2$,   for some $\alpha < 1$.

$$(6.29)$$

Condition (iii) guarantees that, as a function of r,

$$\int_0^1 dr\, r|U_2(r)| < \infty , \quad \sup_{r_0 \geq 2} \int_{|r - r_0| \leq 1} dr\, |U_2(r)| < \infty . \qquad (6.30)$$

Later on, we will discuss the restriction $\alpha < 1$ for (iv).

The conditions (i–iv) ensure that H_l, $\forall l \geq 0$, can be defined through the closure of the form

$$\langle H_l \rangle_\varphi \equiv \int_0^\infty dr\, |\varphi'(r)|^2 + \int_0^\infty dr\, U(r)|\varphi(r)|^2 , \quad \varphi \in C_0^\infty(0, \infty) , \qquad (6.31)$$

and that its domain $D(H_l)$ consists of those reduced radial functions satisfying

1) u, u' absolutely continuous in $(0, \infty)$ and $\in L^2(0, \infty)$

2) $u(0) = 0$

3) $[U_1 + l(l+1)/r^2]^{1/2}u \in L^2(0, \infty)$ (6.32)

4) $H_l u \equiv -u'' + [U + l(l+1)/r^2]u \in L^2(0, \infty)$

(see, for example, [KA 80], [RS 75], [SC 81]).

This result deserves additional comments. Since it is not required that $U \in L^2_{loc}$, the natural domain of H_l does not necessarily contain $C_0^\infty(0, \infty)$, although it is dense, as it consists of all those u functions for which u and u' are absolutely continuous, of compact support in $(0, \infty)$ and $H_l u \in L^2$. In such a domain, although H_l is real, it is not necessarily essentially self-adjoint, and the extension exhibited in (6.32) is the most natural from a physical view-point: in physical states, for which the energy operator has a finite average value, the kinetic energy (plus the potential energies U_2 and U_3, for which average values exist when the kinetic energy's does) must also have a finite average value, as should the repulsive energy $U_1 + l(l+1)/r^2$. In other words, the quadratic form associated with H_l should be the closure of the corresponding form in its natural domain.

As for the explanation of condition $\alpha < 1$, we will consider the potential $U(r) = g/r^2$. If $g > -1/4$, conditions (6.29) are satisfied. Equation (6.25) has two linearly independent solutions, whose behavior at the origin for s-waves is

$$u_1(r) \sim r^{1/2+(g+1/4)^{1/2}}, \qquad u_2(r) \sim r^{1/2-(g+1/4)^{1/2}}, \qquad (6.33)$$

for $g \neq -1/4$, while $u_1(r) \sim r^{1/2}$ and $u_2(r) \sim r^{1/2}\ln r$, if $g = -1/4$. If $g \geq 3/4$, only u_1 is square integrable near the origin. If $-1/4 < g < 3/4$, both are square integrable, but only u_1' is square integrable near the origin. If $g \leq -1/4$, the two solutions fail to have square integrable derivatives near the origin; and (6.29) is not satisfied.

Since the functions u in the domain of the energy operator are to be constructed by a linear superposition of the solutions of (6.25) for different energies, the basis functions are uniquely determined only if $g \geq 3/4$, and this means that $H_{l=0}$ is essentially self-adjoint in C_0^∞. Otherwise, if $-1/4 < g < 3/4$, both u_1 and u_2 can enter into the linear combinations, and only the requirement that the expectation value of the kinetic energy be finite on u fixes the basis functions uniquely, eliminating u_2. Finally, if $g \leq -1/4$, not even the above requirement can be satisfied. It is easily demonstrated that if $g < -1/4$, whichever self-adjoint extension is chosen, it will not be bounded from below: Let $u(r) \in C_0^\infty(0, \infty)$ and be normalized, and consider the function $u_\lambda(r) \equiv \lambda^{1/2}u(\lambda r)$, $\lambda > 0$, also of unit norm. A simple calculation indicates that

$$\langle u_\lambda, H_{l=0}u_\lambda \rangle = \lambda^2 \left[\langle u, -u'' \rangle + g \left\langle u, \frac{1}{r^2}u \right\rangle \right]. \qquad (6.34)$$

On the other hand, starting from the equality

$$(|u|^2/r)' = -|u|^2/r^2 + 2r^{-1}\mathrm{Re}(u^*u'),$$

it suffices to integrate and apply the Schwarz inequality to conclude that, if $u \in C_0^\infty(0, \infty)$, then

$$\left\langle u, \frac{1}{r^2}u \right\rangle \leq 4\langle u, -u'' \rangle , \qquad (6.35)$$

which can be arbitrarily close to being an equality in C_0^∞. Therefore, if $g < -1/4$, there exists a function u such that the r.h.s. of (6.34) is negative; and as a consequence, letting $\lambda \to \infty$, it is clear that the energies are not bounded from below.

Nelson [NE 64] analyzed the time evolution with the potential $U(r) = g/r^2$, by using Feynman's path integral technique, and succeeded in proving that if $g > -l(l+1) - 1/4$, then such time evolution is governed by a unitary group, whose generator is H_l, with continuous spectrum $[0, \infty)$; if $g = -l(l+1) - 1/4$, the time evolution proceeds via unitary operators, and its generator is a semi-bounded extension of the restriction of H_l to $C_0^\infty(0, \infty)$, but with functions in its domain of definition for which the kinetic (potential) energy has a mean value of $\infty (-\infty)$; finally, if $g < -l(l+1) - 1/4$, time evolution is described by a contractive semi-group (that is, the norm of the function can decrease with time), and its generator is not self-adjoint. This last possibility, called the "Nelson phenomenon", is connected with what occurs classically, under the potential $V(r) = \bar{g}/r^2$, when $\bar{g} < -J^2/2\mu$, where J is the angular momentum of the trajectory. Under these circumstances, the particle falls to the center of the potential in a finite time. This connection between the dissipation of quantum mechanical probability and the center acting classically as a sink shows that Feynman's method can provide interesting information in physical problems which are ambiguous by using the operator methods. We should stress, however, that there exist boundary conditions which lead to unitary evolution even if $g < -1/4$, the Nelson contractive evolution being a certain average of unitary evolutions [RA 75].

For potentials more singular than $1/r^2$ at the origin, not only is the energy not semi-bounded in the attractive potential case [it suffices to apply a reasoning similar to that used in (6.34)], but also new types of problems arise. If the potential is of the form g/r^α, $\alpha \geq 4$, $g < 0$, one can prove [SP 67], [FL 71] that no solution of the radial equation is a solution to the three-dimensional Schrödinger equation.

Returning to the reduced radial equation (6.25), the problem of finding the relative bound states of an isolated system of two particles has been translated to finding solutions of (6.25) which satisfy

$$\int_0^\infty dr \, |u_l(r)|^2 < \infty \quad \textit{(integrability} \text{ condition)} ,$$

$$u_l(0) = 0 \quad \textit{(regularity} \text{ condition)} , \qquad (6.36)$$

or equivalently, to finding the bound states in a one-dimensional potential well:

$$U_{\text{eff}}(r) = \begin{cases} \infty \, , & r < 0 \, , \\ U_l(r) = U(r) + l(l+1)/r^2 \, , & r > 0 \, . \end{cases} \qquad (6.37)$$

The graphical representation for a potential well of this type is given in Fig. 6.1.

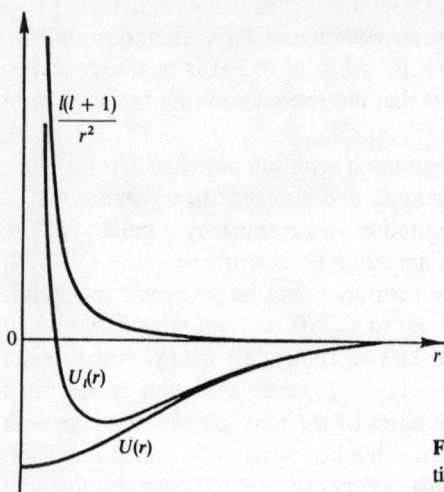

$\dfrac{l(l+1)}{r^2}$

$U_l(r)$

$U(r)$

Fig. 6.1. Representation of the centrifugal potential, the potential $U(r)$ and the effective potential $U_l(r)$

The spectral analysis already done for the one-dimensional potentials in **R** is transferred, with a few modificiations, to the present case. For our purposes, it is sufficient to note that under the conditions [WE 67a], [DS 64]

$$U_l \in L^1_{\text{loc}}(0, \infty), U_l = \tilde{U}_1 + \tilde{U}_2 \, ,$$

$$\tilde{U}_1 \text{ of bounded variation in } (R_0, \infty); \ \tilde{U}_1 \to 0, \ r \to \infty \, ,$$

$$\tilde{U}_2 \in L^1(R_0, \infty) \, , \qquad (6.38)$$

$$U_l(r) \geq -\frac{g}{r^2}, \ g < \frac{1}{4}, \ r \to 0 \, ,$$

one can show that $[0, \infty) \supset \sigma_c(H_l) \supset (0, \infty)$; $\sigma_{\text{a.c.}}(H_l) = [0, \infty)$ and is simple.

With respect to the point spectrum, if $U_l \in L^1_{\text{loc}}(0, \infty)$ satisfies $U_l(r) \geq -g/r^2$, $(g < 1/4)$, as $r \to 0$, and futhermore, $U_l(r) \to \infty$ if $r \to \infty$, then $\sigma(H_l) = \sigma_p(H_l)$, and is discrete. In this case, as in the preceding one, the spectrum is bounded from below, and is also simple for the bound states.

Sturm's theorem remains valid, and allows ordering of the bound states using the number of nodes in $(0, \infty)$.

For a given l value of the orbital angular momentum, it is customary to denote by ε_{nl} and $u_{nl}(r)$ the possible eigenvalues and eigenfunctions, where $(n - 1)$ is the number of nodes in $(0, \infty)$. It is clear that for H, ε_{nl} has at least a degeneracy of $(2l + 1)$ corresponding to the possible different values of the magnetic quantum number m, and that if $n_2 > n_1$, then $\varepsilon_{n_2 l} > \varepsilon_{n_1 l}$.

On the other hand, if one considers the ε_{nl_1} and ε_{nl_2} levels where $l_1 < l_2$, then $u_{nl_1}(r)$ and $u_{nl_2}(r)$ have the same number of nodes in the interval $[0, \infty]$. It is therefore true that either all the nodes of $u_{nl_1}(r)$ coincide with those of $u_{nl_2}(r)$, or there are two adjacent nodes in $u_{nl_2}(r)$, r_1 and r_2, such that $u_{nl_1}(r)$ is non-zero in (r_1, r_2), vanishing at most at one of the endpoints r_1 or r_2. One can always choose the phases in such a way that $u'_{nl_2}(r_1) > 0$, $u'_{nl_2}(r_2) < 0$ and $u_{nl_1}(r) > 0$ in (r_1, r_2). In the second alternative, and for continuous potentials $U(r)$, integrating by parts one obtains

$$\int_{r_1}^{r_2} dr \, [u_{nl_1}(r)u''_{nl_2}(r) - u_{nl_2}(r)u''_{nl_1}(r)] < 0 , \qquad (6.39)$$

and therefore, there exists at least one point $r_0 \in (r_1, r_2)$ such that

$$\frac{u''_{nl_2}(r_0)}{u_{nl_2}(r_0)} < \frac{u''_{nl_1}(r_0)}{u_{nl_1}(r_0)} . \qquad (6.40)$$

Furthermore, from (6.25), one deduces that for all values of r

$$\frac{u''_{nl_2}(r)}{u_{nl_2}(r)} - \frac{u''_{nl_1}(r)}{u_{nl_1}(r)} = \frac{1}{r^2}[l_2(l_2 + 1) - l_1(l_1 + 1)] + (\varepsilon_{nl_1} - \varepsilon_{nl_2}) . \qquad (6.41)$$

Relations (6.40) and (6.41) imply that $\varepsilon_{nl_1} < \varepsilon_{nl_2}$. If, on the contrary, the nodes of $u_{nl_1}(r)$ and $u_{nl_2}(r)$ all coincide and r_1 and r_2 are two adjacent nodes, the integral that appears in (6.39) is zero, and the integrand should at least vanish at some point r_0. Bearing in mind (6.41), one also deduces that in this case $\varepsilon_{nl_1} < \varepsilon_{nl_2}$; and we can, therefore, assert that for radial functions with the same number of nodes, the state corresponding to a smaller angular momentum is always the deepest bound state. In particular, the ground state, should it exist, must have $l = 0$, and therefore, it is spherically symmetric. On the other hand, one can also conclude that the number of bound states with angular momentum $l+1$ is never greater than the number of bound states with angular momentum l.

The same argument allows us to compare the levels ε_{nl}, $\bar{\varepsilon}_{nl}$ corresponding to two continuous potentials $U(r)$ and $\bar{U}(r)$, such that $U(r) \geq \bar{U}(r)$. For this, it suffices to follow the previous reasoning, substituting $\bar{u}_{nl}$ for u_{nl_1} and u_{nl} for u_{nl_2}. Thus, one obtains

$$(\varepsilon_{nl} - \bar{\varepsilon}_{nl}) \int_{r_1}^{r_2} dr \, \bar{u}_{nl}(r)u_{nl}(r) \geq \int_{r_1}^{r_2} dr(U(r) - \bar{U}(r))\bar{u}_{nl}(r)u_{nl}(r) , \quad (6.42)$$

and therefore, $\varepsilon_{nl} \geq \bar{\varepsilon}_{nl}$. In particular, this implies that if we retain only the attractive part of a central potential, the number of bound states for an l wave does not decrease.

The Min-Max principle (Appendix C) allows us to generalize the preceding results: If $U(r) \geq \bar{U}(r)$, the total number of bound states of $U(r)$ with negative energy, including multiplicities, is always less than or equal to that of $\bar{U}(r)$. This is true under very general conditions: it suffices, for example, that $U(r)$ and $\bar{U}(r)$ belong to $L^2(\mathbf{R}^3) + L_\varepsilon^\infty(\mathbf{R}^3)$. These conditions guarantee, as we stated before, that the domains of H and $\bar{H}$ coincide and that their negative spectra consist of isolated points of finite multiplicities.

We have just proved the ordering $\varepsilon_{n-1,l} < \varepsilon_{n,l} < \varepsilon_{n,l+1}$ for very general central potentials. But from this we cannot conclude anything about the ordering of the ε_{nl} levels with constant $n + l$. It is not only pure mathematical interest, but also the physics, which raise the question on what conditions guarantee that $\varepsilon_{nl} > \varepsilon_{n-1,l+1}$ for all n, l (or its opposite). For example, the potential $U(r) = -a/r + br$, $a, b > 0$, appears in the physics of heavy quark-antiquark $(q\bar{q})$ bound systems (charmonium, etc.), and there exists ample numerical evidence for the previous ordering, so that this potential implies the experimentally well-established existence of $q\bar{q}$ bound p-wave states interspersed among other s-wave states. *Martin* and his collaborators [BG 84] have succeeded in proving the following important result: If V is radial, then

$$\Delta V \gtrless 0 , \quad \forall r > 0 \Rightarrow \varepsilon_{nl} \gtrless \varepsilon_{n-1,l+1} , \quad \forall n, l . \tag{6.43}$$

The following inequality is crucial to this proof:

$$-\frac{d^2}{dr^2} \ln u_{1l}(r) \gtrless \frac{l+1}{r^2} \quad \text{if} \quad \Delta V \gtrless 0 , \tag{6.44}$$

which also serves [CO 85] to obtain the relations valid when $\Delta V > 0$:

$$[\langle r \rangle^2_{1l} \le] \langle r^2 \rangle_{1l} \le \frac{2l+4}{2l+3} \langle r_{1l} \rangle^2$$

$$\int_0^\infty dr \, |u'_{1l}(r)|^2 \ge \frac{l+1}{2} \langle r^{-2} \rangle_{1l} . \tag{6.45}$$

Concerning the bounds, it is clear that the results of Sect. 6.1 are also valid for central potentials. But for these potentials, there exist other bounds which exploit their spherical symmetry, and which we now present [SI 76]. Let $N_l \equiv$ rank $E_{H_l}((-\infty, 0))$, i.e. the number of negative eigenvalues of H_l. Recall that these eigenvalues are simple for H_l but have multiplicity at least $2l + 1$ for H. It is also known that N_l coincides with the number of nodes in $(0, \infty)$ of the regular solution to (6.25) for $\varepsilon = 0$ [DS 64]. *Bargmann* [BA 52] proved that if the potential $U(r)$ is such that

$$\int_0^\infty dr \, r|U(r)| < \infty , \quad \int_0^\infty dr \, r^2|U(r)| < \infty , \tag{6.46}$$

then

$$\boxed{N_l \le \frac{1}{2l+1} \int_0^\infty dr \, r|U_-(r)|} , \tag{6.47}$$

where this bound is optimal in the sense that for a given l one can construct potentials for which the integral in (6.47) is arbitrarily close to $(2l+1)N_l$. These potentials, which saturate the Bargmann inequality, have the following form: they consist of N_l very deep and narrow potential wells, which are widely separated, and which degenerate into a set of delta-like potentials in the limit of complete saturation of the inequality.

From the bound in (6.47), one deduces that, given a potential $U(r)$, there always exists a value l_{max} of angular momentum, defined as the greatest integer satisfying

$$2l_{max} + 1 \leq \int_0^\infty dr \ r|U_-(r)| \ , \tag{6.48}$$

and such that for $l > l_{max}$, no bound states exist. Hence, it immediately results that the number of bound states for the potentials which satisfy (6.46), is always finite.

It should be noted that the Bargmann bound requires for its applicability that $U(r)$ be less singular at the origin than $1/r^2$ and decreases faster than $1/r^3$ as $r \to \infty$. Although attractive potentials more singular than $1/r^2$ at the origin have, as we have already stated, energy spectra extending to $-\infty$ and are thus of doubtful physical interest, the condition at large r can be too restrictive. On the other hand, it shares with the Birman-Schwinger bound the inconvenience of scaling incorrectly with the strength of the potential. This is also the case for the family of bounds [GM 76]:

$$N_l \leq (2l+1)^{-(2p-1)} d_p \int_0^\infty dr \ r^{2p-1}|U_-(r)|^p \ , \quad 1 < p < \infty \ ,$$
$$d_p \equiv \frac{(p-1)^{p-1}\Gamma(2p)}{p^p[\Gamma(p)]^2} \ , \tag{6.49}$$

which for $p \downarrow 1$ tend to the Bargmann bound.

A very simple bound, valid for negative central potentials, monotonically increasing with r, is due to *Calogero* [CA 67]:

$$N_l \leq \frac{2}{\pi} \int_0^\infty dr \ |U(r)|^{1/2} \ , \quad l = 0, 1, 2, \dots \ , \tag{6.50}$$

with correct asymptotic behavior for strong potentials.

An approximate formula, easily derivable from the W.B.K. method, and giving usually good results in practical cases, is the following: If $U(r)$ is such that $U(r) + (l + 1/2)^2/r^2$ is monotone or has, at most, one minimum in $r \in (0, \infty)$, and $\lim_{r\to\infty} U(r) = 0$, then

$$N_l \simeq \frac{1}{2} + \frac{1}{\pi} \int dr \left[-U(r) - \frac{(l+1/2)^2}{r^2} \right]^{1/2} \ , \tag{6.51}$$

where the integration is over the region where the argument of the square root is positive.

To finish this discussion on the general features of the reduced radial equation, we will present the example due to *von Neumann-Wigner* [NW 29] already mentioned. It is a very regular central potential, but decreases to zero at infinity only as $1/r$, in an oscillatory manner. These oscillations are such that they prevent, through interference effects, the escape of the particle with a certain positive energy from the attraction of the potential. Concretely, the potential $U(r)$ is

$$a^2 U(r) = \frac{-32 \sin \varrho [f(\varrho)^3 \cos \varrho - 3 f(\varrho)^2 \sin^3 \varrho + f(\varrho) \cos \varrho + \sin^3 \varrho]}{[1 + f(\varrho)^2]^2} ,$$

$$f(\varrho) \equiv 2\varrho - \sin 2\varrho , \quad \varrho \equiv r/a , \tag{6.52}$$

with $a > 0$. For this potential, the radial equation for the $l = 0$ partial wave has an eigenvalue $\varepsilon = 1/a^2$, with normalizable wave function

$$u_0(r) = N \frac{\sin \varrho}{1 + f(\varrho)^2} . \tag{6.53}$$

The graphical representations of $U(r)$ and $u_0(r)$ are given in Fig. 6.2.

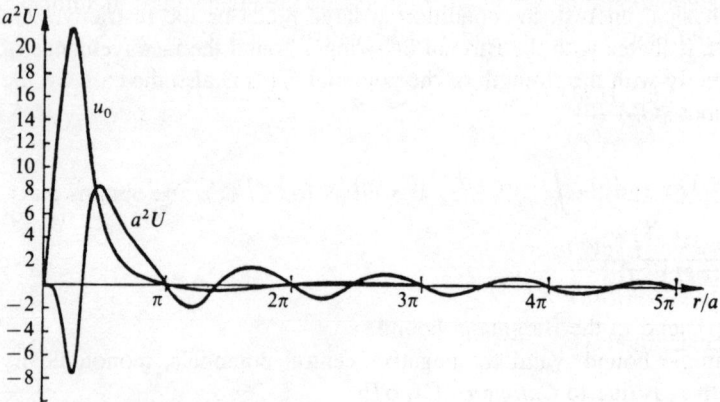

Fig. 6.2. Graphical representation of the von Neumann-Wigner potential, and of the wave function u_0 (with arbitrary normalization)

6.3 Square Wells

As illustrations of the previous results, we will consider some simple examples:

1) The Finite Square Potential Well. Let

$$U(r) = \begin{cases} U_0(< 0) , & 0 \le r \le a \\ 0 , & r > a . \end{cases} \tag{6.54}$$

Then, the radial function for the lth partial wave, for bound states with $U_0 < \varepsilon \le 0$, obeys

$$\frac{d^2 R_l(r)}{dr^2} + \frac{2}{r}\frac{dR_l(r)}{dr} + \left[\alpha^2 - \frac{l(l+1)}{r^2}\right]R_l(r) = 0 , \qquad 0 \le r \le a ,$$

$$\frac{d^2 R_l(r)}{dr^2} + \frac{2}{r}\frac{dR_l(r)}{dr} + \left[-\beta^2 - \frac{l(l+1)}{r^2}\right]R_l(r) = 0 , \qquad r > a ,$$

$$(6.55)$$

where $\alpha = (\varepsilon - U_0)^{1/2}$ and $\beta = |\varepsilon|^{1/2}$. Comparing these equations with (A.115), one sees that

$$R_l(r) = Aj_l(\alpha r) + By_l(\alpha r) , \qquad 0 \le r \le a ,$$
$$R_l(r) = A'j_l(i\beta r) + B'y_l(i\beta r) , \qquad r > a .$$

$$(6.56)$$

Taking into account (A.124) and the regularity condition, we get $B = 0$. On the other hand, from the requirement of square integrability and from (A.125) and (A.118), one derives that $B' = iA'$, and therefore

$$R_l(r) = Aj_l(\alpha r) , \qquad 0 \le r \le a ,$$
$$R_l(r) = A'h_l^{(1)}(i\beta r) , \qquad r > a .$$

$$(6.57)$$

To obtain the equation that permits us to calculate the eigenvalues, it suffices to impose the continuity of the logarithmic derivative at $r = a$:

$$\alpha\frac{j_l'(\alpha a)}{j_l(\alpha a)} = i\beta\frac{h_l^{(1)'}(i\beta a)}{h_l^{(1)}(i\beta a)} .$$

$$(6.58)$$

Using relation (A.129), valid for $j_l(z)$ as well as for $h_L^{(1)}(z)$, one obtains that

$$\alpha\frac{j_{l+1}(\alpha a)}{j_l(\alpha a)} = i\beta\frac{h_{l+1}^{(1)}(i\beta a)}{h_l^{(1)}(i\beta a)} ,$$

$$(6.59)$$

and in particular, for $l = 0$, we have

$$\alpha \cot \alpha a = -\beta ,$$

$$(6.60)$$

as expected, since in the $l = 0$ case, the resulting problem is completely equivalent to that of Example 3 in Sect. 4.3. As in that example, the number of bound states for the s partial wave is given by

$$N_0 = \left[\frac{1}{2} + \frac{a|U_0|^{1/2}}{\pi}\right]_0 ,$$

$$(6.61)$$

where $[x]_0$ indicates the greatest integer smaller than x. If $N_0 = 0$, then there are no bound states for any value of l.

In contrast to the one-dimensional case, in which the square well always has at least one bound state, we see that in three dimensions such potential wells, when they are sufficiently narrow and/or shallow, have no bound states. (In two dimensions, the situation is like that for one dimension. A very general discussion can be found in [SI 76a].)

It is remarkable that, even though for s-waves there are no bound states of zero energy, due to the fact that $U_{l=0}(r)$ is, in this case, of compact support, the repulsive core created by the centrifugal barrier for $l > 0$ allows for the existence of bound states with zero energy in these waves. Letting $\beta \to 0$ in (6.59), and using the behavior at the origin and the recursion formulae given in Sect. A.9, one obtains immediately the existence of a bound state of zero energy in the $l > 0$ partial wave only if

$$j_{l-1}(|U_0|^{1/2}a) = 0 . \tag{6.62}$$

On the other hand, from (6.57) and from the continuity of the wave function, one has

$$R_l(r) \propto \frac{h_l^{(1)}(i\beta r)}{h_l^{(1)}(i\beta a)} , \quad r > a , \tag{6.63}$$

and taking the limit $\beta \to 0$ while holding r fixed results in $R_l(r) \propto r^{-(l+1)}$, so that the solution is square integrable $(l > 0)$, although not with the typical exponential decrease of bound states with negative energy.

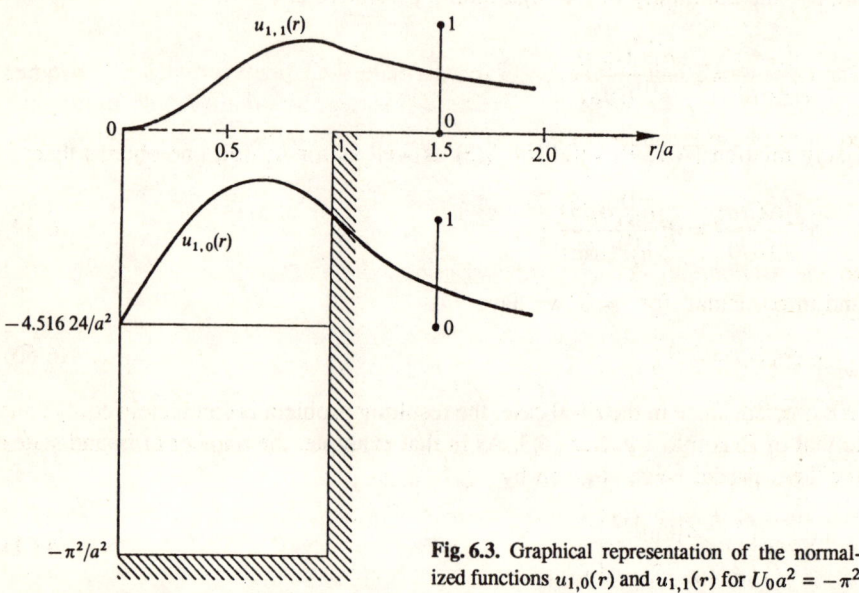

Fig. 6.3. Graphical representation of the normalized functions $u_{1,0}(r)$ and $u_{1,1}(r)$ for $U_0a^2 = -\pi^2$

In Fig. 6.3, one can see the bound wave functions $u_{1,0}(r)$, $l = 0$, and $u_{1,1}(r)$, $l = 1$, for a potential well with $U_0a^2 = -\pi^2$, which has only one s-wave bound state, with $\varepsilon_{1,0}a^2 = -4.51624$, and a p-wave bound state with $\varepsilon_{1,1} = 0$. The functions $u_{1,0}$ and $u_{1,1}$ are normalized, and their scale is indicated by the two vertical marks in the figure.

In nuclear physics, mainly in problems of three or more nucleons, the nucleon-nucleon interaction has been at times approximated using delta-function potentials, to reduce as much as possible the complexity of the associated Schrödinger equation. Although they can be introduced easily in one dimension, either by using quadratic forms, or as limits of square potentials with range tending to zero and intensity inversely proportional to the range, these potentials present certain peculiarities in $\mathbf{R}^3$ (and in general in $\mathbf{R}^n$, $n \geq 2$). Intuitively, one expects that a potential $U(r) \propto \delta(\boldsymbol{x})$ to affect only the s-waves, since waves with $l \geq 1$ have negligible probabilities of being affected by this potential [see (6.28)]. On the other hand, $H_{l=0}$ is of free particle form for $r > 0$, and only a change in the boundary conditions at the origin (that is, the modification of the regularity condition) can simulate the effect of the interaction U. Let us suppose the following boundary condition for s-waves:

$$u(0) + a_0 u'(0) = 0 , \tag{6.64}$$

with a_0 real. We denote by H_{a_0} the Hamiltonian obtained this way [which is that of the free case, but with (6.64) instead of (6.36) when l is zero].

For $a_0 > 0$, there exists a bound state for s-waves, with reduced wave function and energy:

$$u_{1,0}(r) = N \exp\left(-\frac{r}{a_0}\right) , \quad \varepsilon_{1,0} = -\frac{1}{a_0^2} . \tag{6.65}$$

The same result can be obtained formally with the square potential U_a of range a and depth $U_{a,0} = -(\pi/2a)^2 - 2/a_0 a$, and taking the limit $a \downarrow 0$. Requiring the equality of the integrals over $\mathbf{R}^3$ of U_a and of the limit potential, one may write

$$U_a(r) \underset{a \to 0}{\sim} -\frac{\pi^3}{3} a \left[1 + \frac{8}{\pi^2} \frac{a}{a_0}\right] \delta(\boldsymbol{x}) , \tag{6.66}$$

so that the coupling constant of the limiting delta function is infinitesimally small [AG 84a].

For $a_0 < 0$, the simulation of (6.64) as a limit of local repulsive potentials with supports tending to zero does not seem possible. *Friedman* [FR 72] has shown that:

(i) If $a_0 > 0$, then $H_0 + V_a \to H_{a_0}$, as $a \downarrow 0$ (in the strong resolvent sense) and the previous formal results are maintained.

(ii) If $a_0 < 0$, then $H_0 + V_a \to$ free Hamiltonian H_0, as $a \downarrow 0$ (also in the same strong sense), independent of the strength and form of the bounded functions $V_a(\boldsymbol{x})$, with support in the region $|\boldsymbol{x}| \leq a$.

For other methods of approximating these contact δ-like potentials by nonlocal and separable interactions, see, for example, [ZO 80].

2) The Infinite Square Potential Well. We assume that

$$U(r) = \begin{cases} 0, & 0 \leq r \leq a , \\ \infty, & r > a . \end{cases} \tag{6.67}$$

In this case $R_l(r) = 0$ for $r > a$, and in the zone $0 \leq r \leq a$, the radial wave function is a solution of the equation

$$\frac{d^2 R_l(r)}{dr^2} + \frac{2}{r}\frac{dR_l(r)}{dr} + \left[\alpha^2 - \frac{l(l+1)}{r^2}\right] R_l(r) = 0 , \tag{6.68}$$

where $\alpha = \varepsilon^{1/2}$, $\varepsilon > 0$. The regular solution at $r = 0$ is

$$R_l(r) = A j_l(\alpha r) , \tag{6.69}$$

and the condition that the wave function vanishes at $r = a$ implies that the binding energies are determined by the solutions of the equation

$$j_l(\alpha a) = 0 . \tag{6.70}$$

If X_{nl} is the nth positive zero of the function $j_l(x)$, then the energies of the bound states are

$$E_{nl} = \frac{\hbar^2}{2\mu a^2} X_{nl}^2 . \tag{6.71}$$

For an infinite potential well, there exist an infinite number of bound states; the lowest few eigenvalues are given in Table 6.1.

Table 6.1. Energies of the lowest states for the infinite square potential well

States (n, l)	$\dfrac{2\mu a^2}{\hbar^2} E_{nl}$	States (n, l)	$\dfrac{2\mu a^2}{\hbar^2} E_{nl}$
$1s(1,0)$	9.8696	$3s(3,0)$	88.8264
$1p(1,1)$	20.1907	$2f(2,3)$	108.5164
$1d(1,2)$	33.2175	$1i(1,6)$	110.5197
$2s(2,0)$	39.4784	$3p(3,1)$	118.8999
$1f(1,3)$	48.8312	$1j(1,7)$	135.8864
$2p(2,1)$	59.6795	$2g(2,4)$	137.0048
$1g(1,4)$	66.9543	$3d(3,2)$	151.8549
$2d(2,2)$	82.7192	$4s(4,0)$	157.9137
$1h(1,5)$	87.5312	$1k(1,8)$	163.6041

For $l = 0$, equation (6.70) reduces to $\sin \alpha a = 0$ and therefore $X_{n0} = n\pi$; this was completely predictable, since in this case the problem is equivalent to the one-dimensional potential well studied in Example 2 of Sect. 4.3.

In Fig. 6.4, the bound state energies of the first four levels (when they exist) in the potential well $U(r) = U_0 > 0$ for $r > a$, and $U(r) = 0$ for $r \le a$ for different values of U_0, are compared with those corresponding to the same potential well with $U_0 = \infty$. The values of U_0 for which new bound states appear for the lth partial wave are determined by the equations

$$\begin{aligned}\cos(a\sqrt{U_0}) &= 0 , & l &= 0 , \\ j_{l-1}(a\sqrt{U_0}) &= 0 , & l &> 0 .\end{aligned} \tag{6.72}$$

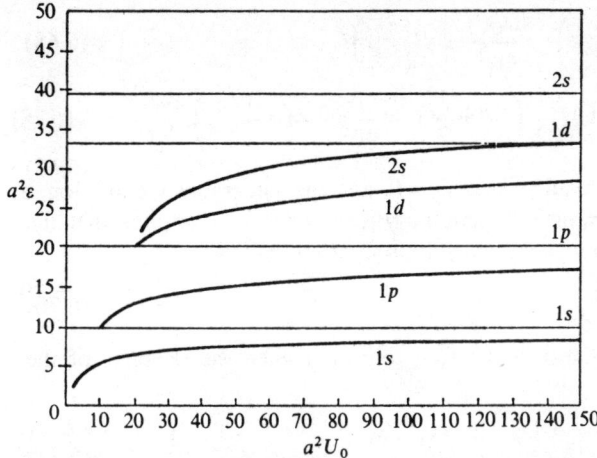

Fig. 6.4. Comparison of the bound state energies of the potential well $U(r) = 0$ for $0 \le r \le a$, and $U(r) = U_0$ for $r > a$, with those of the infinite square potential well (horizontal lines)

6.4 The Three-Dimensional Harmonic Oscillator

For the three-dimensional isotropic harmonic oscillator, the potential is

$$V(r) = \tfrac{1}{2}\mu\omega^2 r^2 \, , \tag{6.73}$$

and the reduced radial equation

$$\frac{d^2 u_l(r)}{dr^2} + \left[\varepsilon - \alpha^4 r^2 - \frac{l(l+1)}{r^2}\right] u_l(r) = 0 \, , \tag{6.74}$$

where $\alpha \equiv \sqrt{\mu\omega/\hbar}$. Therefore, in accordance with the general discussion, it has only a discrete spectrum. The behavior $u_l(r) \sim \text{const} \times r^{l+1}$ as $r \to 0$ and $u_l(r) \sim \text{const} \times \exp(-\alpha^2 r^2/2)$ (in its dominant part) as $r \to \infty$, suggests the change of variables

$$u_l(r) = r^{l+1} e^{-\alpha^2 r^2/2} v_l(r) \, , \tag{6.75}$$

and $v_l(r)$ satisfies the equation

$$\frac{d^2 v_l(r)}{dr^2} + 2\left[\frac{l+1}{r} - \alpha^2 r\right] \frac{dv_l(r)}{dr} + \left[\varepsilon - 2\alpha^2\left(l + \frac{3}{2}\right)\right] v_l(r) = 0 \, . \tag{6.76}$$

Letting $\varrho = \alpha^2 r^2$ and $w_l(\varrho) \equiv v_l(r)$, one obtains

$$\varrho w_l''(\varrho) + \left[\left(l + \frac{3}{2}\right) - \varrho\right] w_l'(\varrho) - \left[\frac{1}{2}\left(l + \frac{3}{2}\right) - \frac{\varepsilon}{4\alpha^2}\right] w_l(\varrho) = 0 \, , \tag{6.77}$$

which is a confluent hypergeometric equation (Sect. A.10). Thus, the most general solution is

$$w_l(\varrho) = AM\left(\frac{1}{2}\left(l+\frac{3}{2}\right) - \frac{\varepsilon}{4\alpha^2}, \; l+\frac{3}{2}, \; \varrho\right)$$
$$+ B\varrho^{-l-1/2}M\left(\frac{1}{2}\left(-l+\frac{1}{2}\right) - \frac{\varepsilon}{4\alpha^2}, \; -l+\frac{1}{2}, \; \varrho\right) . \tag{6.78}$$

The regularity condition requires that $B = 0$, and the integrability condition, as derived from (A.144), is satisfied if and only if

$$\Gamma\left(\frac{1}{2}\left(l+\frac{3}{2}\right) - \frac{\varepsilon}{4\alpha^2}\right) = \infty ; \tag{6.79}$$

that is,

$$\frac{1}{2}\left(l+\frac{3}{2}\right) - \frac{\varepsilon}{4\alpha^2} = -n+1 , \quad n = 1, 2, 3, \ldots . \tag{6.80}$$

Hence the bound state energies are

$$E_{nl} = \hbar\omega\left[2(n-1) + l + \frac{3}{2}\right] . \tag{6.81}$$

Note that the multiplicity of the level corresponding to $2(n-1) + l = N$, with $N = 0, 1, 2, \ldots$, is $(N+1)(N+2)/2$. Thus, one can see that, in addition to the azimuthal degeneracy, there exists an accidental degeneracy, a consequence of the particular dynamical properties of the system.

The corresponding wave functions are

$$\psi_{nlm}(\boldsymbol{r}) = N_{nl}r^l e^{-\alpha^2 r^2/2}M\left(-n+1, \; l+\frac{3}{2}, \; \alpha^2 r^2\right)Y_l^m(\theta, \phi) , \tag{6.82}$$

where N_{nl} is a normalization constant. In this case, the confluent hypergeometric function reduces to a polynomial in r^2 of order $n-1$. Using (A.150), we easily obtain that N_{nl}, apart from a phase factor, is given by

$$N_{nl} = \left[\frac{2\alpha^{2l+3}\Gamma\left(l+\frac{3}{2}+n-1\right)}{(n-1)!\Gamma^2\left(l+\frac{3}{2}\right)}\right]^{1/2} . \tag{6.83}$$

Representations of the normalized wave functions $u_{1,0}$, $u_{1,1}$, $u_{1,2}$ and $u_{2,0}$ can be seen in Fig. 6.5.

It is easy to recognize that for the three-dimensional anisotropic harmonic oscillator, with potential

$$V(r) = \frac{1}{2}\mu(\omega_1^2 x_1^2 + \omega_2^2 x_2^2 + \omega_3^2 x_3^2) \tag{6.84}$$

(which reduces to the previous case if $\omega_1 = \omega_2 = \omega_3 \equiv \omega$), the Schrödinger equation describing the relative motion is separable in Cartesian coordinates, that is, it is equivalent to the study of three uncoupled harmonic oscillators. Therefore, the energy levels are

$$E_{n_1 n_2 n_3} = \hbar\left[\omega_1\left(n_1 + \frac{1}{2}\right) + \omega_2\left(n_2 + \frac{1}{2}\right) + \omega_3\left(n_3 + \frac{1}{2}\right)\right] , \tag{6.85}$$

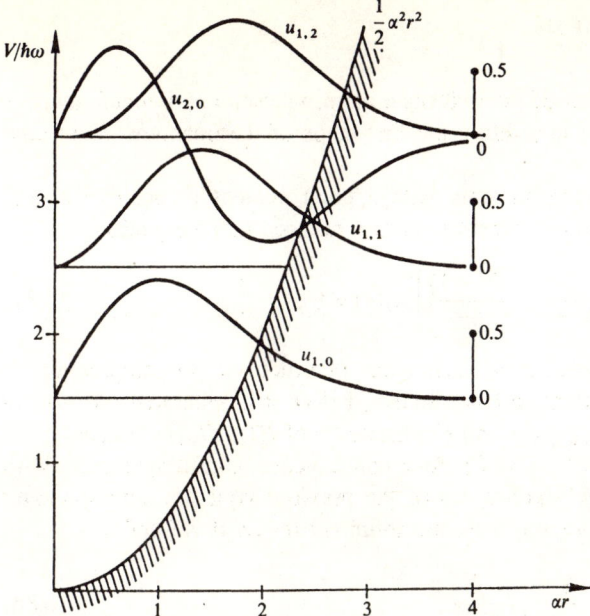

Fig. 6.5. Representation of the normalized wave functions corresponding to the first three energy levels of the isotropic harmonic oscillator

and the corresponding normalized wave functions (4.39) are ($\alpha_i \equiv \sqrt{\mu\omega_i/\hbar}$)

$$\psi_{n_1 n_2 n_3}(\mathbf{r}) = \prod_{i=1}^{3} \left[\frac{\alpha_i}{\sqrt{\pi} 2^{n_i} n_i!} \right]^{1/2} H_{n_i}(\alpha_i x_i) e^{-\alpha_i^2 x_i^2/2} . \tag{6.86}$$

It is evident that the formalism of creation, annihilation and number operators, as well as the notion of coherent states, are easily translated to oscillators in $\mathbb{R}^n$. In this new frame, there appear n copies of these operators $a_j^\dagger$, a_j, N_j, $j = 1, 2, \ldots, n$, with commutation rules identical to (4.53) for operators of equal j, and vanishing commutators in all other cases. The coherent states are now tensorial products of the one-dimensional coherent states. Finally, the wave functions $\psi_{n_1 n_2 \ldots}$ obviously correspond to the states $|n_1 n_2 \ldots\rangle \equiv |n_1\rangle \otimes |n_2\rangle \otimes \ldots$.

6.5 The Hydrogen Atom

We now study the spectrum of the hydrogen atom, whose experimental analysis was of crucial importance in guiding the first steps of the quantum theory and its subsequent developments.

We begin by considering an atom with a point nucleus of positive charge $Z|e|$ and with a single spinless electron, whose reduced radial equation is

$$\frac{d^2 u_l(r)}{dr^2} + \left[\frac{2\mu}{\hbar^2} E + \frac{2\mu Z e^2}{\hbar^2 r} - \frac{l(l+1)}{r^2} \right] u_l(r) = 0 \ . \tag{6.87}$$

Since the Coulomb potential is homogeneous of order -1, the application of the virial theorem to the relative Hamiltonian, before it is separated into partial waves, leads to $E = -\langle H_{0,\mathrm{rel}} \rangle_\psi$ for all eigenstates ψ of $H (\equiv H_{\mathrm{rel}})$ with energy E. Hence, necessarily $\sigma_{\mathrm{p}}(H) \subset (-\infty, 0)$. As a consequence, and in agreement with the general discussion, the eigenvalues of the previous equation corresponding to bound states are negative and there are infinitely many. If we define

$$\varrho = \frac{\sqrt{8\mu |E|}}{\hbar} r \ , \quad \lambda = \frac{Z e^2}{\hbar} \sqrt{\frac{\mu}{2|E|}} \ , \tag{6.88}$$

Eq. (6.87) becomes

$$\frac{d^2 \bar{u}_l(\varrho)}{d\varrho^2} + \left[\frac{\lambda}{\varrho} - \frac{1}{4} - \frac{l(l+1)}{\varrho^2} \right] \bar{u}_l(\varrho) = 0 \ , \tag{6.89}$$

where $\bar{u}_l(\varrho) \equiv u_l(r)$. Given the asymptotic behavior $\bar{u}_l(\varrho) \sim \mathrm{const} \times \varrho^{l+1}$ if $\varrho \to 0$ and $\bar{u}_l(\varrho) \sim \mathrm{const} \times \exp(-\varrho/2)$ (dominant part), if $\varrho \to \infty$, it is convenient to introduce a new function $v_l(\varrho)$, defined as

$$\bar{u}_l(\varrho) = \varrho^{l+1} e^{-\varrho/2} v_l(\varrho) \ . \tag{6.90}$$

This function satisfies

$$\varrho v_l''(\varrho) + [2l + 2 - \varrho] v_l'(\varrho) - [l + 1 - \lambda] v_l(\varrho) = 0 \ , \tag{6.91}$$

which is a confluent hypergeometric equation (Sect. A.10). Bearing in mind the regularity condition, its solution is

$$v_l(\varrho) = A M(l + 1 - \lambda, 2l + 2, \varrho) \ . \tag{6.92}$$

The corresponding function $\bar{u}_l(\varrho)$ is normalizable if and only if $\Gamma(l+1-\lambda) = \infty$, as one immediately deduces from (A.144), and therefore the bound state energies are determined by

$$l + 1 - \lambda = -n_r \ , \quad n_r = 0, 1, 2, \dots \ , \tag{6.93}$$

where n_r is the radial quantum number, and is equal to the number of nodes of the radial wave function in $(0, \infty)$. In this case, instead of following the nomenclature of nodes given in Sect. 6.2, we adopt the notation of Bohr's quantum theory, as usual. With this understanding, the principal quantum number n is defined as $n = n_r + l + 1$, and using (6.93), the energy eigenvalues are

$$E_{nl} = -\frac{1}{2}\mu(Z\alpha c)^2 \frac{1}{n^2}, \quad n = 1,2,3,\dots . \tag{6.94}$$

This formula coincides with that obtained in the Bohr model (1.30). Note that for a given value of n, the possible values of l are $l = 0,1,2,\dots,n-1$, and therefore the degeneracy of the nth level is n^2. Similarly, note that the quantum number n_ψ of the Bohr-Sommerfeld model does not correspond to l, but to $l + 1$. Thus one sees that an accidental degeneracy also appears here due to the particular form of $V(r)$. This has a dynamical origin, as will be discussed later.

If (6.91) and (A.68) are compared, and one takes into account (A.78), then up to phase factors the normalized wave function is

$$\psi_{nlm}(\boldsymbol{r}) = R_{nl}(r)Y_l^m(\hat{\boldsymbol{r}}), \quad \text{with} \tag{6.95}$$

$$\boxed{R_{nl}(r) = \frac{2}{a^{3/2}n^2}\left[\frac{(n-l-1)!}{(n+l)!}\right]^{1/2}\varrho^l e^{-\varrho/2}L_{n-l-1}^{2l+1}(\varrho)}, \tag{6.96}$$

where $\varrho = 2r/an$, and a, the Bohr radius, is given by

$$\boxed{a = \frac{\hbar}{\mu(Z\alpha c)}}. \tag{6.97}$$

In the case $Z = 1$ and $\mu = m_e$ (nucleus of infinite mass), the previous quantity is customarily denoted as

$$a_{\infty\text{Bohr}} = \hbar/m_e(\alpha c) = 0.52917706(44) \times 10^{-8} \text{ cm} .$$

The radial wave functions for the first few bound states are

$$n = 1: \quad \bar{R}_{10}(r) = 2e^{-\bar{r}}$$

$$n = 2: \quad \bar{R}_{20}(r) = \frac{1}{\sqrt{2}}e^{-\bar{r}/2}\left(1 - \frac{1}{2}\bar{r}\right), \quad \bar{R}_{21}(r) = \frac{1}{2\sqrt{6}}e^{-\bar{r}/2}\bar{r}$$

$$n = 3: \quad \bar{R}_{30}(r) = \frac{2}{3\sqrt{3}}e^{-\bar{r}/3}\left(1 - \frac{2}{3}\bar{r} + \frac{2}{27}\bar{r}^2\right), \tag{6.98}$$

$$\bar{R}_{31}(r) = \frac{8}{27\sqrt{6}}e^{-\bar{r}/3}\bar{r}\left(1 - \frac{1}{6}\bar{r}\right), \quad \bar{R}_{32}(r) = \frac{4}{81\sqrt{30}}e^{-\bar{r}/3}\bar{r}^2 ,$$

where $\bar{r} \equiv r/a$ and $\bar{R}_{nl} \equiv a^{3/2}R_{nl}$. Some of the functions $ar^2 R_{nl}^2(r)$ are represented in Fig. 6.6.

Taking into account the definition (A.82), the expectation value of r^p in any arbitrary state (n, l) is given by $\langle r^p \rangle_{nl} = a^p I_{nl}^{(p)}$, and thus, in particular,

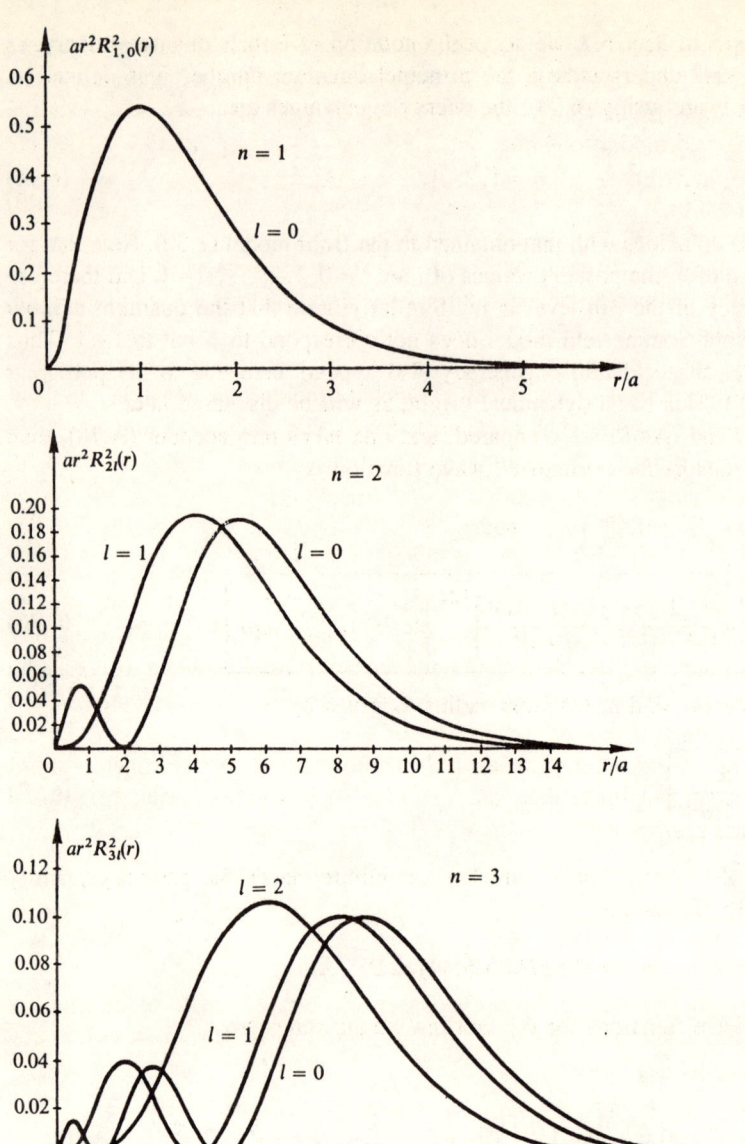

Fig. 6.6. Graphical representation of $ar^2 R_{nl}^2(r)$ as a function of r/a for $n = 1, 2$ and 3

$$\langle r^{-3} \rangle_{nl} = \frac{1}{l(l+1/2)(l+1)n^3} a^{-3}, \quad l \geq 1,$$

$$\langle r^{-2} \rangle_{nl} = \frac{1}{(l+1/2)n^3} a^{-2}, \quad \langle r^{-1} \rangle_{nl} = \frac{1}{n^2} a^{-1}, \tag{6.99}$$

$$\langle r \rangle_{nl} = \tfrac{1}{2}[3n^2 - l(l+1)]a, \quad \langle r^2 \rangle_{nl} = \tfrac{1}{2}[5n^2 + 1 - 3l(l+1)]n^2 a^2.$$

For other values of p, the expectation values can be obtained through Kramers formula (A.84). Note that for $n = l + 1$ [the case where the radial function has no nodes in $(0, \infty)$ and which corresponds to circular orbits ($n = n_\psi$) of the Bohr-Sommerfeld model] one has

$$\langle r \rangle_{n,n-1} = (2n+1)\frac{na}{2} , \quad (\Delta r)_{n,n-1} = \sqrt{2n+1}\frac{na}{2} . \tag{6.100}$$

This allows one to assert that for $n \gg 1$ the corresponding wave functions are concentrated in a spherical shell of radius $n^2 a$ and (relatively) negligible width, as in the classical limit.

Obtaining the expectation values (6.99) through the Kramers relation can generate an appearance of complexity which actually does not exist. In fact, writing the reduced radial Hamiltonian for l waves in convenient units, as $H_l = -D^2 + V_l(r)$, $D \equiv d/dr$, and bearing in mind that $\langle [H_l, Df + fD] \rangle_{nl} = 0$, with f an arbitrary function of r, a simple calculation shows that

$$\langle 4V_l f' + 2V_l' f - f''' \rangle_{nl} = 4E_{nl} \langle f' \rangle_{nl} . \tag{6.101}$$

For the case of the Coulomb potential, the choice $f \equiv r^{N+1}$ in (6.101) produces a linear relation between $\langle r^N \rangle$, $\langle r^{N-1} \rangle$ and $\langle r^{N-2} \rangle$, which is (A.84). Since $\langle 1 \rangle = 1$ and $\langle r^{-1} \rangle$ is easily derivable from the virial theorem, it is evident that by using (6.101) we can calculate the remaining $\langle r^N \rangle$ recursively. Recall that $\langle r^N \rangle_{nl}$ is finite if and only if $N \geq -2(l+1)$.

According to the results obtained up to now, valid for hydrogenic atoms with point nuclei, a spinless electron, no relativistic corrections, etc., the spectral emission lines correspond to the frequencies

$$\boxed{\nu_{n_1 n_2} = \frac{\mu(Z\alpha c)^2}{4\pi\hbar}\left(\frac{1}{n_2^2} - \frac{1}{n_1^2}\right)} \quad n_1 > n_2 = 1, 2, \ldots . \tag{6.102}$$

In reality, formula (6.102) is only approximate, since it neglects the recoil of the atom produced by emission. If this recoil, generally very small, is considered to first order, (6.102) becomes

$$\nu_{n_1 n_2}^{\text{rec}} = \nu_{n_1 n_2}\left[1 - \frac{1}{2}\frac{h\nu_{n_1 n_2}}{Mc^2}\right] , \tag{6.103}$$

where M is the mass of the atom. Note that for atomic systems this correction is of the order of 10^{-8}, though in electromagnetic nuclear transitions it can be much more important.

Equation (6.102) does not give any information concerning the intensity of the corresponding spectral lines. We will see later, when studying radiation theory, that there are lines whose intensity is so small that it is practically impossible to observe them experimentally.

To simplify the formulae that appear in calculations of atomic structure, it is useful to work with the Hartree system of *atomic units* (a.u.), in which m_e, $|e|$

and $\hbar$ are used, respectively, as the units of mass, charge and action. In this system, $a_{\infty\text{Bohr}}$ is the unit of length, the speed of light is $c = \alpha^{-1}$a.u., the unit of momentum is $m_e(\alpha c)$, and the unit of energy is $m_e(\alpha c)^2$. We give next some interesting values related to this system of units which are specially useful in spectroscopy:

$$a_{\infty\text{Bohr}} = \frac{\hbar}{m_e(\alpha c)} = 0.52917706(44) \times 10^{-8}\,\text{cm} \; ;$$

$$\text{Rydberg (energy)} = \frac{m_e(\alpha c)^2}{2} = 13.605804(36)\,\text{eV} \; ;$$

$$\text{Rydberg (frequency)} = \frac{m_e(\alpha c)^2}{4\pi\hbar} = 3.289842(17) \times 10^{15}\,\text{Hz} \; ;$$

$$\text{Rydberg (wave number)} = \frac{m_e(\alpha c)^2}{4\pi\hbar c} = 109737.32(56)\,\text{cm}^{-1} \; .$$

If there is no possibility of misunderstanding, it is usual to represent the last three quantities as R_∞, where the subscript ∞ corresponds to an atom whose nucleus has infinite mass. In practice, μ differs little from m_e, the major difference occurring for the hydrogen atom, for which $\mu = 0.9994556794(30)m_e$.

Table 6.2. Experimental (with respect to vacuum) and theoretical values of the wavelengths of the first few series of the hydrogen atom

n_2	n_1	$\lambda_{n_1 n_2\text{theor}}[\text{Å}]$	$\lambda_{n_1 n_2\text{exp}}[\text{Å}]$	n_2	n_1	$\lambda_{n_1 n_2\text{theor}}[\text{Å}]$	$\lambda_{n_1 n_2\text{exp}}[\text{Å}]$
	2	1 215.684(7)	1 215.68	2	7	3 971.24(2)	3 971.27
1	3	1 025.734(5)	1 025.83		4	18 756.3(1)	18 756.7
	4	972.548(5)	972.54	3	5	12 821.67(7)	12 822.0
	3	6 564.70(3)	6 564.70		6	10 941.16(6)	10 941.3
	4	4 862.74(2)	4 862.76		5	40 522.8(2)	$4.05 \pm 0.03\mu$
2	5	4 341.73(2)	4 341.76	4	6	26 258.8(1)	$2.63 \pm 0.02\mu$
	6	4 102.93(3)	4 102.96	5	6	74 598.8(4)	7.40μ

According to (6.102), the spectral lines of hydrogenic atoms can be classified in series. Each series in the emission spectrum corresponds to a fixed n_2 value, and to n_1 running over the sequence $n_2+1, n_2+2, \ldots$. The typical trait of a series is a set of lines converging to a limit frequency $\nu_{\infty n_2}$. The most characteristic series in the hydrogen atom, whose understanding was of fundamental importance in the beginnings of QM, are those corresponding to $n_2 = 1$ (Lyman series), $n_2 = 2$ (Balmer series), $n_2 = 3$ (Paschen series), $n_2 = 4$ (Brackett series), and $n_2 = 5$ (Pfund series). Their wavelengths are given by

$$\lambda_{n_1 n_2} = \frac{c}{\nu_{n_1 n_2}} = \frac{c}{R_H}\left(\frac{1}{n_2^2} - \frac{1}{n_1^2}\right)^{-1} , \tag{6.104}$$

where R_H is the Rydberg constant for the hydrogen atom, and numerically, $c/R_H = 911.7633(47)$ Å. The limiting wavelengths of these series are, in Å,

$$\lambda_{\infty 1} = 911.763(5), \quad \lambda_{\infty 2} = 3647.05(2), \quad \lambda_{\infty 3} = 8205.87(5),$$
$$\lambda_{\infty 4} = 14588.2(1), \quad \lambda_{\infty 5} = 22794.1(1). \tag{6.105}$$

We thus find that the first series lies in the ultraviolet range; the second, in the visible region, and all the rest, in the infrared. In Table 6.2, we give the first few terms of each series and they are compared with the experimental values [AL 70]. The agreement is extraordinarily good, which makes us expect that all the neglected effects are indeed very small.

Finally, it is clear that the potential in (6.87) satisfies (6.38) and therefore $\sigma_c(H) = [0, \infty)$. Besides, it can be shown that $\sigma_{s.c.}(H) = \emptyset$ [RS 78].

6.6 The Hydrogen Atom: Corrections

The theory just developed deals with a non-existent case: a one-electron atom without spin, without relativistic corrections, with a point nucleus, etc. A comparison of experimental and theoretical values indicates that all these corrections are small. This is not the place for a detailed quantitative treatment of these corrections, but we will discuss them briefly.

In the first place, there are relativistic corrections: From the virial theorem (3.156), one deduces that $\langle v^2/c^2 \rangle_{nl} = (Z\alpha/n)^2$, and therefore the mean speed of the electron in the (n, l) state is $v_{nl}/c \simeq Z\alpha/n$. It is thus expected that the relativistic corrections to the energy are of the order $\Delta E_{nl} \simeq (Z\alpha/n)^2 E_{nl}$, which is small, but not completely negligible. Treating the electron relativistically using the *Dirac* equation [DI 28], [AB 65], and the proton non-relativistically, one can show [GY 69] that in the hydrogen atom, neglecting terms of order $\alpha^6 m_e$ and $\alpha^4 m_e^2/m_p$ and higher, where m_e and m_p are the mass of the electron and proton respectively, the energies corresponding to the bound states are

$$\boxed{E_{nlj} = E_{nl}\left[1 + \frac{\alpha^2}{n}\left(\frac{1}{j+1/2} - \frac{3}{4n}\right)\right].} \tag{6.106}$$

Here j is the total angular momentum of the electron, which can have values $j = 1/2$ if $l = 0$ and $j = l \pm 1/2$ if $l \neq 0$, and E_{nl} is the energy given by (6.94). One sees, therefore, that these corrections make the accidental degeneracy disappear partially. The energy levels are customarily denoted by nl_j; one sees immediately that the old energy level E_{nl} (with degeneracy $2n^2$, if we include the spin of the electron) now splits into n levels. This new structure of the spectrum is called *fine structure*, and is studied in Chap. 12 by perturbation methods. Note that in (6.106) the correction term is always negative, that is, the relativistic corrections tend to increase the binding energy. On the other hand, its modulus decreases as j

increases (with n kept fixed) and therefore the fine structure levels, in increasing order, are

$$(ns_{1/2}, np_{1/2}), \ (np_{3/2}, nd_{3/2}), \ (nd_{5/2}, nf_{5/2}), \ \ldots,$$
$$(n(n-2)_{n-3/2}, \ n(n-1)_{n-3/2}), \ n(n-1)_{n-1/2} \ . \tag{6.107}$$

The fine structure of the first three levels for the hydrogen atom are shown in Fig. 6.7.

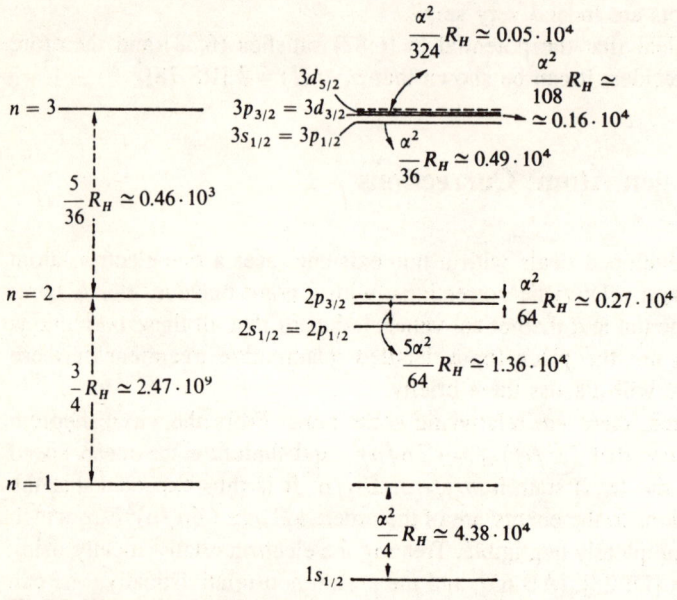

Fig. 6.7. Fine structure of the first three levels of the hydrogen atom. The transition frequencies are given in MHz

Observe that for $j < n - 1/2$, the energy levels $n(j - 1/2)_j$ and $n(j + 1/2)_j$ coincide, and this happens, for example, with the levels $2s_{1/2}$ and $2p_{1/2}$. In practice, on the contrary, one finds that the $2p_{1/2}$ level is below the $2s_{1/2}$ level, and its energy difference, $\Delta E = 2s_{1/2} - 2p_{1/2}$, in the case of the hydrogen atom is 1057.90 ± 0.06 MHz [RS 70]. This level splitting is known as the *Lamb shift*. For a theoretical explanation of this effect it is necessary to resort to quantum electrodynamics, a relativistic theory which describes the interaction of electrons with the electromagnetic field. An exceedingly complex calculation, which incorporates the proton structure and its finite mass, gives $E = 1057.911(12)$ MHz [PE 72], in perfect agreement with the experimental value. A general discussion of the Lamb shift can be found in [LP 72]. It can be shown that the most affected states are those with $l = 0$, whose binding energies are decreased by an amount of order $\alpha(Z\alpha)^4 m_e \ln(Z\alpha)^{-1}$ for hydrogen-like atoms.

Finally, since the proton in the hydrogen atom has spin $1/2$ and a magnetic moment μ_p, there exists an interaction between the proton and electron magnetic moments which we have not yet taken into account. If the electron has a total angular momentum j, the atom has as possible values of the total angular momentum $F = j \pm 1/2$. This interaction produces a further level splitting, called the *hyperfine structure* affecting mainly the levels with $l = 0$. Concretely, the ground state of the hydrogen atom, $1s_{1/2}$, splits into triplet ($F = 1$) and singlet ($F = 0$) states, the singlet level being the most bound. Experimentally, one finds [VP 66]

$$\boxed{\nu_{\text{h.f.s}} \equiv \frac{E_t - E_s}{h} = 1\ 420.405\ 751\ 786\ 4(17)\ \text{MHz}}\ , \qquad (6.108)$$

which is one of the most precise measurements in all of physics. The wavelength for this transition is $21.22\ \text{cm}$, and is of great importance in astrophysics. A simple theoretical calculation due to *Fermi* [FE 30] predicts

$$\nu_{\text{h.f.s.}} \simeq \frac{8}{3}\alpha^2 R_H g_p \frac{m_e}{m_p} \simeq 1\ 421\ \text{MHz}\ , \qquad (6.109)$$

where g_p is the proton gyromagnetic factor, defined as $g_p \equiv 2\mu_p/\mu_N$ with $\mu_N \equiv |e|\hbar/2m_p c$. Quantum electrodynamics allows us to make a more accurate calculation [LP 72], but still very far from achieving the precision of the experimental value; further improvements require a much more complete picture of the strong interactions.

For an extension of the formula (6.106), adding explicit corrections of higher order obtained from an effective potential deduced using quantum electrodynamics, see [AS 83].

Due to the interaction of the electron with the electromagnetic field, only the ground state $1s_{1/2}$ hyperfine singlet ($F = 0$) is really stable; all the rest can decay radiatively to lower levels (see Chap. 15), the typical transition probabilities being of the order of $10^8\ \text{s}^{-1}$. This produces a natural broadening of the lines of the order of $10^{-7}\ \text{eV}$, and as consequence, a masking of some of the smallest displacements considered above.

Finally, it is worth mentioning that nucleus–elementary particle bound systems are currently being studied, in particular, the so-called muonic atoms, where an electron is replaced by a muon, a sort of heavy electron with mass $m_\mu \simeq 207\ m_e$. In this case, in contrast with normal atoms, the hyperfine structure is much more important than the fine structure, since the factor m_e/m_p entering (6.109) is now m_μ/m_p.

6.7 Accidental Degeneracy

In the study of the three-dimensional isotropic harmonic oscillator and the hydrogen atom, it has been observed that the energy levels are more degenerate than expected by the simple fact that the Hamiltonian of the problem is invariant under the rotation group. We will now see that this accidental degeneracy is of dynamical origin, i.e., a direct consequence of the particular form of the interaction potential $V(r)$. It can be easily explained by taking into account the fact that H is invariant under a continuous group of symmetries strictly larger than the rotation group.

In the case of the hydrogen atom, the relative Hamiltonian is

$$H = \frac{p^2}{2\mu} - \frac{e^2}{r} , \tag{6.110}$$

and is invariant under SO(3) $\approx$ SU(2)/Z_2, the rotation group of three-dimensional Euclidean space, whose generators $L = r \times p$ satisfy the well-known commutation rules

$$[L_j, L_k] = i\hbar \varepsilon_{jkl} L_l . \tag{6.111}$$

Defining the Runge-Lenz operator M as

$$M = \frac{1}{2\mu}(p \times L - L \times p) - \frac{e^2}{r} r , \tag{6.112}$$

a long, direct calculation permits us to show that

$$[H, M] = 0 , \quad M \cdot L = L \cdot M = 0 ,$$
$$[L_i, M_j] = i\hbar \varepsilon_{ijk} M_k , \quad [M_i, M_j] = -2i\hbar \mu^{-1} H \varepsilon_{ijk} L_k , \tag{6.113}$$
$$M^2 = 2\mu^{-1} H(L^2 + \hbar^2) + e^4 ,$$

and, in particular, L and M commute with the Hamiltonian due to the particular form of the potential $V(r)$. If the study is limited to a Hilbert subspace corresponding to the bound states of the hydrogen atom, one can define the operators

$$I = \frac{1}{2}\left[L + \sqrt{-\frac{\mu}{2H}} M \right] , \quad K = \frac{1}{2}\left[L - \sqrt{-\frac{\mu}{2H}} M \right] , \tag{6.114}$$

which satisfy

$$[I_j, I_k] = i\hbar \varepsilon_{jkl} I_l , \quad [K_j, K_k] = i\hbar \varepsilon_{jkl} K_l , \quad [I_j, K_k] = 0 . \tag{6.115}$$

Hence, the Lie algebra generated by I and K coincides with that of SO(4) $\approx$ (SU(2)$\otimes$ SU(2))/Z_2. Since $[I, H] = [K, H] = 0$, then due to the form of $V(r)$, the Hamiltonian for the hydrogen atom (in the subspace of its bound states) commutes with a Lie algebra of self-adjoint operators which is isomorphic to the Lie algebra of the group SO(4) $\supset$ SO(3). Furthermore, one can prove [BA 67] that there effectively exists in such a subspace a unitary representation of SO(4) generated by I and K, under whose elements H is invariant. From the expressions in (6.112) and (6.114), one obtains that in the space of bound states,

$$H = -\frac{\mu e^4}{2} \frac{1}{2(I^2 + K^2) + \hbar^2} \ . \tag{6.116}$$

An irreducible SU(2) representation is characterized (Sect. 5.3) by a non-negative integer or half-integer, and therefore irreducible representations of SO(4) are determined by pairs (i, k) of such numbers. If this were all, the energy eigenvalues of the hydrogen atom would be

$$E_{ik} = -\frac{1}{2}\mu(\alpha c)^2 \frac{1}{2i(i+1) + 2k(k+1) + 1} \ . \tag{6.117}$$

However, from $\boldsymbol{L} \cdot \boldsymbol{M} = \boldsymbol{M} \cdot \boldsymbol{L} = 0$, it is deduced that $I^2 = K^2$, and hence $i = k$. Introducing $n \equiv 2i + 1$, where n can take on values $1, 2, 3, \ldots$, one obtains

$$E_n = -\frac{1}{2}\mu(\alpha c)^2 \frac{1}{n^2} \ , \quad n = 1, 2, \ldots \ . \tag{6.118}$$

The irreducibility of the nth level under the SO(4) group (or if one prefers, under $\boldsymbol{L}$ and $\boldsymbol{M}$), can also be seen in the following manner: From the commutation rules (6.113), one immediately obtains that

$$[\boldsymbol{L}^2, M_\pm] = \pm 2\hbar[M_\pm(L_3 \pm \hbar) - M_2 L_\pm] \ ,$$
$$[L_3, M] = \pm\hbar M_\pm \ , \tag{6.119}$$
$$[M_+, M_-] = 4\hbar\mu^{-1} H L_3 \ ,$$

so that

$$M_\pm|nl \pm l\rangle \propto |n, l+1, \pm(l+1)\rangle \ ,$$
$$\|M_\pm|nlm\rangle\|^2 - \|M_-|nlm\rangle\|^2 = \frac{2m}{n^2} \ . \tag{6.120}$$

As a consequence, for $l \neq 0$, $M_\pm|nl \pm l\rangle \neq 0$ and an iterated application of $M_\pm$ permits us to connect the states $|nl \pm l\rangle$, $l \neq 0$, among themselves. Also, if $n > 1$, the state $|n00\rangle$ participates in this connection, since $M^2|n00\rangle = e^4(1 - n^{-2})|n00\rangle$ due to the last equation of (6.113); and hence $M|n00\rangle \neq 0$ if $n > 1$.

The Hamiltonian displays accidental degeneracy even for positive energies. In this case, $\boldsymbol{L}$ and $(\mu/2H)^{1/2}\boldsymbol{M}$ are constants of motion that generate an infinite-dimensional unitary representation of the non-compact group SO(3,1) in the subspace of states with non-negative energy.

The existence of a vector constant of motion $\boldsymbol{M}$ (other than $\boldsymbol{L}$) in the Coulomb problem was expected from the following classical argument: $\boldsymbol{L}$ fixes the orbital plane of the orbit and, given the energy, also the eccentricity. But the direction of the principal axis is not fixed. To determine the major axis, it suffices to give a constant vector directed from the focal point to the perihelion. The Runge-Lenz vector plays precisely this role.

A similar study can be done for the three-dimensional isotropic harmonic oscillator. One can show that H is, in this case, invariant under SU(3) [BA 67], [SC 68a], which completely explains the spectrum of the harmonic oscillator. Furthermore, one can demonstrate that only the attractive potentials of the r^2 and $1/r$ types present accidental degeneracy. It is interesting to remember in this respect a theorem in classical mechanics due to *Bertrand* [BE 73], [WI 41]: Given a point particle which moves under the action of a monotonically increasing and spherically symmetric potential $V(r)$, a continuous function of r and with continuous derivatives up to the third order, a sufficient and necessary condition for all bounded trajectories to be closed is that the potential be proportional to r^2 or $-1/r$.

6.8 The Hydrogen Atom: Parabolic Coordinates

For the Coulomb potential, the Schrödinger equation is separable not only in spherical coordinates, but also in parabolic coordinates. This latter separation of variables is especially useful when there is a privileged direction in space, as we will see in the Stark effect (Chap. 10).

The parabolic coordinates ξ, η, and ϕ are defined as

$$x = \sqrt{\xi\eta}\cos\phi\,, \quad y = \sqrt{\xi\eta}\sin\phi\,, \quad z = \tfrac{1}{2}(\xi - \eta)\,, \tag{6.121}$$

where $0 \le \xi,\ \eta < \infty$, and $0 \le \phi < 2\pi$. Hence one immediately obtains that $r = (\xi + \eta)/2$. The inverse relations of (6.121) are

$$\xi = r + z\,, \quad \eta = r - z\,, \quad \phi = \arctan\frac{y}{x}\,. \tag{6.122}$$

The surfaces $\xi = $ const and $\eta = $ const are paraboloids of revolution with axis Oz and focal point at the origin. In these coordinates, the length and volume elements are

$$ds^2 = \frac{\xi + \eta}{4\xi}d\xi^2 + \frac{\xi + \eta}{4\eta}d\eta^2 + \xi\eta\,d\phi^2\,,$$
$$dx\,dy\,dz = \tfrac{1}{4}(\xi + \eta)\,d\xi\,d\eta\,d\phi\,. \tag{6.123}$$

The Laplacian in these coordinates takes the form given in (3.52):

$$\Delta = \frac{4}{\xi + \eta}\left[\frac{\partial}{\partial\xi}\left(\xi\frac{\partial}{\partial\xi}\right) + \frac{\partial}{\partial\eta}\left(\eta\frac{\partial}{\partial\eta}\right)\right] + \frac{1}{\xi\eta}\frac{\partial^2}{\partial\phi^2}\,, \tag{6.124}$$

so that the Schrödinger equation, for a hydrogenic atom, transforms into

$$\frac{\partial}{\partial \xi}\left(\xi\frac{\partial \psi}{\partial \xi}\right) + \frac{\partial}{\partial \eta}\left(\eta\frac{\partial \psi}{\partial \eta}\right) + \frac{1}{4}\left(\frac{1}{\xi} + \frac{1}{\eta}\right)\frac{\partial^2 \psi}{\partial \phi^2}$$

$$+ \frac{\mu Z e^2}{\hbar^2}\psi + \frac{\mu E}{2\hbar^2}(\xi + \eta)\psi = 0 \ . \tag{6.125}$$

By the change of variables

$$\varrho_1 \equiv \frac{\sqrt{2\mu|E|}}{\hbar}\xi \ , \quad \varrho_2 \equiv \frac{\sqrt{2\mu|E|}}{\hbar}\eta \ , \tag{6.126}$$

the previous equation, for negative energies, becomes

$$\frac{\partial}{\partial \varrho_1}\left(\varrho_1\frac{\partial \bar\psi}{\partial \varrho_1}\right) + \frac{\partial}{\partial \varrho_2}\left(\varrho_2\frac{\partial \bar\psi}{\partial \varrho_2}\right) + \frac{1}{4}\left(\frac{1}{\varrho_1} + \frac{1}{\varrho_2}\right)\frac{\partial^2 \bar\psi}{\partial \phi^2}$$

$$+ \lambda\bar\psi - \frac{1}{4}(\varrho_1 + \varrho_2)\bar\psi = 0 \ , \tag{6.127}$$

where λ is the quantity defined in (6.88) and $\bar\psi(\varrho_1, \varrho_2, \phi) \equiv \psi(\xi, \eta, \phi)$. Equation (6.127) is separable; we thus look for solutions of the form

$$\bar\psi(\varrho_1, \varrho_2, \phi) = f_1(\varrho_1)f_2(\varrho_2)e^{im\phi} \ , \tag{6.128}$$

where m is the magnetic quantum number and $f_i(\varrho_i)$, $i = 1, 2$, are solutions to the equations

$$\frac{d^2 f_i}{d\varrho_i^2} + \frac{1}{\varrho_i}\frac{df_i}{d\varrho_i} + \left[-\frac{1}{4} - \frac{m^2}{4\varrho_i^2} + \frac{1}{\varrho_i}\left(\frac{1 + |m|}{2} + n_i\right)\right]f_i = 0 \ . \tag{6.129}$$

The separation constants n_i satisfy

$$1 + |m| + n_1 + n_2 = \lambda \ . \tag{6.130}$$

Bearing in mind that $f_i \sim \mathrm{const} \times \varrho_i^{|m|/2}$ as $\varrho_i \to 0$ and $f_i \sim \mathrm{const} \times \exp(-\varrho_i/2)$ (dominant part) as $\varrho_i \to \infty$, it is convenient to make the change

$$f_i(\varrho_i) = e^{-\varrho_i/2}\varrho_i^{|m|/2}v_i(\varrho_i) \ , \tag{6.131}$$

so that v_i satisfies

$$\varrho_i v_i'' + [|m| + 1 - \varrho_i]v_i' + n_i v_i = 0 \ . \tag{6.132}$$

This is a confluent hypergeometric equation (Sect. A.10). The regularity condition implies that

$$v_i(\varrho_i) = A_i M(-n_i, |m| + 1, \varrho_i) \ , \tag{6.133}$$

and the integrability condition requires that the n_i be non-negative integers.

In the hydrogenic atom, then, each stationary state of the spectrum is determined in parabolic coordinates by three integers: the parabolic quantum numbers

n_1 and n_2, both non-negative, and the magnetic quantum number m. From relation (6.130), it is evident that λ must be a positive integer, which, from (6.93), coincides with the principal quantum number n:

$$n = n_1 + n_2 + |m| + 1 \ . \tag{6.134}$$

The normalized wave functions for the E_n level are, up to phase factors,

$$\psi_{n_1 n_2 m}(\xi, \eta, \phi) = \frac{\sqrt{2}}{n^2 a^{3/2}} f_{n_1 m}\left(\frac{\xi}{an}\right) f_{n_2 m}\left(\frac{\eta}{an}\right) \frac{1}{\sqrt{2\pi}} e^{im\phi} \ , \tag{6.135}$$

where a is the Bohr radius (6.97) and

$$f_{pm}(x) \equiv \frac{1}{|m|!} \sqrt{\frac{(p+|m|)!}{p!}} M(-p, |m|+1, x) x^{|m|/2} e^{-x/2} \ . \tag{6.136}$$

These functions are also normalized to unity, as can be shown using (A.150). Furthermore, keeping in mind this relation and the orthogonality property of eigenfunctions corresponding to different eigenvalues in (6.129), we can immediately show that

$$\int_0^\infty dx \ f_{pm}(x) f_{p'm}(x) = \delta_{pp'} \ , \tag{6.137}$$

a property we will need later.

6.9 Exactly Solvable Potentials for s-Waves

Only in a few cases (essentially for the square potential well, Coulomb potential, three-dimensional harmonic oscillator and $1/r^4$-type potentials) do we know how to solve completely the radial equation in closed analytic form for all values of l. The last type of potential is too singular at the origin to allow an adequate physical interpretation in the attractive case (the only case of interest at the moment).

There exist interesting potential families, however, whose bound states for the partial wave $l = 0$ can be written in terms of well-known functions. We will now look at a few of them. In the first place there are the *Eckart potentials* [EC 30a], defined as

$$U(r) = U_1 \frac{e^{-r/a}}{1 + ce^{-r/a}} + U_2 \frac{e^{-r/a}}{(1 + ce^{-r/a})^2} \ , \qquad a > 0 \ , \qquad c \geq -1 \ , \tag{6.138}$$

where $U_2 = 0$ if $c = -1$. $U(r)$ exponentially tends to zero as r goes to infinity. If $c(U_2 - U_1)/(U_2 + U_1) > 1$, then $U(r)$ is stationary at the point $r = r_m \equiv a \ln[c(U_2 - U_1)/(U_2 + U_1)]$, with the value $U(r_m) = (U_1 + U_2)^2/4cU_2$. The potential $U(r)$ can sustain a bound state only if $\min\{U(0), U(r_m)\} \leq 0$ and in this case the point spectrum lies in the interval $[\min\{U(0), U(r_m)\}, 0]$.

To find the bound s-states, we consider the corresponding equation (6.25) with $l = 0$ and $\varepsilon_0 \equiv -\lambda$, with $\lambda \geq 0$. Making the changes

$$x \equiv -ce^{-r/a}, \quad u_0(r) \equiv (-x/c)^\sigma (1 - x)^\varrho y(x),$$

$$\sigma \equiv a\sqrt{\lambda}, \quad \varrho \equiv \frac{1}{2}\left[1 - \sqrt{1 - \frac{4a^2 U_2}{c}}\right], \tag{6.139}$$

$y(x)$ is the solution to the equation

$$x(1 - x)\frac{d^2 y(x)}{dx^2} + [(1 + 2\sigma) - (2\sigma + 2\varrho + 1)x]\frac{dy(x)}{dx}$$

$$- \left[2\sigma\varrho - \frac{a^2 U_1}{c} + \varrho^2\right] y(x) = 0, \tag{6.140}$$

which is a hypergeometric equation; therefore, the most general solution of (6.25) for the potential (6.138), with $l = 0$ and $\lambda > 0$, is

$$u_0(r) = e^{-r\sqrt{\lambda}}(1 + ce^{-r/a})^\varrho [A_1 F(\alpha, \beta, \gamma; -ce^{-r/a})$$

$$+ A_2(-ce^{-r/a})^{1-\gamma} F(\gamma - \alpha, \gamma - \beta, \gamma; -ce^{-r/a})], \tag{6.141}$$

$$\alpha \equiv \sigma + \varrho + \sqrt{\sigma^2 + \frac{U_1 a^2}{c}}, \quad \beta \equiv \sigma + \varrho - \sqrt{\sigma^2 + \frac{U_1 a^2}{c}}, \quad \gamma \equiv 1 + 2\sigma,$$

where A_1 and A_2 are arbitrary constants. As $r \to \infty$, the hypergeometric functions tend to one, and

$$u_0(r) \sim A_1 e^{-r\sqrt{\lambda}} + A_2(-c)^{-2a\sqrt{\lambda}} e^{r\sqrt{\lambda}}. \tag{6.142}$$

Square integrability requires that $A_2 = 0$; the correct solution is then

$$u_0(r) = A_1 e^{-r\sqrt{\lambda}}(1 + ce^{-r/a})^\varrho F(\alpha, \beta, \gamma; -ce^{-r/a}). \tag{6.143}$$

Imposing the regularity condition $u_0(0) = 0$, the equation determining the corresponding energies for the bound states is thus

$$F(\alpha, \beta, \gamma; -c) = 0, \tag{6.144}$$

which, in general, must be solved by numerical methods.

For $\lambda = 0$, (6.141) does not represent the most general solution, but following similar arguments, it is easily seen that there do not exist bound states for the s partial wave at zero energy.

We will now examine some particular cases for which (6.144) has explicit solutions.

1) *Hulthén Potential* [HU 42]: $U_2 = 0$, $c = -1$

According to the general discussion, bound states can only exist if $U_1 < 0$. The typical form of this potential for different values of the parameter a can be seen in Fig. 6.8. Since $c = -1$, equation (6.144) can be written [AS 64] as

$$\frac{\Gamma(\gamma)\Gamma(\gamma - \alpha - \beta)}{\Gamma(\gamma - \alpha)\Gamma(\gamma - \beta)} = 0 .$$

(6.145)

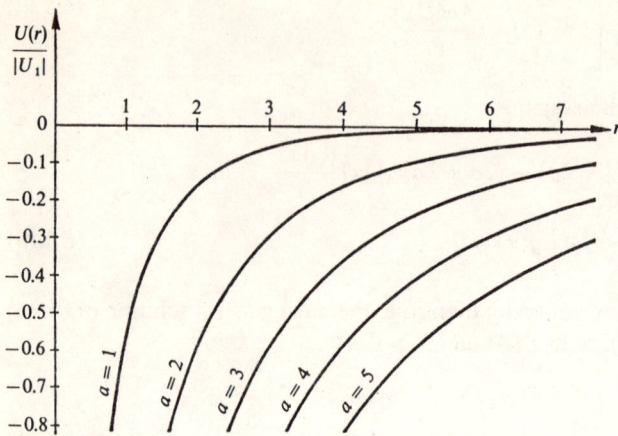

Fig. 6.8. Hulthén potential $U(r) = -|U_1|\exp(-r/a)[1 - \exp(-r/a)]^{-1}$ for different values of the parameter a

Using the definitions of α, β and γ given in (6.141), one now sees that $\gamma - \beta > 0$, and the bound states are obtained by imposing the condition $\Gamma(\gamma - \alpha) = \infty$, that is,

$$a\sqrt{\lambda} - a\sqrt{\lambda + |U_1|} = -n , \quad n = 1, 2, \dots ,$$

(6.146)

so that, since there are no bound states of zero energy for s-waves,

$$-\varepsilon_n = \lambda_n = \frac{(n^2 - a^2|U_1|)^2}{4a^2n^2} , \quad n = 1, 2, \dots < a\sqrt{|U_1|} .$$

(6.147)

2) *Hylleraas Potential* [HY 37]: $U_1 = 0$, $U_2 \equiv 4U_0$, $c = 1$, $a \equiv R/2$

In this case, the potential (6.123) reduces to

$$U(r) = U_0 \cosh^{-2}(r/R) ,$$

(6.148)

which can only have bound states if $U_0 < 0$. The characteristic form of $U(r)$ for distinct values of the parameter is given in Fig. 6.9. Equation (6.144) can be written as $F(2\sigma + \varrho, \varrho, 1 + 2\sigma; -1) = 0$, which is equivalent [AS 64] to

$$\frac{\Gamma(1 + 2\sigma)}{\Gamma(1/2 + \sigma + \varrho/2)\Gamma(1 + \sigma - \varrho/2)} = 0 .$$

(6.149)

Given that $1 + \sigma - \varrho/2 > 0$, the bound states are determined by the equation $\Gamma(1/2 + \sigma + \varrho/2) = \infty$ or, equivalently,

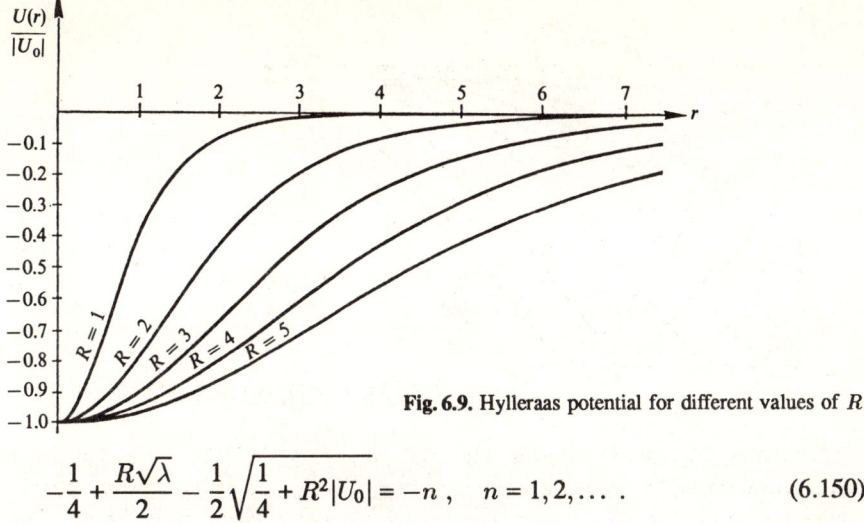

Fig. 6.9. Hylleraas potential for different values of R

$$-\frac{1}{4} + \frac{R\sqrt{\lambda}}{2} - \frac{1}{2}\sqrt{\frac{1}{4} + R^2|U_0|} = -n , \quad n = 1,2,\dots . \tag{6.150}$$

Therefore, since ε_n should be negative,

$$-\varepsilon_n = \lambda_n = \frac{1}{R^2}\left[\sqrt{\frac{1}{4} + R^2|U_0|} - 2n + \frac{1}{2}\right]^2 ,$$

$$n = 1,2,\dots < \left(\frac{1}{2}\sqrt{\frac{1}{4} + R^2|U_0|} + \frac{1}{4}\right) . \tag{6.151}$$

3) *Eckart Potential* [EC 30a]: $U_1 = 0$, $U_2 = -2c/a^2$, $c > 0$

Typical forms of this potential appear in Fig. 6.10. In this case, equation (6.144) reduces to

$$F(2a\sqrt{\lambda} - 1, -1, 1 + 2a\sqrt{\lambda}; -c) = 0 , \tag{6.152}$$

an equation linear in $\sqrt{\lambda}$, whose solution is $\sqrt{\lambda} = (c-1)/2a(c+1)$. For $c \leq 1$, this potential does not have any bound states, and for $c > 1$, it has one bound state with energy

$$-\varepsilon = \lambda = \frac{1}{4a^2}\left(\frac{c-1}{c+1}\right)^2 . \tag{6.153}$$

That this occurs is not surprising, since the potentials in the Eckart family are obtained from the potential in **R**, given by the same analytical formula, but with an infinite barrier located at a point that depends on c. Therefore, all these potentials have at most the same number of bound states as the potential in **R**. But it is easily shown that this has only one bound state. [It always has at least one since it is more attractive than some square well potential; and it cannot have more, for otherwise it would have an odd bound state, that, conveniently translated, would be bound for the Hylleraas potential with $U_0 =$

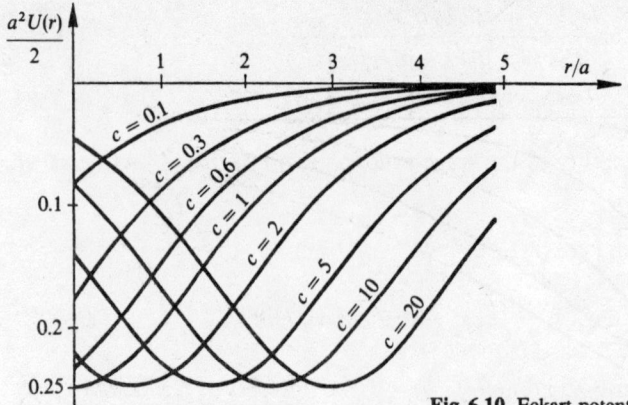

Fig. 6.10. Eckart potential for different values of c

$-2/R^2$; however, that is impossible according to (6.151).] As the barrier is displaced to the right, the Eckart potential becomes less attractive until the barrier reaches the minimum of the potential ($c = 1$), the point at which the bound state disappears.

Another family of potentials which is solvable in closed form for s-waves is [GP 67]

$$U(r) = U_1 e^{-2r/a} + U_2 e^{-r/a} \,, \quad a > 0 \,. \tag{6.154}$$

These also tend exponentially to zero when $r \to \infty$. If $(-2U_1/U_2) > 1$, then $U(r)$ is stationary at the point $r = r_m = a \ln(-2U_1/U_2)$, at which $U(r_m) = -U_2^2/4U_1$. Again bound states exist only if $\min\{U(0), U(r_m)\} < 0$, and they lie in $[\min\{U(0), U(r_m)\}, 0]$. The bound states for s-waves are determined by solving (6.25) for $l = 0$. Defining

$$x \equiv 2a\sqrt{U_1}e^{-r/a} \,, \quad u_0(r) \equiv x^{a\sqrt{\lambda}}e^{-x/2}y(x) \,, \quad \varepsilon \equiv -\lambda \,, \tag{6.155}$$

equation (6.25) is converted into

$$x\frac{d^2y(x)}{dx^2} + [(1 + 2a\sqrt{\lambda}) - x]\frac{dy(x)}{dx} - \left[\frac{1}{2} + a\sqrt{\lambda} + \frac{aU_2}{2\sqrt{U_1}}\right]y(x) = 0 \,, \tag{6.156}$$

which is a confluent hypergeometric equation. Using the results of Appendix A, the most general solution of (6.25) for s-waves is

$$\begin{aligned}
u_0(r) = e^{-r\sqrt{\lambda}}\exp(-a\sqrt{U_1}e^{-r/a}) \\
\times \left[A_1 M\left(\frac{1}{2} + a\sqrt{\lambda} + \frac{aU_2}{2\sqrt{U_1}}, 1 + 2a\sqrt{\lambda}, 2a\sqrt{U_1}e^{-r/a}\right)\right. \\
\left. + A_2 U\left(\frac{1}{2} + a\sqrt{\lambda} + \frac{aU_2}{2\sqrt{U_1}}, 1 + 2a\sqrt{\lambda}, 2a\sqrt{U_1}e^{-r/a}\right)\right] \,,
\end{aligned} \tag{6.157}$$

where A_1 and A_2 are arbitrary constants. Keeping in mind the results of Appendix A, it is immediately shown that as $r \to \infty$, and $\lambda > 0$,

$$u_0(r) \sim A_1 e^{-r\sqrt{\lambda}} + A_2 \frac{\Gamma(2a\sqrt{\lambda})}{\Gamma(1/2 + a\sqrt{\lambda} + aU_2/2\sqrt{U_1})}$$
$$\times \, (2a\sqrt{U_1})^{-2a\sqrt{\lambda}} e^{r\sqrt{\lambda}} \,. \tag{6.158}$$

Therefore, the normalizability of the wave function requires that $A_2 = 0$, and, as a consequence,

$$u_0(r) = A_1 e^{-r\sqrt{\lambda}} \exp\left(-a\sqrt{U_1} e^{-r/a}\right)$$
$$\times \, M\left(\frac{1}{2} + a\sqrt{\lambda} + \frac{aU_2}{2\sqrt{U_1}}, 1 + 2a\sqrt{\lambda}, 2a\sqrt{U_1} e^{-r/a}\right) \,. \tag{6.159}$$

By imposing the regularity condition $u_0(0) = 0$, the energies of the bound states with negative energy are the solutions of

$$M\left(\frac{1}{2} + a\sqrt{\lambda} + \frac{aU_2}{2\sqrt{U_1}}, 1 + 2a\sqrt{\lambda}, 2a\sqrt{U_1}\right) = 0 \,, \tag{6.160}$$

which, in general, must be solved numerically.

From (6.157) with $\lambda = 0$, and a study of the asymptotic behavior of $u_0(r)$, it is seen that it is never normalizable, and therefore there are no bound states for s-waves at zero energy.

Finally, we will discuss some particular cases of the potential family in (6.154).

1) *Exponential Potential: $U_2 = 0$*

Bound states can only exist if $U_1 < 0$. Equation (6.160) becomes

$$M\left(\tfrac{1}{2} + a\sqrt{\lambda}, 1 + 2a\sqrt{\lambda}, 2ia\sqrt{|U_1|}\right) = 0 \,. \tag{6.161}$$

Bearing in mind the properties of the confluent hypergeometric function of the type $M(\alpha, 2\alpha, z)$ [AS 64], the previous equation is equivalent to

$$J_{a\sqrt{\lambda}}(a\sqrt{|U_1|}) = 0 \,. \tag{6.162}$$

In Fig. 6.11, the graphical representation of the positive real solutions of this equation [GU 66] are given.

2) *Morse Potential* [MO 29]: $U_1 = U_0 e^{4b}$, $U_2 = -2U_0 e^{2b}$, $a \equiv R/2b$, $b, R > 0$, $U_0 > 0$

For this case the potential can be written as

$$U(r) = U_0 \left[\left(1 - \exp\left[-2b\left(\frac{r-R}{R}\right)\right]\right)^2 - 1\right] \,. \tag{6.163}$$

This potential, for some values of the parameter b, is represented in Fig. 6.12. The equation determining the energies of the bound states is deduced from (6.160):

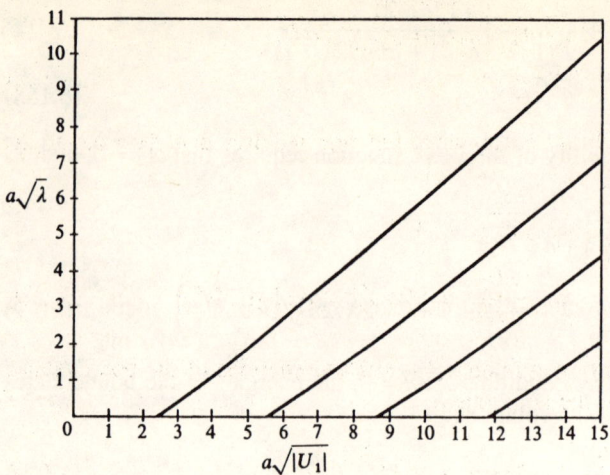

Fig. 6.11. Graphical representations of the positive real solutions of $J_{a\sqrt{\lambda}}(a\sqrt{|U_1|}) = 0$

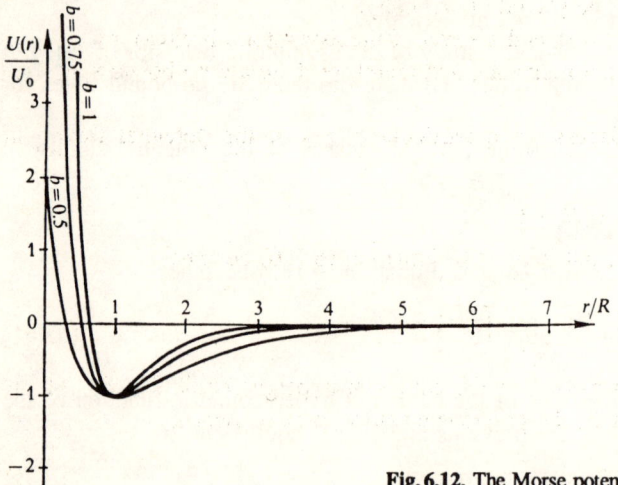

Fig. 6.12. The Morse potential for different values of b

$$M\left(\frac{1}{2} + \frac{R}{2b}(\sqrt{\lambda} - \sqrt{U_0}), 1 + \frac{R\sqrt{\lambda}}{b}, \frac{R}{b}\sqrt{U_0}e^{2b}\right) = 0, \qquad (6.164)$$

to be solved numerically.

The potential has been heavily used in molecular physics calculations to describe the interaction potential between two molecules. For the H_2 molecule [SL 39], the value of the parameters are $U_0 = 1074\,\text{Å}^{-2}$, $R = 0.75\,\text{Å}$, and $b = 0.728$ and thus, $R\sqrt{U_0}\exp(2b)/b \simeq 145$. In the cases of interest in molecular physics, $1 \ll R\sqrt{U_0}/2b \ll R\sqrt{U_0}\exp(2b)/b$, so that, using an appropriate asymptotic expansion [AS 64], the equation determining the bound state energies is in this approximation

$$\sin\left[\pi\left(\frac{1}{2}+\frac{R}{2b}(\sqrt{\lambda}-\sqrt{U_0})\right)\right]=0\,,\tag{6.165}$$

and hence,

$$-\varepsilon_n=\lambda_n\simeq\frac{4b^2}{R^2}\left[n-\frac{1}{2}-\frac{R}{2b}\sqrt{U_0}\right]^2\,,$$

$$n=1,2,\ldots<\frac{R}{2b}\sqrt{U_0}+\frac{1}{2}\,.\tag{6.166}$$

There exist other potential families whose reduced radial equation can be solved in closed form for $l=0$ or for other values of l. The reader interested in this type of problems can consult [BA 49] and [NE 82].

7. Symmetry Transformations

7.1 Introduction

In this chapter we will discuss symmetry transformations, in order to introduce the symmetry principles; namely, the invariance properties which the equations of motion of a quantum mechanical system can display with respect to a given symmetry. Although the interest in these transformations already appears in classical physics, it is in quantum physics that they acquire greatest relevance, not only because of the physical characteristics of the problems studied, but also due to the mathematical structure of the theory.

It frequently occurs that the dynamics of a quantum mechanical system is not correctly known, but experimental data clearly show the existence of a conservation law. As we will see, this implies the existence of a symmetry principle, the knowledge of which may help in the correct formulation of the equations of motion. Sometimes, it is the proper understanding, either abstract or intuitive, of a physical problem that suggests the convenience of imposing the invariance under a certain set of transformations. The fact that the dynamics is perfectly understood does not imply that symmetry principles lose their utility, since knowledge of their existence allows us, among other things, to establish selection rules and relationships among the different matrix elements of a given operator, and to classify the solutions of the equations of motion.

7.2 Symmetry Transformations: Wigner's Theorem

Let S be a physical system and $\mathcal{H}$ be the Hilbert space of its (pure) states; every transformation G which: (i) establishes a one-to-one correspondence between the different unit rays $|\xi\rangle_R$ of $\mathcal{H}$; (ii) transforms the rays corresponding to the physically realizable pure states among themselves and (iii) conserves its transition probabilities

$$_R\langle\psi|\phi\rangle_R = {_R\langle\psi_G|\phi_G\rangle_R} \,, \quad |\xi\rangle_R \xrightarrow{\;G\;} |\xi_G\rangle_R \equiv G|\xi\rangle_R \tag{7.1}$$

is called a *symmetry transformation*. (It is customary to define a "scalar" product between rays $|\phi\rangle_R$ and $|\psi\rangle_R$ as $|\langle\phi|\psi\rangle|$, $|\phi\rangle \in |\phi\rangle_R$ and $|\psi\rangle \in |\psi\rangle_R$. We extend G to any arbitrary ray in the natural way: $G\{|\alpha|\xi\rangle\}_R \equiv |\alpha|G|\xi\rangle_R$, where by $|a| \, |\eta\rangle_R$ we understand the ray $\{|\alpha| \, |\eta\rangle : |\eta\rangle \in |\eta\rangle_R\}$.)

Should superselection rules (Sect. 2.15) exist, not all unit rays of $\mathcal{H}$ correspond to physically realizable pure states. In general, $\mathcal{H}$ is a direct sum of coherent Hilbert spaces $\mathcal{H}_c$ such that in each one of them the superselection operators have well-defined eigenvalues, and the only physically realizable pure states are those that belong to one of these coherent spaces. The symmetry transformation G maps a coherent space $\mathcal{H}_c$ (as a collection of rays) onto another coherent space $G\mathcal{H}_c$ because if $|\psi\rangle_R$ and $|\phi\rangle_R \in \mathcal{H}_c$ were mapped into $|\psi_G\rangle_R$ and $|\phi_G\rangle_R$ belonging to different coherent spaces, then the physically realizable pure state $(\alpha|\psi\rangle + \beta|\phi\rangle)_R$ (with $\alpha\beta \neq 0$) with nonzero projections over $|\psi\rangle_R$ and $|\phi\rangle_R$ would be transformed (G being a symmetry) into other physically realizable pure state with nonzero projections over $|\psi_G\rangle_R$ and $|\phi_G\rangle_R$. Evidently, this is impossible if these belong to different coherent spaces. On the other hand, every state $|\phi'\rangle_R$ coherent with $|\phi_G\rangle_R$ should be the image under G of some pure state, and as a consequence, $G^{-1}|\phi'\rangle_R$ is coherent with $|\phi\rangle_R$; that is, it belongs to $\mathcal{H}_c$. Hence, the image of $\mathcal{H}_c$ under G fills a whole coherent space. Furthermore, $\dim(\mathcal{H}_c) = \dim(G\mathcal{H}_c)$. It could however happen that $\mathcal{H}_c \neq G\mathcal{H}_c$.

On the other hand, if the correspondence $|\xi\rangle_R \to |\xi_G\rangle_R$ is implementable by a linear or anti-linear operator U_G (that is, there exist $|\xi\rangle \in |\xi\rangle_R$ and $|\xi\rangle_G \in |\xi_G\rangle_R$ such that $|\xi_G\rangle = U_G|\xi\rangle$), this operator U_G is not necessarily unique; it may be altered by phase changes in the representatives selected in each ray.

It is evident that given a unitary or anti-unitary operator U which transforms state vectors (physically admissible) amongst themselves, the correspondence $|\phi\rangle_R \to (U|\phi\rangle)_R$ satisfies (7.1) and defines a symmetry transformation.

The converse statement constitutes a famous and important theorem called *Wigner's theorem* [WI 31]: "Every symmetry transformation G between coherent spaces is implementable by a one-to-one linear or anti-linear isometry U_G". That is, given $|\xi\rangle_R$ and $|\xi_G\rangle_R$, it is possible to choose certain representatives $|\xi\rangle$ and $|\xi_G\rangle$ such that $|\xi_G\rangle = U_G|\xi\rangle$, and therefore,

$$
\boxed{
\begin{aligned}
\langle\psi_G|\phi_G\rangle &= \langle\psi|\phi\rangle && \text{if } U_G \text{ is unitary}, \\
\langle\psi_G|\phi_G\rangle &= \langle\psi|\phi\rangle^* && \text{if } U_G \text{ is anti-unitary}.
\end{aligned}
}
\tag{7.2}
$$

Even though it is an abuse of language in the case of $\mathcal{H}_c \neq G\mathcal{H}_c$, we will retain the usual physics nomenclature of linear and anti-linear unitary operators (Appendix G) for these isometries.

Given the importance of this theorem, we will give its proof, which is not excessively difficult but is long and delicate. We will closely follow the proof given in [BA 64]:

Fix a unit ray $|e\rangle_R$ in $\mathcal{H}_c$ and let $|e\rangle \in |e\rangle_R$. Similarly, choose a fixed vector $|\bar{e}\rangle \in |e_G\rangle_R$. The vectors $|e\rangle$ and $|\bar{e}\rangle$ have unit norm, and we will write $|\bar{e}\rangle \equiv U|e\rangle$.

1) If $\dim(\mathcal{H}_c) = 1$, and $\alpha|e\rangle$ is any vector of $\mathcal{H}_c$, defining $U(\alpha|e\rangle) \equiv \chi(\alpha)|\bar{e}\rangle$ with $\chi(\alpha) = \alpha$ or α^* (a choice independent of α), it is obvious that U implements

G. Thus, in one-dimensional coherent spaces the symmetry transformation is always implementable, both through linear or anti-linear maps.

2) Let $\dim(\mathcal{H}_c) \geq 2$, and let $\mathcal{H}'_c$ denote the subspace of $\mathcal{H}_c$ orthogonal to $|e\rangle$. From (7.1) it follows that $G\mathcal{H}'_c$ is also orthogonal to $|\bar{e}\rangle$. If $|z\rangle \in \mathcal{H}'_c$, then it is possible to find $|\bar{z}\rangle \in |z_G\rangle_R$ such that

$$|\bar{e}\rangle + |\bar{z}\rangle \in G(|e\rangle + |z\rangle)_R \ . \tag{7.3}$$

In fact, (7.1) implies that $G(|e\rangle + |z\rangle)_R$ is in the plane spanned by $|e_G\rangle_R$ and $|z_G\rangle_R$, forming with these the same "angles" that $(|e\rangle + |z\rangle)_R$ makes with respect to $|e\rangle_R$ and $|z\rangle_R$. This means that $G(|e\rangle + |z\rangle)_R$ contains vectors of the form $|\bar{e}\rangle + e^{i\alpha}|\bar{z}_1\rangle)$, with $|\bar{z}_1\rangle \in |z_G\rangle_R$. It suffices now to let $|\bar{z}\rangle = e^{i\alpha}|\bar{z}_1\rangle$. Evidently, $|\bar{z}\rangle$ is unique. We will write $|\bar{z}\rangle \equiv V|z\rangle$, and prove that V defines a linear or anti-linear isometry from $\mathcal{H}'_c$ onto $G\mathcal{H}'_c$; i.e.

$$\begin{aligned} V(\alpha_1|z_1\rangle + \alpha_2|z_2\rangle) &= \chi(\alpha_1)V|z_1\rangle + \chi(\alpha_2)V|z_2\rangle \ , \\ \langle Vz_1|Vz_2\rangle &= \chi(\langle z_1|z_2\rangle) \ , \end{aligned} \tag{7.4}$$

where χ is either the identity operation or the complex conjugation.

3) Supposing that (7.4) is proved, we will see how to achieve the implementation of G in $\mathcal{H}_c$. It suffices to define

$$U(\alpha|e\rangle + |z\rangle) \equiv \chi(\alpha)|\bar{e}\rangle + V|z\rangle \ . \tag{7.5}$$

In fact, if $\alpha = 0$, then by construction $U|z\rangle = V|z\rangle \in |z_G\rangle_R$. If $\alpha \neq 0$, we have

$$(\alpha|e\rangle + |z\rangle)_R = |\alpha| \, (|e\rangle + \alpha^{-1}|z\rangle)_R \ , \tag{7.6}$$

and, because G is a symmetry, it follows that

$$G(\alpha|e\rangle + |z\rangle)_R = |\alpha| \, G(|e\rangle + \alpha^{-1}|z\rangle)_R \ . \tag{7.7}$$

By the construction of V, and using (7.4),

$$|\bar{e}\rangle + \chi(\alpha^{-1})V|z\rangle \in G(|e\rangle + \alpha^{-1}|z\rangle)_R = \frac{1}{|\alpha|} G(\alpha|e\rangle + |z\rangle)_R \ , \tag{7.8}$$

so that

$$\begin{aligned} U(\alpha|e\rangle + z)) &\equiv \chi(\alpha)|\bar{e}\rangle + V|z\rangle \in \frac{|\chi(\alpha)|}{|\alpha|} G(\alpha|e\rangle + |z\rangle)_R \\ &= G(\alpha|e\rangle + |z\rangle)_R \ , \end{aligned} \tag{7.9}$$

which proves that U, defined in (7.5), implements G in $\mathcal{H}_c$.

This operator has the same linear or anti-linear character as V.

4) Finally, we now proceed to prove the non-trivial part, i.e., the content of (7.4). In the first place, note that if $|z_1\rangle$ and $|z_2\rangle \in \mathcal{H}'_c$, from the equalities

$$\begin{aligned} |\langle e + z_1|e + z_2\rangle| &= |\langle \bar{e} + \bar{z}_1|\bar{e} + \bar{z}_2\rangle| \ , \\ |\langle z_1|z_2\rangle| &= |\langle \bar{z}_1|\bar{z}_2\rangle| \ , \end{aligned} \tag{7.10}$$

which are consequences of (7.1) and of the construction of $|\bar{z}\rangle$, one sees that

$$|\langle z_1|z_2\rangle| = |\langle Vz_1|Vz_2\rangle| \,,$$
$$\text{Re}\langle z_1|z_2\rangle = \text{Re}\langle Vz_1|Vz_2\rangle \,, \tag{7.11}$$

and in particular,

$$\langle z_1|z_2\rangle \quad \text{real} \Rightarrow \quad \langle z_1|z_2\rangle = \langle Vz_1|Vz_2\rangle \,. \tag{7.12}$$

Given $|e\rangle + \alpha|z\rangle$, $|z\rangle \neq 0$, there exists a unique complex number $\chi_z(\alpha)$ such that

$$|\bar{e}\rangle + \chi_z(\alpha)V|z\rangle \in G(|e\rangle + \alpha|z\rangle)_R \,, \tag{7.13}$$

as inferred from the construction of V. From (7.3) and (7.11), it is easily shown gradually that

a) $\chi_z(1) = 1$;
b) $\chi_z(\alpha) = \alpha$, $\forall \alpha$ real;
c) $\chi_z(i) = \pm i$;
d) $\chi_z(\alpha + i\beta) = \alpha + \chi_z(i)\beta$, $\forall \alpha, \beta$ real,

and as a result, χ_z is either the identity or the complex conjugate operation. Furthermore, let us see that χ_z does not depend on z:
If $\dim(\mathcal{H}_c) = 2$, it suffices to show that $\chi_{\lambda z} = \chi_z$, for $\lambda|z\rangle \neq 0$. Since

$$V(\alpha\lambda|z\rangle) = \begin{cases} \chi_z(\alpha\lambda)V|z\rangle \\ \chi_{\lambda z}(\alpha)V(\lambda|z\rangle) = \chi_{\lambda z}(\alpha)\chi_z(\lambda)V|z\rangle \end{cases} \,, \tag{7.14}$$

it follows that $\chi_z(\alpha\lambda) = \chi_{\lambda z}(\alpha)\chi_z(\lambda)$. But $\chi_z(\alpha\lambda) = \chi_z(\alpha)\chi_z(\lambda)$, and therefore $\chi_{\lambda z} = \chi_z$. That is, χ_z is a constant on each one-dimensional subspace of $\mathcal{H}'_c$.
If $\dim(\mathcal{H}_c) \geq 3$, it just remains for us to show that χ_z does not depend upon the one-dimensional subspace containing $|z\rangle$. Let $|\varphi_1\rangle$ and $|\varphi_2\rangle$ be two orthonormal vectors in $\mathcal{H}'_c$, and also let $|z\rangle = \alpha_1|\varphi_1\rangle + \alpha_2|\varphi_2\rangle$.
As a new consequence of (7.1), $V|z\rangle = \alpha'_1 V|\varphi_1\rangle + \alpha'_2 V|\varphi_2\rangle$, with $|\alpha'_i| = |\alpha_i|$. If $\alpha_i \neq 0$, then

$$1 = \langle \alpha_i^{*-1}\varphi_i|\alpha_i\varphi_i\rangle = \langle \alpha_i^{*-1}\varphi_i|z\rangle \,. \tag{7.15}$$

Applying (7.12),

$$\chi_{\varphi_i}^*(\alpha_i^{*-1})\chi_{\varphi_i}(\alpha_i) = \chi_{\varphi_i}^*(\alpha_i^{*-1})\alpha'_i \tag{7.16}$$

so that $\alpha'_i = \chi_{\varphi_i}(\alpha_i)$, and hence

$$V|z\rangle = \chi_{\varphi_i}(\alpha_1)V|\varphi_1\rangle + \chi_{\varphi_2}(\alpha_2)V|\varphi_2\rangle \,. \tag{7.17}$$

In particular, if $|w\rangle = |\varphi_1\rangle + |\varphi_2\rangle$, we have, using property (a), that $V|w\rangle = V|\varphi_1\rangle + V|\varphi_2\rangle$. By (7.17),

$$V(\lambda|w\rangle) = \begin{cases} \chi_w(\lambda)V|w\rangle = \chi_w(\lambda)(V|\varphi_1\rangle + V|\varphi_2\rangle) \\ \chi_{\varphi_1}(\lambda)V|\varphi_1\rangle + \chi_{\varphi_2}(\lambda)V|\varphi_2\rangle \end{cases} \tag{7.18}$$

and therefore $\chi_{\varphi_1} = \chi_{\varphi_2}(= \chi_w)$, that is, χ_z coincides on orthogonal one-dimensional subspaces. Finally, using (7.17)

$$V(\lambda|z)) = \begin{cases} \chi_z(\lambda)V|z) = \chi_z(\lambda)[\chi_{\varphi_1}(\alpha_1)V|\varphi_1) + \chi_{\varphi_1}(\alpha_2)V|\varphi_2)] \\ V(\lambda\alpha_1|\varphi_1) + \lambda\alpha_2|\varphi_2)) = \chi_{\varphi_1}(\lambda\alpha_1)V|\varphi_1) + \chi_{\varphi_1}(\lambda\alpha_2)V|\varphi_2) \end{cases}$$

$$(7.19)$$

whence it turns out that $\chi_z = \chi_{\varphi_1}$, for every $|z) \neq 0$ in the $|\varphi_1), |\varphi_2)$ plane. Given two non-zero arbitrary vectors $|z)$ and $|z')$ in $\mathcal{H}'_c$, there always exists a plane which contains them, and therefore $\chi_z = \chi_{z'} \equiv \chi$, which demonstrates the independence of χ from z. Equation (7.17) produces the first relation in (7.4), and in order to prove the second, it suffices to express $|z_1)$ and $|z_2)$ as linear combinations of two orthonormal vectors in the plane defined by them, to recall that V preserves orthonormality, and to use the first of the relations (7.4).

This concludes the proof of Wigner's theorem: the symmetry map $G : \mathcal{H}_c \rightarrow G\mathcal{H}_c$ is implementable by an isometry U which is either linear or anti-linear, according to whether χ is the identity operation or the complex conjugation.

With respect to the uniqueness of U, we have already shown at the beginning of this proof that if $\dim(\mathcal{H}_c) = 1$, U can be chosen at will to be either linear or anti-linear. Otherwise, we will see that if $\dim(\mathcal{H}_c) \geq 2$, and if U_1 and U_2 implement G, then both must necessarily be either linear or anti-linear and $U_2 = e^{i\theta}U_1$, θ real. In other words, U is unique ($\equiv U_G$), apart from global phases. In fact, given two linearly independent non-zero vectors $|a)$ and $|b)$ in $\mathcal{H}_c$, we know that since U_1 and U_2 implement G, then

$$U_2|a) = e^{i\theta}U_1|a) \quad \text{and}$$

$$U_2(|a) + |b)) = e^{i\theta}U_1|a) + U_2|b) = e^{i\theta}(U_1|a) + e^{-i\theta}U_2|b)) \ .$$

But, we also have $U_2(|a) + |b)) = e^{i\gamma}U_1(|a) + |b))$. Comparing, we have $U_2|b) = e^{i\theta}U_1|b)$, as was to be shown.

Returning to the initial Hilbert space of the physical system $\mathcal{H} = \oplus_c \mathcal{H}_c$, the symmetry G is represented on each $\mathcal{H}_c$ through the map $U_{G,c}$ which implements it. Given that for our purposes it suffices to deal with *coherent Hilbert spaces* (the only ones that we will consider from now on unless otherwise stated), in which no superselection rules operate, and since we do not pretend to treat symmetries such as particle-antiparticle conjugation, which interchange coherent spaces, in all that follows our *symmetries* will always be *implemented* by one-to-one linear or anti-linear isometries of $\mathcal{H}$ onto $\mathcal{H}$, i.e., by *unitary* or *anti-unitary operators*.

7.3 Transformation Properties of Operators

Given a linear operator A, it is of interest to know its transform A_G under a symmetry G implemented by a unitary or anti-unitary operator U_G. The transform A_G is defined, by analogy with (7.1), such that its matrix elements fulfill

$$|\langle\phi_G|A_G|\psi_G\rangle| = |\langle\phi|A|\psi\rangle| \qquad (7.20)$$

for all pairs $|\phi\rangle$ and $|\psi\rangle$ of elements in $\mathcal{H}$.

Equation (7.20) is evidently satisfied by taking

$$\boxed{A_G = U_G A U_G^\dagger} \;. \qquad (7.21)$$

We will see that (7.21), except for phases, is the only possible solution for (7.20), and A_G defined by (7.21) is the operator we will take as the transform of A. This essential uniqueness is proved if we show that given two operators B and C such that $|\langle\phi|B|\psi\rangle| = |\langle\phi|C|\psi\rangle|$ for all $|\phi\rangle \in \mathcal{H}$ and $|\psi\rangle \in D(B) = D(C)$, then $C = e^{i\alpha} B$. Letting $|\phi\rangle$ run over an orthonormal basis, the condition imposed implies that $\|B\psi\| = \|C\psi\|$, so that taking $|\phi\rangle = B|\psi\rangle$ results in

$$\|C\psi\|^2 = \|B\psi\|^2 = \langle B\psi|B\psi\rangle = |\langle B\psi|C\psi\rangle| \leq \|B\psi\|\|C\psi\| \;. \qquad (7.22)$$

Therefore, the Schwarz inequality must be saturated, and as a consequence, $C|\psi\rangle = \exp[i\alpha(\psi)]B|\psi\rangle$. We will show, finally, that α does not depend on $|\psi\rangle$. To do this, it suffices to take a real linear combination of two arbitrary vectors $|\psi_1\rangle$ and $|\psi_2\rangle$ and notice that

$$C(\lambda_1|\psi_1\rangle + \lambda_2|\psi_2\rangle) = \begin{cases} \exp[i\alpha(\lambda_1\psi_1 + \lambda_2\psi_2)](\lambda_1 B|\psi_1\rangle + \lambda_2 B|\psi_2\rangle)\,, \\ \lambda_1\exp[i\alpha(\psi_1)]B|\psi_1\rangle + \lambda_2\exp[i\alpha(\psi_2)]B|\psi_2\rangle\,, \end{cases} \qquad (7.23)$$

whence $\exp[i\alpha(\psi_1)] = \exp[i\alpha(\psi_2)]$. Note, furthermore, that we assumed nothing about the linear or anti-linear character of B and C, and therefore the argument presented also shows that the starting equality implies that both should be simultaneously linear or anti-linear whenever $\dim(\mathcal{H}) \geq 2$.

From (7.21), it follows that

$$\boxed{\begin{aligned} \langle\phi_G|A_G|\psi_G\rangle &= \langle\phi|A|\psi\rangle && \text{if } U_G \text{ is unitary} \\ \langle\phi_G|A_G|\psi_G\rangle &= \langle\phi|A|\psi\rangle^* && \text{if } U_G \text{ is anti-unitary}\,. \end{aligned}} \qquad (7.24)$$

If A is an observable, equation (7.21) allows us to assert that A_G is also an observable. It has the same spectrum as A, and the subspace spanned by the eigenvectors of A_G, for a given eigenvalue, is just the transform under U_G of the eigensubspace of A for the same eigenvalue.

Similarly, (7.21) implies that every algebraic relation amongst the observables of the system is form-invariant under a symmetry transformation if U_G is linear or is brought over its complex conjugate if U_G is anti-linear.

If $\{A_i\}$ is an irreducible set of observables, in the sense that every linear operator which commutes with these observables is necessarily a multiple of the identity, then the symmetry G will be determined upon giving the transformation law for each A_i, i.e., specifying the $\{A_{iG}\}$; since if U and V are two unitary (or anti-unitary) operators which satisfy $U A_i U^\dagger = V A_i V^\dagger, \forall A_i$, the unitary operator $V^{-1}U$ will commute with all operators A_i, and, the system being irreducible, $V^{-1}U = \alpha I$, with $|\alpha| = 1$.

For example, for a system of particles with spin, the position, momentum and spin of each particle can be taken as the set $\{A_i\}$.

In practice, symmetries are generally defined by their action on the basic observables, and not by their explicit and direct action on the states, so that their linear or anti-linear character is not always evident. However, in the cases of interest, the transforms of each basic observable are simple functions of the irreducible set of such observables:

$$U_G A_i U_G^\dagger = f_i(A_1, A_2, \ldots) . \tag{7.25}$$

As mentioned above, knowing f_i fixes U_G, up to a phase. In general, the structure of these functions immediately reveals the linear or anti-linear character of U_G: from the commutators between the basic observables,

$$[A_i, A_j] = i g_{ij}(A_1, A_2, \ldots) , \tag{7.26}$$

here expressed as "functions" of the irreducible set $\{A_k\}$ (in practice they are simple functions), and applying U_G to both sides of (7.26) we obtain

$$[f_i(A_1, A_2, \ldots), f_j(A_1, A_2, \ldots)] = \pm i \hat{g}_{ij}(f_1(A_1, A_2, \ldots), \ldots) , \tag{7.27}$$

where $\hat{g} = g$ if U_G is linear and $\hat{g} = g^*$ if U_G is anti-linear. Calculating the l.h.s. using (7.26), and comparing it with the r.h.s. reveals, in general, the linear or anti-linear character of U_G.

7.4 Symmetry Groups

Bearing in mind the definition of a symmetry transformation, it is easily shown that the set of all symmetry transformations forms a group. In practice, only a small subgroup of symmetries is of interest, by virtue of being intimately connected to feasible operations in the laboratory, and leading to dynamical invariances, which we consider next.

Given a symmetry group $\mathcal{G}$ (a subgroup of the whole group), we already know that each element $G_i \in \mathcal{G}$ is implementable by a unitary or anti-unitary operator U_{G_i} uniquely determined up to a phase. Due to this ambiguity, it cannot be ensured that the chosen U_G's have the group property, we can only guarantee that

$$\boxed{U_{G_i G_j} = \omega(G_i, G_j) U_{G_i} U_{G_j}} , \tag{7.28}$$

where $\omega(G_i, G_j)$ has modulus equal to one. In other words, $G \to U_G$ is a unitary or anti-unitary projective representation of $\mathcal{G}$. For $G \to U_G$ to be a true representation, it is necessary that all phase factors ω be equal to one, which, although desirable, is not always possible.

A particularly interesting case is when the subgroup $\mathcal{G}$ of symmetry transformations is a connected Lie group which locally depends on n normal parameters $\alpha_1, \alpha_2, \ldots, \alpha_n$. Now any finite transformation can be generated from infinitesimal transformations. The operator corresponding to such a finite transformation is necessarily unitary (if the action of the transformation group is continuous on rays). To see this, it suffices to keep in mind that the identity transformation is always implementable by the identity unitary operator, and that any other transformation of this group is connected to the identity transformation in a continuous manner. Hence, it is necessarily represented by a unitary operator, since a continuous change from unitary to anti-unitary is impossible. Given a group element G in a neighborhood of the identity characterized by the normal parameters $\alpha_1, \ldots, \alpha_n$, then, from among the corresponding U_G's, we can select one of the form

$$U(\alpha_1, \ldots, \alpha_n) = \exp\left[-i \sum_{j=1}^{n} \alpha_j I_j \right] , \tag{7.29}$$

where the operators $I_j (= I_j^\dagger)$ are the infinitesimal generators of the group.

If, under an infinitesimal symmetry transformation defined by the normal parameters $(0, 0, \ldots \delta\alpha_i, \ldots 0)$, the observable A changes to $A_m + \delta_i A_m$, then (7.29) and (7.25) lead to

$$[A_m, I_i] = -i \frac{\delta_i A_m}{\delta \alpha_i} . \tag{7.30}$$

The set of these relations for an irreducible set of observables allows us to determine the generators, up to additive constants.

7.5 Space Translations

We now consider translations for a system of N particles of given types. The space of states is therefore coherent. We will take as an irreducible set of observables the one formed by their positions r_i, momenta p_i and spins S_i, $1 \le i \le N$. Under a translation $G(a)$, associated with the vector a, the observables are transformed according to the laws

$$\boxed{\begin{aligned} U(a) r_i U^\dagger(a) &= r_i - a , \\ U(a) p_i U^\dagger(a) &= p_i , \\ U(a) S_i U^\dagger(a) &= S_i , \end{aligned}} \tag{7.31}$$

where $U(a), a \in \mathbf{R}^3$, is a unitary single valued representation of the connected, and simply connected Lie group of space translations. Its existence and explicit expression will be explained presently.

The first of these relations is a consequence of (7.24) and the requirement that a localized state at the point r transforms into another localized state, with the same "norm" at $r + a$. The second is obtained in the same manner, assuming, by classical analogy, that the momentum of a particle is invariant under translations. From the first two relations in (7.31), the orbital angular momentum L_i, under a translation, becomes $U(a)L_iU^\dagger(a) = L_i - a \times p_i$. This implies that the intrinsic orbital angular momentum (that is, with respect to the center of mass of the system) is invariant, and we assume that the same occurs with the spin operator.

The point of view adopted here corresponds to translating by a the experimental set-up necessary for preparing the system. In this set-up, we shall include, of course, all the external agents that may influence the preparation of the state within the required precision. Otherwise, displacing the set-up to a region in which, for example, there are additional external fields, these fields could influence the dynamics of the preparation, given the finite time that such a preparation requires. This operational viewpoint is called the *active interpretation* of the symmetry transformation. Once we have eliminated all the external agents that could affect the preparation of the states, the homogeneity assumed for space is what guarantees that the operational translation of these states, as previously discussed, is implemented by the operator $U(a)$ as given in (7.31). There exists another interpretation, the *passive interpretation*, in which there are two distinct observers, related by a symmetry transformation (in the translation case, observers with reference systems translated one from the other), who contemplate the *same* experimental set-up, attributing to the outcome of a unique preparation generally different state vectors. The relation between the state vectors that each of the observers assign to the same preparation is given by the unitary or anti-unitary operator U_G^{passive}, such that

$$U_G^{\text{passive}} = (U_G^{\text{active}})^{-1} . \tag{7.32}$$

Unless otherwise stated, we will adopt the active point of view.

If we denote by D the translation generator, then, in accordance with (7.30) and (7.31), this operator is determined by the equations

$$[(r_i)_j, D_k] = i\delta_{jk} , \quad [p_i, D] = [S_i, D] = 0 , \tag{7.33}$$

and one immediately derives:

$$D = \hbar^{-1} \sum_{i=1}^{N} p_i + kI ,$$

where k is an arbitrary constant vector chosen equal to zero, according to (3.17). (This is evident for a single particle. When there are several particles, we have to assume that the effect of the translation on the whole system is obtained by applying it to each individual particle.)

Introducing the total momentum $P = \sum_{i=1}^{N} p_i$, we have that

$$D = \hbar^{-1} P , \tag{7.34}$$

that is, the translation generator is proportional to the total momentum operator. The unitary operator corresponding to the translation $G(a)$ is, therefore,

$$\boxed{U(a) = \exp(-\mathrm{i}a \cdot P/\hbar)} , \tag{7.35}$$

which produces a faithful representation of the translation group and clearly leads to (7.31).

With this choice, the wave functions transform under translations as

$$\boxed{\begin{aligned} &(U(a)\Psi)_{m_1,\dots,m_N}(r_1,\dots,r_N; t) \\ &= \Psi_{m_1,\dots,m_N}(r_1 - a,\dots,r_N - a; t) , \end{aligned}} \tag{7.36}$$

where the m_i's are the projections of the individual spins on a fixed direction of space. Equation (7.36) generalizes (3.18), with $U(a) \equiv V_a$.

7.6 Rotations

Let $G(\hat{n}, \theta)$ be a rotation defined by the axis $\hat{n}$ and angle θ between 0 and π, measured according to the right-hand rule. Under this rotation of the experimental set-up, a point of the system with coordinates r rotates to a new position r', given by

$$\begin{aligned} r' &= \mathcal{R}(\hat{n}, \theta)r , \\ \mathcal{R}_{ij}(\hat{n}, \theta) &\equiv \delta_{ij} \cos\theta + (1 - \cos\theta)n_i n_j - \varepsilon_{ijk} n_k \sin\theta . \end{aligned} \tag{7.37}$$

Assuming the space to be isotropic, after elimination of external fields which might dynamically influence the preparation of the states, the effect of such a rotation in its active interpretation is a symmetry of the quantum mechanical system. The unitary operator $U(\hat{n}, \theta)$ associated with this rotation, acting on the system of N particles introduced earlier, transforms the observables as

$$\boxed{\begin{aligned} U(\hat{n}, \theta)r_i U^\dagger(\hat{n}, \theta) &= \mathcal{R}^{-1}(\hat{n}, \theta)r_i , \\ U(\hat{n}, \theta)p_i U^\dagger(\hat{n}, \theta) &= \mathcal{R}^{-1}(\hat{n}, \theta)p_i , \\ U(\hat{n}, \theta)S_i U^\dagger(\hat{n}, \theta) &= \mathcal{R}^{-1}(\hat{n}, \theta)S_i . \end{aligned}} \tag{7.38}$$

The first relation in (7.38) is a consequence of (7.37) and (7.24), while the second relation is obtained from (7.24) by accepting the classical transformation law under rotations for the momentum of a particle. The transformation law for the spin operator has been taken to correspond to that for the orbital angular momentum, which can be deduced from the first two relations in (7.38). If we let I be the rotation generator, we have locally

$$U(\hat{n}, \theta) = \exp(-i\theta\hat{n} \cdot I) , \tag{7.39}$$

and the relations equivalent to (7.30) are written as

$$[A_j, I_i] = i\varepsilon_{jik}A_k , \tag{7.40}$$

where A is any of the observables r_i, p_i and S_i. From this and from (5.123), one immediately sees that the rotation generator is given, within an additive constant, by

$$I = \hbar^{-1}J = \hbar^{-1}(L + S) = \hbar^{-1}\sum_{i=1}^{N}(r_i \times p_i + S_i) , \tag{7.41}$$

i.e., it is proportional to the total angular momentum. The arbitrary additive constant vector in I is zero if (5.8) is to hold true and if the global effect of the rotation is obtained from its application to each of the particles of the system. The unitary operator corresponding to the rotation $\mathcal{R}(\hat{n}, \theta)$ and leading to (7.38) is therefore

$$\boxed{U(\hat{n}, \theta) = \pm\exp(-i\theta\hat{n} \cdot J/\hbar)} . \tag{7.42}$$

We have already explained the significance of the $\pm$ sign in Chap. 5; there, it was shown that these operators form a group isomorphic to the rotation group if the system contains an even number of spin one-half particles. When the number of such particles is odd, then there only exists a two-to-one homomorphism between the two groups.

The transformation laws for the wave functions under rotations were already given in (5.30), where $U(\hat{n}, \theta) \equiv R(\hat{n}, \theta)$.

7.7 Parity

In the previous two cases we have considered continuous symmetry transformations, in the sense that they could be generated through successive infinitesimal transformations. In this and the following section, we will analyze symmetries for which this is not the case, generally called "discrete symmetries". First, we consider the parity transformation P which "reflects" with respect to the coordinate origin the experimental equipment that prepares the states. This reflection can be decomposed into a product of a rotation and a mirror reflection. Having already described how the rotations act, it suffices to clarify how one performs a mirror reflection of an experimental set-up. In such set-ups, we could have, for example, electric currents, electromagnetic fields, etc. The mirror images of such classical parameters change in the usual way, dictated by the classical transformation laws under reflection. Thus, if an apparatus prepares a neutron with spin in a certain direction, determined by a magnetic field, in the mirror image with respect to a plane parallel to the field, this appears inverted, and thus it will prepare the neutron with spin in the opposite sense. It is convenient to take into account, however, that there could be physical situations whose mirror image is not physically realizable. The known experimental data support the hypothesis that the neutrinos are particles of helicity $-\hbar/2$, so that their mirror images are not realizable as neutrinos. In all other cases, the operator U_P, corresponding to the parity transformation, must be such that

$$\boxed{U_P r_i U_P^\dagger = -r_i\,, \quad U_P p_i U_P^\dagger = -p_i\,, \quad U_P S_i U_P^\dagger = S_i\,.} \tag{7.43}$$

Here the transformation law for S_i has been taken analogous to that for L_i, which follows from the first two relations in (7.43). In accordance with the general discussion, the fact that r_i and p_i simultaneously change sign under a parity transformation, and thus that the commutation law between these two observables is preserved, implies that the operator U_P must be unitary.

The operator U_P defined in this way is such that U_P^2 commutes with every operator of an irreducible set of observables, and therefore, it is a multiple of the identity. Moreover, since U_P is unitary, it is obvious that $U_P^2 = \exp(i\alpha)I$, where α is a real number. We choose the phase of U_P such that $U_P^2 = I$, and $U_P = U_P^\dagger$, i.e., U_P is self-adjoint. With this choice of phase, the operator U_P is an observable with eigenvalues ± 1. The observable U_P is called *parity*.

If, for simplicity, a particle A with spin s is considered in the state $|r, m\rangle$, where m is the projection (in units of $\hbar$) of the spin in a fixed direction, then the operator U_P acts in the following manner:

$$U_P|r, m\rangle = \eta_A|-r, m\rangle\,. \tag{7.44}$$

Therefore,

$$(U_P\psi)_m(r) = \eta_A\psi_m(-r)\,, \tag{7.45}$$

where $\eta_A = \pm 1$ and is called the *intrinsic parity* of A. The non-dependence of the phase factor η_A on r, m is an immediate consequence of the second and third relations in (7.43).

If $\psi_m(-r) = \eta_\psi\psi_m(r)$, then $\eta_A\eta_\psi$ is called the parity of the state considered, and as mentioned before, its possible values are ± 1. These considerations can be extended without difficulty to a system of N particles. The action of parity on a wave function of such a system is

$$\boxed{(U_P\psi)_{m_1 m_2\dots}(r_1, r_2, \dots) = \eta_1\cdots\eta_N\psi_{m_1 m_2\dots}(-r_1, -r_2, \dots)}\,, \tag{7.46}$$

where η_i is the intrinsic parity of the ith particle. Such an operator U_P implements effectively the relations (7.43).

If we use helicity for the description of the wave functions, and since the helicity is a pseudoscalar, the transformation law is

$$(U_P\psi)_{\lambda_1\lambda_2...}(\boldsymbol{p}_1,\boldsymbol{p}_2,\ldots) = \eta_1\cdots\eta_N\hat{\psi}_{-\lambda_1,-\lambda_2,...}(-\boldsymbol{p}_1,-\boldsymbol{p}_2,\ldots)\ . \tag{7.47}$$

Under those circumstances in which the number and type of particles remain constant throughout a process, the intrinsic parities are irrelevant, since their product affects every state in the same way. However, in processes in which the number and/or the characteristics of the particles change, the intrinsic parities assigned are, in general, important. The same question also arises even if the number and type of constituent particles are maintained but their initial and final clusterings are different, and we want to describe such clusters as particles, without details of internal structure. In this latter case, if, for example, we consider a particle A that is a bound state of two other particles B and C, with relative orbital angular momentum L, then the intrinsic parity of A is $\eta_A = (-1)^L\eta_B\eta_C$, since $Y_L^M(-\hat{\boldsymbol{r}}) = (-1)^L Y_L^M(\hat{\boldsymbol{r}})$ [see (A.34)].

7.8 Time Reversal

A symmetry transformation, T, that changes one physical system into another with an inverted sense of time evolution is called time reversal. In classical mechanics, this corresponds to substituting for each trajectory $\boldsymbol{r}_i = \boldsymbol{r}_i(t), i = 1,\ldots, N$, of the system the trajectory $\boldsymbol{r}_i = \boldsymbol{r}_i(-t), i = 1,\ldots, N$, i.e., to moving along the given trajectories with the opposite velocities at each point. It should be noted that if the original trajectories are dynamically possible, it is not necessary, in general, that the time inverted trajectories be so for the same dynamics, unless the Hamiltonian of the system is invariant under the transformations $t \to -t, \boldsymbol{r}_i \to \boldsymbol{r}_i$ and $\boldsymbol{p}_i \to -\boldsymbol{p}_i$. Given that, classically, time reversal changes $\boldsymbol{r}_i \to \boldsymbol{r}_i$ and $\boldsymbol{p}_i \to -\boldsymbol{p}_i$, this suggests that the operator U_T which implements such symmetry in quantum mechanics must be such that

$$\boxed{\begin{aligned} U_T\boldsymbol{r}_i U_T^{\dagger} &= \boldsymbol{r}_i\ , \\ U_T\boldsymbol{p}_i U_T^{\dagger} &= -\boldsymbol{p}_i\ , \\ U_T\boldsymbol{S}_i U_T^{\dagger} &= -\boldsymbol{S}_i\ , \end{aligned}} \tag{7.48}$$

where, as usual, the transformation law of the orbital angular momentum, derived from the first two relations, is taken for $\boldsymbol{S}_i$. From (7.48) and from our general discussion, one can immediately obtain, by considering the commutator $[\boldsymbol{r}_i, \boldsymbol{p}_i]$, that the operator U_T must be anti-unitary. As in the previous cases, we accept for the moment the existence of such a symmetry in order to arrive at the explicit form of the operator which implements it.

The operator U_T defined in this way is such that U_T^2 commutes with every operator of an irreducible set of observables and is thus a multiple of the identity.

Moreover, since it is unitary, we have $U_T^2 = \exp(i\alpha)I$, with α real. From $U_T U_T^2 = U_T^2 U_T$ it results that $\exp(-i\alpha)U_T = \exp(i\alpha)U_T$, and hence that $U_T^2 = \pm I$. In order to fix the sign of this operator, we will consider a rotation through angle π around the Oy-axis, whose corresponding operator is denoted by $U(\hat{e}_2, \pi)$. From (7.38), it follows that

$$U(\hat{e}_2, \pi)J_i U^\dagger(\hat{e}_2, \pi) = (-1)^{1+\delta_{i2}} J_i , \quad \text{and thus} \tag{7.49}$$

$$[U(\hat{e}_2, \pi)U_T, J_3] = [U(\hat{e}_2, \pi)U_T, \boldsymbol{J}^2] = 0 . \tag{7.50}$$

The equalities in (7.50) also hold if they are restricted to any subsystem of the given system. Since the state of the system can always be characterized by the kinetic energies of its particles, the angular momenta of the subsystems and the third component of the total angular momentum, there will exist a basis of states $(|\alpha JM\rangle)$, where α are those other quantum numbers just cited, such that

$$U(\hat{e}_2, \pi)U_T|\alpha JM\rangle = e^{i\gamma}|\alpha JM\rangle . \tag{7.51}$$

Since $U(\hat{e}_2, \pi)U_T$ is anti-unitary,

$$U(\hat{e}_2, \pi)U_T e^{i\gamma/2}|\alpha JM\rangle = e^{i\gamma/2}|\alpha JM\rangle , \tag{7.52}$$

and with a convenient modification of the phase of $|\alpha JM\rangle$, we can write

$$U(\hat{e}_2, \pi)U_T|\alpha JM\rangle = |\alpha JM\rangle . \tag{7.53}$$

Applying the operator $U(\hat{e}_2, \pi)U_T$ to both sides of (7.53) and bearing in mind that $[U(\hat{e}_2, \pi), U_T] = 0$, we see that

$$U_T^2 U^2(\hat{e}_2, \pi)|\alpha JM\rangle = |\alpha JM\rangle . \tag{7.54}$$

On the other hand, from (B.22) (or from Sect. 5.6), one immediately obtains that $U^2(\hat{e}_2, \pi)|\alpha JM\rangle = (-1)^{2J}|\alpha JM\rangle$. As a consequence, we have

$$U_T^2 = (-1)^n , \tag{7.55}$$

where n is the number of spin one-half particles in the system.

In view of (7.55), and by analogy with what was said about parity, one might think that it is possible to prepare eigenstates of U_T, with eigenvalues ± 1 (n even) and $\pm i$ (n odd), the two alternatives being physically distinguishable in each case. That this is not possible, can be seen as follows: If

$$U_T|a\rangle = e^{i\alpha}|a\rangle , \quad \text{then} \tag{7.56}$$

$$U_T e^{i\alpha/2}|a\rangle = e^{i\alpha/2}|a\rangle ; \tag{7.57}$$

thus, the ray $|a\rangle_R$ contains eigenvectors of U_T with any complex number of modulus one as eigenvalue. [For more details, see Appendix G.]

From (7.53), and making use of (B.21) and (B.22), one derives that

$$\overline{|\alpha JM\rangle} \equiv U_T|\alpha JM\rangle = (-1)^{J+M}|\alpha J - M\rangle , \tag{7.58}$$

where, as usual, $|\overline{\alpha JM}\rangle$ is the state obtained from $|\alpha JM\rangle$ by time reversal.

For a particle of spin s, the time reversal operator can be constructed as follows: Let K be the anti-unitary operator which performs complex conjugation in the representation associated with the complete set of compatible observables $\{r, S_3\}$. If the phases of the states in the basis for this representation are taken according to the usual specifications, then the matrix representation of the spin operator is such that S_1 and S_3 have all their components real, and S_2 has only pure imaginary matrix elements. Thus

$$KrK = r , \quad KpK = -p ,$$
$$KS_1K = S_1 , \quad KS_2K = -S_2 , \quad KS_3K = S_3 , \tag{7.59}$$

where one has taken into account that $K = K^\dagger$, since $K^2 = I$. Comparing (7.59) and (7.48), it becomes evident that the action of $\exp(-i\pi S_2/\hbar)K$ on an irreducible set of observables is identical with the action of U_T, and thus, selecting the arbitrary phase conveniently, one can write

$$U_T = e^{-i\pi S_2/\hbar} K , \tag{7.60}$$

which for spin one-half particles reduces to

$$U_T = -i\sigma_2 K , \tag{7.61}$$

where σ_2 is the Pauli matrix. In a similar way, one can derive that for a system of N particles with spin operators $S_1, \ldots, S_N$, the time reversal operator U_T, which leads to (7.48), is

$$U_T = \exp\left[-\frac{i\pi}{\hbar}(S_{1,2} + \ldots + S_{N,2})\right] K , \tag{7.62}$$

where $S_{i,2}$ is the second component of the spin operator of the ith particle.

Note, finally, that in the basis satisfying (7.53), the relation

$$\langle \alpha_2 J_2 M_2 | A | \alpha_1 J_1 M_1 \rangle$$
$$= \langle \alpha_2 J_2 M_2 | U(\hat{e}_2, \pi) U_T A (U(\hat{e}_2, \pi) U_T)^{-1} | \alpha_1 J_1 M_1 \rangle^* \tag{7.63}$$

holds. Here A is an arbitrary operator. If A is invariant under $U(\hat{e}_2, \pi)U_T$, then all these matrix elements are real, while, if A changes sign, all the matrix elements are pure imaginary.

7.9 Invariances and Conservation Laws

Let $\mathcal{G}$ be a group of symmetry transformations, and $\hat{\mathcal{G}}$ be the group of associated operators, which for the moment we suppose to be linear. We denote by U_G an arbitrary element of $\hat{\mathcal{G}}$. If we suppose that the Hamiltonian $H(t)$ is invariant under the operators $U_G \in \hat{\mathcal{G}}$, then according to (7.21) we can write

$$[H(t), U_G] = 0 , \quad \forall U_G \in \hat{\mathcal{G}} . \tag{7.64}$$

For conservative systems, the eigenstates of H can be characterized by $|\alpha j \mu\rangle$, where j indicates the set of numbers necessary to characterize an irreducible representation of $\hat{\mathcal{G}}$, μ the set of numbers necessary to distinguish the different vectors in the basis for such an irreducible representation, and α the additional quantum numbers characterizing the state, among which might be the energy. An immediate consequence of (7.64) is that

$$H|\alpha j \mu\rangle = E(\alpha j)|\alpha j \mu\rangle , \tag{7.65}$$

i.e., the energy level characterized by (αj) is d_j-fold degenerate, d_j being the dimension of the irreducible representation characterized by the numbers j. In fact, in the subspace spanned by all the vectors with the quantum numbers α and j, corresponding to an irreducible representation of $\hat{\mathcal{G}}$, the eigenvectors of H with a definite energy form a stable subspace under this representation. Since it is irreducible, this subspace fills the entire space of the representation if it contains any non-null vectors. Thus, for example, if H is invariant under the group of rotations, the eigenstates of H can be characterized by $|\alpha J M\rangle$ and the degeneracy of a level with fixed α and J is $2J + 1$.

We know that, for a conservative system, if A is an observable which does not depend explicitly on time and such that $[A, H] = 0$, then this observable is a constant of motion. Thus, we can assert that if H is conservative and invariant under the group $\hat{\mathcal{G}}$, then all observables without explicit time dependence which can be constructed from the group operators will be constants of motion. It can be seen, therefore, that each group of symmetry transformations leaving H invariant and which are implemented by unitary operators, have associated a certain number of conservation laws.

We now demonstrate that the invariance of $H(t)$ under such a group $\hat{\mathcal{G}}$ implies that

$$[U(t, t_0), U_G] = 0 , \quad \forall U_G \in \hat{\mathcal{G}} , \tag{7.66}$$

where $U(t, t_0)$ is the time evolution operator. Indeed, this operator is the solution of the integral equation (2.127)

$$U(t, t_0) = I - \frac{i}{\hbar} \int_{t_0}^{t} dt' \, H(t')U(t', t_0) . \tag{7.67}$$

If one multiplies this equation by U_G from the left and by $U_G^\dagger$ from the right, and takes into account that $[U_G, H(t')] = 0$, $t_0 \leq t' \leq t$, one immediately obtains

$$U_G U(t, t_0) U_G^\dagger = I - \frac{i}{\hbar} \int_{t_0}^{t} dt' \, H(t) U_G U(t', t_0) U_G^\dagger . \tag{7.68}$$

Since $U(t, t_0)$ and $U_G U(t, t_0) U_G^\dagger$ satisfy the same integral equation, and the solution is unique, they are equal, thus proving (7.66). Hence, two states which can be obtained from each other by a transformation $U_G \in \hat{\mathcal{G}}$ retain this property in the course of time. Note also that if the system is prepared in the pure state represented by the vector $|\varphi\rangle$ at time t_0, and then at time $t > t_0$ the probability of the system being in the state $|\psi\rangle$ is measured, we deduce from (7.66) that

$$|\langle \psi | U(t, t_0) | \varphi \rangle|^2 = |\langle \psi | U_G^\dagger U(t, t_0) U_G | \varphi \rangle|^2 , \tag{7.69}$$

i.e., the result of this measurement coincides with that which would be obtained starting at t_0 with the state $U_G|\varphi\rangle$ and asking for the probability of finding the system at time t in the state $U_G|\psi\rangle$. An identical conclusion is true for mixed states.

7.10 Invariance Under Translations

Invariance under spatial or temporal translations corresponds to the accepted fact that the localization in space or time of an isolated system plays no essential role in the formulation of its dynamical laws. One direct way of testing the homogeneity of space-time consists of analyzing possible variations of fundamental constants by comparing processes well separated in both space and time. Thus, from the observation of spectra of galaxies separated by as much as 2×10^9 light years, it can be concluded that the value of the fine structure constant in these galaxies, at the time the light was emitted, differs at most by 3×10^{-3} parts from its value measured in our laboratories. The Oklo "nuclear reactor", which reached criticality in a natural way some 2×10^9 years ago, has permitted inference of a much more restrictive bound on the variation of α with time:

$$|\alpha_{\text{Oklo}} - \alpha_{\text{now}}| \lesssim 10^{-8} \alpha_{\text{now}} \tag{7.70}$$

(for more details concerning this and the possible variations of other constants see [DY 72], [WI 84]).

On the other hand, we proved that the generator of translations is the total momentum operator of the system, so that the homogeneity of space implies that for an isolated system $[H, P] = 0$; it follows from this that for such a system the total momentum should be conserved. A test of this conservation law thus provides an indirect test of the homogeneity of space.

In exactly the same way in which the invariance of dynamics under spatial translations leads to the conservation of momentum, the homogeneity of time in the description of fundamental physical laws, that is, the independence of these laws of the time origin chosen, leads to the conservation of energy in isolated systems.

These energy-momentum conservation laws have been accurately tested experimentally, and are universally accepted. One of their immediate consequences is to forbid processes such as $e^- + e^+ \rightarrow \gamma$, $p \rightarrow p + \gamma$, etc., none of which have been observed.

7.11 Invariance Under Rotations

Invariance under rotations expresses the belief that physical laws for isolated systems are also independent of their orientation in space. It could be thought that even if space were isotropic for purposes of description of these laws, the position of the earth in the galaxy could give rise to a local anisotropy. *Hughes* et al. [HR 60] have measured the dependence of the inertial mass M on the direction of acceleration, and have found, using nuclear resonance methods, that measurements of the mass in the direction of the center of the galaxy and in a perpendicular direction give rise to a fractional anisotropy of $\Delta M / M \leq 10^{-20}$; this appears to be a good direct test of the isotropy of space as well. Nevertheless, there are authors [DI 61] who doubt this interpretation and consider the experiment to be a "null" experiment.

Since the total angular momentum J is the generator of rotations, the isotropy of space implies that $[H, J] = 0$ for an isolated system, and thus implies the law of conservation of angular momentum for these systems. A measurement of this conservation law consequently gives indirect evidence for the isotropy of space.

We discuss next a pair of experiments which confirm the validity of the law of conservation of total angular momentum.

One decay forbidden by conservation of total angular momentum is $K^+ \rightarrow \pi^+ + \gamma$; the K^+ and π^+ mesons have spin and parity 0^- (i.e., spin 0 and intrinsic parity -1). If one chooses a system of reference in which the K^+ is at rest, then the π^+ and the γ should emerge in opposite directions along, say, the Oz-axis, here chosen as the axis of quantization. In the initial state, $(J_i)_3 = 0$, and in the final state $(J_f)_3 = \pm \hbar$, depending upon the polarization state of the photon, since the projection of the orbital angular momentum along the direction of relative motion is zero, and it is known that the helicity of the photon is always $\pm \hbar$ (corresponding to right- and left-handed circular polarization).

Experimentally, it has been observed [AC 84] that

$$\frac{\Gamma(K^+ \rightarrow \pi^+ \gamma)}{\Gamma(K^+ \rightarrow \text{all})} < 1.4 \times 10^{-6} , \tag{7.71}$$

where $\Gamma(A \rightarrow B)$ denotes $\hbar$ times the transition probability from A to B per unit time, or the width of the transition, and has the dimensions of energy. If angular momentum conservation were not operative, the ratio on the left could be estimated to be 10^{-2}.

An analysis parallel to the one above permits us to show that the law of conservation of total angular momentum forbids the electromagnetic transition

$X^* \rightarrow X + \gamma$, where X and X^* are two states of the same nucleus, both with spin 0. Several such transitions have been studied, and no experimental evidence for them has been found [SI 68]. From this type of experiment, it has been concluded [FG 59] that a possible amplitude violating the conservation of angular momentum should have a modulus $< 0.3 \times 10^{-8}$ times the modulus of the conserving amplitude.

7.12 Invariance Under Parity

According to the development thus far, the fact that the behavior of a physical system is invariant under spatial reflection or under parity means that if we prepare a state $|\Psi(t_0)\rangle$ at some initial instant t_0, and its image under parity is $U_P|\Psi(t_0)\rangle$, these will evolve in time in such a way that $U(t,t_0)U_P|\Psi(t_0)\rangle$ will continue to be the image of $|\Psi(t)\rangle$ under P for all time. Due to the law of conservation of angular momentum, this is equivalent to saying that the mirror image of the time evolution of a state is physically realizable with the same dynamics, since the operation of reflection through a plane is the equivalent to a parity operation followed by a rotation through π around an axis perpendicular to the plane.

For a further discussion of parity (and the fact that it is not, as we shall see, a universal invariance) it is necessary to recall a few facts concerning the basic forces of nature. Until the beginning of the 70s, four types of fundamental interactions between elementary particles were postulated: *gravitational, weak, electromagnetic* and *strong*, ordered according to increasing strength. The strong interaction is necessary to explain, among other things, the stability of atomic nuclei; the weak is responsible for processes such as $\beta^{\pm}$-decay; and the electromagnetic and gravitational – the only two which manifest themselves ostensibly at the macroscopic level – account for the existence of atoms, molecules, etc., and of gravity, respectively. These four classes of interactions differ enormously at low energies in intensity and/or range.

A myriad of experimental results over the last twenty years have shown that the fundamental components of matter are *leptons* (electron, muon, tau and associated neutrinos) and *quarks* (u, d, s, c, b, t (?), each one in three different "color" states). These leptons and quarks, all of spin 1/2, interact with each other via the emission and absorption of particles of integer spin (photon, $W^{\pm}$, Z^0, gluons, graviton (?), etc.). Motivated by a simplifying impulse, Glashow, Salam and Weinberg proposed a model, known as the *electroweak* theory, in which the electromagnetic and weak interactions appear as part of the same global framework, and only at energies much lower than $m_W \simeq m_Z \sim 10^2 m_p$ do they become differentiated and acquire their ordinarily observed characteristics. Their different ranges and strengths are understood in terms of the zero mass of the photon (carrier of the electromagnetic interactions) and the large masses of the W and Z (carriers of the weak interaction).

There exists, on the other hand, the theory called *quantum chromodynamics*, which describes the strong interactions among the quarks via the exchange of an octet of gluons. While this theory has had experimental successes, it is still far from possible to use it to deduce the phenomenological strong interactions between *hadrons* ("white" or colorless systems of quarks and/or antiquarks, such as the nucleons, pions, etc.). The search for theories which will unify the electroweak interactions with quantum chromodynamics and even with gravitation is currently being pursued with great vigor. *Grand unification* models have been proposed (one of whose most notable predictions is the instability of protons; they also point to the solution to important cosmological problems), some of which include a new type of symmetry, called *supersymmetry*. These supersymmetries open a way for the inclusion of gravitational interactions; however, the actual situation is uncertain at this time. Another promising path toward *total unification* appears to be through the theory of *superstrings*.

Of fundamental interest for this book are the problems dealing with low energies, so that the residual effects of any unification would be very small (even though some would be observable in very high precision measurements). We will thus retain the traditional classifications of the four interactions. The strengths are usually characterized by the coupling constants written in dimensionless form: gravitation in terms of the gravitational fine structure constant $\alpha_G = Gm_p^2/\hbar c \sim 10^{-38}$; electromagnetism via $\alpha \sim 1/137$, the strong interaction via $g^2/4\pi\hbar c \sim 1 - 10$, and the weak force through $G_F(m_p c^2)/(\hbar c)^3 \sim 10^{-5}$. The small value of α_G explains the irrelevance of the gravitational interaction in nuclei, atoms and molecules; however, this does not mean that gravity is not observable at the quantum level. In a beautiful experiment using neutron interferometry, *Colella, Overhauser* and *Werner* [CO 75] have detected the influence of terrestrial gravity on the neutron wave function, and verified that the correct way of introducing this interaction in the Schrödinger equation is the usual one, $V = Mgz$ [GR 83].

We continue now with a discussion of some of the experimental evidence relative to the question of the conservation or violation of parity. In the first place, we consider a nuclear system, whose structure, as we know, is basically governed by the strong and electromagnetic interactions. We will suppose that for this system $[H, U_P] = 0$, which implies that for each nuclear level there exist states of well-defined parity $\Pi(= \pm 1)$. The complexity of H for a nuclear system favors that there is no energy level degeneracy with respect to parity if it is conserved. Finding a level with no definite parity would thus require attributing this to a violation of P. To see the extent to which this is true, we consider a nuclear state with spin-parity J^Π which can undergo an α-decay to a nuclear state whose spin-parity is 0^+. As an α-particle has spin-parity 0^+, the law of conservation of angular momentum implies that the angular momentum of the final state should be $L = J$, so that the wave function which describes the relative motion of the two final nuclei has the form $R(r)Y_J^M(\theta, \varphi)$. Taking into account that the spherical coordinates of $(-r)$ are $(r, \pi - \theta, \varphi + \pi)$, the effect of the parity operator on the wave function of relative motion is

$$U_P : R(r)Y_J^M(\theta, \varphi) \rightarrow R(r)Y_J^M(\pi - \theta, \varphi + \pi)$$
$$= (-1)^J R(r)Y_J^M(\theta, \varphi) , \qquad (7.72)$$

where we have made use of [A.34]. The initial parity is $\Pi_i = \Pi$; and the final one is $\Pi_f = (+1)(+1)(-1)^J = (-1)^J$. Thus the conservation of parity implies that these α-decays are only possible if $\Pi = (-1)^J$, i.e., if the initial state is $J^\Pi = 0^+, 1^-, 2^+, \ldots$. According to this, the transition

$$^{16}O(2^-; 8.88) \rightarrow {}^{12}C(0^+; \text{g.s.}) + \alpha + 7.148 \,\text{MeV} \qquad (7.73)$$

should be forbidden, where g.s. stands for the ground state and 8.88 is the energy of excitation, in MeV, of the excited state of ^{16}O being considered. The width Γ_α for this process has been measured, and it is found [HH 70] that $\Gamma_\alpha = (1.8 \pm 0.8) \times 10^{-10}$ eV. If the initial state had analogous characteristics, but its parity were positive, it has been estimated [BM 69] that the corresponding width would be $\Gamma_\alpha^{(+)} \simeq 6 \times 10^4$ eV. Now if the initial state did not have a well-defined parity, its wave function could be written $\psi = (\psi_- + \varepsilon\psi_+)/\sqrt{1 + |\varepsilon|^2}$, where ψ_Π are normalized wave functions with parity Π and ε is a parameter of the order of magnitude of the ratio of the strength of the parity-violating forces to those conserving it. Then $\Gamma_\alpha \simeq [|\varepsilon|^2/(1 + |\varepsilon|^2)]\Gamma_\alpha^{(+)}$, whence $|\varepsilon| \simeq 6 \times 10^{-8}$. Note that even if parity were exactly conserved in strong and electromagnetic interactions, it is not to be expected that ε be zero, since the weak interactions also enter into nuclear structure and violate parity, as we will see, and can generate ε values which are small, but not zero. *Gari* [GA 70] has carried out calculations of Γ_α assuming that only the weak interactions violate parity, and arrives at theoretical values in good agreement with the experimental data. From arguments of this type one can infer that the strong and electromagnetic interactions could conserve parity exactly, but that it is not possible to discount the presence of a part of the strong (electromagnetic) interaction which violates parity if it is smaller by a factor of the order of $10^{-8}(10^{-3})$ than the conserving part.

One could object, in principle, to the previous argument by imagining a nuclear level accidentally degenerate under parity, in such a way that it would have even and odd states, and it would thus not be necessary to have recourse to violation of parity to explain the presence of ψ_+ and ψ_-. However, under such circumstances, the excitation of the nucleus to such a level would produce a statistical mixture of even or odd states of similar proportions, and thus its α-decay width would be $\sim (1/2)\Gamma_\alpha^{(+)}$, contrary to experimental evidence. Similar results obtain from analysis of the circular polarization of γ radiation. A more detailed discussion can be found in [VI 72].

We go on to discuss an experiment which demonstrates non-conservation of parity in the weak interaction. *Lee* and *Yang* [LY 56] were the first to postulate that, contrary to what was then believed, parity was not only not conserved in weak interactions, it was in fact maximally violated; i.e., for weak interactions, the transition amplitudes which conserve and those which violate parity have the same intensity. To confirm this postulate, *Wu* et al. [WA 57] studied the β-decay

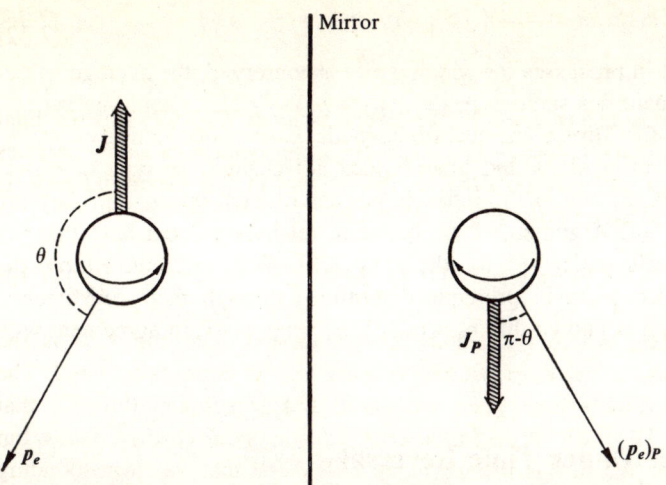

Fig. 7.1. Schematic representation of the angular correlation experiment with ^{60}Co and its image in a mirror perpendicular to the plane of the drawing

$$^{60}\text{Co}(5^+; \text{g.s.}) \rightarrow {}^{60}\text{Ni}(4^+; 2.505) + e^- + \bar{\nu}_e \ , \tag{7.74}$$

measuring the correlation between the direction of emission of the electrons and the direction of polarization of the original nucleus. In Fig. 7.1 we represent the process and its mirror image. If parity were conserved, the mirror image would be dynamically possible, and thus the number of electrons emitted with an angle θ between the direction of the electron's path and the direction of the original nuclear spin should be equal to the number emitted with angle $\pi - \theta$. In the model proposed by *Lee* and *Yang* [LY 56], in which parity is maximally violated, the angular distribution of the emitted electrons is given by

$$W(\theta) = 1 - \frac{\langle J_{i3} \rangle}{J_i} \frac{v}{c} \cos \theta \ , \tag{7.75}$$

where J_i is the spin of the original nucleus, v is the velocity of the outgoing electron and $\langle J_{i3} \rangle / J_i$ measures the amount of polarization of the original sample. The results of the experiment mentioned above, which revealed the $\cos \theta$ dependence of the angular distribution, constituted the first evidence that the law of conservation of parity was violated in the weak interaction. Subsequent experiments in nuclear physics and with elementary particles have established this fact clearly and in such way that it can now be affirmed to within an error of a part in 10^3 that the violation is the largest possible [HE 69a], [SA 64].

In the interpretation of all these experiments, it should be kept in mind that if N particles participate (including the initial and final ones) with momenta $p_1, p_2, \ldots p_N$ and spins $s_1, s_2, \ldots s_N$, and if parity is conserved, then the distribution function $W(p_1, p_2, \ldots p_N; s_1, s_2, \ldots s_N)$, which represents the transition probability of an initial state to a different final state, satisfies the relation

$$W(\boldsymbol{p}_1, \ldots \boldsymbol{p}_N; \boldsymbol{s}_1, \ldots \boldsymbol{s}_N) = W(-\boldsymbol{p}_1, \ldots, -\boldsymbol{p}_N; \boldsymbol{s}_1, \ldots \boldsymbol{s}_N) \,, \tag{7.76}$$

which implies that in processes for which parity is conserved, the average values of pseudoscalar quantities such as $\boldsymbol{p}_i \cdot \boldsymbol{s}_j$, $(\boldsymbol{p}_i \times \boldsymbol{p}_j) \cdot \boldsymbol{p}_k$ etc., should be zero.

Finally, since the strong and electromagnetic interactions conserve parity, and the exchange of photons and gluons does not change the type of quark, it is possible to fix arbitrarily the intrinsic parities of each quark, and they are usually chosen as +1. Arguments from quantum relativistic local field theories then require intrinsic parity -1 for the antiquarks. With this convention, the parities of all hadrons are in principle determined through their constituents; thus, for example, it is known that $\eta_{\rm p} = \eta_{\rm n} = 1, \eta_\pi = \eta_{\rm K} = -1$, in agreement with experiment.

7.13 Invariance Under Time Reversal

As we have already seen, the operator U_T, corresponding to the operation of time reversal, is anti-unitary, so that the results obtained in Sect. 7.9 will not be applicable in general. In particular, the assumption of invariance under temporal inversion does not lead to a new conserved quantum number, as was the case with parity.

If $U_T H(t) U_T^{-1} = H(-t)$, then the application of the operator U_T to both sides of the Schrödinger equation which describes the time evolution of the system

$$i\hbar \frac{d}{dt}|\Psi(t)\rangle = H(t)|\Psi(t)\rangle \tag{7.77}$$

gives

$$-i\hbar \frac{d}{dt} U_T |\Psi(t)\rangle = H(-t) U_T |\Psi(t)\rangle \,, \tag{7.78}$$

from which

$$i\hbar \frac{d}{dt} U_T |\Psi(-t)\rangle = H(t) U_T |\Psi(-t)\rangle \,. \tag{7.79}$$

Thus, if $|\Psi(t)\rangle$ represents a possible trajectory of the system being considered, the vector

$$|\overline{\Psi(t)}\rangle \equiv U_T |\Psi(-t)\rangle \tag{7.80}$$

also represents another possible quantum trajectory of the system. The state represented by the vector $|\overline{\Psi}\rangle$ at any instant t is the transform by time reversal of the vector $|\Psi\rangle$ at the instant $(-t)$.

Conversely, let us now suppose that the evolution of the system under consideration is invariant under time reversal, in the sense that if $|\Psi(t)\rangle = U(t, t_0)|\Psi(t_0)\rangle$ represents a possible motion of the system, then the vector $|\overline{\Psi(t)}\rangle = U_T U(-t, -t_0)|\Psi(-t_0)\rangle$ equally represents a possible motion, for all vectors $|\Psi(t_0)\rangle$. This requires that, except for a phase, $U_T U(-t, -t_0)|\Psi(-t_0)\rangle$ should coincide with $U(t, t_0)U_T|\Psi(-t_0)\rangle$ so that

$$U(t, t_0)U_T = e^{i\alpha(t)}U_T U(-t, -t_0) . \tag{7.81}$$

If the system is conservative,

$$U(t, t_0) = \exp\left[-\frac{i}{\hbar}(t - t_0)H\right] , \quad U(-t, -t_0) = U^\dagger(t, t_0) , \tag{7.82}$$

and thus, (7.81) is equivalent to

$$U(t, t_0) = e^{i\alpha(t)}U_T U^\dagger(t, t_0)U_T^\dagger . \tag{7.83}$$

Multiplying (7.83) from the right by $U_T^\dagger$ and from the left by U_T, and using (7.55), we obtain

$$U^\dagger(t, t_0) = e^{i\alpha(t)}U_T U(t, t_0)U_T^\dagger . \tag{7.84}$$

Comparing (7.84) with the adjoint of (7.83), we deduce that $\exp[i\alpha(t)] = \pm 1$. Nevertheless, since $\alpha(t)$ should be a continuous function of t, and for $t = t_0$ we must have $\exp[i\alpha(t_0)] = 1$, then $\exp[i\alpha(t)] = 1$, and thus

$$U^\dagger(t, t_0) = U_T U(t, t_0)U_T^\dagger , \tag{7.85}$$

an equation from which we immediately get $H = U_T H U_T^\dagger$; that is, $[H, U_T] = 0$. Thus one sees that for conservative systems the fact that $[H, U_T] = 0$ is equivalent to the statement that the evolution law of the system is invariant under time reversal.

For conservative physical systems, another way of postulating the invariance under time reversal is to accept the *principle of microreversibility*; the probability that at a particular instant t the system can be found in the state $|\varphi\rangle$ if it was prepared in the state $|\psi\rangle$ at time t_0 coincides with the probability that at time t the system be found in the state $U_T|\psi\rangle$ if at t_0 it was prepared in the state $U_T|\varphi\rangle$. This is equivalent to saying:

$$\boxed{|\langle\varphi|U(t, t_0)|\psi\rangle|^2 = |\langle U_T\psi|U(t, t_0)U_T|\varphi\rangle|^2} . \tag{7.86}$$

Since

$$\langle U_T\psi|U(t, t_0)U_T|\varphi\rangle = \langle\psi|U_T^\dagger U(t, t_0)U_T|\varphi\rangle^*$$
$$= \langle\varphi|U_T^\dagger U^\dagger(t, t_0)U_T|\psi\rangle , \tag{7.87}$$

the equality (7.86) is true if and only if

$$|\langle\varphi|U(t, t_0)|\psi\rangle|^2 = |\langle\varphi|U_T^\dagger U^\dagger(t, t_0)U_T|\varphi\rangle|^2 . \tag{7.88}$$

If this holds for every pair of states $|\psi\rangle$ and $|\varphi\rangle$, then it implies (see Sect. 7.3) a relation equivalent to (7.84); repeating the reasoning developed there, we obtain thus an identical conclusion. On the other hand, it is easy to prove, starting with (7.81), the validity of (7.86) for conservative systems, i.e., the principle of microreversibility.

An important consequence of the fact that $[H, U_T] = 0$ is the following. If $|\psi\rangle$ is a solution of the equation $H|\psi\rangle = E|\psi\rangle$, then also $H(U_T|\psi\rangle) = E(U_T|\psi\rangle)$, and thus we can say that the subspace $\mathcal{H}_E$ generated by the eigenvectors of H with eigenvalue E is stable under time reversal. Using the results shown in Appendix G, if the number of fermions (particles of half-integer spin) is even, the vectors in an orthonormal basis in $\mathcal{H}_E$ can always be chosen in such a way that $U_T|\varphi\rangle = |\varphi\rangle$. On the other hand, if the number of fermions is odd, then an orthonormal basis in $\mathcal{H}_E$ can be found, such that if $|\varphi\rangle$ belongs to the basis, then $U_T|\varphi\rangle$ is another, different member of that basis; this implies that the dimension of $\mathcal{H}_E$ is even. This leads to the famous *Kramers' theorem*: if $[H, U_T] = 0$, and the system has an odd number of fermions, every eigenvalue of H is at least doubly degenerate, and its degeneracy is always even. *Kramers* [KR 30] proved this theorem for the case of an odd number of fermions with a mutual interaction which depended only on their relative positions, and in the presence of an external, possibly inhomogeneous, electrostatic field $E(x)$; to do so, he did not need to refer to the concept of time reversal. That this special case is encompassed in our previous theorem can be inferred from the fact that under time reversal $E(x) \xrightarrow{T} E(x)$, and thus $[H, U_T] = 0$.

One consequence of this theorem is that Curie's law, dealing with the variation of paramagnetic susceptibility as a function of temperature, fails for solutions of rare earth salts with trivalent ions having an even number of electrons outside closed shells, even for moderately low temperatures. Thus, for example, the law is no longer correct for Pr^{3+}, which has the configuration $4f^2 5s^2 6p^6$, while it is correct for Nd^{3+}, whose configuration is $4f^3 5s^2 6p^6$ [WA 72].

We finally consider a few experimental tests to verify the invariance (or lack thereof) of the different basic interactions under time reversal.

In the same way that the conservation of parity implies the equality

$$W(p_1 s_1, \ldots \to p_1', s_1', \ldots) = W(-p_1, s_1, \ldots \to -p_1', s_1', \ldots) \tag{7.89}$$

between the distribution functions of the final products of a collision,

$$A_1 + \ldots \to A_1' + \ldots , \tag{7.90}$$

it is a simple consequence of (7.86) that the invariance of dynamics under time reversal implies an equality of the type

$$W(p_1, s_1, \ldots \to p_1', s_1', \ldots) = W(-p_1', -s_1', \ldots \to -p_1, -s_1, \ldots) \tag{7.91}$$

between the direct reaction (7.90) and its inverse

$$A_1' + \ldots \to A_1 + \ldots . \tag{7.92}$$

When the dynamics of the reaction is sufficiently weak that the Born approximation is applicable (see Chap. 8), one can deduce from (7.91) that

$$W(p_1 s_1, \ldots \to p_1', s_1', \ldots) = W(-p_1, -s_1, \ldots \to -p_1', -s_1', \ldots) . \qquad (7.93)$$

One possible way of investigating whether a given interaction violates invariance under time reversal is to measure the differential cross section in the center-of-mass system of a reaction $A_1 + A_2 \to A_3 + A_4$ mediated by this interaction, for unpolarized incident particles, not measuring the final polarizations, and comparing this to the analogous quantity for the reaction $A_4 + A_3 \to A_2 + A_1$. If the comparison is made for the same scattering angle and the same total center-of-mass energy for both reactions, it can be shown (see Chap. 8) that time reversal invariance implies

$$\frac{d\sigma(A_1 + A_2 \to A_3 + A_4)/d\Omega}{d\sigma(A_4 + A_3 \to A_2 + A_1)/d\Omega} = \left(\frac{p'}{p}\right)^2 \frac{(2J_3 + 1)(2J_4 + 1)}{(2J_1 + 1)(2J_2 + 1)} , \qquad (7.94)$$

where J_i is the spin of particle A_i, p and p' are respectively the magnitudes of the 3-momenta (in the center-of-mass system) of the initial and final particles in the first reaction. This result is known as the *reciprocity theorem* or *detailed balance theorem*. It should be noted that these experiments can only give a clean test of the invariance under time reversal if a number of requirements are satisfied. Here we mention only that if the Born approximation is valid, the hermiticity of the Hamiltonian guarantees (7.94). A more detailed study of this problem can be found in [HE 69a].

One of the clearest experiments of this type is that described in [WR 68], in which the reactions

$$\alpha(0^+) + {}^{24}\text{Mg}(0^+) \leftrightarrows {}^{27}\text{Al}(5/2^+) + \text{p}(1/2^+) \qquad (7.95)$$

are compared for a given angle near 180° in the center-of-mass system, and for a proton kinetic energy in the laboratory of 10.10 to 10.60 MeV. The relation (7.94) is very well satisfied, and the analysis of the results obtained permit the assertion that if the amplitude F for the above process is written as $F = f + f'$, where f indicates the part of the amplitude which is invariant under time reversal and f' the violating part, and we define the ratio

$$\xi = \left(\frac{\langle|f'|^2\rangle}{\langle|f^2|\rangle}\right)^{1/2} , \qquad (7.96)$$

where $\langle \ldots \rangle$ indicates an average over the energies involved, then

$$\xi \leq 3 \times 10^{-3} \qquad (7.97)$$

with a confidence level of 85 %. Since the strong interactions are the dominant ones in the process considered, this result translates into the statement that if there is time reversal violation in the strong interactions, it cannot exceed a few parts in a thousand. Other results of the same type confirming the above conclusion are discussed in [HE 69a].

Possible violations of time reversal invariance have been sought, without apparent success, in the strong reactions of the type ${}^{7}\text{Li} + {}^{3}\text{He} \leftrightarrows \text{p} + {}^{9}\text{Be}$, by measuring the polarization of the protons in the direct reaction, and then using polarized protons in the inverse process [TB 84], [MK 84].

Another typical class of experiments which can be done to detect possible violations of time reversal invariance is the following: consider the collision $A_1 + A_2 \rightarrow A_3 + A_4$, with unpolarized initial particles. Let p and p' be the initial and final momenta (say of A_1 and A_3) in the center of mass; the unitary vector $\hat{n} = (p \times p')/|p \times p'|$ is perpendicular to the plane of the reaction. If the final particle A_i ($i = 3$ or 4) has spin J_i, and if J_i is the corresponding operator, the polarization vector for this particle is defined as $P_i = \langle J_i \rangle / J_i$, where $\langle J_i \rangle$ is the average value of J_i for the state of that particle. If parity is conserved in this process, the values $P_i \cdot p$ and $P_i \cdot p'$ should be zero, but $P_i \cdot \hat{n}$ can be different from zero, i.e., parity permits the existence of a polarization perpendicular to the plane of the process. If, in addition, the process is mediated by an interaction that is invariant under time reversal, and for which the Born approximation is valid, then the expectation value of $J_i \cdot \hat{n}$ should also be zero from (7.93). One experiment of this type has been carried out [PS 68] using the process

$$e^- + d \rightarrow e^- + d \tag{7.98}$$

for 1 GeV electrons, and a deuteron polarization of $|P_d| = 0.075 \pm 0.088$ is found, where the error is purely statistical in origin. Since the reaction (7.98) is mediated basically by electromagnetic interactions, it can be concluded that there is no significant violation of T in these interactions. From this and similar results [HE 69a], we conclude that the violation of T in electromagnetic interactions is at most a few percent.

Possible T violations have also been sought for in weak interactions [HE 69a]. An attempt has been made to determine the existence of a term of the type $J_i \cdot (p_e \times p_\nu)$ that is odd under time reversal in the decay $^{19}\text{Ne} \rightarrow {}^{19}\text{F} + e^+ + \nu_e$, with negative results [HC 84].

Summarizing: all direct experimental evidence leads to the conclusion that the contribution to the amplitude for processes involving strong, electromagnetic and weak interactions of a term which violates time reversal is at the most of order of a few parts in a thousand. There exist, nevertheless, powerful reasons which indirectly support that violation, possibly at the electroweak level (we refer to the CP violation in the physics of neutral kaons, which together with the famous CPT theorem implies the violation of T [MR 69]).

7.14 Galilean Transformations

Spatial translations and rotations form the Euclidean group, which is a subgroup of the Galilei group, whose central role in Newtonian mechanics is well known as the associated relativity group. This group consists of the transformations

$$r \xrightarrow{g} Rr + vt + a , \quad t \xrightarrow{g} t + \tau , \tag{7.99}$$

where a is the spatial translation vector, v is the relative velocity of the two inertial systems in question, $\mathcal{R}$ is a rotation, and τ a displacement of the time origin. Such a group is a 10-parameter connected Lie group.

Representing the transformation (7.99) symbolically by $g \equiv (\mathcal{R}, v, a, \tau)$ we have immediately the product law

$$(\mathcal{R}_2, v_2, a_2, \tau_2)(\mathcal{R}_1, v_1, a_1, \tau_1)$$
$$= (\mathcal{R}_2 \mathcal{R}_1, v_2 + \mathcal{R}_2 v_1, a_2 + \mathcal{R}_2 a_1 + \tau_1 v_2, \tau_1 + \tau_2) \qquad (7.100)$$

It is convenient to introduce appropriate notation before defining operationally the action of a Galilean transformation g on a physical system. Considering for simplicity the case of a single spinless particle, $|x; t\rangle$ will denote an eigenstate of the position operator $X(t)$ in the Heisenberg picture (leaving understood the subscript "H") with eigenvalue x, i.e.,

$$|x; t\rangle \equiv U^\dagger(t, t_0)|x\rangle , \qquad (7.101)$$

where t_0 is the instant of time at which the Schrödinger and Heisenberg pictures coincide, and $U(t, t_0)$ the time evolution operator in the Schrödinger picture (2.201). The equations (7.99) suggest the operational definition of the action of g on a physical state

$$|x; t\rangle \xrightarrow{g} \exp[\mathrm{i}\varphi_g(x, t)/\hbar]|\mathcal{R}x + vt + a; t + \tau\rangle , \qquad (7.102)$$

where φ_g is a local phase to be determined. Availing ourselves of measuring devices which prepare localized states at a given time, the state $|x; t\rangle$ is prepared by such an apparatus at rest, which at $t = 0$ is at x and has instructions to prepare the state at time t. Similarly, the final state in (7.102) is produced at time $t + \tau$ with another such device which at $t = 0$ is located at the position $x - \mathcal{R}^{-1}v\tau$, is turned through $\mathcal{R}$ and translated by a, and moves with a constant velocity v.

The specification of the φ_g phase in (7.102) requires the analysis of what happens to the momentum under the action of g. Introducing in the same way the eigenstate $|p; t\rangle$ of $P(t)$ with eigenvalue p, it is clear that

$$|p; t\rangle \xrightarrow{g} \exp[\mathrm{i}\psi_g(p, t)/\hbar]|\mathcal{R}p + Mv; t + \tau\rangle , \qquad (7.103)$$

as suggested by the classical transformation law for momentum, M being the mass of the particle.

The action of g, given by (7.102) and (7.103), is a symmetry transformation due to the assumed equivalence of inertial frames. Thus, given the connected nature of the Galilei group, this action can be implemented by a unitary operator $U_t(g)$, which in principle can depend on the instant t of the initial states. The invariance of the scalar product $\langle x; t|p; t\rangle$ under $U_t(g)$ leads to

$$\varphi_g(x, t) = Mv \cdot (\mathcal{R}x) + \alpha_g(t) ,$$
$$\psi_g(p, t) = -(vt + a) \cdot \mathcal{R}p - \beta_g(t) , \qquad (7.104)$$
$$\alpha_g(t) + \beta_g(t) = Mv \cdot (vt + a) .$$

We can already determine $U_t(g)$ explicitly. For this it suffices to observe that for any characterization of the states by quantum numbers a,

$$|a; t + \tau\rangle = U_{\mathrm{H}}(t + \tau, t)|a; t\rangle , \quad \text{where} \tag{7.105}$$

$$U_{\mathrm{H}}(t_2, t_1) \equiv U(t_0, t_2)U(t_1, t_0) . \tag{7.106}$$

Then

$$|\mathcal{R}\boldsymbol{x}{+}\boldsymbol{v}t + \boldsymbol{a}; t + \tau\rangle = U_{\mathrm{H}}(t + \tau, t)|\mathcal{R}\boldsymbol{x} + \boldsymbol{v}t + \boldsymbol{a}; t\rangle$$

$$= U_{\mathrm{H}}(t + \tau, t) \exp\left[-\frac{\mathrm{i}}{\hbar}\boldsymbol{P}(t) \cdot [\boldsymbol{v}t + \boldsymbol{a}]\right] U_t(\mathcal{R})|\boldsymbol{x}; t\rangle , \tag{7.107}$$

where the operator implementing the rotation at time t is indicated by $U_t(\mathcal{R})$, having generators $\boldsymbol{J}(t)$. As a consequence,

$$U_t(g) = \exp\left[\frac{\mathrm{i}}{\hbar}\alpha_g(t)\right] U_{\mathrm{H}}(t + \tau, t) \exp\left[-\frac{\mathrm{i}}{\hbar}\boldsymbol{P}(t) \cdot [\boldsymbol{v}t + \boldsymbol{a}]\right]$$

$$\times \exp\left[\frac{\mathrm{i}}{\hbar}M\boldsymbol{v} \cdot \boldsymbol{X}(t)\right] U_t(\mathcal{R})$$

$$= \exp\left\{\frac{\mathrm{i}}{\hbar}[\alpha_g(t) - Mv^2t/2]\right\} U_{\mathrm{H}}(t + \tau, t) \tag{7.108}$$

$$\times \exp\left[-\frac{\mathrm{i}}{\hbar}\boldsymbol{P}(t) \cdot \boldsymbol{a}\right] \exp\left[-\frac{\mathrm{i}}{\hbar}[t\boldsymbol{P}(t) - M\boldsymbol{X}(t)] \cdot \boldsymbol{v}\right] U_t(\mathcal{R}) .$$

Note thus that $\boldsymbol{K}(t) \equiv t\boldsymbol{P}(t) - M\boldsymbol{X}(t)$ is the generator of the pure Galilean transformation.

From (7.108), or directly from (7.102) and (7.103), we get the expected transformation laws of the position and momentum observables:

$$\boldsymbol{X}(t) \overset{g}{\longrightarrow} \mathcal{R}^{-1}[\boldsymbol{X}(t + \tau) - \boldsymbol{v}t - \boldsymbol{a}] ,$$

$$\boldsymbol{P}(t) \overset{g}{\longrightarrow} \mathcal{R}^{-1}[\boldsymbol{P}(t + \tau) - M\boldsymbol{v}] . \tag{7.109}$$

It is clear that these same expressions (7.109) are applicable to the operators $\boldsymbol{X}_i, \boldsymbol{P}_i$ for each particle of a composite system.

Through the operational definition given for $U_t(g)$ it is evident that

$$U_t(g_2 g_1) = \omega_t(g_2, g_1)U_{t+\tau_1}(g_2)U_t(g_1) , \tag{7.110}$$

where $\omega_t(g_2, g_1)$ is a global phase which may be non-trivial. Using (7.108) it is easy to show that

$$\omega_t(g_2, t_1) = \exp\left(\frac{\mathrm{i}}{\hbar}[\alpha_{g_2 g_1}(t) - \alpha_{g_2}(t + \tau_1) - \alpha_{g_1}(t)\right.$$

$$\left. - M(\mathcal{R}_2^{-1}\boldsymbol{v}_2) \cdot (\boldsymbol{v}_1 t + \boldsymbol{a}_1)]\right) . \tag{7.111}$$

The extension of all of the above to the case of systems of N particles with or without spin is straightforward. We note first that for a system of N spinless particles the relative orbital angular momentum is

$$L_{\text{rel}}(t) \equiv \sum_{1}^{N} (X_i(t) - X(t)) \times \left(P_i(t) - \frac{M_i}{M} P(t) \right) , \qquad (7.112)$$

where

$$X(t) \equiv \frac{1}{M} \sum_{1}^{N} M_i X_i(t), \; P(t) \equiv \sum_{1}^{N} P_i(t) \; \text{and} \; M \equiv \sum_{1}^{N} M_i$$

denote the position and momentum of the center of mass and the total mass of the system, respectively. It changes under a Galilean transformation according to

$$L_{\text{rel}}(t) \xrightarrow{g} \mathcal{R}^{-1} L_{\text{rel}}(t + \tau) , \qquad (7.113)$$

as can be seen from (7.109). This suggests that we should accept for the spin

$$S_i(t) \xrightarrow{g} \mathcal{R}^{-1} S_i(t + \tau) , \quad i = 1, \ldots, N . \qquad (7.114)$$

From this we see that the operator $U_t(g)$ which implements g for an arbitrary system is given by (7.108), interpreting $X(t), P(t), J(t)$ as the position and momentum observables of the center of mass and the total angular momentum, respectively.

In general, the equality (7.110) does not correspond to a projective representation of the Galilei group, given that the operators which enter there refer to different times. If our physical system is closed, however, and we admit the principle of Galilean invariance, the generators $P(t), K(t), J(t)$ and $H(t)$ are constants of motion; thus, choosing $\alpha_g(t) \equiv Mv^2 t/2$, we obtain

$$U(g) = e^{iH\tau/\hbar} e^{-iP \cdot a/\hbar} e^{-iK \cdot v/\hbar} U(\mathcal{R}) , \qquad (7.115)$$

where we have omitted the subscript t in $U(g)$ since the r.h.s. is independent of t. With this, (7.110) and (7.111) become

$$U(g_2 g_1) = \omega(g_2, g_1) U(g_2) U(g_1) ,$$
$$\omega(g_2 g_1) = \exp \left\{ -\frac{i}{\hbar} M \left[\frac{1}{2} \tau_1 v_2^2 + v_2 \cdot (\mathcal{R}_2 a_1) \right] \right\} . \qquad (7.116)$$

Starting with (7.115), one can also directly prove (7.116), using (2.178) and the relations

$$e^A f(B) e^{-A} = f(e^A B e^{-A}) ,$$
$$e^A B e^{-A} = B + [A, B] + \frac{1}{2!}[A, [A, B]] + \cdots \equiv e^{[A}, B] . \qquad (7.117)$$

The operators $U(g)$ provide a projective representation of the Galilei group. In contrast to what occurs with the rotation group, for example, the phase ω is not locally removable via a global change of phase for each $U(g)$, but is instead essential.

To see how a wave function transforms, we will discuss it in terms of a particle of spin s since the extension to composite systems is straightforward using tensor products. Given at the instant t a state $|\boldsymbol{x}, m; t\rangle$ of this particle, localized at $\boldsymbol{x}$ and with $S_z(t)$ fixed and equal to $m\hbar$, we have

$$|\boldsymbol{x}, m; t\rangle \xrightarrow{g} \pm \exp\left\{\frac{i}{\hbar} M \left[\boldsymbol{v} \cdot \mathcal{R}\boldsymbol{x} + \frac{1}{2}v^2 t\right]\right\}$$

$$\times \sum_{m'} D_{m'm}^s(\mathcal{R})|\mathcal{R}\boldsymbol{x} + \boldsymbol{v}t + \boldsymbol{a}, m'; t + \tau\rangle, \qquad (7.118)$$

where the matrix element $D_{m'm}^s(\mathcal{R}) = \langle m'; t_1 | \exp[-i\hat{\boldsymbol{n}} \cdot \boldsymbol{S}(t_1)\theta/\hbar]|m; t_1\rangle$ for the rotation $\mathcal{R}(\hat{\boldsymbol{n}}, \theta)$ is independent of the time t_1 at which it is calculated, and thus we can adopt for it the value in the standard representation of SU(2). Given a state $|\Psi\rangle$ for the particle, i.e., $|\Psi\rangle$ in the Heisenberg picture and $|\Psi_S(t)\rangle$ in the Schrödinger picture, its wave function is simply $\Psi_m(\boldsymbol{x}; t) \equiv \langle \boldsymbol{x}, m|\Psi_S(t)\rangle = \langle \boldsymbol{x}, m; t|\Psi\rangle$. Then, using (7.118) one sees that the wave function of the state transformed under g would be, at time t:

$$(U_t(g)\Psi)_m(\boldsymbol{x}; t) = \langle \boldsymbol{x}, m; t|U_t(g)|\Psi\rangle = \pm \exp\left[-\frac{i}{\hbar} M \left(-\boldsymbol{v} \cdot \boldsymbol{x} + \frac{1}{2}v^2 t\right)\right]$$

$$\times \sum_{m'} D_{m'm}^s(\mathcal{R})\Psi_{m'}(\mathcal{R}^{-1}(\boldsymbol{x} - (t - \tau)\boldsymbol{v} - \boldsymbol{a}), t - \tau).$$

$$(7.119)$$

In particular, if the wave function is a plane wave $\exp[i(\boldsymbol{p} \cdot \boldsymbol{x} - Et)/\hbar]$, and we suppress the spin for simplicity, the transformed wave function is, aside from a global phase, also a plane wave with momentum and energy

$$\boldsymbol{p}' = \mathcal{R}\boldsymbol{p} + M\boldsymbol{v}, \qquad E' = E + \mathcal{R}\boldsymbol{p} \cdot \boldsymbol{v} + \tfrac{1}{2}Mv^2. \qquad (7.120)$$

These formulae, in which the terms $M\boldsymbol{v}$ and $Mv^2/2$ stem from the essential phase appearing in (7.119), indicate that the reduced wavelength $\lambdabar$ and the angular frequency ω associated with a plane wave change according to

$$\frac{1}{\lambdabar'} = \frac{1}{\lambdabar} + \frac{Mv}{\hbar}, \qquad \omega' = \omega + \frac{v}{\lambdabar} + \frac{1}{2}\frac{Mv^2}{\hbar} \qquad (7.121)$$

under Galilean transformations having velocities parallel to $\boldsymbol{p}$.

Another important consequence of the essential phase which occurs in (7.119) is the existence of a superselection rule associated with the mass (the Bargmann superselection rule). In fact, let us consider the identity transformation, written as

$$(I, 0, 0, 0) = (I, -\boldsymbol{v}, 0, 0)(I, 0, -\boldsymbol{a}, 0)(I, \boldsymbol{v}, 0, 0)(I, 0, \boldsymbol{a}, 0). \qquad (7.122)$$

Using (7.108) with $\alpha_g(t) = Mv^2t/2$, the product of the operators implementing the transformations on the r.h.s. is $I\exp[iM\boldsymbol{a} \cdot \boldsymbol{v}/\hbar]$, where M is the total mass of the system. On the other hand, it is clear that the physical observables must not change under (7.122), since the resultant geometric transformation is the identity. Then if $|\Psi_1\rangle, |\Psi_2\rangle$ are states of systems with two different masses M_1 and M_2, their linear superposition cannot be coherent, since under the r.h.s. of (7.122), calculated in successive steps,

$$\alpha_1|\Psi_1\rangle + \alpha_2|\Psi_2\rangle \longrightarrow \alpha_1 \exp\left[\frac{i}{\hbar}M_1\boldsymbol{a}\cdot\boldsymbol{v}\right]|\Psi_1\rangle + \alpha_2 \exp\left[\frac{i}{\hbar}M_2\boldsymbol{a}\cdot\boldsymbol{v}\right]|\Psi_2\rangle ,$$

(7.123)

and, since the relative phase of the amplitudes has changed, there can be no observables which connect $|\Psi_1\rangle$ and $|\Psi_2\rangle$; in other words, the mass operator defines a superselection rule if one accepts that basic observables transform according to the Galilei group in going from one inertial frame to another. Of course, this is an approximate superselection rule, valid only in the Galilean (non-relativistic) framework.

Since a free particle forms a closed system, invariant under Galilean transformations, it is to be expected that the free-particle Schrödinger equation should have the same differential expression in every inertial frame. It is precisely the essential phase in (7.119) which guarantees this. Indeed, taking spin 0 for simplicity,

$$\Psi'(\boldsymbol{x};t) \equiv (U(g)\Psi)(\boldsymbol{x};t)$$
$$= \exp\left[-\frac{i}{\hbar}M\left(-\boldsymbol{v}\cdot\boldsymbol{x}+\frac{1}{2}v^2t\right)\right]\Psi(\boldsymbol{x}';t') ,$$

(7.124)

with $\boldsymbol{x}' \equiv \mathcal{R}^{-1}(\boldsymbol{x} - \boldsymbol{v}t' - \boldsymbol{a})$, $t' \equiv t - \tau$. A direct calculation shows that if

$$i\hbar\frac{\partial\Psi(\boldsymbol{x}';t')}{\partial t'} = -\frac{\hbar^2}{2M}\Delta'\Psi(\boldsymbol{x}';t') ,$$

(7.125)

then also

$$i\hbar\frac{\partial\Psi'(\boldsymbol{x};t)}{\partial t} = -\frac{\hbar^2}{2M}\Delta\Psi'(\boldsymbol{x};t) .$$

(7.126)

For more general systems, the Schrödinger equation

$$i\hbar\frac{d}{dt}|\Psi_S(t)\rangle = H_S(t)|\Psi_S(t)\rangle$$

(7.127)

becomes

$$i\hbar\frac{d}{dt}|\Psi'_S(t)\rangle = H'_S(t)|\Psi'_S(t)\rangle , \quad \text{where}$$

(7.128)

$$H'_S(t) = H_S(t) + \left[i\hbar\frac{dU_t(g)}{dt}U_t^\dagger(g)\right]_S .$$

(7.129)

Thus the invariance of the Schrödinger equation is assured if and only if $U_t(g)$ does not depend on time t. This requirement, with the usual choice

$\alpha_g(t) = Mv^2t/2$, leads, upon requiring it for all Galilean transformations, to the conditions

$$\frac{d}{dt}J(t) = \frac{d}{dt}K(t) = \frac{d}{dt}P(t) = \frac{d}{dt}H(t) = 0 . \tag{7.130}$$

In particular, from (7.130), we deduce that for a single particle immersed in a potential $V(\boldsymbol{x}; t)$, the only choice for Galilean invariance is $V(\boldsymbol{x}; t) = $ constant. For a system of two spinless particles with a mutual interaction $V(\boldsymbol{x}_1, \boldsymbol{x}_2; t)$ the conditions (7.130) will be satisfied if and only if V depends only on the relative distance $|\boldsymbol{x}_1 - \boldsymbol{x}_2|$.

It can be interesting, nevertheless, to observe that there exists an extensive family of Hamiltonians which preserve a weak Galilean invariance in the following sense [JA 68]: consider a subgroup of the Galilei group formed by the transformations $g_v \equiv (I, \boldsymbol{v}, -\boldsymbol{v}t, 0)$, at some time t. Under these transformations, the position and momentum observables (suppressing the spin and considering only one particle) change as follows:

$$X(t) \xrightarrow{g_v} X(t) , \qquad P(t) \xrightarrow{g_v} P(t) - M\boldsymbol{v} , \tag{7.131}$$

and the operators which implement these transformations are

$$U_t(g_v) = \exp\left[\frac{\mathrm{i}}{\hbar}M\boldsymbol{v} \cdot X(t)\right] . \tag{7.132}$$

Classically it is to be expected that the velocities change according to

$$\frac{dX(t)}{dt} \xrightarrow{g_v} \frac{dX(t)}{dt} - \boldsymbol{v} . \tag{7.133}$$

The velocity operators being related to the momentum operators via the Heisenberg equations of motion, it is not strange that the requirement (7.133) imposes conditions on $H(t)$ in our case. Indeed: from the second of equations (7.131), together with (7.133), we see that $P(t) - M\ dX(t)/dt$ should commute with $U_t(g_v)$ for all $\boldsymbol{v}$. This requires that

$$P(t) - M\frac{dX(t)}{dt} = A(X(t); t) . \tag{7.134}$$

Taking into account that (7.134) implies

$$\left[X_j(t), M\frac{dX_k(t)}{dt}\right] = \mathrm{i}\hbar\delta_{jk}I , \tag{7.135}$$

and using the Heisenberg equation for time evolution, we immediately obtain

$$\left[H(t) - \frac{1}{2M}(P(t) - A(X(t); t))^2, X(t)\right] = 0 , \tag{7.136}$$

and thus

$$H(t) = \frac{1}{2M}(P(t) - A(X(t); t))^2 + V(X(t); t) , \tag{7.137}$$

which is the standard form of the Hamiltonian for a charged particle in an electromagnetic field, as we shall discuss in Chap. 12.

We will not expand further on the Galilei group, whose projective unitary representations require a mathematical treatment beyond the scope of this book. There exists abundant literature on this subject [LE 74], [LO 71], and it is shown that these representations for cases of physical interest are characterized, if they are irreducible, by two parameters: the mass and the spin of the system. We should indicate here that although historically the spin appeared first as a relativistic phenomenon (in the sense of Poincaré), it already occurs naturally within the Galilean framework. We return to this question when we analyze the interaction of a particle with an electromagnetic field in Chap. 12.

7.15 Isospin

Independent of the space-time symmetries we have considered thus far, there exist in physics the so-called *internal* symmetries, very important in the development of nuclear physics and in the theory of elementary particles. Of all the internal symmetries, we discuss only isospin; the reader interested in increasing his or her understanding of this general area should consult [GN 64], [DY 66]. The two particles which are the constituents of the atomic nucleus are the neutron and the proton, whose mass, spin-parity, electric charge, magnetic moment and mean lifetime are [AC 84]:

$$
\begin{aligned}
&m_p = 938.279\ 6(27)\,\text{MeV}/c^2 , &&J_p^{\Pi} = 1/2^+ , &&Q_p = |e| , \\
&m_n = 939.573\ 1(27)\,\text{MeV}/c^2 , &&J_n^{\Pi} = 1/2^+ , &&Q_n = 0 , \\
&\mu_p = 2.792\ 845\ 6(11)\mu_N , &&\tau_p > 10^{32}\ \text{years} , \\
&\mu_n = -1.913\ 041\ 84(88)\mu_N , &&\tau_n = 0.898(16) \times 10^3\ \text{s} ,
\end{aligned}
\tag{7.138}
$$

where e is the electric charge of the electron, and $\mu_N = |e|\hbar/2m_p c = 3.1524515(53)$ $\times 10^{-18}\,\text{MeV}\,\text{G}^{-1}$. The properties of these particles are essentially due to the joint effects of the strong, electromagnetic, and weak interactions. *Heisenberg* [HE 32] was the first to pose the problem of how these properties would be modified if only the strong interactions existed in nature. Since the neutron decays only through the weak interaction $n \rightarrow p + e^- + \bar{\nu}_e$, upon disappearance of this interaction both the proton and the neutron would be stable (we ignore their possible instability due to processes typical of grand unification models). On the other hand, the concepts of electrical charge and magnetic moment cease to make sense if there is no electromagnetic interaction. If we suppose that the small difference between the masses of the proton and the neutron, $(m_n - m_p)/m_p \simeq 1.4 \times 10^{-3}$, is of electromagnetic origin, we can already assert that in the absence of weak and electromagnetic interactions, the neutron and the proton are indistinguishable particles; the word *nucleon* is used to denote either of them.

It is known that in the absence of external electromagnetic fields which determine some privileged direction in space, there can be no observable phenomena which depend upon whether the third component of the spin of the electron be $+1/2$ or $-1/2$ along some arbitrarily selected quantization direction; this is due to the invariance of the Hamiltonian under the rotation group. In a totally analogous way, it can be argued that the fact that the neutron and the proton are indistinguishable in the absence of electromagnetic and weak interactions is due to the Hamiltonian being invariant with respect to rotations in some internal space, which has come to be called *isospin space*. The algebra of this group of rotations is that of $SU_T(2)$ where T indicates that the rotation takes place in the internal space. Thus, the Hamiltonian with only the strong interactions, H_s, is such that if $G \in SU_T(2)$, then $[H_s, U_G] = 0$, while the entire Hamiltonian, which includes the electromagnetic and weak interactions, $H = H_s + H_{em} + H_w$, should satisfy $[H, U_G] \neq 0$, since it is known that there are marked differences between real neutrons and protons.

The proton and the neutron correspond to the $\mathcal{D}^{(1/2)}$ representation of $SU_T(2)$, and we say that the nucleon has isospin $1/2$. We identify the proton with the state $|T = 1/2, T_3 = +1/2\rangle$, and the neutron with the state $|T = 1/2, T_3 = -1/2\rangle$. (Notice that this identification is totally arbitrary, and in some nuclear physics books they are reversed.)

One consequence of H_s being invariant under $SU_T(2)$ is that for each energy level of a nucleus (A, Z) (A nucleons, Z protons), there exist in the absence of the weak and electromagnetic interactions states with well-defined values of the isospin T and its third component T_3. In general, given the complex dynamics of a nucleus, its levels will not show accidental degeneracy; they will have a definite value of T. For a given nucleus, the value of T_3 is fixed: $T_3 =$ (number of protons $-$ number of neutrons)$/2 = Z - A/2$. On the other hand, if the level $(T, T_3 = Z - A/2)$ exists in a given nucleus (A, Z), all the other levels $(T, T), (T, T - 1), \ldots, (T, -T + 1), (T, -T)$ should exist in the nuclei $(A, A/2 + T), (A, A/2 + T - 1), \ldots, (A, A/2 - T + 1), (A, A/2 - T)$, respectively, with the same energy and identical values of spin-parity, since the $SU_T(2)$ operators cannot change the values of these quantities. The set of these states is said to form an *isospin multiplet* or *isomultiplet*. To see the extent to which this is true, we consider some isospin multiplets which have been observed and we compare their energies.

The ground state of ^{14}C, the state with excitation energy of $2.312\,\text{MeV}$ in ^{14}N, and the ground state of ^{14}O form an isospin triplet, with $J^\Pi = 0^+$, and T values of $T_3 = -1, 0, +1$, respectively. From [AL 59] we obtain that the differences in energies between the corresponding states are $E(^{14}O) - E(^{14}N) = 2.835\,\text{MeV}$, $E(^{14}N) - E(^{14}C) = 2.157\,\text{MeV}$. Taking into account that $m_n - m_p = 1.293323(16)\,\text{MeV}/c^2$ and that the mass of the electron is $m_e = 0.5110034(14)\,\text{MeV}/c^2$, the correction needed to account for the neutron-proton mass difference, as well as the presence of extra electrons in each case, is to add to each of the above quantities $(1.2933 - 0.5110)\,\text{MeV} = 0.7823\,\text{MeV}$. The results for the differences in energy between those nuclear states, corrected for the neutron-proton mass difference are: $E_N(^{14}O) - E_N(^{14}N) = 3.617$ MeV and $E_N(^{14}N) - E_N(^{14}C) = 2.939$ MeV. These differences must be mostly due to the electrostatic repulsions between the protons, provided that H_s is invariant under $SU_T(2)$. The calculation of this energy is not simple. Approximating the total Coulomb energy of a nucleus by the formula [BM 69]

$$E_C(A, Z) \simeq \frac{3Z^2\alpha}{5R} \left[1 - 5 \left(\frac{3}{16\pi Z} \right)^{2/3} \right], \tag{7.139}$$

where the nuclear radius R is given by $R = 1.25A^{1/3}$ fm, it is found that $E_C(^{14}O) - E_C(^{14}N) = 3.730$ MeV and $E_C(^{14}N) - E_C(^{14}C) = 3.183$ MeV, in good agreement with the experimental results if we take into account the uncertainty of the latter and the approximations used to find E_C.

Another typical example is the isospin quartet $(T = 3/2)$ formed by: the ground state of ^{33}P, the 5.48 MeV state of ^{33}S, the 5.56 MeV state of ^{33}Cl and the ground state of ^{33}Ar, all with spin-parity $1/2^+$ and with T_3 values, respectively, of $-3/2, -1/2, 1/2$, and $3/2$. From the tables in [ML 67] we obtain that the differences in energies between the nuclear states corrected for the neutron-proton mass difference are $E_N(^{33}Ar) - E_N(^{33}Cl) = 6.82$ MeV, $E_N(^{33}Cl) - E_N(^{33}S) = 6.43$ MeV and $E_N(^{33}S) - E_N(^{33}P) = 6.014$ MeV. The differences between the Coulomb energies calculated from (7.139) are $E_C(^{33}Ar) - E_C(^{33}Cl) = 6.973$ MeV, $E_C(^{33}Cl) - E_C(^{33}S) = 6.553$ MeV, and $E_C(^{33}S) - E_C(^{33}P) = 6.133$ MeV, again in good agreement with the previous data, taking into account the uncertainties.

Another test of $SU_T(2)$ symmetry is the study of nuclear reactions, which, being mediated primarily by the strong interaction, should conserve isospin. The reactions [CP 63]

$$\begin{aligned} d(1^+; \text{g.s.}; T = 0) &+ {}^{16}O(0^+; \text{g.s.}; T = 0) \\ &\rightarrow {}^{14}N + \alpha(0^+; \text{g.s.}; T = 0) \end{aligned} \tag{7.140}$$

have been studied for different possible states of ^{14}N, in particular the production of the states $^{14}N(1^+; \text{g.s.}; T = 0)$ and $^{14}N(0^+; 2.312; T = 1)$. The conservation of isospin implies that this last state cannot be produced. Using 24 MeV deuterons it was found that the excited state in question was produced only with a rate of (0.7 ± 0.6) percent with respect to the production rate of the ground state. It is entirely possible that the low production of the excited state be in large measure due to the selection rule considered here, but it cannot be ruled out that a part of this suppression is due to the detailed mechanism causing the transition.

One possible additional test concerning the validity of this symmetry is to compare, for given kinematic conditions, the cross sections for the two processes

a) $p + d \rightarrow \pi^+ + {}^3H$,

b) $p + d \rightarrow \pi^0 + {}^3He$. $\tag{7.141}$

It is known that the proton is a state of isospin $|1/2, 1/2\rangle$; on the other hand, the deuteron is $|0, 0\rangle$. We also know that ^{3}He and ^{3}H form a doublet and correspond to the states $|1/2, 1/2\rangle$ and $|1/2, -1/2\rangle$, respectively. Moreover, there occur in nature three pions, π^+, π^0, π^-, which form an isospin triplet and which can be identified with the states of the basis of a $T = 1$ representation of $SU_T(2)$ as follows: $|\pi^+\rangle = -|1, 1\rangle, |\pi^0\rangle = |1, 0\rangle$ and $|\pi^-\rangle = |1, -1\rangle$, where the origin of the minus sign is due to the use of the Condon-Shortley convention for the Clebsch-Gordan coefficients and the requirement that the quantum fields associated with π^+ and π^- be hermitian conjugates of each other, as can be demonstrated. In this type of test, two or more reactions are compared under identical kinematic conditions, their only difference being that some of the interacting particles in one reaction are replaced by others in the same isomultiplets. In the case considered here (7.141), the initial state is a state of pure isospin $|1/2, 1/2\rangle$. The final isospin states for the two reactions are

a) $\quad |f\rangle = -\sqrt{\tfrac{1}{3}}|3/2, 1/2\rangle - \sqrt{\tfrac{2}{3}}|1/2, 1/2\rangle$,

b) $\quad |f\rangle = \sqrt{\tfrac{2}{3}}|3/2, 1/2\rangle - \sqrt{\tfrac{1}{3}}|1/2, 1/2\rangle$.

$$(7.142)$$

Conservation of isospin does not permit the final state to have $T = 3/2$. As a consequence, if the isospin is conserved, then the isospin amplitude of the final state for the first reaction is $\sqrt{2}$ times the corresponding amplitude for the second, and thus the cross section for identical kinematic configurations for the two processes differ by the factor $(\sqrt{2})^2 = 2$:

$$R \equiv \frac{\sigma(\mathrm{p+d} \to \pi^+ + {}^3\mathrm{H})}{\sigma(\mathrm{p+d} \to \pi^0 + {}^3\mathrm{He})} = 2 . \qquad (7.143)$$

In reality, this is valid only if isomultiplet mass differences and electromagnetic interactions are neglected. For incident proton kinetic energies of 590 MeV, it has been estimated [KO 60] that these corrections produce $R = 2.20 \pm 0.07$, to be compared to the experimental value [HK 60] of $R_{\mathrm{exp}} = 2.13 \pm 0.06$; this agreement with the theoretical prediction is quite satisfactory.

In the state with isospin 1 and with $L = J = 0$, the phase shifts δ (Chap. 8) of strong interaction origin at low energies for the nucleon-nucleon system can be parametrized, as we shall see in general later on, in the form

$$k \cot \delta(k) = \frac{-1}{a} + \frac{1}{2}rk^2 + \dots , \qquad (7.144)$$

where $\hbar k$ is the relative momentum of the two nucleons in the center-of-mass system and a, r are constants, the so-called scattering length and effective range, respectively. A careful and non-trivial analysis of the experimental data gives the following results [HE 69b]:

$$
\begin{array}{ll}
a_{\mathrm{pp}} = -17.0 \pm 0.2 \,\mathrm{fm} , & r_{\mathrm{pp}} = 2.83 \pm 0.03 \,\mathrm{fm} , \\
a_{\mathrm{nn}} = -17.6 \pm 1.5 \,\mathrm{fm} , & r_{\mathrm{nn}} = 3.2 \pm 1.6 \,\mathrm{fm} , \\
a_{\mathrm{np}} = -23.175 \pm 0.013 \,\mathrm{fm} , & r_{\mathrm{np}} = 2.76 \pm 0.07 \,\mathrm{fm} .
\end{array}
\qquad (7.145)
$$

If the nucleon-nucleon strong interactions are invariant under $SU_T(2)$, then in a state of given relative energy, spin, orbital and total angular momentum, and isospin 1, the forces in the (pp), (nn) and (np) systems are equal (charge independence of nuclear forces); thus, $\delta_{pp}(k) = \delta_{nn}(k) = \delta_{np}(k)$, and in particular, $a_{pp} = a_{nn} = a_{np}$ and $r_{pp} = r_{nn} = r_{np}$. Although the effective range equalities are well satisfied, the scattering legnth a_{np} is markedly different from that for $a_{pp} \simeq a_{nn}$. Though the discrepancy appears to be very large, it can be explained completely assuming that the (np) potential is more attractive than the (nn) or (pp) by about 2% [HE 69b], [BS 73], and it cannot be ascertained that this difference is not due to the mass difference between the $\pi^{\pm}$ and the π^0 [HE 69b], which are the particles mediating the nuclear force.

Finally, we note that the concept of isospin can also be useful in the study of the electromagnetic and weak interactions, since it appears that the Hamiltonians which describe these interactions consist of only a few terms, each one of which has well-defined $SU_T(2)$ tensorial character. For example, there is strong experimental evidence which indicates that H_{em} is the sum of two terms: an isoscalar and the zero component of an isovector term.

As a consequence of this fact, we mention the possibility of establishing the so-called electromagnetic mass formulae. It is supposed that the masses of the components of an isomultiplet are due to the joint action of the strong and electromagnetic interactions, and that in particular the difference in mass among the members of an isomultiplet is basically due to a second order effect of the electromagnetic interaction. Taking into account the transformation properties of H_s and H_{em} under $SU_T(2)$ and the above hypotheses, the mass operator M should be of the form

$$M = M_0^0 + M_0^1 + M_0^2 , \tag{7.146}$$

where $M_{T_3}^T$ indicates an irreducible tensor under $SU_T(2)$. The expectation value of this operator for a given state in an isomultiplet will give us its mass.

According to the Wigner-Eckart theorem, we can write

$$M(T T_3) = \alpha + C(T1T; T_30T_3)\beta + C(T2T; T_30T_3)\gamma , \tag{7.147}$$

where α, β and γ are the reduced matrix elements depending upon the isomultiplet considered, but not on the value of T_3, and whose evaluation would require a much greater understanding of the dynamics of the strong and electromagnetic interactions than is available today. If $T \geq 3/2$, then one obtains, eliminating the quantities α, β and γ, a set of $2(T-1)$ independent relations among the masses of the states of the isomultiplet. If, for example, $T = 3/2$, there results

$$[M(3/2, 3/2) - M(3/2, 1/2)] + [M(3/2, -1/2) - M(3/2, -3/2)]$$
$$= 2[M(3/2, 1/2) - M(3/2, -1/2)] . \tag{7.148}$$

Thus, the isoquartet of nuclear states considered earlier should satisfy

$$[E(^{33}Ar) - E(^{33}Cl)] + [E(^{33}S) - E(^{33}P)] = 2[E(^{33}Cl) - E(^{33}S)] . \tag{7.149}$$

The values 12.83 MeV and 12.86 MeV, respectively, can be deduced from the experimental data for each side of (7.149). For a much more detailed discussion of this point, [BL 73] and [CE 68] can be consulted.

Quantum chromodynamics permits an understanding of this isospin symmetry. If the masses of all types of quarks occurring in such a theory were equal, the equations of motion would be invariant under a group SU(N), where N is the number of different quark types (= "flavor"), and thus the strong interactions between hadrons would also present this invariance. In reality this is not the case, but nevertheless the u and d quark masses are very much smaller than the hadron masses and we still have, to a good approximation, invariance under the group SU(2), which is the isospin group [PT 84].

Appendix A: Special Functions

When trying to solve the Schrödinger equation for specific cases, it is frequently found that the problem is reduced to finding the solution of a differential equation of second order with only regular singular points at finite distance. The general theory for solving these equations, as well as the properties of the solutions for some of the particularly interesting cases, can be found in many books, among which the classic one by *Whittaker* and *Watson* [WW 69] is especially recommendable. The books by *Erdélyi* [ER 53] and *Abramowitz* and *Stegun* [AS 64], the latter having extensive numerical tables, are both extraordinarily useful for the large number of properties of the solutions to particular cases.

In this appendix we summarize the properties of those special functions which have been most frequently used in the development of this book, without giving proofs. Except when otherwise noted, the notation followed is that of *Abramowitz* and *Stegun* [AS 64].

A.1 Legendre Polynomials

The Legendre polynomials $P_L(x)$, with $L = 0, 1, 2, \ldots$, are the solutions of polynomial type, conveniently normalized, of the equations

$$\left[(1 - x^2)\frac{d^2}{dx^2} - 2x\frac{d}{dx} + L(L+1) \right] P_L(x) = 0 . \tag{A.1}$$

The general expression for them is

$$P_L(x) = 2^{-L} \sum_{n=0}^{[L/2]} \frac{(-1)^n}{n!} \frac{(2L - 2n)!}{(L - n)!(L - 2n)!} x^{L-2n} , \tag{A.2}$$

where $[L/2]$ indicates the integer part of $L/2$. The first few lowest order Legendre polynomials are

$$
\begin{aligned}
&P_0(x) = 1 , &&P_1(x) = x , \\
&P_2(x) = \tfrac{1}{2}(3x^2 - 1) , &&P_3(x) = \tfrac{1}{2}(5x^3 - 3x) , \\
&P_4(x) = \tfrac{1}{8}(35x^4 - 30x^2 + 3) , &&P_5(x) = \tfrac{1}{8}(63x^5 - 70x^3 + 15x) .
\end{aligned}
\tag{A.3}
$$

Some interesting properties of these polynomials are:

1) $P_L(1) = 1$, $P_L(-1) = (-1)^L$.   (A.4)

2) All zeros of $P_L(x)$ are simple and lie in the interval $(-1, +1)$; furthermore,

$$|P_L(x)| \leq 1 , \quad x \in [-1, +1] .$$ (A.5)

3) $P_L(-x) = (-1)^L P_L(x)$.   (A.6)

4) $P_{2L+1}(0) = 0$, $P_{2L}(0) = (-1)^L \dfrac{(2L)!}{2^{2L}(L!)^2}$.   (A.7)

5) Rodrigues' formula:

$$P_L(x) = \frac{1}{2^L L!} \frac{d^L}{dx^L} (x^2 - 1)^L .$$ (A.8)

6) Schläfli's formula:

$$P_L(x) = \frac{1}{2^{L+1}\pi i} \int_C \frac{(z^2 - 1)^L}{(z - x)^{L+1}} dz ,$$ (A.9)

where C is a simple closed contour in the complex z-plane which contains in its interior the point $z = x$ and is traversed in a positive or counterclockwise sense.

7) The Legendre polynomials form an orthogonal, complete set in $L^2[-1, +1]$ and satisfy the normalization condition

$$\int_{-1}^{+1} dx \, P_L(x) P_{L'}(x) = \frac{2}{2L + 1} \delta_{LL'} .$$ (A.10)

If $f(x) \in L^2[-1, +1]$, then

$$f(x) = \sum_{L=0}^{\infty} a_L P_L(x) , \quad \text{where}$$ (A.11)

$$a_L = \frac{2L + 1}{2} \int_{-1}^{+1} dx \, f(x) P_L(x) = \frac{2L + 1}{L! 2^{L+1}} \int_{-1}^{+1} dx \, (1 - x^2)^L \frac{d^L f(x)}{dx^L} .$$

(A.12)

In particular, if $|h| < 1$, one has

$$[1 - 2hx + h^2]^{-1/2} = \sum_{L=0}^{\infty} h^L P_L(x) ,$$ (A.13)

where the function on the l.h.s. is called the generating function. Furthermore,

$$e^{i\mathbf{k} \cdot \mathbf{r}} = \sum_{L=0}^{\infty} i^L (2L + 1) j_L(kr) P_L(\cos \theta) ,$$ (A.14)

where $j_L(kr)$ are the spherical Bessel functions and θ is the angle formed by the vectors $\mathbf{r}$ and $\mathbf{k}$.

8) Recursion formulae:

$$P'_{L+1}(x) - xP'_L(x) = (L+1)P_L(x) \, ; \tag{A.15}$$

$$(L+1)P_{L+1}(x) = (2L+1)xP_L(x) - LP_{L-1}(x) \, ; \tag{A.16}$$

$$xP'_L(x) - P'_{L-1}(x) = LP_L(x) \, ; \tag{A.17}$$

$$P'_{L+1}(x) - P'_{L-1}(x) = (2L+1)P_L(x) \, ; \tag{A.18}$$

$$(x^2 - 1)P'_L(x) = LxP_L(x) - LP_{L-1}(x) \, . \tag{A.19}$$

It is interesting to note for numerical evaluations that the iterative use of (A.16) permits the calculation of $P_L(x)$, starting with $P_0(x)$ and $P_1(x)$, without appreciable loss of significant digits.

A.2 Associated Legendre Functions

The associated Legendre functions $P_L^M(x)$, with $L = 0, 1, 2, \ldots$, and $M = -L, -L+1, \ldots, L-1, L$, are defined as the, appropriately normalized, regular solutions for $x \in [-1, +1]$ of the equations

$$\left[(1 - x^2)\frac{d^2}{dx^2} - 2x\frac{d}{dx} + L(L+1) - \frac{M^2}{1 - x^2} \right] P_L^M(x) = 0 \, , \tag{A.20}$$

and they are given by

$$P_L^M(x) = (1 - x^2)^{M/2}\frac{(-1)^M}{L!2^L}\frac{d^{L+M}}{dx^{L+M}}(x^2 - 1)^L \, . \tag{A.21}$$

Equivalent expressions for them are

$$P_L^M(x) = (-1)^M(1 - x^2)^{M/2}\frac{d^M}{dx^M}P_L(x) \, , \quad M \geq 0 \, ;$$

$$P_L^{-M}(x) = (-1)^M\frac{(L-M)!}{(L+M)!}P_L^M(x) \, . \tag{A.22}$$

Interesting properties of these functions are:

1) $P_L^0(x) = P_L(x)$. \hfill (A.23)

2) Orthogonality and normalization properties:

$$\int_{-1}^{+1} dx \, P_L^M(x)P_{L'}^M(x) = \frac{2}{(2L+1)}\frac{(L+M)!}{(L-M)!}\delta_{LL'} \, . \tag{A.24}$$

3) Recursion formulae:

$$P_L^{M-1}(x) = \frac{2Mx}{(1 - x^2)^{1/2}}P_L^M(x) - [L(L+1) - M(M-1)]P_L^{M-1} \, ; \tag{A.25}$$

$$(L - M + 1)P_{L+1}^M(x) = (2L+1)xP_L^M(x) - (L+M)P_{L-1}^M(x) \, ; \tag{A.26}$$

$$(2L + 1)(1 - x^2)^{1/2} P_L^M(x) = P_{L+1}^{M+1}(x) - P_{L-1}^{M+1}(x) \; ; \tag{A.27}$$

$$(2L + 1)(1 - x^2)^{1/2} P_L^M(x) = (L + M)(L + M - 1)P_{L-1}^{M-1}(x)$$
$$- (L - M + 1)(L - M + 2)P_{L+1}^{M-1}(x) \; . \tag{A.28}$$

4) Addition formula: if x_1, x_2, x lie in the interval $[-1, +1]$ and are related by

$$x = x_1 x_2 - \sqrt{(1 - x_1^2)(1 - x_2^2)} \cos \omega \; , \quad \text{then} \tag{A.29}$$

$$P_L(x) = P_L(x_1)P_L(x_2) + 2 \sum_{M=1}^{L} \frac{(L - M)!}{(L + M)!} P_L^M(x_1)P_L^M(x_2) \cos M\omega \; .$$

$$\tag{A.30}$$

A.3 Spherical Harmonics

The spherical harmonics are defined to be the functions

$$Y_L^M(\theta, \phi) = \left[\frac{(2L + 1)}{4\pi} \frac{(L - M)!}{(L + M)!} \right]^{1/2} P_L^M(\cos \theta) \, e^{iM\phi} \; , \tag{A.31}$$

$$0 \le \theta \le \pi \; , \quad 0 \le \phi < 2\pi \; ,$$

and, aside from multiplicative constants, they are the regular solutions of the equations

$$\left[\frac{1}{\sin \theta} \frac{\partial}{\partial \theta} \left(\sin \theta \frac{\partial}{\partial \theta} \right) + \frac{M^2}{\sin^2 \theta} + L(L + 1) \right] Y_L^M(\theta, \phi) = 0 \; ,$$

$$\left[i \frac{\partial}{\partial \phi} + M \right] Y_L^M(\theta, \phi) = 0 \; . \tag{A.32}$$

From their definition we can derive that

$$Y_L^{M*}(\theta, \phi) = (-1)^M Y_L^{-M}(\theta, \phi) \; , \tag{A.33}$$

$$Y_L^M(\pi - \theta, \phi + \pi) = (-1)^L Y_L^M(\theta, \phi) \; . \tag{A.34}$$

Explicit expressions for the spherical harmonics of lowest order are:

$$Y_0^0(\theta, \phi) = \frac{1}{\sqrt{4\pi}} \; ,$$

$$Y_1^0(\theta, \phi) = \sqrt{\frac{3}{4\pi}} \cos \theta \; , \qquad Y_1^1(\theta, \phi) = -\sqrt{\frac{3}{8\pi}} \sin \theta \, e^{i\phi} \; ,$$

$$Y_2^0(\theta, \phi) = \sqrt{\frac{5}{16\pi}} (3 \cos^2 \theta - 1) \; , \quad Y_2^1(\theta, \phi) = -\sqrt{\frac{15}{8\pi}} \sin \theta \cos \theta \, e^{i\phi}$$

$$Y_2^2(\theta, \phi) = \sqrt{\frac{15}{32\pi}} \sin^2 \theta \, e^{2i\phi} \; . \tag{A.35}$$

If r is a vector with Cartesian components (x_1, x_2, x_3) and with spherical components $(r \sin \theta \cos \phi, r \sin \theta \sin \phi, r \cos \theta)$, it is of interest to introduce the quantities

$$x_{+1} = -\frac{1}{\sqrt{2}}(x_1 + ix_2) , \qquad x_0 = x_3 , \qquad x_{-1} = +\frac{1}{\sqrt{2}}(x_1 - ix_2) . \qquad (A.36)$$

It can then be immediately shown that

$$Y_1^M(\theta, \phi) = \frac{1}{r}\sqrt{\frac{3}{4\pi}} x_M . \qquad (A.37)$$

Spherical harmonics are orthonormal:

$$\int d\Omega \, Y_{L'}^{M'*}(\theta, \phi) Y_L^M(\theta, \phi) = \delta_{L'L}\delta_{M'M} ,$$

$$\int d\Omega \dots \equiv \int_0^{2\pi} d\phi \int_0^\pi d\theta \, \sin\theta \dots , \qquad (A.38)$$

and in addition they form a complete set in $L^2(S_2)$, $S_2 \equiv \{|r| = 1\}$, with measure $d\Omega$.

Additional interesting properties of these functions are:

1) $Y_L^M(0, \phi) = \delta_{M0}\sqrt{\dfrac{2L+1}{4\pi}}$, $\displaystyle\sum_M Y_L^{M*}(\theta, \phi)Y_L^M(\theta, \phi) = \dfrac{2L+1}{4\pi}$,   (A.39)

$$Y_L^0(\theta, \phi) = \sqrt{\frac{2L+1}{4\pi}} P_L(\cos\theta) . \qquad (A.40)$$

2) If we have two directions in space characterized by the spherical coordinates (θ_1, ϕ_1) and (θ_2, ϕ_2), respectively, and if θ is the angle between them, then

$$P_L(\cos\theta) = \frac{4\pi}{2L+1} \sum_{M=-L}^{+L} Y_L^{M*}(\theta_1, \phi_1)Y_L^M(\theta_2, \phi_2) , \qquad (A.41)$$

which permits (A.14) to be rewritten in the following form:

$$e^{i\boldsymbol{k}\cdot\boldsymbol{r}} = 4\pi \sum_{L=0}^{\infty} i^L j_L(kr) \sum_{M=-L}^{+L} Y_L^{M*}(\hat{\boldsymbol{k}})Y_L^M(\hat{\boldsymbol{r}}) . \qquad (A.42)$$

3) The coupling rules for two spherical harmonics are:

$$Y_{L_1}^{M_1}(\theta, \phi)Y_{L_2}^{M_2}(\theta, \phi) = \sum_{L=0}^{\infty} \left[\frac{(2L_1+1)(2L_2+1)}{4\pi(2L+1)}\right]^{1/2}$$

$$\times C(L_1 L_2 L; M_1 M_2)C(L_1 L_2 L; 00)Y_L^{M_1+M_2}(\theta, \phi) , \qquad (A.43)$$

where $C(L_1 L_2 L; M_1 M_2 M)$ are the Clebsch-Gordan coefficients. Starting with (A.43) and using (A.38) we can immediately find:

$$\int d\Omega \, Y_{L_3}^{M_3*}(\theta, \phi) Y_{L_2}^{M_2}(\theta, \phi) Y_{L_1}^{M_1}(\theta, \phi)$$

$$= \left[\frac{(2L_1 + 1)(2L_2 + 1)}{4\pi(2L_3 + 1)} \right]^{1/2} C(L_1 L_2 L_3; M_1 M_2 M_3) C(L_1 L_2 L_3; 000) \, .$$

$$(A.44)$$

A.4 Hermite Polynomials

The Hermite polynomials $H_n(x)$, with $n = 0, 1, 2, \ldots$, are the polynomial-type solutions, with appropriate normalization, of the equations

$$\left[\frac{d^2}{dx^2} - 2x\frac{d}{dx} + 2n \right] H_n(x) = 0 \, . \tag{A.45}$$

The general expression for them is

$$H_n(x) = \sum_{r=0}^{[n/2]} \frac{(-1)^r n! (2x)^{n-2r}}{r!(n - 2r)!} \, . \tag{A.46}$$

The first few of them are:

$$H_0(x) = 1 \, , \qquad\qquad H_1(x) = 2x \, ,$$
$$H_2(x) = 4x^2 - 2 \, , \qquad\qquad H_3(x) = 8x^3 - 12x \, , \tag{A.47}$$
$$H_4(x) = 16x^4 - 48x^2 + 12 \, , \qquad H_5(x) = 32x^5 - 160x^3 + 120x \, .$$

Frequently used properties of the Hermite polynomials are the following:

1) If C is a simple closed contour in the complex z-plane which encloses the origin, and is traversed in a positive sense (counterclockwise), then

$$H_n(x) = \frac{n!}{2\pi i} \int_C \frac{\exp(2xz - z^2)}{z^{n+1}} dz \, . \tag{A.48}$$

2) The generating function is

$$e^{2xy-y^2} = \sum_{n=0}^{\infty} \frac{1}{n!} H_n(x) y^n \, . \tag{A.49}$$

3) Recursion formulae:

$$H_n'(x) = 2n H_{n-1}(x) \, , \tag{A.50}$$

$$x H_n'(x) = n H_{n-1}'(x) + n H_n(x) \, , \tag{A.51}$$

$$H_n(x) = 2x H_{n-1}(x) - 2(n - 1) H_{n-2}(x) \, . \tag{A.52}$$

4) Rodrigues' formula:

$$H_n(x) = (-1)^n e^{x^2} \frac{d^n}{dx^n} e^{-x^2} . \tag{A.53}$$

5) These polynomials are orthogonal according to the following relation:

$$\int_{-\infty}^{+\infty} dx\, e^{-x^2} H_n(x) H_m(x) = \sqrt{\pi} 2^n n! \delta_{nm} . \tag{A.54}$$

6) The integrals

$$I_{nm}^k \equiv [2^{n+m} \pi n! m!]^{-1/2} \int_{-\infty}^{+\infty} dx\, e^{-x^2} x^k H_n(x) H_m(x) \tag{A.55}$$

are particularly interesting. The relation (A.54) is equivalent to $I_{nm}^0 = \delta_{nm}$. Using the generating function, we can obtain

$$I_{nm}^1 = \sqrt{\frac{n+1}{2}} \delta_{m,n+1} + \sqrt{\frac{n}{2}} \delta_{m,n-1} , \tag{A.56}$$

and then with Parseval's theorem we can prove that

$$I_{nm}^k = \sum_{l=0}^{\infty} I_{nl}^1 I_{lm}^{k-1} = \sqrt{\frac{n+1}{2}} I_{n+1,m}^{k-1} + \sqrt{\frac{n}{2}} I_{n-1,m}^{k-1} . \tag{A.57}$$

Thus,

$$I_{nm}^2 = \tfrac{1}{2}\sqrt{(n+1)(n+2)}\delta_{m,n+2} + \tfrac{1}{2}(2n+1)\delta_{mn} + \tfrac{1}{2}\sqrt{(n-1)n}\delta_{m,n-2} ,$$

$$I_{nm}^3 = \frac{1}{2\sqrt{2}}\sqrt{(n+1)(n+2)(n+3)}\delta_{m,n+3} + \frac{3}{2\sqrt{2}}(n+1)^{3/2}\delta_{m,n+1}$$

$$\qquad + \frac{3}{2\sqrt{2}}n^{3/2}\delta_{m,n-1} + \frac{1}{2\sqrt{2}}\sqrt{(n-2)(n-1)n}\delta_{m,n-3} ,$$

$$I_{mn}^4 = \tfrac{1}{4}\sqrt{(n+1)(n+2)(n+3)(n+4)}\delta_{m,n+4}$$

$$\qquad + \tfrac{1}{2}(2n+3)\sqrt{(n+1)(n+2)}\delta_{m,n+2}$$

$$\qquad + \tfrac{3}{4}[(n+1)^2 + n^2]\delta_{mn} + \tfrac{1}{2}(2n-1)\sqrt{(n-1)n}\delta_{m,n-2}$$

$$\qquad + \tfrac{1}{4}\sqrt{(n-3)(n-2)(n-1)n}\delta_{m,n-4} . \tag{A.58}$$

A.5 Laguerre Polynomials

The Laguerre polynomials $L_n(x)$, with $n = 0, 1, 2, \ldots$, are appropriately normalized polynomial solutions of the equations

$$\left[x\frac{d^2}{dx^2} + (1-x)\frac{d}{dx} + n\right] L_n(x) = 0 , \tag{A.59}$$

whose general expression is

$$L_n(x) = \sum_{k=0}^{n} \frac{(-1)^k n!}{(k!)^2(n-k)!} x^k . \tag{A.60}$$

The lowest order ones are

$$
\begin{aligned}
&L_0(x) = 1 , && L_1(x) = 1 - x , \\
&L_2(x) = \tfrac{1}{2}[2 - 4x + x^2] , && L_3(x) = \tfrac{1}{6}[6 - 18x + 9x^2 - x^3] , \\
&L_4(x) = \tfrac{1}{24}[24 - 96x + 72x^2 - 16x^3 + x^4] , \\
&L_5(x) = \tfrac{1}{120}[120 - 600x + 600x^2 - 200x^3 + 25x^4 - x^5] .
\end{aligned}
\tag{A.61}
$$

A few frequently used properties are the following:

1) Generating function:

$$\frac{1}{1-y} \exp\left[-\frac{xy}{1-y}\right] = \sum_{n=0}^{\infty} y^n L_n(x) , \quad |y| < 1 . \tag{A.62}$$

2) Rodrigues' formula:

$$L_n(x) = \frac{1}{n!} e^x \frac{d^n}{dx^n}(x^n e^{-x}) . \tag{A.63}$$

3) Recursion formulae:

$$(n+1)L_{n+1}(x) = (2n+1-x)L_n(x) - nL_{n-1}(x) , \tag{A.64}$$

$$xL_n'(x) = n[L_n(x) - L_{n-1}(x)] . \tag{A.65}$$

4) These polynomials satisfy the following orthonormality relation:

$$\int_0^{\infty} dx\, e^{-x} L_n(x) L_m(x) = \delta_{nm} . \tag{A.66}$$

A.6 Generalized Laguerre Polynomials

The generalized Laguerre polynomials $L_n^k(x)$ are defined by

$$L_n^k(x) = (-1)^k \frac{d^k}{dx^k} L_{n+k}(x) , \quad n, k = 0, 1, 2, \dots , \tag{A.67}$$

and they are the regular solutions, aside from multiplicative constants, of the equations

$$\left[x \frac{d^2}{dx^2} + (k + 1 - x) \frac{d}{dx} + n \right] L_n^k(x) = 0 . \tag{A.68}$$

The explicit expression for them is

$$L_n^k(x) = \sum_{r=0}^{n} \frac{(-1)^r (n + k)!}{(n - r)!(k + r)!r!} x^r . \tag{A.69}$$

In particular,

$$L_0^k(x) = 1 . \tag{A.70}$$

The lowest order generalized Laguerre polynomials which appear more frequently in quantum mechanical problems are:

$$
\begin{aligned}
&L_1^1(x) = 2 - x , && L_2^1(x) = 3 - 3x + \tfrac{1}{2}x^2 , \\
&L_1^3(x) = 4 - x , && L_3^1(x) = 4 - 6x + 2x^2 - \tfrac{1}{6}x^3 , \\
&L_2^3(x) = 10 - 5x + \tfrac{1}{2}x^2 , && L_1^5(x) = 6 - x .
\end{aligned}
\tag{A.71}
$$

Some of their interesting properties are:

1) Generating function:

$$\frac{1}{(1-y)^{1+k}} \exp \left[-\frac{xy}{1-y} \right] = \sum_{n=0}^{\infty} y^n L_n^k(x) . \tag{A.72}$$

2) Rodrigues' formula:

$$L_n^k(x) = \frac{e^x}{n! x^k} \frac{d^n}{dx^n} [x^{n+k} e^{-x}] . \tag{A.73}$$

3) Recursion relations:

$$x L_n^{k+1}(x) = (x - n) L_n^k(x) + (n + k) L_{n-1}^k(x) , \tag{A.74}$$

$$L_n^{k-1}(x) = L_n^k(x) - L_{n-1}^k(x) , \tag{A.75}$$

$$x L_n^{k+1}(x) = (n + k + 1) L_n^k(x) - (n + 1) L_{n+1}^k(x) , \tag{A.76}$$

$$(n + k) L_n^{k-1}(x) = (n + 1) L_{n+1}^k(x) - (n + 1 - x) L_n^k(x) . \tag{A.77}$$

4) These polynomials satisfy the following orthonormality relation:

$$\int_0^\infty dx \, e^{-x} x^k L_n^k(x) L_m^k(x) = \frac{(n+k)!}{n!} \delta_{nm} \,. \tag{A.78}$$

Integrals which appear frequently in applications to quantum mechanics are:

$$I_s(nk; n'k') \equiv \int_0^\infty dx \, e^{-x} x^s L_n^k(x) L_{n'}^{k'}(x) \,, \tag{A.79}$$

which were calculated by *Schrödinger* [SC 26] to be

$$I_s(nk; n'k') = s! \sum_{r=0}^b (-1)^{n+n'+r} \binom{s-k}{n-r} \binom{s-k'}{n'-r} \binom{-s-1}{r} \,, \tag{A.80}$$

where $b = \min(n, n')$ and the binomial coefficients are

$$\binom{n}{k} = \frac{n!}{n!(n-k)!} \,, \quad n \ge k \ge 0 \,;$$
$$\binom{-n}{k} = (-1)^k \binom{n+k-1}{k} \,; \quad \binom{n}{k} = 0 \,, \quad k > n > 0 \,. \tag{A.81}$$

Similarly, the integrals

$$I_{nl}^{(p)} \equiv \frac{n^{p-1}}{2^{p+1}} \frac{(n-l-1)!}{(n+l)!} \int_0^\infty dx \, x^{2l+2+p} e^{-x} [L_{n-l-1}^{2l+1}(x)]^2 \,, \tag{A.82}$$

can easily be computed using (A.80). They are found to be

$$I_{nl}^{(2)} = \frac{n^2}{2} [5n^2 + 1 - 3l(l+1)] \,, \quad I_{nl}^{(1)} = \frac{1}{2} [3n^2 - l(l+1)] \,,$$
$$I_{nl}^{(0)} = 1 \,, \quad I_{nl}^{(-1)} = \frac{1}{n^2} \,, \quad I_{nl}^{(-2)} = \frac{2}{(2l+1)n^3} \,. \tag{A.83}$$

If one starts with the equation satisfied by $y_{nl} \equiv x^{l+1} e^{-x/2} L_{n-l-1}^{2l+1}(x)$, after multiplying by $x^{p+1} y_{nl}' - \frac{1}{2}(p+1) x^p y_{nl}$ and integrating by parts, one easily obtains the so-called *Kramers relation*, which can be written as

$$\frac{p+1}{n^2} I_{nl}^{(p)} - (2p+1) I_{nl}^{(p-1)} + \frac{p}{4} [(2l+1)^2 - p^2] I_{nl}^{(p-2)} = 0 \tag{A.84}$$

and is valid for $p > -(2l+1)$.

A.7 The Euler Gamma Function

The Euler gamma function, $\Gamma(z)$, is defined as

$$\Gamma(z) = \lim_{n \to \infty} \frac{(n-1)!}{z(z+1) \cdots (z+n-1)} n^z \, , \tag{A.85}$$

where z is any arbitrary complex number. $\Gamma(z)$ is an analytic function of z in the entire complex plane except at the points $z = -n$, where n is a non-negative integer; at these points $\Gamma(z)$ has simple poles. Some interesting properties of this function are the following:

1) Integral representation:

$$\Gamma(z) = \int_0^\infty dt\, e^{-t} t^{z-1} \, , \quad \mathrm{Re}\, z > 0 \, . \tag{A.86}$$

2) Euler infinite product:

$$\frac{1}{\Gamma(z)} = z e^{\gamma z} \prod_{n=1}^{\infty} \left[\left(1 + \frac{z}{n} \right) e^{-z/n} \right] \, , \quad |z| < \infty \, , \tag{A.87}$$

where the Euler constant, γ, is given by

$$\gamma = \lim_{n \to \infty} \left[\sum_{k=1}^{n} \frac{1}{k} - \ln n \right] = 0.577\,215\,664\,9\ldots \, . \tag{A.88}$$

3) Frequently used formulae:

$$\Gamma(1+z) = z\Gamma(z) \tag{A.89}$$

$$\Gamma(z)\Gamma(1-z) = \frac{\pi}{\sin \pi z} \, . \tag{A.90}$$

From the first of these expressions we see that if n is a non-negative integer,

$$\Gamma(n+1) = n! \, . \tag{A.91}$$

4) A few values of particular interest are:

$$\Gamma(1) = 1 \, , \qquad\qquad \Gamma(1/2) = \sqrt{\pi} \, ,$$
$$\Gamma(1/3) = 2.678\,938\,534\,7\ldots \, , \quad \Gamma(1/4) = 3.625\,609\,908\,2\ldots \, . \tag{A.92}$$

5) A function related to $\Gamma(z)$ is the function $\psi(z)$, defined by

$$\psi(z) = \frac{\Gamma'(z)}{\Gamma(z)} \tag{A.93}$$

for which the following relations hold:

$$\psi(1/2) = -\gamma - 2\ln 2 , \quad \psi(1) = -\gamma ,$$

$$\psi(z+1) = \psi(z) + \frac{1}{z} ,$$

$$\psi(n) = -\gamma + \sum_{k=1}^{n-1} \frac{1}{k} , \quad \text{integer } n \geq 2 . \tag{A.94}$$

A.8 Bessel Functions

The Bessel equation is

$$\left[\frac{d^2}{dz^2} + \frac{1}{z}\frac{d}{dz} + \left(1 - \frac{\nu^2}{z^2}\right) \right] w(z) = 0 ; \tag{A.95}$$

where ν is a complex number and z is a complex variable. This equation has a solution

$$J_\nu(z) = \sum_{k=0}^{\infty} \frac{(-1)^k}{k!\,\Gamma(\nu + k + 1)} \left(\frac{z}{2}\right)^{\nu+2k} , \tag{A.96}$$

which is called a Bessel function of the first kind, or simply a Bessel function. The function $J_\nu(z)$ is holomorphic in the entire complex z-plane except for a cut along the negative real axis when ν is not an integer; otherwise, $J_n(z)$ is analytic everywhere. Note that if $J_\nu(z)$ is a solution, so is $J_{-\nu}(z)$. If $\nu \neq n$, then $J_\nu(z)$ and $J_{-\nu}(z)$ are two independent functions and thus, a linear combination of these two is the general solution. When $\nu = n$,

$$J_{-n}(z) = (-1)^n J_n(z) . \tag{A.97}$$

We sometimes need Bessel functions of the second kind, or Weber functions, defined as

$$Y_\nu(z) = \frac{1}{\sin \pi\nu} [J_\nu(z)\cos \pi\nu - J_{-\nu}(z)] . \tag{A.98}$$

The right-hand side is replaced by its limit value when $\nu = n$. This is a solution to Bessel's equation which is analytic in $\mathbb{C} - [0, \infty)$ and is always independent of $J_\nu(z)$, so that the general solution can always be written as a linear combination of $J_\nu(z)$ and $Y_\nu(z)$. For $\nu = n$ the explicit expression for $Y_n(z)$ is

$$Y_n(z) = -\frac{1}{\pi} \sum_{k=0}^{n-1} \frac{(n-k-1)!}{k!} \left(\frac{z}{2}\right)^{2k-n} + \frac{2}{\pi} J_n(z) \ln\frac{z}{2}$$

$$- \frac{1}{\pi} \sum_{k=0}^{\infty} [\psi(k+1) + \psi(n+k+1)] \frac{(-1)^k}{k!(n+k)!} \left(\frac{z}{2}\right)^{2k+n} . \tag{A.99}$$

Often used are the Bessel functions of the third kind, or Hankel functions, defined by

$$H_\nu^{(1)}(z) = J_\nu(z) + iY_\nu(z) , \quad H_\nu^{(2)}(z) = J_\nu(z) - iY_\nu(z) , \tag{A.100}$$

obviously linearly independent for all ν.

The modified Bessel functions, $I_\nu(z)$, are also frequently used. They are defined by:

$$\begin{aligned}
I_\nu(z) &= e^{-i\pi\nu/2} J_\nu(ze^{i\pi/2}) , & -\pi < \arg z \le \pi/2 , \\
I_\nu(z) &= e^{3i\pi\nu/2} J_\nu(ze^{-3i\pi/2}) , & \pi/2 < \arg z \le \pi ,
\end{aligned} \tag{A.101}$$

and whose explicit expression is

$$I_\nu(z) = \sum_{k=0}^{\infty} \frac{1}{k!\,\Gamma(\nu + k + 1)} \left(\frac{z}{2}\right)^{2k+\nu} \tag{A.102}$$

We continue by listing several interesting properties of these functions:

1) Recursion formulae:

$$\frac{2\nu}{z} J_\nu(z) = J_{\nu-1}(z) + J_{\nu+1}(z) ; \tag{A.103}$$

$$2J_\nu'(z) = J_{\nu-1}(z) - J_{\nu+1}(z) ; \tag{A.104}$$

$$J_\nu'(z) = J_{\nu-1}(z) - \frac{\nu}{z} J_\nu(z) ; \tag{A.105}$$

$$J_\nu'(z) = -J_{\nu+1}(z) + \frac{\nu}{z} J_\nu(z) . \tag{A.106}$$

These same recursion formulae are valid for Y_ν, $H_\nu^{(1)}$, and $H_\nu^{(2)}$, or for any linear combination of these whose coefficients are independent of ν and of z.

2) Asymptotic expansions: As $|z| \to \infty$ and $|\arg z| < \pi$, we have

$$J_\nu(z) \sim \sqrt{\frac{2}{\pi z}} \cos\left(z - \frac{\pi\nu}{2} - \frac{\pi}{4}\right) , \tag{A.107}$$

$$Y_\nu(z) \sim \sqrt{\frac{2}{\pi z}} \sin\left(z - \frac{\pi\nu}{2} - \frac{\pi}{4}\right) . \tag{A.108}$$

3) The following relations hold for Bessel functions of integer order:

$$e^{z(t-1/t)/2} = \sum_{n=-\infty}^{+\infty} t^n J_n(z) , \quad t \ne 0 ; \tag{A.109}$$

$$J_n(z) = \frac{1}{\pi i^n} \int_0^\pi d\theta\, e^{iz\cos\theta} \cos n\theta ; \tag{A.110}$$

$$J_0(z) + 2 \sum_{n=1}^{\infty} J_{2n}(z) = 1 \; ; \tag{A.111}$$

$$J_0^2(z) + 2 \sum_{n=1}^{\infty} J_n^2(z) = 1 \; . \tag{A.112}$$

Relations (A.103) and (A.111) are particularly interesting for numerical evaluations of $J_n(x)$ for different values of n and for a given range of x values. For this we proceed in the following way: choose an $n = N$ sufficiently large that $J_N(x) \simeq 0$ for those values of x of interest, to the level of approximation desired. Then set $J_N(x) = 0$ and $J_{N-1}(x) = 1$; and use (A.103) to calculate $J_{N-2}(x), \ldots J_0(x)$ and normalize the results obtained so that they satisfy (A.111), thus solving the problem. For example, if we wish to calculate the values of $J_n(1)$ to six significant figures, the expression (A.96) allows us to write that $J_8(1) = 0$ within the desired limits. Starting from this and using $J_7(1) = 1$, we calculate $J_6(1), \ldots, J_0(1)$, which when properly normalized give

$$J_0(1) = 0.765\ 198 \;, \quad J_1(1) = 0.440\ 051 \;, \quad J_2(1) = 0.114\ 903 \;,$$

$$J_3(1) = 0.019\ 563 \;, \quad J_4(1) = 0.002\ 477 \;, \quad J_5(1) = 0.000\ 250 \;, \tag{A.113}$$

$$J_6(1) = 0.000\ 021 \;, \quad J_7(1) = 0.000\ 002 \;; \quad J_n(1) = 0.000\ 000 \;, \quad n \geq 8 \;,$$

coinciding with the values appearing in tables.

Finally, we immediately derive from (A.112) that

$$|J_0(x)| \leq 1 \; ; \quad |J_n(x)| \leq \frac{1}{\sqrt{2}} \;, \quad n \geq 1 \;, \tag{A.114}$$

for real arguments.

A.9 Spherical Bessel Functions

In the development of quantum mechanics we frequently encounter the differential equation

$$\left[\frac{d^2}{dz^2} + \frac{2}{z} \frac{d}{dz} + 1 - \frac{n(n+1)}{z^2} \right] w(z) = 0 \;, \tag{A.115}$$

with n an integer. Two linearly independent, particular solutions are

$$j_n(z) = \sqrt{\frac{\pi}{2z}} J_{n+1/2}(z) \;, \tag{A.116}$$

$$y_n(z) = \sqrt{\frac{\pi}{2z}} Y_{n+1/2}(z) \;, \tag{A.117}$$

which are called, respectively, spherical Bessel functions of the first kind, or simply spherical Bessel functions, and spherical Bessel functions of the second kind or spherical Neumann functions. Hankel functions, or spherical Bessel functions of the third kind, form another pair of linearly independent solutions, defined as

$$h_n^{(1)}(z) = j_n(z) + iy_n(z) = \sqrt{\frac{\pi}{2z}} H_{n+1/2}^{(1)}(z) \,,$$

$$h_n^{(2)}(z) = j_n(z) - iy_n(z) = \sqrt{\frac{\pi}{2z}} H_{n+1/2}^{(2)}(z) \,.$$

(A.118)

The explicit form of the functions defined by (A.116) and (A.117) for $n \geq 0$ is

$$j_n(z) = z^n \left[-\frac{1}{z}\frac{d}{dz} \right]^n \frac{\sin z}{z} \,, \quad y_n(z) = -z^n \left[-\frac{1}{z}\frac{d}{dz} \right]^n \frac{\cos z}{z} \,,$$

(A.119)

or equivalently,

$$j_n(z) = \frac{1}{z}\left[P\left(n+\frac{1}{2},z\right)\sin\left(z-\frac{n\pi}{2}\right) + Q\left(n+\frac{1}{2},z\right)\cos\left(z-\frac{n\pi}{2}\right) \right] \,,$$

$$y_n(z) = \frac{(-1)^{n+1}}{z}\left[P\left(n+\frac{1}{2},z\right)\cos\left(z+\frac{n\pi}{2}\right) \right.$$

$$\left. -Q\left(n+\frac{1}{2},z\right)\sin\left(z+\frac{n\pi}{2}\right) \right] \,,$$

(A.120)

where

$$P\left(n+\frac{1}{2},z\right) = \sum_{k=0}^{[n/2]}(-1)^k\frac{(n+2k)!}{(2k)!(n-2k)!}\frac{1}{(2z)^{2k}} \,,$$

$$Q\left(n+\frac{1}{2},z\right) = \sum_{k=0}^{[(n-1)/2]}(-1)^k\frac{(n+2k+1)!}{(2k+1)!(n-2k-1)!}\frac{1}{(2z)^{2k+1}} \,.$$

(A.121)

From these relations we easily see that the lowest order functions are

$$j_0(z) = \frac{\sin z}{z} \,; \qquad\qquad y_0(z) = -\frac{\cos z}{z} \,;$$

$$j_1(z) = \frac{\sin z}{z^2} - \frac{\cos z}{z} \,; \qquad y_1(z) = -\frac{\cos z}{z^2} - \frac{\sin z}{z} \,; \quad \text{(A.122)}$$

$$j_2(z) = \left(\frac{3}{z^3} - \frac{1}{z}\right)\sin z - \frac{3}{z^2}\cos z \,; \quad y_2(z) = \left(-\frac{3}{z^3} + \frac{1}{z}\right)\cos z - \frac{3}{z^2}\sin z \,.$$

Other expressions for these functions are ($n \geq 0$)

$$j_n(z) = z^n \sum_{k=0}^{\infty} \frac{(-1)^k}{k!(2n+2k+1)!!}\left(\frac{z^2}{2}\right)^k \,,$$

(A.123)

$$y_n(z) = -\frac{(2n-1)!!}{z^{n+1}}\left[1 + \sum_{k=1}^{\infty}\frac{(-z^2/2)^k}{k!(1-2n)(3-2n)\cdots(2k-1-2n)}\right].$$

Some useful properties are:

1) Asymptotic forms: For $z \to 0$ we have

$$j_n(z) \sim \frac{z^n}{(2n+1)!!} \;,\quad y_n(z) \sim -\frac{(2n-1)!!}{z^{n+1}}\;. \tag{A.124}$$

For $|z| \to \infty$ with $|\arg z| < \pi$ we find

$$j_n(z) \sim \frac{1}{z}\cos\left[z - \frac{(n+1)\pi}{2}\right]\;,\quad y_n(z) \sim \frac{1}{z}\sin\left[z - \frac{(n+1)\pi}{2}\right]\;;$$

$$h_n^{(1)}(z) \sim \frac{1}{z}\exp\left[\mathrm{i}\left(z - \frac{(n+1)\pi}{2}\right)\right]\;, \tag{A.125}$$

$$h_n^{(2)}(z) \sim \frac{1}{z}\exp\left[-\mathrm{i}\left(z - \frac{(n+1)\pi}{2}\right)\right]\;.$$

2) Recursion relations: The relations

$$j_{n-1}(z) + j_{n+1}(z) = \frac{2n+1}{z}j_n(z)\;, \tag{A.126}$$

$$(2n+1)j_n'(z) = nj_{n-1}(z) - (n+1)j_{n+1}(z)\;, \tag{A.127}$$

$$j_n'(z) = j_{n-1}(z) - \frac{n+1}{z}j_n(z)\;, \tag{A.128}$$

$$j_n'(z) = -j_{n+1}(z) + \frac{n}{z}j_n(z)\;, \tag{A.129}$$

are true for arbitrary integer values of n. The same relations hold for $y_n(z), h_n^{(1)}(z)$, and $h_n^{(2)}(z)$. When using them it is useful to keep in mind that

$$y_n(z) = (-1)^{n+1}j_{-n-1}(z)\;,\quad n = 0, \pm 1, \pm 2, \ldots\;; \tag{A.130}$$

$$h_{-n-1}^{(1)}(z) = \mathrm{i}(-1)^n h_n^{(1)}(z)\;,\quad h_{-n-1}^{(2)}(z) = -\mathrm{i}(-1)^n h_n^{(2)}(z)\;,$$
$$n = 0, 1, \ldots\;. \tag{A.131}$$

3) Another interesting relation is

$$\sum_{n=0}^{\infty}(2n+1)j_n^2(z) = 1\;, \tag{A.132}$$

which permits us to write for real values of z:

$$|j_n(x)| \leq \frac{1}{\sqrt{2n+1}}\;. \tag{A.133}$$

For numerical evaluation of the functions $j_n(z)$ one can proceed as in the evaluation of $J_n(z)$, using (A.126) and determining the normalization factor by means either of (A.132) or of the explicit expression for $j_0(z)$.

4) The orthogonality relations are:

$$\int_0^\infty dr\, r^2 j_n(kr) j_n(k'r) = \frac{\pi}{2k^2} \delta(k - k') .$$ (A.134)

5) Let r_1 and r_2 be two vectors and ω be the angle between them. If $r_<$ ($r_>$) designates the smaller (larger) of the two moduli of r_1 and r_2, then

$$\frac{1}{|r_1 - r_2|} = \sum_{L=0}^\infty \frac{r_<^L}{r_>^{L+1}} P_L(\cos\omega)$$

$$= \sum_{L=0}^\infty \sum_{M=-L}^{+L} \frac{4\pi}{2L+1} \frac{r_<^L}{r_>^{L+1}} Y_L^{M*}(\hat{r}_1) Y_L^M(\hat{r}_2) ;$$ (A.135)

$$\frac{\exp(ik|r_1 - r_2|)}{|r_1 - r_2|} = ik \sum_{L=0}^\infty (2L+1) j_L(kr_<) h_L^{(1)}(kr_>) P_l(\cos\omega) .$$

This last formula is valid for all k, real or complex.

A.10 Confluent Hypergeometric Functions

Consider the equation

$$\left[z\frac{d^2}{dz^2} + (b - z)\frac{d}{dz} - a \right] w(z) = 0 ,$$ (A.136)

where a and b are arbitrary complex constants, the only limitation being that $b \neq -n$, and n is a non-negative integer. This equation has a regular singular point at $z = 0$ and an irregular singular point at $z = \infty$.

One solution of this equation is the Kummer function, defined by

$$M(a, b, z) = \sum_{n=0}^\infty \frac{(a)_n z^n}{(b)_n n!} ,$$ (A.137)

where $(a)_n = a(a + 1)(a + 2)\ldots(a + n - 1)$ and $(a)_0 = 1$. This is a convergent series for all values of a, b and z and it reduces to a polynomial if $a = -n$. A second solution [lineary independent of (A.137) if $b \neq -n$] is

$$U(a, b, z) = \frac{\pi}{\sin \pi b} \left[\frac{M(a, b, z)}{\Gamma(1 + a - b)\Gamma(b)} - z^{1-b} \frac{M(1 + a - b, 2 - b, z)}{\Gamma(a)\Gamma(2 - b)} \right] .$$ (A.138)

This is a multivalued function whose principal branch is defined by choosing $-\pi < \arg z \leq \pi$. The r.h.s. should be replaced by its limit when $b = n + 1, n = 0, 1, 2, \ldots$, and in this case the explicit expression of the function $U(a, n + 1, z)$ is

$$U(a, n+1, z) = \frac{(-1)^{n+1}}{n! \, \Gamma(a-n)} \left[M(a, n+1, z) \ln z \right.$$

$$\left. + \sum_{r=0}^{\infty} [\psi(a+r) - \psi(1+r) - \psi(1+n+r)] \frac{(a)_r z^r}{(1+n)_r r!} \right]$$

$$+ \frac{(n-1)!}{\Gamma(a)} z^{-n} \sum_{r=0}^{n-1} \frac{(a-n)_r z^r}{(1-n)_r r!} , \qquad (A.139)$$

where the last sum vanishes if $n = 0$. Evidently, if b is not an integer, the most general solution can be written as a linear combination of $M(a, b, z)$ and $z^{1-b} M(1 + a - b, 2 - b, z)$. Some interesting properties of these functions are:

1) Asymptotic expansions: When $|z| \to 0$ one finds

$$M(a, b, 0) = 1 ; \qquad (A.140)$$

$$U(a, b, z) = \frac{\Gamma(b-1)}{\Gamma(a)} z^{1-b} + O(|z|^{\mathrm{Re}\,b-2}), \quad \mathrm{Re}\, b \geq 2 , \quad b \neq 2 ; \qquad (A.141)$$

$$U(a, b, z) = \frac{\Gamma(b-1)}{\Gamma(a)} z^{1-b} + O(|\ln z|) , \quad b = 2 ; \qquad (A.142)$$

$$U(a, b, z) = -\frac{1}{\Gamma(a)} [\ln z + \psi(a)] + O(|z \ln z|) , b = 1 . \qquad (A.143)$$

Here, we have limited ourselves to the most interesting cases.
When $|z| \to \infty$

$$M(a, b, z) = \frac{\Gamma(b)}{\Gamma(b-a)} e^{i \varepsilon \pi a} z^{-a} g(a, a - b + 1, -z)$$

$$+ \frac{\Gamma(b)}{\Gamma(a)} e^z z^{a-b} g(1 - a, b - a, z) , \qquad (A.144)$$

where

$$\varepsilon = +1 \quad \text{if} \quad -\frac{\pi}{2} < \arg z < \frac{3\pi}{2} , \quad \varepsilon = -1 \quad \text{if} \quad -\frac{3\pi}{2} < \arg z \leq -\frac{\pi}{2} ,$$

$$g(a, b, z) = \sum_{n=0}^{\infty} \frac{(a)_n (b)_n}{n! z^n} . \qquad (A.145)$$

Furthermore, if $-3\pi/2 < \arg z < 3\pi/2$, then

$$U(a, b, z) = z^{-a} g(a, a - b + 1, -z) . \qquad (A.146)$$

2) It is interesting to note that the confluent hypergeometric functions are intimately related to the functions defined in the previous sections. Thus, for example

$$M(-n, k+1, z) = \frac{n!}{(k+1)_n} L_n^k(z) ;$$

$$M\left(\nu + \frac{1}{2}, 2\nu + 1, 2iz\right) = \Gamma(1+\nu)e^{iz}\left(\frac{z}{2}\right)^{-\nu} J_\nu(z) .$$

(A.147)

3) Some useful integrals containing confluent hypergeometric functions are [LL 65]:

a) If $\mathrm{Re}\,\nu > 0$, and $\mathrm{Re}\,\lambda > |\mathrm{Re}\,k|$ (or $\mathrm{Re}\,\lambda > 0$ if a is integer ≤ 0), then

$$\int_0^\infty dx\, e^{-\lambda x} x^{\nu-1} M(a, b, kx) = \Gamma(\nu)\lambda^{-\nu} F(a, \nu, b; k/\lambda)$$

(A.148)

where $F(\alpha, \beta, \gamma; z)$ is the hypergeometric function defined by

$$F(\alpha, \beta, \gamma; z) = \sum \frac{(\alpha)_n (\beta)_n}{n!(\gamma)_n} z^n .$$

(A.149)

b) If n is an integer ≥ 0, $\mathrm{Re}\,k > 0$ and $\mathrm{Re}\,\nu > 0$,

$$\int_0^\infty dx\, e^{-kx} x^{\nu-1} [M(-n, \beta, kx)]^2$$

$$= \frac{\Gamma(\nu)n!}{k^\nu(\beta)_n}\left\{1 + \frac{n(\gamma-1)\gamma}{1^2 \cdot \beta} + \frac{n(n-1)(\gamma-2)(\gamma-1)\gamma(\gamma+1)}{(2!)^2 \beta(\beta+1)}\right.$$

$$\left. + \cdots + \frac{n(n-1)\cdots 1 (\gamma-n)\cdots(\gamma+n-1)}{(n!)^2 \beta(\beta+1)\cdots(\beta+n-1)}\right\}$$

(A.150)

where $\gamma \equiv \beta - \nu$.

c) Furthermore, under conditions similar to those in (a),

$$\int_0^\infty dx\, e^{-\lambda x} x^{\beta-1} M(\alpha, \beta, kx) M(\alpha', \beta, k'x)$$

$$= \Gamma(\beta)\lambda^{\alpha+\alpha'-\beta}(\lambda - k)^{-\alpha}(\lambda - k')^{-\alpha'} F\left(\alpha, \alpha', \beta; \frac{kk'}{(\lambda - k)(\lambda - k')}\right) .$$

(A.151)

A.11 Coulomb Wave Functions

In the study of scattering states of a particle moving in a Coulomb potential, the following radial equation is encountered:

$$\left[\frac{d^2}{d\varrho^2} + 1 - \frac{2\eta}{\varrho} - \frac{L(L+1)}{\varrho^2}\right] w(\varrho) = 0 , \qquad (A.152)$$

where $\varrho > 0, -\infty < \eta < \infty$ and L is a non-negative integer. If we change variables to

$$w(\varrho) = \varrho^{L+1} e^{-i\varrho} y(\varrho) , \quad z = 2i\varrho , \qquad (A.153)$$

the equation reduces to

$$\left[z\frac{d^2}{dz^2} + (2L + 2 - z)\frac{d}{dz} - (L + 1 - i\eta)\right] y(z) = 0 , \qquad (A.154)$$

which is simply the differential equation for the confluent hypergeometric functions. We have then that the solution to (A.152) is given by the so-called regular Coulomb function, defined as

$$F_L(\eta, \varrho) = C_L(\eta)\varrho^{L+1} e^{-i\varrho} M(L + 1 - i\eta, 2L + 2, 2i\varrho) , \qquad (A.155)$$

where

$$C_L(\eta) = \frac{2^L e^{-\pi\eta/2} |\Gamma(L + 1 + i\eta)|}{\Gamma(2L + 2)} . \qquad (A.156)$$

It is interesting to note that

$$C_0^2(\eta) = 2\pi\eta[e^{2\pi\eta} - 1]^{-1} , \quad C_L(\eta) = \frac{(L^2 + \eta^2)^{1/2}}{L(2L + 1)} C_{L-1}(\eta) , \qquad (A.157)$$

where we have used (A.89) and (A.90). From this we immediately obtain

$$C_L^2(\eta) = \frac{2^{2L}(1 + \eta^2)(2^2 + \eta^2)\dots(L^2 + \eta^2)}{[(2L + 1)!]^2} C_0^2(\eta) , \quad L > 0 . \qquad (A.158)$$

Another solution of (A.152), linearly independent of that in (A.155), is the so-called irregular Coulomb function

$$G_L(\eta, \varrho) = 2^L i e^{\pi\eta/2} \varrho^{L+1} e^{-i\varrho} [e^{-i(\sigma_L - \pi L)} U(L + 1 - i\eta, 2L + 2, 2i\varrho)$$
$$- e^{+i(\sigma_L - \pi L)} e^{2i\varrho} U(L + 1 + i\eta, 2L + 2, -2i\varrho)] , \qquad (A.159)$$

as results from the fact that if $U(a, b, z)$ is a solution of the confluent hypergeometric function, then so is $e^z U(b - a, b, -z)$. The quantity σ_L is defined by

$$\sigma_L = \arg \Gamma(L + 1 + i\eta) . \qquad (A.160)$$

It is of interest to observe that

$$\sigma_{L+1} = \sigma_L + \arctan \frac{\eta}{L+1} \,,$$

$$\sigma_0 = -\eta\gamma + \sum_{n=0}^{\infty} \left(\frac{\eta}{1+n} - \arctan \frac{\eta}{1+n} \right)$$

(A.161)

where γ is Euler's constant.

Interesting properties of these functions are:

a) Asymptotic behaviors: If $\varrho \rightarrow +\infty$,

$$F_L(\eta, \varrho) \sim \sin \left[\varrho - \eta \ln 2\varrho - \frac{\pi L}{2} + \sigma_L \right] \,,$$

$$G_L(\eta, \varrho) \sim \cos \left[\varrho - \eta \ln 2\varrho - \frac{\pi L}{2} + \sigma_L \right] \,.$$

(A.162)

If $\varrho \rightarrow 0$,

$$F_L(\eta, \varrho) \sim C_L(\eta)\varrho^{L+1} \,,$$

$$G_L(\eta, \varrho) \sim \frac{\varrho^{-L}}{(2L+1)C_L(\eta)} \,.$$

(A.163)

b) Note that if $\eta = 0$,

$$F_L(0, \varrho) = \varrho j_L(\varrho) \,, \quad G_L(0, \varrho) = -\varrho y_L(\varrho) \,,$$

(A.164)

as expected.

Appendix B: Angular Momentum

The treatment of angular momentum within the framework of quantum mechanics has been discussed in Chap. 5 of this book. In this appendix we present some of the more frequently used formulae associated with this subject. We also include tables of Clebsch-Gordan and Racah coefficients of lowest order.

There are many books dedicated to the study of angular momentum. Among the better known ones, we wish to cite those of *Rose* [RO 57] (whose notation we adopt in the majority of cases), *Edmonds* [ED 57], *Fano* and *Racah* [FR 59], *Biedenharn* and *Louck* [BL 81] and *Wigner* [WI 31]. For Clebsch-Gordan and Racah coefficients not tabulated in this appendix, the tables of *Rotenberg* et al. [RB 59] can be consulted.

B.1 Angular Momentum

Throughout this appendix we will use a system of units in which $\hbar = 1$. In quantum mechanics any self-adjoint operator $\boldsymbol{J}$ with components J_1, J_2, and J_3 satisfying the commutation relations

$$[J_j, J_k] = i\varepsilon_{jkl}J_l \tag{B.1}$$

is called an angular momentum. Introducing the operators

$$\boldsymbol{J}^2 = J_1^2 + J_2^2 + J_3^2 , \quad J_\pm = J_1 \pm iJ_2 , \tag{B.2}$$

it is straightforward to see that

$$[\boldsymbol{J}^2, J_k] = 0 , \tag{B.3}$$

$$J_\pm^\dagger = J_\mp , \quad J_1 = \frac{1}{2}(J_+ + J_-) , \quad J_2 = \frac{1}{2i}(J_+ - J_-) , \tag{B.4}$$

$$[J_3, J_\pm] = \pm J_\pm , \quad [J_+, J_-] = 2J_3 . \tag{B.5}$$

The operator $\boldsymbol{J}^2$ and one of the components of $\boldsymbol{J}$, e.g., J_3, can be part of a complete set of commuting observables $\{\boldsymbol{J}^2, J_3, A, \ldots\}$ chosen such that $A, \ldots$ are compatible with $\boldsymbol{J}$. The eigenstates of this complete set will be denoted by $|JM\alpha\rangle$, where the indices α refer to all observables other than $\boldsymbol{J}^2$ and J_3 (and will be omitted in what follows when they are not necessary); they obey the relations

$$J^2|JM\alpha\rangle = J(J+1)|JM\alpha\rangle \, , \tag{B.6}$$

$$J_3|JM\alpha\rangle = M|JM\alpha\rangle \, , \tag{B.7}$$

$$\langle J'M'\alpha'|JM\alpha\rangle = \delta_{JJ'}\delta_{MM'}\delta_{\alpha\alpha'} \, . \tag{B.8}$$

The possible values of J are $J = 0, 1/2, 1, 3/2, \ldots$, that is, any non-negative integer or half-integer. For a given J the possible values of M are $M = J, J - 1, \ldots, -J + 1, -J$, and the $(2J + 1)$ vectors $|JM\rangle$ (for fixed α, here omitted) form the basis for an irreducible representation of the rotation group denoted by $\mathcal{D}^{(J)}$.

The relative phases of the different states of a basis for a representation $\mathcal{D}^{(J)}$ are fixed using the Condon-Shortley convention [CS 57], for which the following relation is valid

$$J_\pm|JM\rangle = [(J \mp M)(J \pm M + 1)]^{1/2}|JM \pm 1\rangle \, . \tag{B.9}$$

The matrices corresponding to J_1, J_2, and J_3 for two irreducible representations $\mathcal{D}^{(J)}$ of more than usual interest are:

a) $\mathcal{D}^{(1/2)}$

$$J = \tfrac{1}{2}\sigma \, ,$$

$$\sigma_1 = \begin{pmatrix} 0 & 1 \\ 1 & 0 \end{pmatrix}, \quad \sigma_2 = \begin{pmatrix} 0 & -i \\ i & 0 \end{pmatrix}, \quad \sigma_3 = \begin{pmatrix} 1 & 0 \\ 0 & -1 \end{pmatrix}; \tag{B.10}$$

the matrices σ are called the Pauli spin matrices and they satisfy the following relations:

$$[\sigma_j, \sigma_k] = 2i\varepsilon_{jkl}\sigma_l \, ,$$
$$\sigma_i\sigma_j + \sigma_j\sigma_i = 2\delta_{ij} \, , \quad \sigma_j\sigma_k = \delta_{jk} + i\varepsilon_{jkl}\sigma_l \, , \tag{B.11}$$
$$(\sigma_j)_{\alpha\beta}(\sigma_j)_{\gamma\delta} = 2\delta_{\alpha\delta}\delta_{\beta\gamma} - \delta_{\alpha\beta}\delta_{\gamma\delta} \, ,$$

where there is an implicit sum over j in the last relation. If $A, B, \ldots$ are arbitrary vectors, it can be easily shown that

$$(A \cdot \sigma)\sigma = A - i(A \times \sigma) \, , \quad \sigma(A \cdot \sigma) = A + i(A \times \sigma) \, ,$$
$$(A \cdot \sigma)(B \cdot \sigma) = (A \cdot B) + i(A \times B) \cdot \sigma \, ,$$
$$(A \cdot \sigma)\sigma(B \cdot \sigma) = A(B \cdot \sigma) + (A \cdot \sigma)B - (A \cdot B)\sigma - i(A \times B) \, ,$$
$$\sigma_k(A \cdot \sigma)\sigma_k = -(A \cdot \sigma) \, . \tag{B.12}$$

The traces of these matrices obey the following relations:

$$\mathrm{Tr}\{\sigma_i\} = 0 \, , \quad \mathrm{Tr}\{\sigma_i\sigma_j\} = 2\delta_{ij} \, , \quad \mathrm{Tr}\{\sigma_j\sigma_k\sigma_l\} = 2i\varepsilon_{jkl} \, ,$$
$$\mathrm{Tr}\{\sigma_i\sigma_j\sigma_k\sigma_l\} = 2[\delta_{ij}\delta_{kl} - \delta_{ik}\delta_{jl} + \delta_{il}\delta_{jk}] \, . \tag{B.13}$$

b) $\mathcal{D}^{(1)}$

$$J_1 = \frac{1}{\sqrt{2}} \begin{pmatrix} 0 & 1 & 0 \\ 1 & 0 & 1 \\ 0 & 1 & 0 \end{pmatrix} , \quad J_2 = \frac{i}{\sqrt{2}} \begin{pmatrix} 0 & -1 & 0 \\ 1 & 0 & -1 \\ 0 & 1 & 0 \end{pmatrix} ,$$

$$J_3 = \begin{pmatrix} 1 & 0 & 0 \\ 0 & 0 & 0 \\ 0 & 0 & -1 \end{pmatrix} . \tag{B.14}$$

B.2 Matrix Representation of the Rotation Operators

The operator corresponding to an active rotation characterized by the Euler angles $\alpha, \beta,$ and γ is

$$R(\alpha, \beta, \gamma) = e^{-i\alpha J_3} e^{-i\beta J_2} e^{-i\gamma J_3} . \tag{B.15}$$

Under this rotation, the basis vectors $|JM\rangle$ of the irreducible representation $\mathcal{D}^{(J)}$ transform among themselves according to

$$R(\alpha, \beta, \gamma)|JM\rangle = \sum_{M'} D^J_{M'M}(\alpha, \beta, \gamma)|JM'\rangle , \tag{B.16}$$

and thus the matrix representation of this rotation operator in the irreducible representation under consideration is

$$D^J_{M'M}(\alpha, \beta, \gamma) = \langle JM'|e^{-i\alpha J_3} e^{-i\beta J_2} e^{-i\gamma J_3}|JM\rangle , \tag{B.17}$$

which can be written as

$$D^J_{M'M}(\alpha, \beta, \gamma) = e^{-i\alpha M'} d^J_{M'M}(\beta) e^{-i\gamma M} ,$$
$$d^J_{M'M}(\beta) = \langle JM'|e^{-i\beta J_2}|JM\rangle . \tag{B.18}$$

Wigner [WI 31] has obtained an explicit expression for these functions:

$$d^J_{M'M}(\beta) = [(J + M)!(J - M)!(J + M')!(J - M')!]^{1/2}$$
$$\times \sum_\nu \frac{(-1)^\nu}{(J - M' - \nu)!(J + M - \nu)!(\nu + M' - M)!\nu!}$$
$$\times \left(\cos \frac{\beta}{2} \right)^{2J+M-M'-2\nu} \left(-\sin \frac{\beta}{2} \right)^{M'-M+2\nu} , \tag{B.19}$$

where the sum over ν extends over all integer values for which the arguments of all factorials are larger than or equal to zero.

Some interesting properties of these functions are the following:

$$d^J_{M'M}(\beta) = (-1)^{M'-M} d^J_{MM'}(\beta) , \quad d^J_{M'M}(\beta) = d^J_{-M-M'}(\beta) , \qquad (\text{B.20})$$

$$d^J_{M'M}(\beta) = d^J_{MM'}(-\beta) , \quad d^J_{M'M}(\beta) = (-1)^{J+M'} d^J_{M'-M}(\pi - \beta) , \qquad (\text{B.21})$$

$$d^J_{M'M}(0) = \delta_{M'M} , \quad d^J_{M'M}(\pi) = (-1)^{J-M} \delta_{M'-M} ,$$

$$d^J_{M'M}(\beta + 2\pi) = (-1)^{2J} d^J_{M'M}(\beta) , \qquad (\text{B.22})$$

$$\int_0^{\pi} d\beta \, \sin\beta \, d^J_{MM'}(\beta) d^{J'}_{MM'}(\beta) = \frac{2}{2J+1} \delta_{JJ'} , \qquad (\text{B.23})$$

$$\frac{1}{2} \sum_J (2J+1) d^J_{MM'}(\beta) d^J_{MM'}(\beta') = \delta(\cos\beta - \cos\beta') , \qquad (\text{B.24})$$

$$d^{J_1}_{M_1 M_1'}(\beta) d^{J_2}_{M_2 M_2'}(\beta) = \sum_{J_3} C(J_1 J_2 J_3; M_1, -M_2)$$

$$\times C(J_1 J_2 J_3; M_1', -M_2')(-1)^{M_2-M_2'} d^{J_3}_{M_1-M_2, M_1'-M_2'}(\beta) , \qquad (\text{B.25})$$

$$d^{J_1+J_2}_{k\lambda}(\beta) = \sum_\nu \frac{(2J_1)!}{(J_1+\nu)!(J_1-\nu)!}$$

$$\times \left[\frac{(J_1+J_2-k)!(J_1+J_2+k)!(J_2-J_1-\lambda)!}{(J_2-k-\nu)!(J_2+k+\nu)!(J_1+J_2+\lambda)!}\right]^{1/2}$$

$$\times \left(\cos\frac{\beta}{2}\right)^{J_1+\nu} \left(-\sin\frac{\beta}{2}\right)^{J_1-\nu} d^{J_2}_{k+\nu, J_1+\lambda}(\beta) , \qquad (\text{B.26})$$

$$\left.\begin{array}{l} \displaystyle\sum_M d^J_{MM'}(\beta) d^J_{MM''}(\beta) = \delta_{M'M''} , \\[3mm] d^J_{M0}(\beta) = \sqrt{\dfrac{4\pi}{2J+1}} Y^M_J(\beta,0) , \\[4mm] d^J_{JM}(\beta) = \sqrt{\dfrac{(2J)!}{(J+M)!(J-M)!}} \left(\cos\dfrac{\beta}{2}\right)^{J+M} \left(-\sin\dfrac{\beta}{2}\right)^{J-M} , \end{array}\right\} \qquad (\text{B.27})$$

where the second relation in (B.27) is true only for non-negative integer J.
Taking into account (B.17) and (B.20)–(B.27) one can easily obtain:

$$D^{J*}_{M'M}(\alpha,\beta,\gamma) = (-1)^{M'-M} D^J_{-M'-M}(\alpha,\beta,\gamma) ,$$

$$D^{J*}_{M'M}(\alpha,\beta,\gamma) = D^J_{MM'}(-\gamma,-\beta,-\alpha) , \qquad (\text{B.28})$$

$$D^J_{M'M}(0,0,0) = \delta_{MM'} ,$$

$$D^J_{M'M}(\alpha,\beta+2\pi,\gamma) = (-1)^{2J} D^J_{M'M}(\alpha,\beta,\gamma) , \qquad (\text{B.29})$$

$$\sum_M D^{J*}_{M'M}(\alpha,\beta,\gamma) D^J_{M''M}(\alpha,\beta,\gamma) = \delta_{M'M''} ,$$

$$\sum_M D^{J*}_{MM'}(\alpha,\beta,\gamma) D^J_{MM''}(\alpha,\beta,\gamma) = \delta_{M'M''} , \qquad (\text{B.30})$$

$$D_{M_1 M_1'}^{J_1}(\alpha, \beta, \gamma) D_{M_2 M_2'}^{J_2}(\alpha, \beta, \gamma) = \sum_{J_3} C(J_1 J_2 J_3; M_1 M_2)$$

$$\times C(J_1 J_2 J_3; M_1' M_2') D_{M_1 + M_2, M_1' + M_2'}^{J_3}(\alpha, \beta, \gamma) , \tag{B.31}$$

$$C(J_1 J_2 J_3; M_1' M_2' M_3') D_{M_3 M_3'}^{J_3}(\alpha, \beta, \gamma)$$

$$= \sum_{M_1} C(J_1 J_2 J_3; M_1, M_3 - M_1, M_3) D_{M_1 M_1'}^{J_1}(\alpha, \beta, \gamma) D_{M_3 - M_1, M_2'}^{J_2}(\alpha, \beta, \gamma),$$

$$D_{M0}^{J}(\alpha, \beta, 0) = \sqrt{\frac{4\pi}{2J+1}} Y_J^{M*}(\beta, \alpha) . \tag{B.32}$$

In (B.32) J should be a non-negative integer. In addition,

$$\int dU \, D_{M_1 M_1'}^{J_1 *}(\alpha, \beta, \gamma) D_{M_2 M_2'}^{J_2}(\alpha, \beta, \gamma) = \frac{8\pi^2}{2J_1 + 1} \delta_{J_1 J_2} \delta_{M_1 M_2} \delta_{M_1' M_2'} , \tag{B.33}$$

with

$$\int dU \ldots \equiv \frac{1}{4} \int_0^{4\pi} d\alpha \int_0^{\pi} d\beta \, \sin \beta \int_0^{4\pi} d\gamma \ldots , \tag{B.34}$$

where the integrations over α and γ run from 0 to 4π due to the double covering of SO(3) by SU(2). Finally,

$$\int dU \, D_{M_3 M_3'}^{J_3 *}(\alpha, \beta, \gamma) D_{M_2 M_2'}^{J_2}(\alpha, \beta, \gamma) D_{M_1 M_1'}^{J_1}(\alpha, \beta, \gamma)$$

$$= \frac{8\pi^2}{2J_3 + 1} \delta_{M_1 + M_2, M_3} \delta_{M_1' + M_2', M_3'} C(J_1 J_2 J_3; M_1 M_2) C(J_1 J_2 J_3; M_1' M_2') . \tag{B.35}$$

The explicit expressions of these matrices for $J = 1/2$ and $J = 1$ are

$$D^{1/2}(\alpha, \beta, \gamma) = \begin{pmatrix} \cos \frac{\beta}{2} e^{-i(\alpha+\gamma)/2} & -\sin \frac{\beta}{2} e^{-i(\alpha-\gamma)/2} \\ \sin \frac{\beta}{2} e^{i(\alpha-\gamma)/2} & \cos \frac{\beta}{2} e^{i(\alpha+\gamma)/2} \end{pmatrix} , \tag{B.36}$$

$$D^{1}(\alpha, \beta, \gamma) = \begin{pmatrix} \frac{1+\cos \beta}{2} e^{-i(\alpha+\gamma)} & -\frac{\sin \beta}{\sqrt{2}} e^{-i\alpha} & \frac{1-\cos \beta}{2} e^{-i(\alpha-\gamma)} \\ \frac{\sin \beta}{\sqrt{2}} e^{-i\gamma} & \cos \beta & -\frac{\sin \beta}{\sqrt{2}} e^{i\gamma} \\ \frac{1-\cos \beta}{2} e^{i(\alpha-\gamma)} & \frac{\sin \beta}{\sqrt{2}} e^{i\alpha} & \frac{1+\cos \beta}{2} e^{i(\alpha+\gamma)} \end{pmatrix} . \tag{B.37}$$

B.3 Clebsch-Gordan Coefficients

If J_1 and J_2 are the angular momentum operators of two kinematically independent subsystems 1 and 2, i.e., such that $[J_1, J_2] = 0$, then $J_3 = J_1 + J_2$ is an angular momentum operator which we call the total angular momentum of the system $1+2$. The operators $J_1^2, (J_1)_3, J_2^2, (J_2)_3$, together with some other self-adjoint operators A such that $[A, \boldsymbol{J}_1] = 0$ and $[A, \boldsymbol{J}_2] = 0$, form a complete set of commuting observables for the total system. The elements $|J_1 M_1 J_2 M_2 \alpha\rangle$ of the complete system of simultaneous eigenvectors satisfy

$$\boldsymbol{J}_i^2 |J_1 M_1 J_2 M_2 \alpha\rangle = J_i(J_i + 1)|J_1 M_1 J_2 M_2 \alpha\rangle \,,$$
$$(J_i)_3 |J_1 M_1 J_2 M_2 \alpha\rangle = M_i |J_1 M_1 J_2 M_2 \alpha\rangle \,, \tag{B.38}$$
$$\langle J_1' M_1' J_2' M_2' \alpha' | J_1 M_1 J_2 M_2 \alpha\rangle = \delta_{J_1 J_1'} \delta_{J_2 J_2'} \delta_{M_1 M_1'} \delta_{M_2 M_2'} \delta_{\alpha \alpha'} \,,$$

with $i = 1, 2$. For given values of J_1, J_2 and α the relative phases are determined using the Condon-Shortley convention. Another complete set of commuting observables for the total system, fully equivalent to the one above, is $\boldsymbol{J}_1^2, \boldsymbol{J}_2^2, \boldsymbol{J}_3^2, (J_3)_3, A$, and the elements $|J_1 J_2 J_3 M_3 \alpha\rangle$ of the system of simultaneous eigenvectors for this set satisfy

$$\boldsymbol{J}_i^2 |J_1 J_2 J_3 M_3 \alpha\rangle = J_i(J_i + 1)|J_1 J_2 J_3 M_3 \alpha\rangle \,,$$
$$(J_3)_3 |J_1 J_2 J_3 M_3 \alpha\rangle = M_3 |J_1 J_2 J_3 M_3 \alpha\rangle \,, \tag{B.39}$$
$$\langle J_1' J_2' J_3' M_3' \alpha' | J_1 J_2 J_3 M_3 \alpha\rangle = \delta_{J_1 J_1'} \delta_{J_2 J_2'} \delta_{J_3 J_3'} \delta_{M_3 M_3'} \delta_{\alpha \alpha'} \,,$$

with $i = 1, 2, 3$. For given values of J_1, J_2, J_3 and α the relative phases are again determined through the Condon-Shortley convention.

For fixed J_1, J_2 and α, the systems of vectors $|J_1 M_1 J_2 M_2 \alpha\rangle$ and $|J_1 J_2 J_3 M_3 \alpha\rangle$ are complete orthonormal systems; each vector of one can be expressed as a linear combination of the vectors of the other, and we can write

$$|J_1 J_2 J_3 M_3 \alpha\rangle = \sum_{M_1 M_2} C(J_1 J_2 J_3; M_1 M_2 M_3)|J_1 M_1 J_2 M_2 \alpha\rangle \,,$$
$$|J_1 M_1 J_2 M_2 \alpha\rangle = \sum_{J_3 M_3} C(J_1 J_2 J_3; M_1 M_2 M_3)|J_1 J_2 J_3 M_3 \alpha\rangle \,. \tag{B.40}$$

The coefficients $C(J_1 J_2 J_3; M_1 M_2 M_3)$, which can always be chosen real with appropriate selection of the relative phases of both orthonormal systems, are called the Clebsch-Gordan coefficients. Their general expression is [WI 31]

$$C(J_1 J_2 J_3; M_1 M_2 M_3) = \delta_{M_1 + M_2, M_3} \Delta(J_1 J_2 J_3) \Big[(2J_3 + 1)$$

$$\times \frac{(J_1 + J_2 - J_3)!(J_2 + J_3 - J_1)!(J_3 + J_1 - J_2)!(J_3 + M_3)!(J_3 - M_3)!}{(J_1 + J_2 + J_3 + 1)!(J_1 - M_1)!(J_1 + M_1)!(J_2 - M_2)!(J_2 + M_2)!} \Big]^{1/2}$$

$$\times \sum_\nu \frac{(-1)^{\nu + J_2 + M_2}}{\nu!} \frac{(J_2 + J_3 + M_1 - \nu)!(J_1 - M_1 + \nu)!}{(J_2 + J_3 - J_1 - \nu)!(J_3 + M_3 - \nu)!(J_1 - J_2 - M_3 + \nu)!} \,, \tag{B.41}$$

where the triangle function $\Delta(J_1 J_2 J_3)$ is zero if J_3 does not appear in the series $|J_1 - J_2|, |J_1 - J_2| + 1, |J_1 - J_2| + 2, \dots, J_1 + J_2 - 1, J_1 + J_2$; otherwise the function is one. The sum on ν is over all integer values such that none of the arguments of the factorials become negative. It should be noted that since $M_3 = M_1 + M_2$, the index M_3 which appears in the Clebsch-Gordan coefficients is redundant and may be omitted, writing instead

$$C(J_1 J_2 J_3; M_1 M_2) \equiv C(J_1 J_2 J_3; M_1 M_2 M_3) . \tag{B.42}$$

The general formula is not very manageable; we give tables at the end of this appendix for those coefficients which are most frequently used.

Important properties of these coefficients are:

1) Orthogonality relations:

$$\sum_{M_1} C(J_1 J_2 J_3; M_1, M_3 - M_1) C(J_1 J_2 J_3'; M_1, M_3 - M_1) = \delta_{J_3 J_3'} ,$$

$$\sum_{J_3} C(J_1 J_2 J_3; M_1, M_3 - M_1) C(J_1 J_2 J_3; M_1', M_3' - M_1') \tag{B.43}$$

$$= \delta_{M_1 M_1'} \delta_{M_3 M_3'} .$$

2) $C(J_1 J_2 J_1 + J_2; J_1 J_2) = 1 , \quad C(J_1 0 J_3; M_1 0 M_3) = \delta_{J_1 J_3} \delta_{M_1 M_3},$

$$C(J_1 J_2 J_1 + J_2; M_1 M_2) \tag{B.44}$$

$$= \left[\frac{(2J_1)!(2J_2)!(J_1 + J_2 + M_1 + M_2)!(J_1 + J_2 - M_1 - M_2)!}{(2J_1 + 2J_2)!(J_1 + M_1)!(J_1 - M_1)!(J_2 + M_2)!(J_2 - M_2)!} \right]^{1/2} ;$$

$$C(J_1 J_2 J_3; 00) = 0 \quad \text{if} \quad J_1 + J_2 + J_3 = \text{odd} ;$$

$$C(J_1 J_2 J_3; 00) = (-1)^{(J_1 + J_2 - J_3)/2} \sqrt{\frac{2J_3 + 1}{J_1 + J_2 + J_3 + 1}}$$

$$\times \frac{\tau(J_1 + J_2 + J_3)}{\tau(J_1 + J_2 - J_3)\tau(J_1 - J_2 + J_3)\tau(-J_1 + J_2 + J_3)} \tag{B.45}$$

if $J_1 + J_2 + J_3 =$ even number, with $\tau(x) \equiv (x/2)!(x!)^{-1/2}$;

$$C\left(J_1 J_2 J_3; \tfrac{1}{2} 0 \tfrac{1}{2}\right)$$

$$= (-1)^{J_1 + J_2 - J_3} \sqrt{\frac{(J_1 + J_2 + J_3 + 2)(J_1 - J_2 + J_3 + 1)}{(2J_1 + 1)(2J_3 + 2)}}$$

$$\times C\left(J_1 + \tfrac{1}{2}, J_2, J_3 + \tfrac{1}{2}; 000\right)$$

when $J_1 + J_2 + J_3$ is odd.

3) Symmetry properties:

$$C(J_1 J_2 J_3; M_1 M_2 M_3) = (-1)^{J_1 + J_2 - J_3}$$

$$\times C(J_1 J_2 J_3; -M_1 - M_2 - M_3) , \tag{B.46}$$

$$C(J_1 J_2 J_3; M_1 M_2 M_3) = (-1)^{J_1 + J_2 - J_3} C(J_2 J_1 J_3; M_2 M_1 M_3) , \tag{B.47}$$

$$C(J_1 J_2 J_3; M_1 M_2 M_3) = (-1)^{J_1 - M_1} \sqrt{\frac{2J_3 + 1}{2J_2 + 1}}$$

$$\times C(J_1 J_3 J_2; M_1 - M_3 - M_2) . \tag{B.48}$$

Immediate consequences of these relations are

$$C(J_1 J_2 J_3; M_1 M_2 M_3) = (-1)^{J_2+M_2} \sqrt{\frac{2J_3+1}{2J_1+1}}$$
$$\times C(J_3 J_2 J_1; -M_3 M_2 - M_1), \qquad (B.49)$$

$$C(J_1 J_2 J_3; M_1 M_2 M_3) = (-1)^{J_1-M_1} \sqrt{\frac{2J_3+1}{2J_2+1}}$$
$$\times C(J_3 J_1 J_2; M_3 - M_1 M_2), \qquad (B.50)$$

$$C(J_1 J_2 J_3; M_1 M_2 M_3) = (-1)^{J_2+M_2} \sqrt{\frac{2J_3+1}{2J_1+1}}$$
$$\times C(J_2 J_3 J_1; -M_2 M_3 M_1). \qquad (B.51)$$

B.4 Racah Coefficients

Let J_1, J_2 and J_3 be the angular momentum operators for three kinematically independent subsystems; then $J = J_1+J_2+J_3$, $J' = J_1+J_2$ and $J'' = J_2+J_3$ are the angular momentum operators corresponding to the systems $1+2+3$, $1+2$, $2+3$, respectively. The operators $J_1^2, (J_1)_3, J_2^2, (J_2)_3, J_3^2, (J_3)_3$ together, as before, with some other operators A, form a complete set of commuting observables for the total system. Other complete sets equivalent to this one are

$$J_1^2, J_2^2, J_3^2, J'^2, J^2, J_3, A; \qquad (B.52)$$

$$J_1^2, J_2^2, J_3^2, J''^2, J^2, J_3, A. \qquad (B.53)$$

We denote by $|J_1 J_2 J_3 J' J M \alpha\rangle$ and $|J_1 J_2 J_3 J'' J M \alpha\rangle$ the simultaneous eigenvectors of the sets (B.52) and (B.53), respectively. Each of the sets of simultaneous eigenvectors form a complete orthogonal system, which we assume to be normalized. We will use the Condon-Shortley convention to determine the relative phases of the vectors of a given system differing only by their M values.

Given that, for fixed $J_1 J_2 J_3 J M$ and α, the sets $|J_1 J_2 J_3 J' J M \alpha\rangle$ and $|J_1 J_2 J_3 J'' J M \alpha\rangle$ form independent complete orthonormal systems, each vector of one set can be expressed as a linear combination of the vectors of the other:

$$|J_1 J_2 J_3 J' J M \alpha\rangle$$
$$= \sum_{J''} \sqrt{(2J'+1)(2J''+1)} W(J_1 J_2 J J_3; J' J'') |J_1 J_2 J_3 J'' J M \alpha\rangle. \qquad (B.54)$$

The coefficients $W(J_1 J_2 J J_3; J' J'')$, which can be chosen to be real by appropriate choices of the relative phases for both orthonormal systems, are called *Racah coefficients* [RA 42], [RA 43]. Their general expression is

$$W(abcd; ef) = \Delta_R(abe)\Delta_R(cde)\Delta_R(acf)\Delta_R(bdf)$$

$$\times \sum_\nu (-1)^{a+b+c+d+\nu}(\nu+1)![(\nu-a-b-e)!(\nu-c-d-e)!$$

$$\times (\nu-a-c-f)!(\nu-b-d-f)!(a+b+c+d-\nu)!$$

$$\times (a+d+e+f-\nu)!(b+c+e+f-\nu)!]^{-1}, \tag{B.55}$$

where the sum over ν is over all integers such that none of the arguments of the factorials become negative, and $\Delta_R(abc)$ is the Racah triangle function, equal to

$$\Delta_R(abc) = \left[\frac{(a+b-c)!(b+c-a)!(c+a-b)!}{(a+b+c+1)!}\right]^{1/2} \tag{B.56}$$

if all arguments of the factorials are non-negative integers, and is zero otherwise. The appearance of Δ_R functions in (B.55) implies that

$$W(abcd; ef) = 0 \tag{B.57}$$

unless $\Delta(abe) = \Delta(cde) = \Delta(acf) = \Delta(bdf) = 1$, with Δ defined as in Sect. B.3. The Racah and Clebsch-Gordan coefficients are related by

$$C(J_1 J_2 J'; M_1 M_2)C(J' J_3 J; M_1 + M_2 M_3)$$

$$= \sum_{J''} \sqrt{(2J'+1)(2J''+1)}W(J_1 J_2 J J_3; J' J'')$$

$$\times C(J_2 J_3 J''; M_2 M_3)C(J_1 J'' J; M_1 M_2 + M_3). \tag{B.58}$$

Taking into account the orthogonality relations (B.43), we can show that

$$\sqrt{(2J'+1)(2J''+1)}W(J_1 J_2 J J_3; J' J'')C(J_1 J'' J; M_1 M_2 + M_3)$$

$$= \sum_{\substack{M_2 M_3 \\ M_2+M_3=\text{const}}} C(J_1 J_2 J'; M_1 M_2)C(J' J_3 J; M_1 + M_2 M_3)C(J_2 J_3 J''; M_2 M_3),$$

$$\sqrt{(2J'+1)(2J''+1)}W(J_1 J_2 J J_3; J' J'') \tag{B.59}$$

$$= \sum_{\substack{M_1 M_2 M_3 \\ M_1+M_2+M_3=\text{const}}} C(J_1 J_2 J'; M_1 M_2)C(J' J_3 J; M_1 + M_2 M_3)$$

$$\times C(J_1 J'' J; M_1 M_2 + M_3)C(J_2 J_3 J''; M_2 M_3). $$

Another relation easily obtained from (B.59) is

$$\sqrt{(2L'+1)(2J'+1)}C(L'lL; 000)W(L'J'LJ; \tfrac{1}{2}l)$$

$$= (-1)^{J-L+l-1/2}C(JlJ'; 1/2\,0\,1/2), \tag{B.60}$$

valid only if $L + L' + l$ is even.

The orthogonality relations for these coefficients are

$$\sum_e (2e+1)(2f+1)W(abcd; ef)W(abcd; eg) = \delta_{fg}. \tag{B.61}$$

The Racah coefficients satisfy the following 24 symmetry properties:

$$W(abcd; ef) = W(badc; ef) = W(cdab; ef) = W(dcba; ef)$$
$$= W(acbd; fe) = W(cadb; fe) = W(bdac; fe)$$
$$= W(dbca; fe) \; ; \tag{B.62}$$

$$(-1)^{b+c-e-f} \, W(abcd; ef) = W(aefd; bc) = W(eadf; bc)$$
$$= W(fdae; bc) = W(dfea; bc) = W(afed; cb)$$
$$= W(fade; cb) = W(edaf; cb) = W(defa; cb) \; ; \tag{B.63}$$

$$(-1)^{a+d-e-f} W(abcd; ef) = W(ebcf; ad) = W(befc; ad)$$
$$= W(cfeb; ad) = W(fcbe; ad) = W(ecbf; da)$$
$$= W(cefb; da) = W(bfec; da) = W(fbce; da) \; . \tag{B.64}$$

It should be observed that any variable in a Racah coefficient can be moved to any predetermined position using these symmetry relations.

The case in which one of the coefficients, e.g., e, takes on the values $0, 1$, and 2 appears often. For zero, we have

$$W(abcd; 0f) = \frac{(-1)^{f-b-d} \delta_{ab} \delta_{cd}}{\sqrt{(2b+1)(2d+1)}} \tag{B.65}$$

and for the two other cases the explicit expression is given in Table B.3.

Remarks similar to those in the last paragraphs of Sect. 5.7 and Sect. 5.8 can be repeated now with respect to notations $|J_1 J_2 J_3 J' \, JM\alpha\rangle$ and $|J_1 J_2 J_3 J'' \, JM\alpha\rangle$ and to the validity of relations like (B.61), (B.65) and others, which presuppose for instance the fulfilment of triangle relations [as $\Delta(acf) = 1$ for (B.65)].

B.5 Irreducible Tensors

An irreducible tensor of order J with respect to the rotation group is defined as (a linear combination of) the set of $(2J + 1)$ operators T_M^J, where $M = -J, -J+1, \ldots, J-1, J$, transforming under the rotation group according to the rule

$$R(\alpha, \beta, \gamma) T_M^J R^{-1}(\alpha, \beta, \gamma) = \sum_{M'} D_{M'M}^J(\alpha, \beta, \gamma) T_{M'}^J \; . \tag{B.66}$$

The operators T_M^J constitute the basis or standard components of the tensorial operator under consideration.

This definition is equivalent to requiring

$$[J_\pm, T_M^J] = \sqrt{(J \mp M)(J \pm M + 1)} T_{M\pm 1}^J \; , \tag{B.67}$$
$$[J_3, T_M^J] = M T_M^J \; .$$

The following properties are very useful:

1) A vector operator A, with self-adjoint Cartesian components A_1, A_2, and A_3 is an irreducible tensor of order 1, with standard (or spherical) components

$$A_{+1} = -\frac{1}{\sqrt{2}}(A_1 + iA_2) , \quad A_0 = A_3 , \quad A_{-1} = +\frac{1}{\sqrt{2}}(A_1 - iA_2) . \tag{B.68}$$

From this we deduce that the scalar product of two vectors can be written in terms of their standard components as

$$A \cdot B = \sum_M (-1)^M A_{-M} B_M = \sum_M A_M^\dagger B_M . \tag{B.69}$$

2) The spherical harmonics $Y_L^M(\theta, \varphi)$, acting as multiplicative operators, are the standard basis of an irreducible tensor of order L.

3) If A_M^J and B_M^J are two irreducible tensors of order J, then $\alpha A_M^J + \beta B_M^J$, where α and β are arbitrary complex numbers, is also an irreducible tensor of the same order.

4) If $A_{M_1}^{J_1}$ and $B_{M_2}^{J_2}$ are irreducible tensors, then so is

$$T_M^J = \sum_{M_1} C(J_1 J_2 J; M_1, M - M_1) A_{M_1}^{J_1} B_{M-M_1}^{J_2} . \tag{B.70}$$

If T_M^J is an irreducible tensor, the calculation of its matrix elements for states with well-defined angular momentum is extraordinarily simplified by the Wigner-Eckart theorem [WI 31], [EC 30]:

$$\langle J_2 M_2 \alpha_2 | T_M^J | J_1 M_1 \alpha_1 \rangle = C(J_1 J J_2; M_1 M M_2) \langle J_2 \alpha_2 \| T^J \| J_1 \alpha_1 \rangle , \tag{B.71}$$

where $\langle J_2 \alpha_2 \| T^J \| J_1 \alpha_1 \rangle$ is called the reduced matrix element, and its value does not depend upon M_1, M_2 or M. Note that the importance of (B.71) stems from the fact that it permits factorization of the dependence on the third component of the angular momentum and therefore in particular the matrix element being considered is zero unless $M = M_2 - M_1$ and $\Delta(J_1 J_2 J) = 1$.

Some reduced matrix elements of particular importance are

$$\langle J_2 \alpha_2 \| J \| J_1 \alpha_1 \rangle = \delta_{\alpha_1 \alpha_2} \delta_{J_1 J_2} \sqrt{J_1(J_1 + 1)} , \tag{B.72}$$

$$\langle L_2 \alpha_2 \| Y_L \| L_1 \alpha_1 \rangle = \delta_{\alpha_1 \alpha_2} C(L_1 L L_2; 00) \sqrt{\frac{(2L_1 + 1)(2L + 1)}{4\pi(2L_2 + 1)}} . \tag{B.73}$$

The following relations, valid only for first order irreducible tensors, are consequences of the Wigner-Eckart theorem:

$$\langle J_1 M_2 \alpha_2 | T_M^1 | J_1 M_1 \alpha_1 \rangle = [J_1(J_1 + 1)]^{-1} \langle J_1 M_2 \alpha_2 | J_M(J \cdot T) | J_1 M_1 \alpha_1 \rangle ,$$
$$\langle J_2 M_2 \alpha_2 | J_M(J \cdot T) | J_1 M_1 \alpha_1 \rangle \tag{B.74}$$
$$= \delta_{J_1 J_2} \langle J_1 M_2 | J_M | J_1 M_1 \rangle \langle J_1 \alpha_2 \| (J \cdot T) \| J_1 \alpha_1 \rangle .$$

If 1 and 2 are kinematically independent systems and $T_M^K(i)$ is an irreducible tensor which acts only on the system i, then

$$\langle J_1' J_2' J' \alpha_1' \alpha_2' \| T^K(1) \| J_1 J_2 J \alpha_1 \alpha_2 \rangle$$
$$= \delta_{J_2 J_2'} \delta_{\alpha_2 \alpha_2'} (-1)^{J_2 + K - J_1 - J'}$$
$$\times W(J_1 J J_1' J'; J_2 K) \sqrt{(2J + 1)(2J_1' + 1)} \langle J_1' \alpha_1' \| T^K(1) \| J_1 \alpha_1 \rangle . \quad \text{(B.75)}$$

If

$$T_M^K(1,2) = \sum_{M_1 M_2} C(K_1 K_2 K; M_1 M_2 M) T_{M_1}^{K_1}(1) T_{M_2}^{K_2}(2) , \quad \text{(B.76)}$$

then we have

$$\langle J_1' J_2' J' \alpha_1' \alpha_2' \| T^K(1,2) \| J_1 J_2 J \alpha_1 \alpha_2 \rangle$$
$$= [(2J + 1)(2J_1' + 1)(2J_2' + 1)(2K + 1)]^{1/2} X(J_1 J_2 J; K_1 K_2 K; J_1' J_2' J')$$
$$\times \langle J_1' \alpha_1' \| T^{K_1}(1) \| J_1 \alpha_1 \rangle \langle J_2' \alpha_2' \| T^{K_2}(2) \| J_2 \alpha_2 \rangle , \quad \text{(B.77)}$$

where the coefficient X is defined as

$$X(l_1 s_1 j_1; l_2 s_2 j_2; LSJ) = (-1)^\sigma \sum_g (2g + 1) W(s_1 l_2 j_1 L; g l_1)$$
$$\times W(l_2 s_1 j_2 S; g s_2) W(L j_1 S j_2; g J) , \quad \text{(B.78)}$$

σ being the sum $\sigma = l_1 + s_1 + j_1 + l_2 + s_2 + j_2 + L + S + J$. The coefficient X is known as the Rose coefficient [RO 57], and appears frequently in the coupling of four angular momenta. One of the principal properties of X is that if its arguments are written in the form of a 3×3 matrix with rows $l_1 s_1 j_1$, $l_2 s_2 j_2$, and LSJ, then the interchange of any pair of rows or columns changes the corresponding X coefficient only by a sign $(-1)^\sigma$.

Other interesting properties of these coefficients are

1) $$\sum_{j_1 j_2} (2j_1 + 1)(2j_2 + 1)\sqrt{2L + 1}\sqrt{2L' + 1}\sqrt{2s + 1}\sqrt{2s' + 1}$$
$$\times X(l_1 s_1 j_1; l_2 s_2 j_2; LSJ) X(l_1 s_1 j_1; l_2 s_2 j_2; L'S'J) = \delta_{LL'} \delta_{SS'} ; \quad \text{(B.79)}$$

2) $$X(l_1 s_1 j_1; l_2 s_2 j_1; LL0)$$
$$= (-1)^{j_1 + L - l_1 - s_2} \frac{1}{\sqrt{2j_1 + 1}\sqrt{2L + 1}} W(l_1 s_1 l_2 s_2; j_1 L) ; \quad \text{(B.80)}$$

3) $$\sqrt{2}\sqrt{2l_1 + 1}\sqrt{2l_2 + 1}\sqrt{2L + 1}\sqrt{2S + 1}$$
$$\times C(l_1 l_2 L; 000) X(l_1 \tfrac{1}{2} j_1; l_2 \tfrac{1}{2} j_2; LSJ)$$
$$= (-1)^{j_1 + l_1 + 1/2} C(j_1 j_2 J; 1/2 - 1/2 0) f_{12}(LSJ) , \quad \text{(B.81)}$$

where $l_1 + l_2 + L = $ even, and

$$f_{12}(L0J) = -\delta_{LJ}, \quad f_{12}(J1J) = \frac{k_1 - k_2}{\sqrt{J(J+1)}},$$

$$f_{12}(J-11J) = \frac{1}{2J+1}\sqrt{\frac{2J-1}{J}}(k_1 + k_2 - J),$$

$$f_{12}(J+11J) = \frac{1}{2J+1}\sqrt{\frac{2J+3}{J+1}}(k_1 + k_2 + J + 1),$$

$$k_i = (2j_i + 1)(l_i - j_i), \quad i = 1,2.$$

Particular cases of (B.77) are (B.75) and the relation

$$\langle J_1' J_2' J' \alpha_1' \alpha_2' \| \sum_M (-1)^M T_M^K(1) T_{-M}^K(2) \| J_1 J_2 J \alpha_1 \alpha_2 \rangle$$

$$= \delta_{JJ'}(-1)^{J_1'+J_2-J} W(J_1 J_2 J_1' J_2'; JK)\sqrt{(2J_1'+1)(2J_2'+1)}$$

$$\times \langle J_1' \alpha_1' \| T^K(1) \| J_1 \alpha_1 \rangle \langle J_2' \alpha_2' \| T^K(2) \| J_2 \alpha_2 \rangle. \tag{B.82}$$

B.6 Irreducible Vector Tensors

Let $V(r)$ be a vector field and consider a rotation characterized by an axis $\hat{n}$ and a rotation angle θ in the positive sense. The application of this rotation in an active sense to the experimental device which served to prepare the vector field $V(r)$ produces a transformed apparatus which sets up a new vector field $V'(r)$. If the rotation is $r \to r' = \mathcal{R}r$, then $V'(r') = \mathcal{R}V(r)$, and there exists a unitary operator $R(\hat{n}, \theta)$ such that $V'(r) = \mathcal{R}(R(\hat{n}, \theta)V)(r)$.

We can write

$$\mathcal{R}R(\hat{n}, \theta) = \exp[-i\hat{n} \cdot (L + S)], \tag{B.83}$$

where $L = -ir \times \nabla$ is the orbital angular momentum operator and S is the spin operator, whose components, for the basis $e_1 = (1,0,0), e_2 = (0,1,0)$ and $e_3 = (0,0,1)$, are

$$S_1 = \begin{pmatrix} 0 & 0 & 0 \\ 0 & 0 & -i \\ 0 & i & 0 \end{pmatrix}, \quad S_2 = \begin{pmatrix} 0 & 0 & i \\ 0 & 0 & 0 \\ -i & 0 & 0 \end{pmatrix}, \quad S_3 = \begin{pmatrix} 0 & -i & 0 \\ i & 0 & 0 \\ 0 & 0 & 0 \end{pmatrix}, \tag{B.84}$$

and they generate a $\mathcal{D}^{(1)}$ representation.

For orbital angular momentum, the eigenfunctions of L^2 and L_3 are the spherical harmonics $Y_L^M(\theta, \phi)$ obeying

$$L^2 Y_L^M(\theta, \phi) = L(L+1) Y_L^M(\theta, \phi), \quad L_3 Y_L^M(\theta, \phi) = M Y_L^M(\theta, \phi). \tag{B.85}$$

For S^2 and S_3 the simultaneous eigenvectors

$$S^2 \chi_\mu = 2\chi_\mu, \quad S_3 \chi_\mu = \mu \chi_\mu \tag{B.86}$$

are given by

$$\chi_{\pm 1} = \mp \frac{1}{\sqrt{2}}(e_1 \pm ie_2) \quad \chi_0 = e_3 . \tag{B.87}$$

These vectors have been chosen in such a way that they are normalized according to

$$\chi_\mu^* \cdot \chi_{\mu'} = \delta_{\mu\mu'} , \tag{B.88}$$

and their relative phases agree with the Condon-Shortley convention, so that

$$S_\mu \chi_\lambda = (-1)^\mu \sqrt{2} C(111; \mu + \lambda, -\mu, \lambda) \chi_{\mu+\lambda} . \tag{B.89}$$

Equation (B.89) is a particular case of the more general relation

$$J_\mu |JM\rangle = (-1)^\mu \sqrt{J(J+1)} C(J1J; M+\mu, -\mu, M) |JM+\mu\rangle , \tag{B.90}$$

which itself is merely a compact way of writing (B.7) and (B.9) for J_μ.

If the total angular momentum is defined as $J = L + S$, it is of interest to find the simultaneous eigenvectors of L^2, S^2, J^2 and J_3, which we will denote by $\mathcal{Y}_{JM}^L(\theta, \phi)$, and which must satisfy the equations

$$L^2 \mathcal{Y}_{JM}^L(\theta, \phi) = L(L+1) \mathcal{Y}_{JM}^L(\theta, \phi) , \quad S^2 \mathcal{Y}_{JM}^L(\theta, \phi) = 2 \mathcal{Y}_{JM}^L(\theta, \phi) ,$$
$$J^2 \mathcal{Y}_{JM}^L(\theta, \phi) = J(J+1) \mathcal{Y}_{JM}^L(\theta, \phi) , \quad J_3 \mathcal{Y}_{JM}^L(\theta, \phi) = M \mathcal{Y}_{JM}^L(\theta, \phi) . \tag{B.91}$$

According to the general theory, these are

$$\mathcal{Y}_{JM}^L(\theta, \phi) = \sum_\mu C(L1J; M-\mu, \mu) Y_L^{M-\mu}(\theta, \phi) \chi_\mu . \tag{B.92}$$

For a fixed value of L the $(2J+1)$ functions $\mathcal{Y}_{JM}^L$ constitute an irreducible tensor of order J, which is customarily called an irreducible vector tensor or a vector spherical harmonic.

Note that the possible values of L are generically $|J-1|, J, J+1$. Taking into account (B.43) and (B.68) it is easy to show that the spherical components of $\mathcal{Y}_{JM}^L(\theta, \phi)$ are

$$[\mathcal{Y}_{JM}^L(\theta, \phi)]_\mu = (-1)^\mu C(L1J; M+\mu, -\mu) Y_L^{M+\mu}(\theta, \phi) . \tag{B.93}$$

On the other hand, using (A.38), (B.43) and (B.88) we obtain

$$\int d\Omega \, \mathcal{Y}_{J'M'}^{L'*}(\theta, \phi) \cdot \mathcal{Y}_{JM}^L(\theta, \phi) = \delta_{JJ'} \delta_{LL'} \delta_{MM'} . \tag{B.94}$$

The notation

$$X_{JM}(\theta, \phi) \equiv \mathcal{Y}_{JM}^J(\theta, \phi) \tag{B.95}$$

is frequently introduced. Taking into account that $L = \sum_\mu (-1)^\mu \chi_{-\mu} L_\mu$ and using (B.90) and (B.92) it is easily shown that

$$X_{JM}(\theta,\phi) = \frac{LY_J^M(\theta,\phi)}{\sqrt{J(J+1)}} .$$ (B.96)

In applications, certain properties of the vector spherical harmonics occur frequently; we collect them here:

1) Through (B.93), (A.37), (A.43) and the orthogonality and symmetry properties of the Clebsch-Gordan coefficients it is easily shown that

$$\hat{\mathbf{r}} \cdot \mathcal{Y}_{JM}^L(\hat{\mathbf{r}}) = -C(J1L;000)Y_J^M(\hat{\mathbf{r}}) ,$$ (B.97)

from which, together with (B.45), one obtains

$$\hat{\mathbf{r}} \cdot X_{JM}(\hat{\mathbf{r}}) = 0 .$$ (B.98)

2) Examination of (A.34), plus the vector character of the vector spherical harmonics show that the parity of $\mathcal{Y}_{JM}^L(\theta,\phi)$ is $(-1)^L$, and thus for $X_{JM}(\theta,\phi)$ the parity is $(-1)^J$.

3) Another interesting relation is

$$i(\hat{\mathbf{r}} \times L)Y_L^M(\hat{\mathbf{r}}) = -L\sqrt{\frac{L+1}{2L+1}}\mathcal{Y}_{LM}^{L+1}(\hat{\mathbf{r}}) - (L+1)\sqrt{\frac{L}{2L+1}}\mathcal{Y}_{LM}^{L-1}(\hat{\mathbf{r}}) .$$ (B.99)

To show this, it suffices to start with the first term in (B.99), take into account that both $\hat{\mathbf{r}}$ and L can be expressed in the form $A = \sum_\mu (-1)^\mu A_\mu \chi_{-\mu}$, and use (B.90) to find $L_\mu Y_L^M(\hat{\mathbf{r}})$. There occurs in the above a vector product $\chi_{\mu'} \times \chi_\mu$, which can be expressed as

$$\chi_{\mu'} \times \chi_\mu = i\sqrt{2}C(111; \mu'\mu)\chi_{\mu'+\mu} .$$ (B.100)

On the other hand, $\hat{r}_\mu$ can be written in terms of $Y_1^\mu(\hat{\mathbf{r}})$, and the two spherical harmonics which appear can be combined using (A.43). The sum over the third component of the angular momentum involves only the Clebsch-Gordan coefficients and can be carried out using (B.59), and, recalling the definition of the irreducible vector tensors, we have

$$i(\hat{\mathbf{r}} \times L)Y_L^M(\hat{\mathbf{r}}) = [6L(L+1)(2L+1)]^{1/2}$$
$$\times \sum_{L'} C(L1L';00)W(LL'11;1L)\mathcal{Y}_{LM}^{L'}(\hat{\mathbf{r}}) .$$ (B.101)

Finally, with the explicit expression for the Clebsch-Gordan and Racah coefficients given in the tables at the end of this appendix we obtain (B.99).

4) Starting from $\hat{\mathbf{r}}Y_L^M(\hat{\mathbf{r}})$ and writing $\hat{\mathbf{r}}$ in terms of χ_μ and the spherical harmonics of order 1, we obtain a product of two spherical harmonics which can be combined using (A.43), and taking into account the symmetry properties of the Clebsch-Gordan coefficients and the definition of irreducible vector tensors finally gives

$$\hat{\mathbf{r}}Y_L^M(\hat{\mathbf{r}}) = -\sum_{L'} C(L1L';00)\mathcal{Y}_{LM}^{L'}(\hat{\mathbf{r}}) .$$ (B.102)

5) The irreducible vector tensors form an orthonormal basis for vector fields $V(r)$, complete in $L^2(S_2, \mathbb{C}^3)$ so that

$$V(r) = \sum_{JLM} C_{JM}^L(r) \mathcal{Y}_{JM}^L(\hat{r}) ,$$

(B.103)

where the coefficients, using (B.94), are

$$C_{JM}^L(r) = \int d\Omega\, \mathcal{Y}_{JM}^{L*}(\hat{r}) \cdot V(r) .$$

(B.104)

In particular, if $R(r)$ is a scalar function, then

$$\nabla[R(r)Y_L^M(\hat{r})] = -\sqrt{\frac{L+1}{2L+1}}\left[\frac{dR(r)}{dr} - \frac{L}{r}R(r)\right] \mathcal{Y}_{LM}^{L+1}(\hat{r})$$
$$+ \sqrt{\frac{L}{2L+1}}\left[\frac{dR(r)}{dr} + \frac{L+1}{r}R(r)\right] \mathcal{Y}_{LM}^{L-1}(\hat{r}) ,$$

(B.105)

which is the famous gradient formula, with a myriad of applications. To prove it, it suffices to take into account that

$$\nabla = \hat{r} \cdot (\hat{r} \cdot \nabla) - \hat{r} \times (\hat{r} \times \nabla) = \hat{r}\frac{\partial}{\partial r} - \frac{i}{r}(\hat{r} \times L) ,$$

(B.106)

and using (B.104), (B.99) and the orthogonality relation for the irreducible vector tensors we obtain the desired relation.

It is also of interest to calculate $\nabla \cdot [R(r)\mathcal{Y}_{JM}^L(\hat{r})]$. Given that $\nabla = \sum_\mu (-1)^\mu \nabla_\mu \chi_{-\mu}$ and also (B.92) we can write

$$\nabla \cdot [R(r)\mathcal{Y}_{JM}^L(\hat{r})] = \sum_\mu C(L1J; M-\mu, \mu)\nabla_\mu R(r)Y_L^{M-\mu}(\hat{r}) .$$

(B.107)

With the help of (B.105), (B.49), the fact that $(\chi_\lambda)_\mu = (-1)^\lambda \delta_{\mu,-\lambda}$, and the orthogonality relations for the Clebsch-Gordan coefficients, we obtain

$$\nabla \cdot (R(r)\mathcal{Y}_{JM}^L(\hat{r})) = \sqrt{\frac{L+1}{2L+3}}\delta_{J,L+1}Y_{L+1}^M(\hat{r})\left[\frac{dR(r)}{dr} - \frac{L}{r}R(r)\right]$$
$$- \sqrt{\frac{L}{2L-1}}\delta_{J,L-1}Y_{L-1}^M(\hat{r})\left[\frac{dR(r)}{dr} + \frac{L+1}{r}R(r)\right] ,$$

(B.108)

the required expression.

In a similar vein, using

$$\nabla \times L = -i\left[r\Delta - \left(2 + r\frac{\partial}{\partial r}\right)\nabla\right] ,$$

(B.109)

a simple, but lengthy, calculation permits us to show that

$$\nabla \times [j_J(\omega r) \boldsymbol{X}_{JM}(\hat{\boldsymbol{r}})] = \mathrm{i}\omega \left[-\sqrt{\frac{J}{2J+1}} j_{J+1}(\omega r) \mathcal{Y}_{JM}^{J+1}(\hat{\boldsymbol{r}}) \right.$$

$$\left. + \sqrt{\frac{J+1}{2J+1}} j_{J-1}(\omega r) \mathcal{Y}_{JM}^{J-1}(\hat{\boldsymbol{r}}) \right] . \qquad \text{(B.110)}$$

6) Another expansion of some interest is that for $\chi_\mu \exp(\mathrm{i}kz)$, with $\mu = \pm 1$, in terms of vector spherical harmonics. Using (A.14), (A.40) and the inverse of (B.92) we have

$$\chi_\mu \mathrm{e}^{\mathrm{i}kz} = \sqrt{4\pi} \sum_{L=0}^{\infty} \mathrm{i}^L \sqrt{2L+1} j_L(kr) \sum_J C(L1J; 0\mu) \mathcal{Y}_{J\mu}^L(\hat{\boldsymbol{r}}) . \qquad \text{(B.111)}$$

Rearranging the terms and using Table B.2 permits writing this as

$$\chi_\mu \mathrm{e}^{\mathrm{i}kz} = \sqrt{2\pi} \sum_{L=1}^{\infty} \mathrm{i}^L \sqrt{2L+1} \left[\mathrm{i}\sqrt{\frac{L}{2L+1}} j_{L+1}(kr) \mathcal{Y}_{L\mu}^{L+1}(\hat{\boldsymbol{r}}) \right.$$

$$\left. -\mathrm{i}\sqrt{\frac{L+1}{2L+1}} j_{L-1}(kr) \mathcal{Y}_{L\mu}^{L-1}(\hat{\boldsymbol{r}}) - \mu j_L(kr) \boldsymbol{X}_{L\mu}(\hat{\boldsymbol{r}}) \right] , \qquad \text{(B.112)}$$

the required formula. For $\varepsilon_\mu \exp(\mathrm{i}\boldsymbol{k} \cdot \boldsymbol{r})$, where ε_μ, $\hat{\boldsymbol{k}}$ are obtained from χ_μ, χ_0 by a rotation $\mathcal{R}$, (B.112) immediately gives

$$\varepsilon_\mu \mathrm{e}^{\mathrm{i}\boldsymbol{k}\cdot\boldsymbol{r}} = \sqrt{2\pi} \sum_{L=1}^{\infty} \sum_{\nu=-L}^{+L} \mathrm{i}^L \sqrt{2L+1} D_{\nu\mu}^L(\mathcal{R}) \left[\mathrm{i}\sqrt{\frac{L}{2L+1}} j_{L+1}(kr) \mathcal{Y}_{L\nu}^{L+1}(\hat{\boldsymbol{r}}) \right.$$

$$\left. -\mathrm{i}\sqrt{\frac{L+1}{2L+1}} j_{L-1}(kr) \mathcal{Y}_{L\nu}^{L-1}(\hat{\boldsymbol{r}}) - \mu j_L(kr) \boldsymbol{X}_{L\nu}(\hat{\boldsymbol{r}}) \right] . \qquad \text{(B.113)}$$

B.7 Tables of Clebsch-Gordan and Racah Coefficients

The entries in Table B.1 are

$J_1 J_2 J_3$	$M_1 M_2 M_3$	$C(J_1 J_2 J_3; M_1 M_2 M_3)$

.

A square root has been omitted in all Clebsch-Gordan coefficients, which affects their moduli. (For instance, $-1/3$ should be read $-\sqrt{1/3}$.) The symmetry relations (B.46–47) allow us to limit the tabulation to the cases $J_1 \geq J_2, M_3 \geq 0$.

Tables B.2 and B.3 are self-explanatory.

Table B.1. Clebsch-Gordan coefficients

1/2 1/2 0	1/2 −1/2 0	1/2		5/2 1/2 3	3/2 1/2 2	5/6
1/2 1/2 1	1/2 1/2 1	1		5/2 1/2 3	3/2 −1/2 1	1/3
1/2 1/2 1	1/2 −1/2 0	1/2		5/2 1/2 3	1/2 1/2 1	2/3
1 1/2 1/2	1 −1/2 1/2	2/3		5/2 1/2 3	1/2 −1/2 0	1/2
1 1/2 1/2	0 1/2 1/2	−1/3		3 1/2 5/2	3 −1/2 5/2	6/7
1 1/2 3/2	1 1/2 3/2	1		3 1/2 5/2	2 1/2 5/2	−1/7
1 1/2 3/2	1 −1/2 1/2	1/3		3 1/2 5/2	2 −1/2 3/2	5/7
1 1/2 3/2	0 1/2 1/2	2/3		3 1/2 5/2	1 1/2 3/2	−2/7
3/2 1/2 1	3/2 −1/2 1	3/4		3 1/2 5/2	1 −1/2 1/2	4/7
3/2 1/2 1	1/2 1/2 1	−1/4		3 1/2 5/2	0 1/2 1/2	−3/7
3/2 1/2 1	1/2 −1/2 0	1/2		3 1/2 7/2	3 1/2 7/2	1
3/2 1/2 2	3/2 1/2 2	1		3 1/2 7/2	3 −1/2 5/2	1/7
3/2 1/2 2	3/2 −1/2 1	1/4		3 1/2 7/2	2 1/2 5/2	6/7
3/2 1/2 2	1/2 1/2 1	3/4		3 1/2 7/2	2 −1/2 3/2	2/7
3/2 1/2 2	1/2 −1/2 0	1/2		3 1/2 7/2	1 1/2 3/2	5/7
2 1/2 3/2	2 −1/2 3/2	4/5		3 1/2 7/2	1 −1/2 1/2	3/7
2 1/2 3/2	1 1/2 3/2	−1/5		3 1/2 7/2	0 1/2 1/2	4/7
2 1/2 3/2	1 −1/2 1/2	3/5		7/2 1/2 3	7/2 −1/2 3	7/8
2 1/2 3/2	0 1/2 1/2	−2/5		7/2 1/2 3	5/2 1/2 3	−1/8
2 1/2 5/2	2 1/2 5/2	1		7/2 1/2 3	5/2 −1/2 2	3/4
2 1/2 5/2	2 −1/2 3/2	1/5		7/2 1/2 3	3/2 1/2 2	−1/4
2 1/2 5/2	1 1/2 3/2	4/5		7/2 1/2 3	3/2 −1/2 1	5/8
2 1/2 5/2	1 −1/2 1/2	2/5		7/2 1/2 3	1/2 1/2 1	−3/8
2 1/2 5/2	0 1/2 1/2	3/5		7/2 1/2 3	1/2 −1/2 0	1/2
5/2 1/2 2	5/2 −1/2 2	5/6		7/2 1/2 4	7/2 1/2 4	1
5/2 1/2 2	3/2 1/2 2	−1/6		7/2 1/2 4	7/2 −1/2 3	1/8
5/2 1/2 2	3/2 −1/2 1	2/3		7/2 1/2 4	5/2 1/2 3	7/8
5/2 1/2 2	1/2 1/2 1	−1/3		7/2 1/2 4	5/2 −1/2 2	1/4
5/2 1/2 2	1/2 −1/2 0	1/2		7/2 1/2 4	3/2 1/2 2	3/4
5/2 1/2 3	5/2 1/2 3	1		7/2 1/2 4	3/2 −1/2 1	3/8
5/2 1/2 3	5/2 −1/2 2	1/6		7/2 1/2 4	1/2 1/2 1	5/8

Table B.1. (cont.)

7/2	1/2	4	1/2	-1/2	0	1/2
1	1	0	1	-1	0	1/3
1	1	0	0	0	0	-1/3
1	1	1	1	0	1	1/2
1	1	1	0	1	1	-1/2
1	1	1	1	-1	0	1/2
1	1	1	0	0	0	0
1	1	2	1	1	2	1
1	1	2	1	0	1	1/2
1	1	2	0	1	1	1/2
1	1	2	1	-1	0	1/6
1	1	2	0	0	0	2/3
3/2	1	1/2	3/2	-1	1/2	1/2
3/2	1	1/2	1/2	0	1/2	-1/3
3/2	1	1/2	-1/2	1	1/2	1/6
3/2	1	3/2	3/2	0	3/2	3/5
3/2	1	3/2	1/2	1	3/2	-2/5
3/2	1	3/2	3/2	-1	1/2	2/5
3/2	1	3/2	1/2	0	1/2	1/15
3/2	1	3/2	-1/2	1	1/2	-8/15
3/2	1	5/2	3/2	1	5/2	1
3/2	1	5/2	3/2	0	3/2	2/5
3/2	1	5/2	1/2	1	3/2	3/5
3/2	1	5/2	3/2	-1	1/2	1/10
3/2	1	5/2	1/2	0	1/2	3/5
3/2	1	5/2	-1/2	1	1/2	3/10
2	1	1	2	-1	1	3/5
2	1	1	1	0	1	-3/10
2	1	1	0	1	1	1/10
2	1	1	1	-1	0	3/10
2	1	1	0	0	0	-2/5

2	1	2	2	0	2	2/3
2	1	2	1	1	2	-1/3
2	1	2	2	-1	1	1/3
2	1	2	1	0	1	1/6
2	1	2	0	1	1	-1/2
2	1	2	1	-1	0	1/2
2	1	2	0	0	0	0
2	1	3	2	1	3	1
2	1	3	2	0	2	1/3
2	1	3	1	1	2	2/3
2	1	3	2	-1	1	1/15
2	1	3	1	0	1	8/15
2	1	3	0	1	1	2/5
2	1	3	1	-1	0	1/5
2	1	3	0	0	0	3/5
5/2	1	3/2	5/2	-1	3/2	2/3
5/2	1	3/2	3/2	0	3/2	-4/15
5/2	1	3/2	1/2	1	3/2	1/15
5/2	1	3/2	3/2	-1	1/2	2/5
5/2	1	3/2	1/2	0	1/2	-2/5
5/2	1	3/2	-1/2	1	1/2	1/5
5/2	1	5/2	5/2	0	5/2	5/7
5/2	1	5/2	3/2	1	5/2	-2/7
5/2	1	5/2	5/2	-1	3/2	2/7
5/2	1	5/2	3/2	0	3/2	9/35
5/2	1	5/2	1/2	1	3/2	-16/35
5/2	1	5/2	3/2	-1	1/2	16/35
5/2	1	5/2	1/2	0	1/2	1/35
5/2	1	5/2	-1/2	1	1/2	-18/35
5/2	1	7/2	5/2	1	7/2	1
5/2	1	7/2	5/2	0	5/2	2/7

Table B.1. (cont.)

5/2	1	7/2	3/2	1	5/2	5/7	3	1	4	2	−1	1	3/28
5/2	1	7/2	5/2	−1	3/2	1/21	3	1	4	1	0	1	15/28
5/2	1	7/2	3/2	0	3/2	10/21	3	1	4	0	1	1	5/14
5/2	1	7/2	1/2	1	3/2	10/21	3	1	4	1	−1	0	3/14
5/2	1	7/2	3/2	−1	1/2	1/7	3	1	4	0	0	0	4/7
5/2	1	7/2	1/2	0	1/2	4/7	3/2	3/2	0	3/2	−3/2	0	1/4
5/2	1	7/2	−1/2	1	1/2	2/7	3/2	3/2	0	1/2	−1/2	0	−1/4
3	1	2	3	−1	2	5/7	3/2	3/2	1	3/2	−1/2	1	3/10
3	1	2	2	0	2	−5/21	3/2	3/2	1	1/2	1/2	1	−2/5
3	1	2	1	1	2	1/21	3/2	3/2	1	−1/2	3/2	1	3/10
3	1	2	2	−1	1	10/21	3/2	3/2	1	3/2	−3/2	0	9/20
3	1	2	1	0	1	−8/21	3/2	3/2	1	1/2	−1/2	0	−1/20
3	1	2	0	1	1	1/7	3/2	3/2	2	3/2	1/2	2	1/2
3	1	2	1	−1	0	2/7	3/2	3/2	2	1/2	3/2	2	−1/2
3	1	2	0	0	0	−3/7	3/2	3/2	2	3/2	−1/2	1	1/2
3	1	3	3	0	3	3/4	3/2	3/2	2	1/2	1/2	1	0
3	1	3	2	1	3	−1/2	3/2	3/2	2	−1/2	3/2	1	−1/2
3	1	3	3	−1	2	1/4	3/2	3/2	2	3/2	−3/2	0	1/4
3	1	3	2	0	2	1/3	3/2	3/2	2	1/2	−1/2	0	1/4
3	1	3	1	1	2	−5/12	3/2	3/2	3	3/2	3/2	3	1
3	1	3	2	−1	1	5/12	3/2	3/2	3	3/2	1/2	2	1/2
3	1	3	1	0	1	1/12	3/2	3/2	3	1/2	3/2	2	1/2
3	1	3	0	1	1	−1/2	3/2	3/2	3	3/2	−1/2	1	1/5
3	1	3	1	−1	0	1/2	3/2	3/2	3	1/2	1/2	1	3/5
3	1	3	0	0	0	0	3/2	3/2	3	−1/2	3/2	1	1/5
3	1	4	3	1	4	1	3/2	3/2	3	3/2	−3/2	0	1/20
3	1	4	3	0	3	1/4	3/2	3/2	3	1/2	−1/2	0	9/20
3	1	4	2	1	3	3/4	2	3/2	1/2	2	−3/2	1/2	2/5
3	1	4	3	−1	2	1/28	2	3/2	1/2	1	−1/2	1/2	−3/10
3	1	4	2	0	2	3/7	2	3/2	1/2	0	1/2	1/2	1/5
3	1	4	1	1	2	15/28	2	3/2	1/2	−1	3/2	1/2	−1/10

Table B.1. (cont.)

2	3/2	3/2	2	-1/2	3/2	2/5	5/2	3/2	1	1/2	-1/2	0	-3/10
2	3/2	3/2	1	1/2	3/2	-2/5	5/2	3/2	2	5/2	-1/2	2	10/21
2	3/2	3/2	0	3/2	3/2	1/5	5/2	3/2	2	3/2	1/2	2	-8/21
2	3/2	3/2	2	-3/2	1/2	2/5	5/2	3/2	2	1/2	3/2	2	1/7
2	3/2	3/2	1	-1/2	1/2	0	5/2	3/2	2	5/2	-3/2	1	5/14
2	3/2	3/2	0	1/2	1/2	-1/5	5/2	3/2	2	3/2	-1/2	1	1/42
2	3/2	3/2	-1	3/2	1/2	2/5	5/2	3/2	2	1/2	1/2	1	-25/84
2	3/2	5/2	2	1/2	5/2	4/7	5/2	3/2	2	-1/2	3/2	1	9/28
2	3/2	5/2	1	3/2	5/2	-3/7	5/2	3/2	2	3/2	-3/2	0	3/7
2	3/2	5/2	2	-1/2	3/2	16/35	5/2	3/2	2	1/2	-1/2	0	-1/14
2	3/2	5/2	1	1/2	3/2	1/35	5/2	3/2	3	5/2	1/2	3	5/8
2	3/2	5/2	0	3/2	3/2	-18/35	5/2	3/2	3	3/2	3/2	3	-3/8
2	3/2	5/2	2	-3/2	1/2	6/35	5/2	3/2	3	5/2	-1/2	2	5/12
2	3/2	5/2	1	-1/2	1/2	5/14	5/2	3/2	3	3/2	1/2	2	1/12
2	3/2	5/2	0	1/2	1/2	-3/35	5/2	3/2	3	1/2	3/2	2	-1/2
2	3/2	5/2	-1	3/2	1/2	-27/70	5/2	3/2	3	5/2	-3/2	1	1/8
2	3/2	7/2	2	3/2	7/2	1	5/2	3/2	3	3/2	-1/2	1	49/120
2	3/2	7/2	2	1/2	5/2	3/7	5/2	3/2	3	1/2	1/2	1	-1/60
2	3/2	7/2	1	3/2	5/2	4/7	5/2	3/2	3	-1/2	3/2	1	-9/20
2	3/2	7/2	2	-1/2	3/2	1/7	5/2	3/2	3	3/2	-3/2	0	3/10
2	3/2	7/2	1	1/2	3/2	4/7	5/2	3/2	3	1/2	-1/2	0	1/5
2	3/2	7/2	0	3/2	3/2	2/7	5/2	3/2	4	5/2	3/2	4	1
2	3/2	7/2	2	-3/2	1/2	1/35	5/2	3/2	4	5/2	1/2	3	3/8
2	3/2	7/2	1	-1/2	1/2	12/35	5/2	3/2	4	3/2	3/2	3	5/8
2	3/2	7/2	0	1/2	1/2	18/35	5/2	3/2	4	5/2	-1/2	2	3/28
2	3/2	7/2	-1	3/2	1/2	4/35	5/2	3/2	4	3/2	1/2	2	15/28
5/2	3/2	1	5/2	-3/2	1	1/2	5/2	3/2	4	1/2	3/2	2	5/14
5/2	3/2	1	3/2	-1/2	1	-3/10	5/2	3/2	4	5/2	-3/2	1	1/56
5/2	3/2	1	1/2	1/2	1	3/20	5/2	3/2	4	3/2	-1/2	1	15/56
5/2	3/2	1	-1/2	3/2	1	-1/20	5/2	3/2	4	1/2	1/2	1	15/28
5/2	3/2	1	3/2	-3/2	0	1/5	5/2	3/2	4	-1/2	3/2	1	5/28

Table B.1. (cont.)

5/2	3/2	4	3/2	−3/2	0	1/14	2	2	3	2	0	2	1/2
5/2	3/2	4	1/2	−1/2	0	3/7	2	2	3	1	1	2	0
2	2	0	2	−2	0	1/5	2	2	3	0	2	2	−1/2
2	2	0	1	−1	0	−1/5	2	2	3	2	−1	1	3/10
2	2	0	0	0	0	1/5	2	2	3	1	0	1	1/5
2	2	1	2	−1	1	1/5	2	2	3	0	1	1	−1/5
2	2	1	1	0	1	−3/10	2	2	3	−1	2	1	−3/10
2	2	1	0	1	1	3/10	2	2	3	2	−2	0	1/10
2	2	1	−1	2	1	−1/5	2	2	3	1	−1	0	2/5
2	2	1	2	−2	0	2/5	2	2	3	0	0	0	0
2	2	1	1	−1	0	−1/10	2	2	4	2	2	4	1
2	2	1	0	0	0	0	2	2	4	2	1	3	1/2
2	2	2	2	0	2	2/7	2	2	4	1	2	3	1/2
2	2	2	1	1	2	−3/7	2	2	4	2	0	2	3/14
2	2	2	0	2	2	2/7	2	2	4	1	1	2	4/7
2	2	2	2	−1	1	3/7	2	2	4	0	2	2	3/14
2	2	2	1	0	1	−1/14	2	2	4	2	−1	1	1/14
2	2	2	0	1	1	−1/14	2	2	4	1	0	1	3/7
2	2	2	−1	2	1	3/7	2	2	4	0	1	1	3/7
2	2	2	2	−2	0	2/7	2	2	4	−1	2	1	1/14
2	2	2	1	−1	0	1/14	2	2	4	2	−2	0	1/70
2	2	2	0	0	0	−2/7	2	2	4	1	−1	0	8/35
2	2	3	2	1	3	1/2	2	2	4	0	0	0	18/35
2	2	3	1	2	3	−1/2							

Table B.2. Clebsch-Gordan coefficients

$\mathcal{D}^{(J)} \otimes \mathcal{D}^{(1/2)}$	$\|J\,1/2\,J+1/2\,M\rangle$	$\|J\,1/2\,J-1/2\,M\rangle$
$\|J\,M+1/2\,1/2\,-1/2\rangle$	$\sqrt{\dfrac{J-M+1/2}{2J+1}}$	$\sqrt{\dfrac{J+M+1/2}{2J+1}}$
$\|J\,M-1/2\,1/2\,1/2\rangle$	$\sqrt{\dfrac{J+M+1/2}{2J+1}}$	$-\sqrt{\dfrac{J-M+1/2}{2J+1}}$

$\mathcal{D}^{(J)} \otimes \mathcal{D}^{(1)}$	$\|J\,1\,J+1\,M\rangle$	$\|J1JM\rangle$	$\|J\,1\,J-1\,M\rangle$
$\|JM+1\,1\,-1\rangle$	$\sqrt{\dfrac{(J-M)(J-M+1)}{(2J+1)(2J+2)}}$	$\sqrt{\dfrac{(J-M)(J+M+1)}{2J(J+1)}}$	$\sqrt{\dfrac{(J+M)(J+M+1)}{2J(2J+1)}}$
$\|JM10\rangle$	$\sqrt{\dfrac{(J-M+1)(J+M+1)}{(J+1)(2J+1)}}$	$\dfrac{M}{\sqrt{J(J+1)}}$	$-\sqrt{\dfrac{(J-M)(J+M)}{J(2J+1)}}$
$\|JM-1\,1\,1\rangle$	$\sqrt{\dfrac{(J+M)(J+M+1)}{(2J+1)(2J+2)}}$	$-\sqrt{\dfrac{(J-M+1)(J+M)}{2J(J+1)}}$	$\sqrt{\dfrac{(J-M)(J-M+1)}{2J(2J+1)}}$

Table B.2. (cont.)

$\mathscr{D}^{(J)} \otimes \mathscr{D}^{(3/2)}$	$\lvert J\,3/2\,J+3/2\,M\rangle$	$\lvert J\,3/2\,J+1/2\,M\rangle$
$\lvert J\,M+3/2\,3/2-3/2\rangle$	$\sqrt{\dfrac{(J-M-1/2)(J-M+1/2)(J-M+3/2)}{(2J+1)(2J+2)(2J+3)}}$	$\sqrt{\dfrac{3(J-M-1/2)(J+M+1/2)(J+M+3/2)}{2J(2J+1)(2J+3)}}$
$\lvert J\,M+1/2\,3/2-1/2\rangle$	$\sqrt{\dfrac{3(J-M+1/2)(J-M+3/2)(J+M+3/2)}{(2J+1)(2J+2)(2J+3)}}$	$(J+3M+3/2)\sqrt{\dfrac{(J-M+1/2)}{2J(2J+1)(2J+3)}}$
$\lvert J\,M-1/2\,3/2\,1/2\rangle$	$\sqrt{\dfrac{3(J-M+3/2)(J+M+1/2)(J+M+3/2)}{(2J+1)(2J+2)(2J+3)}}$	$-(J-3M+3/2)\sqrt{\dfrac{(J+M+1/2)}{2J(2J+1)(2J+3)}}$
$\lvert J\,M-3/2\,3/2\,3/2\rangle$	$\sqrt{\dfrac{(J+M-1/2)(J+M+1/2)(J+M+3/2)}{(2J+1)(2J+2)(2J+3)}}$	$-\sqrt{\dfrac{3(J-M+3/2)(J+M-1/2)(J+M+1/2)}{2J(2J+1)(2J+3)}}$

$\mathscr{D}^{(J)} \otimes \mathscr{D}^{(3/2)}$	$\lvert J\,3/2\,J-1/2\,M\rangle$	$\lvert J\,3/2\,J-3/2\,M\rangle$
$\lvert J\,M+3/2\,3/2-3/2\rangle$	$\sqrt{\dfrac{3(J-M-1/2)(J+M+1/2)(J+M+3/2)}{(2J-1)(2J+1)(2J+2)}}$	$\sqrt{\dfrac{(J+M-1/2)(J+M+1/2)(J+M+3/2)}{(2J-1)(2J)(2J+1)}}$
$\lvert J\,M+1/2\,3/2-1/2\rangle$	$-(J-3M-1/2)\sqrt{\dfrac{(J-M+1/2)}{(2J-1)(2J+1)(2J+2)}}$	$-\sqrt{\dfrac{3(J-M-1/2)(J+M-1/2)(J+M+1/2)}{(2J-1)(2J)(2J+1)}}$
$\lvert J\,M-1/2\,3/2\,1/2\rangle$	$-(J+3M-1/2)\sqrt{\dfrac{(J-M+1/2)}{(2J-1)(2J+1)(2J+2)}}$	$\sqrt{\dfrac{3(J-M-1/2)(J-M+1/2)(J+M-1/2)}{(2J-1)(2J)(2J+1)}}$
$\lvert J\,M-3/2\,3/2\,3/2\rangle$	$\sqrt{\dfrac{3(J-M+1/2)(J-M+3/2)(J+M-1/2)}{(2J-1)(2J+1)(2J+2)}}$	$-\sqrt{\dfrac{(J-M-1/2)(J-M+1/2)(J-M+3/2)}{(2J-1)(2J)(2J+1)}}$

Table B.3. Racah coefficients

$$W(b+1/2\,b\,d+1/2\,d;1/2\,f) = (-1)^{b+d-f}\left[\frac{(b+d+f+2)(b+d-f+1)}{(2b+1)(2b+2)(2d+1)(2d+2)}\right]^{1/2}$$

$$W(b+1/2\,b\,d-1/2\,d;1/2\,f) = (-1)^{b+d-f}\left[\frac{(f+b-d+1)(f-b+d)}{(2b+1)(2b+2)(2d)(2d+1)}\right]^{1/2}$$

$$W(b-1/2\,b\,d+1/2\,d;1/2\,f) = (-1)^{b+d-f}\left[\frac{(f-b+d+1)(f+b-d)}{2b(2b+1)(2d+1)(2d+2)}\right]^{1/2}$$

$$W(b-1/2\,b\,d-1/2\,d;1/2\,f) = (-1)^{b+d-f-1}\left[\frac{(b+d+f+1)(b+d-f)}{2b(2b+1)2d(2d+1)}\right]^{1/2}$$

$$W(b+1\,b\,d+1\,d;1\,f) = (-1)^{b+d-f}\left[\frac{(f+b+d+3)(f+b+d+2)(-f+b+d+2)(-f+b+d+1)}{4(2b+3)(b+1)(2b+1)(2d+3)(d+1)(2d+1)}\right]^{1/2}$$

$$W(b\,b\,d+1\,d;1\,f) = (-1)^{b+d-f}\left[\frac{(f+b+d+2)(-f+b+d+1)(f-b+d+1)(f+b-d)}{4b(2b+1)(b+1)(2d+1)(d+1)(2d+3)}\right]^{1/2}$$

$$W(b-1\,b\,d+1\,d;1\,f) = (-1)^{b+d-f}\left[\frac{(f+b-d)(f+b-d-1)(f-b+d+2)(f-b+d+1)}{4b(2b-1)(2b+1)(d+1)(2d+1)(2d+3)}\right]^{1/2}$$

$$W(b+1\,b\,d\,d;1\,f) = (-1)^{b+d-f}\left[\frac{(f+b+d+2)(f+b-d+1)(b+d-f+1)(f-b+d)}{4(2b+1)(b+1)(2b+3)d(d+1)(2d+1)}\right]^{1/2}$$

$$W(b\,b\,d\,d;1\,f) = (-1)^{b+d-f-1}\frac{b(b+1)+d(d+1)-f(f+1)}{[4b(b+1)(2b+1)d(d+1)(2d+1)]^{1/2}}$$

$$W(b-1\,b\,d\,d;1\,f) = (-1)^{b+d-f-1}\left[\frac{(b+d+f+1)(b+d-f)(f+b-d)(f-b+d+1)}{4(2b+1)b(2b-1)d(2d+1)(d+1)}\right]^{1/2}$$

$$W(b+1\,b\,d-1\,d;1\,f) = (-1)^{b+d-f}\left[\frac{(f-b+d)(f-b+d-1)(f+b-d+2)(f+b-d+1)}{4(2b+1)(b+1)(2b+3)(2d-1)d(2d+1)}\right]^{1/2}$$

$$W(b\,b\,d-1\,d;1\,f) = (-1)^{b+d-f-1}\left[\frac{(f+b+d+1)(f+b-d+1)(f+d-b)(b+d-f)}{4b(2b+1)(b+1)d(2d+1)(2d-1)}\right]^{1/2}$$

$$W(b-1\,b\,d-1\,d;1\,f) = (-1)^{b+d-f}\left[\frac{(f+b+d+1)(f+b+d)(-f+b+d)(-f+b+d-1)}{4(2b+1)b(2b-1)(2d+1)d(2d-1)}\right]^{1/2}$$

Appendix C: Summary of Operator Theory

We bring together here some of the results of the theory of linear operators in a Hilbert space which are of direct interest for quantum mechanics. We suppose a certain familiarity with the elements of the theory, at perhaps the undergraduate level. For a more complete development of the concepts and properties sketched here the reader can, for example, consult the references: [DS 64], [KA 80], [SI 71], [RS 72], [SI 72], [RS 75], [RS 78], [RS 79], [SC 81], [TH 81] and [AG 88] among others.

C.1 Notation and Basic Definitions

We will use $\mathcal{H}$, with subscripts if necessary, to denote a *complex* and *separable Hilbert space* with a scalar product $\langle\cdot|\cdot\rangle$ or $\langle\cdot,\cdot\rangle$ and associated norm $\||\cdot\rangle\|$ or $\|\cdot\|$. The notation $|\varphi\rangle_R$ will indicate the ray $\{\exp(i\alpha)|\varphi\rangle : \alpha \in \mathbf{R}\} \subset \mathcal{H}$. It is said that $|\varphi\rangle_R$ is a unit ray if $\|\varphi\| = 1$. The set of bounded linear operators defined over all of $\mathcal{H}$ will be denoted by $\mathcal{A}(\mathcal{H})$.

Example: $L^2(\mathbf{R}^N, d\mu)$, μ being a non-negative measure (σ-additive, σ-finite, and separable). We will write $L^2(\mathbf{R}^N) \equiv L(\mathbf{R}^N, d^N x)$ when μ is the Lebesgue measure.

If $A : D(A)(\subset \mathcal{H}) \to \mathcal{H}$ is a linear operator defined in the subspace $D(A)$, we say that $D(A)$ is its domain and $R(A) \equiv AD(A)$ is its range. The operator A is characterized by its graph

$$\Gamma(A) \equiv \{(\varphi, \psi) \in \mathcal{H} \oplus \mathcal{H} : \varphi \in D(A), \psi = A\varphi\} \, .$$

A is said to be *closed* if the subspace $\Gamma(A)$ is closed. A is said to be *closable* if $\overline{\Gamma(A)}$, the closure of $\Gamma(A)$, is a graph: $\overline{\Gamma(A)} = \Gamma(\bar{A})$. $\bar{A}$ is called the closure of A, and is its minimal closed extension. [Given A_1, A_2, we say that A_2 is an extension of A_1 and we will write $A_1 \subset A_2$, or $A_2 \supset A_1$ if $\Gamma(A_1) \subset \Gamma(A_2)$.]

Given A, and $\lambda \in \mathbf{C}$ we say that λ belongs to the *resolvent set* $\varrho(A)$ if: $\overline{R(\lambda I - A)} = \mathcal{H}$, $(\lambda I - A)^{-1} \equiv R_\lambda(A)$ exists (the *resolvent operator* for A in λ) and is bounded. The set $\varrho(A)$ is open. Its complement $\mathbf{C} - \varrho(A) \equiv \sigma(A)$ is closed and is called the *spectrum* of A.

In turn, the points in $\sigma(A)$ can be classified as follows:

$\sigma_c(A)$ (*continuous* spectrum): $\lambda \in \sigma_c(A)$ if $\overline{R(\lambda I - A)} = \mathcal{H}$, and $R_\lambda(A)$ exists and is not bounded.

$\sigma_r(A)$ (*residual* spectrum): $\lambda \in \sigma_r(A)$ if $\overline{R(\lambda I - A)} \neq \mathcal{H}$, and $R_\lambda(A)$ exists.

$\sigma_p(A)$ (*point* spectrum): $R_\lambda(A)$ does not exist. [Thus $(A - \lambda I)\varphi = 0$ has a non-trivial solution φ; λ is then called an *eigenvalue* of A and φ an *eigenvector*.]

$\sigma_c, \sigma_r, \sigma_p$ are pairwise disjoint and $\sigma(A) = \sigma_c(A) \cup \sigma_r(A) \cup \sigma_p(A)$.

If A is closable, $\varrho(A) = \varrho(\bar{A}), \sigma(A) = \sigma(\bar{A})$ and $R(\lambda - \bar{A})$ is closed for $\lambda \in \varrho(\bar{A})$. (We write on occasion $\lambda - B$ instead of $\lambda I - B$.)

It is well known that if A is bounded, with $\overline{D(A)} = \mathcal{H}$, then $\varrho(A) \supset \{\lambda : |\lambda| > \|A\|\}$, and $\sigma(A)$ is not the empty set. When A is not bounded, even if $D(A)$ be dense in $\mathcal{H}$, it can occur that $\varrho(A)$ be empty [as in the case of the annihilation operator a for the harmonic oscillator, since $\sigma_p(a) = \mathbb{C}$] or that $\sigma(A)$ be empty (for example if $A : \varphi(x) \rightarrow d\varphi/dx$, in $L^2[0, 1]$, with φ absolutely continuous, $\varphi(0) = 0, d\varphi/dx \in L^2[0, 1]$.)

Finally, if $\overline{D(A)} = \mathcal{H}$, the adjoint $A^\dagger$ exists: $\langle \varphi, A\psi \rangle = \langle A^\dagger \varphi, \psi \rangle$ for $\forall \psi \in D(A), \forall \varphi \in D(A^\dagger)$. Its graph $\Gamma(A^\dagger) = (\mathcal{H} \oplus \mathcal{H}) \ominus \{(A\psi, -\psi) : \psi \in D(A)\}$. $A^\dagger$ is closed, and $A^{\dagger\dagger}$ exists if and only if, in addition, A is closable, in which case $(\bar{A})^\dagger = A^\dagger, \bar{A} = A^{\dagger\dagger}$.

C.2 Symmetric, Self-Adjoint, and Essentially Self-Adjoint Operators

A is said to be *symmetric* or *Hermitian* if $A \subset A^\dagger$. When $A = A^\dagger$, we say that A is *self-adjoint*, and if $\bar{A} = A^\dagger$, we say it is *essentially self-adjoint*. σ_p and σ_c lie on the real axis $\mathbb{R}$ for every symmetric operator, and if $\lambda \in \sigma_r$, with $\text{Im}\,\lambda \neq 0$, then the entire half-plane $\mathbb{C}_\lambda \equiv \{\mu : (\text{Im}\,\mu)(\text{Im}\,\lambda) > 0\} \subset \sigma_r$, and in particular, $\mathbb{R} \subset \sigma$. Given a symmetric operator A, its spectrum $\sigma(A) \subset \mathbb{R}$ if and only if A is essentially self-adjoint. Finally, if A is self-adjoint, $\sigma_r(A) = \emptyset$.

Given a symmetric operator A and non-real λ, the cardinal number $d_\lambda(A) \equiv \dim[\mathcal{H} \ominus R(\lambda I - A)]$ is constant over the half-plane $\mathbb{C}_\lambda$. The cardinal numbers $d_\pm(A) \equiv d_{\pm i}(A)$ are called *deficiency indices* of A. It is clear that $d_\pm(A) = \dim\{\varphi \in \mathcal{H} : A^\dagger \varphi = \pm i\varphi\}$. The symmetric operator A has a self-adjoint extension if and only if $d_+(A) = d_-(A)$, and is essentially self-adjoint if and only if $d_\pm(A) = 0$. The equality of the deficiency indices is assured, for example, if there exists a complex conjugation J (= antilinear, isometric, involutive, one-to-one mapping of $\mathcal{H}$ onto $\mathcal{H}$) such that $JA = AJ$. Another important case in which A can be extended to a self-adjoint operator is the following: if $A \geq -\alpha I$, that is, if $\langle \varphi, A\varphi \rangle \geq -\alpha\|\varphi\|^2, \forall \varphi \in D(A)$, in which case we say that A is bounded from below or semibounded, then $d_+(A) = d_-(A)$, and there exists a self-adjoint *extension* A_F, called the *Friedrichs* extension of A, characterized uniquely as the restriction $A^\dagger \upharpoonright \mathcal{H}_A \cap D(A^\dagger)$, $\mathcal{H}_A$ being the subspace completion of $D(A)$ under the norm $\|\varphi\|^2_{+1,A} \equiv \langle \varphi, (A + (\alpha + 1)I)\varphi \rangle$. This extension A_F has the same lower bound as A.

Examples:

1) $\mathcal{H} = L^2(I)$, where I is an interval, finite or not, in $\mathbf{R}$. Given a real and locally L^2 Borel function $f(x)$, the operator

$$f : \varphi(\cdot) \to f(\cdot)\varphi(\cdot) ,$$

is essentially self-adjoint in $C_0^\infty(I)$ (infinitely differentiable functions having compact support in the interior of I) and self-adjoint in the subspace $\{\varphi \in \mathcal{H} : f(\cdot)\varphi(\cdot) \in \mathcal{H}\}$.

2) The operator $P_{(0)} : \varphi(\cdot) \to -i\varphi'(\cdot)$ is symmetric in $C_0^\infty(I)$. But $d_\pm(P_{(0)}) = 1$ if I is finite, $d_+(P_{(0)}) = 0$, $d_-(P_{(0)}) = 1$ if I is semi-infinite to the right (vice versa if to the left), and $d_\pm(P_{(0)}) = 0$ if $I = \mathbf{R}$. In the first case it will have self-adjoint extensions; in the second case it will not and in the third case it is essentially self-adjoint.

The self-adjoint extensions of $P_{(0)}$ when I is finite with end-points a, b are characterized by *boundary conditions*, and are always of the form $(P_{(0)})^\dagger \upharpoonright \{\varphi \in D(P_{(0)}^\dagger) : \varphi(b) = \exp(i\theta)\varphi(a))\}$, with fixed real θ. Calling them P_θ, with $0 \le \theta < 2\pi$, $D(P_\theta) = \{\varphi \in L^2(I) : \varphi$ absolutely continuous, $\varphi' \in L^2(I), \varphi(b) = e^{i\theta}\varphi(a)\}$.

In the case for which $I = \mathbf{R}$, $\bar{P}_{(0)} = P$ is self-adjoint, and its domain $D(P) = \{\varphi \in L^2(\mathbf{R}) : \varphi$ absolutely continuous, $\varphi' \in L^2(\mathbf{R})\}$.

3) The operator $-\Delta : \varphi(\cdot) \to -\varphi''(\cdot)$ is symmetric in $C_0^\infty(I)$, and satisfies $-\Delta \ge 0$.

If I is finite, with end-points a, b, $d_\pm(-\Delta) = 2$, and its self-adjoint extensions are fixed by appropriate boundary conditions such as $\varphi(a) = \varphi(b) = 0$, or $\varphi'(a) = \varphi'(b) = 0$, or $\varphi(a) = \varphi(b), \varphi'(a) = \varphi'(b)$. The Friedrichs extension of $-\Delta$ is precisely the one corresponding to the boundary conditions $\varphi(a) = \varphi(b) = 0$.

If the interval I is semi-infinite, $d_\pm(-\Delta) = 1$, and each self-adjoint extension is determined by a boundary condition of the type $\varphi(c)\cos \alpha + \varphi'(c)\sin \alpha = 0$, where c is the finite boundary point of I, and α is real. The Friedrichs extension corresponds to the case $\alpha = 0$. When $I = \mathbf{R}$, $-\Delta$ is essentially self-adjoint.

In all three cases, every function in the domain of a self-adjoint extension is absolutely continuous, as is its first derivative. When I is not finite, it can be shown that the functions must be bounded, uniformly continuous and Hölder continuous.

4) The operators $-i\nabla$ are symmetric in $C_0^\infty(\mathbf{R}^3)$. Their closures are definable in $\mathcal{S}(\mathbf{R}^3)$ (those C^∞ functions which decrease rapidly). Since $\mathcal{S}$ is stable under Fourier transformations, and the operation of multiplication by p is essentially self-adjoint in $\mathcal{S}$, then $-i\nabla$ is essentially self-adjoint in C_0^∞, and, a fortiori, in $\mathcal{S}$. If P denotes its self-adjoint extension,

$$D(P) = \{\varphi \in L^2 : p\hat{\varphi}(p) \in L^2\} , \tag{C.1}$$

$\hat{\varphi}$ being the Fourier transform $\mathcal{F}\varphi$ of φ. Note that φ does not have to be differentiable in the usual sense.

Analogously, $-\Delta = -\nabla \cdot \nabla$ is essentially self-adjoint in C_0^∞, and in $\mathcal{S}$, and continuing to use $-\Delta$ to denote the self-adjoint extension,

$$D(-\Delta) = \{\varphi \in L^2 : p^2\hat{\varphi}(p) \in L^2\}. \tag{C.2}$$

The functions in $D(-\Delta)$ are bounded, uniformly continuous and Hölder continuous.

C.3 Spectral Theory of Self-Adjoint Operators

Given a self-adjoint operator A in $\mathcal{H}$, one can prove the *spectral theorem*: there exists a (finite or countably infinite) partition

$$\mathbf{R} = \cup_1^N B_i\,, \quad N \leq \aleph_0\,,$$

of $\mathbf{R}$ into disjoint Borel sets B_i and a regular non-negative Borel measure μ such that A is equivalent (= isometrically isomorphic) to the operator A', in

$$\mathcal{H}' \equiv \oplus_{n=1}^N [\underbrace{L^2(B_n, d\mu) \oplus \ldots \oplus L^2(B_n, d\mu)}_{n \text{ summands}}]\,, \tag{C.3}$$

defined as "multiplication by the variable" in each summand and with domain

$$D(A') \equiv \left\{ (f_1, (f_{21}, f_{22}), \ldots) \in \mathcal{H}' : \sum_{n=1}^N \sum_{i=1}^n \int_{B_n} x^2 |f_{ni}(x)|^2 d\mu < \infty \right\}. \tag{C.4}$$

This realization or *spectral representation* of A is unique in the sense that given another with $B_1', \ldots, B_{N'}', \mu'$, then $\mu \simeq \mu'$ (equivalent or absolutely continuous with each other), $N = N'$, and $B_i' = B_i$ almost everywhere with respect to μ. This determines A, aside from unitary equivalences.

The support supp μ of μ is $\sigma(A)$. The spectral theorem permits immediately the introduction of a functional calculus using A which parallels the one based on its equivalent A'. Given a complex Borel function φ, finite μ-almost everywhere, the operator $\varphi(A)$, which is equivalent to multiplication by $\varphi(\cdot)$ in $\mathcal{H}'$, is closed, with a dense domain formed by those vectors whose images in $\mathcal{H}'$ satisfy

$$\sum_{n=1}^N \sum_{i=1}^n \int_{B_n} |\varphi(x)|^2 |f_{ni}(x)|^2 d\mu < \infty\,.$$

In particular, if $\|\varphi\|_\infty < \infty$ (the norm $\|\cdot\|_\infty$ being understood to be with respect to μ), $\varphi(A) \in \mathcal{A}(\mathcal{H})$ and $\|\varphi(A)\| = \|\varphi\|_\infty$. The correspondence $\varphi \to \varphi(A), \varphi \in L^\infty(\mathbf{R}, d\mu)$, is an isometric *-homomorphism:

1) $(\alpha\varphi)(A) = \alpha\varphi(A), \alpha \in \mathbb{C}$.
2) $(\varphi_1 + \varphi_2)(A) = \varphi_1(A) + \varphi_2(A)$.
3) $(\varphi_1\varphi_2)(A) = \varphi_1(A)\varphi_2(A)$.   (C.5)
4) $(\varphi^*)(A) = \varphi^\dagger(A)$.
5) $\|\varphi(A)\| = \|\varphi\|_\infty$.

For general Borel functions φ which are finite μ-almost everywhere, properties (1) and (4) are fulfilled, and (2) and (3) will be as well, if we substitute $\supset$ for =. In particular, φ real $\Rightarrow \varphi(A)$ is self-adjoint, $|\varphi(x)| = 1, \forall x \Rightarrow \varphi(A)$ is unitary. For any Borel set B having characteristic function $\chi_B, \chi_B(A)$ is an orthogonal projection which we will denote by $E_A(B)$, or simply by $E(B)$ if A is understood. The association $B \rightarrow E(B)$ defines the *spectral measure* associated with A, and $E(B)$ is called the *spectral projection* of A associated with B. The family $E(\cdot)$ uniquely determines A by means of

$$A = \int_{\mathbb{R}} \lambda \, dE_\lambda , \qquad\qquad (C.6)$$

where $E_\lambda \equiv E((-\infty, \lambda])$, and the integral is understood in the sense of

$$\langle\psi_1, A\psi_2\rangle = \int_{\mathbb{R}} \lambda \, d\langle\psi_1, E_\lambda\psi_2\rangle \quad \text{for} \qquad\qquad (C.7)$$

$$\forall\psi_1 \in \mathcal{H}, \forall\psi_2 \in D(A) \left(= \left\{\psi : \int_{\mathbb{R}} \lambda^2 d\|E_\lambda\psi\|^2 < \infty\right\}\right) . \qquad (C.8)$$

The function $\lambda \rightarrow E_\lambda$ is non-decreasing and strongly continuous to the right, and is known as the *resolution of the identity* for $A : I = \int_{\mathbb{R}} dE_\lambda$. In terms of E_λ, and in the same sense as (C.6):

$$\varphi(A) = \int_{\mathbb{R}} \varphi(\lambda)dE_\lambda , \quad D(\varphi(A)) = \left\{\psi : \int_{\mathbb{R}} |\varphi(\lambda)^2 d\|E_\lambda\psi\|^2 < \infty\right\} \quad (C.9)$$

for all Borel φ which are E-finite (i.e., finite except on at most a subset of a Borel set on which the spectral projection is null; this is equivalent to μ-finite).
Finally,

$$\sigma(\varphi(A)) = \bigcap_{E(B)=I} \overline{\varphi(B)} ,$$

and if φ is real, the spectral measure $E_\varphi(\cdot)$ associated with $\varphi(A)$ obeys $E_\varphi(B) = E(\varphi^{-1}(B))$. If φ is complex, then writing $\varphi(A) = (\text{Re } \varphi)(A)+\text{i}(\text{Im } \varphi)(A)$ we will obtain in the same way the spectral decomposition of its self-adjoint and its anti-self-adjoint parts, and thus any two functions $\varphi_1(A)$ and $\varphi_2(A)$ will commute in the sense that their spectral measures do.

C.4 The Spectrum of a Self-Adjoint Operator

Given a self-adjoint A, we know that $\sigma(A) = \sigma_p(A) \cup \sigma_c(A) \subset \mathbf{R}$. Moreover, λ real $\in \sigma(A) \Leftrightarrow E_\lambda$ is not constant in any neighborhood of λ : $E_{\lambda+\varepsilon} - E_{\lambda-\varepsilon} \neq 0, \forall \varepsilon > 0$. And, λ real $\in \sigma_p(A) \Leftrightarrow E_\lambda \neq E_{\lambda-0}$, i.e., if E_μ is discontinuous at $\mu = \lambda$. All of this is an immediate consequence of the functional calculus discussed in Sect. C.3.

Equivalently, in terms of the measure μ given by the spectral theorem,

$$\lambda \text{ real } \in \sigma(A) \Leftrightarrow \mu((\lambda - \varepsilon, \lambda + \varepsilon)) \neq 0, \quad \forall \varepsilon > 0$$
$$\lambda \text{ real } \in \sigma_p(A) \Leftrightarrow \mu(\{\lambda\}) \neq 0.$$

If we remove the pure point part from μ, $\mu_{\text{p.p.}} \equiv \sum_\lambda \mu(\{\lambda\})\delta_\lambda$, where $\delta_\lambda : B \to 0(1)$ if $\lambda \notin B (\lambda \in B)$, there remains $\mu_c = \mu - \mu_{\text{p.p.}}$, where μ_c is *continuous*, i.e., it has no pure points ($\mu_c(\{\lambda\}) = 0, \forall \lambda$). In turn, μ_c can be decomposed (using the Lebesgue decomposition theorem) into a part $\mu_{\text{a.c.}}$ which is *absolutely continuous* (Lebesgue null $B \Rightarrow \mu_{\text{a.c.}}(B) = 0$) and another part $\mu_{\text{s.c.}}$ which is *continuous singular* [i.e. it is continuous and there exists a Lebesgue null B such that $\mu_{\text{s.c.}}(\mathbf{R} - B) = 0$]. In summary, $\mu = \mu_{\text{p.p.}} + \mu_{\text{s.c.}} + \mu_{\text{a.c.}}$, and correspondingly

$$\mathcal{H} = \mathcal{H}_{\text{p.p.}} \oplus \mathcal{H}_{\text{s.c.}} \oplus \mathcal{H}_{\text{a.c.}}, \tag{C.10}$$

these subspaces being stable under A and such that

$$A_{\text{p.p.}} \equiv A \restriction \mathcal{H}_{\text{p.p.}}, \quad A_{\text{s.c.}} \equiv A \restriction \mathcal{H}_{\text{s.c.}}, \quad A_{\text{a.c.}} \equiv A \restriction \mathcal{H}_{\text{a.c.}}, \tag{C.11}$$

are all self-adjoint, with associated measures $\mu_{\text{p.p.}}, \mu_{\text{s.c.}}$, and $\mu_{\text{a.c.}}$, according to the spectral theorem. [In terms of E_λ, ψ belongs to one of the subspaces $\mathcal{H}_{\text{p.p.}}$, etc., if and only if the measure $d\|E_\lambda\psi\|^2$ is pure point, etc.]

We define $\sigma_{\text{p.p.}}(A) \equiv \sigma(A_{\text{p.p.}}), \sigma_{\text{s.c.}}(A) \equiv \sigma(A_{\text{s.c.}}), \sigma_{\text{a.c.}}(A) \equiv \sigma(A_{\text{a.c.}})$. Observe that $\sigma_{\text{p.p.}} = \overline{\sigma_p}$ and that $\sigma = \sigma_{\text{p.p.}} \cup \sigma_{\text{s.c.}} \cup \sigma_{\text{a.c.}}$, even though these subsets are not necessarily disjoint. There exists an orthonormal basis of eigenvectors for the operator $A_{\text{p.p.}}$. If $A = A_{\text{p.p.}}$, we say simply that the spectrum of A is a pure point spectrum.

Another useful decomposition of the spectrum of A is the following: $\sigma(A) = \sigma_{\text{disc}}(A) \cup \sigma_{\text{ess}}(A)$, $\sigma_{\text{disc}}(A)$ being the *discrete* spectrum, $\sigma_{\text{disc}}(A) \equiv \{\lambda \in \sigma_p(A) : \lambda$ has finite multiplicity and is isolated in $\sigma(A)\}$ and $\sigma_{\text{ess}}(A)$ being the *essential* spectrum: $\sigma_{\text{ess}}(A) \equiv \sigma(A) - \sigma_{\text{disc}}(A)$. It is clear that $\sigma_{\text{ess}}(A) = \overline{\sigma_{\text{ess}}(A)}$.

We shall see that $\sigma_{\text{a.c.}}$ and σ_{ess} are stable under certain perturbations of A.

C.5 One-Parameter Unitary Groups

Given a self-adjoint A, $t \rightarrow U(t) \equiv \exp(-itA)$ defines a one-parameter group of unitary operators, continuous in the strong sense (and in the norm topology if A is bounded). Conversely, we have *Stone's theorem*:

If $U(t)$ is a one-parameter group of unitary operators $[U(t_1 + t_2) = U(t_1)U(t_2)]$ continuous in the strong topology, the operator A, defined by

$$A\psi \equiv \lim_{t \rightarrow 0} \frac{i}{t}(U(t) - I)\psi \qquad (C.12)$$

in a domain in which this limit exists in the strong sense, is self-adjoint and *generates* $U(t) : U(t) = \exp(-itA)$.

[Actually, the continuity of $U(t)$ is assured by the condition that $\langle \psi_1, U(t)\psi_2 \rangle$ be Lebesgue measurable, as a function of t, $\forall \psi_1, \psi_2 \in \mathcal{H}$ (we have supposed $\mathcal{H}$ separable).]

C.6 Quadratic Forms

Given a self-adjoint operator A, we shall denote by $Q(A)$ the subspace $D(|A|^{1/2})$. It is clear that $\varphi, \psi \rightarrow \langle \varphi | A | \psi \rangle$ is meaningful for $\forall \varphi, \psi \in Q(A)$. This map is called the *quadratic form* associated with A; we shall say that it is a *self-adjoint* form.

In general, a quadratic form is a map $a : Q(a) \times Q(a) \rightarrow \mathbb{C}$, where $Q(a)$ is a dense subspace in $\mathcal{H}$, called the domain of a, such that a is linear in its second argument, and antilinear in the first. The form is said to be symmetric if $a(\psi_1, \psi_2) = a(\psi_2, \psi_1)^*$, and *semi-bounded* (or *bounded from below*) if in addition $a(\psi, \psi) \geq -\alpha \|\psi\|^2$, for a fixed $\alpha \in \mathbf{R}$. If $\alpha = 0$, a is called *positive*.

Given a semi-bounded symmetric form a, the map

$$\psi \rightarrow \|\psi\|_{+1,a} \equiv a(\psi, \psi) + (\alpha + 1)\|\psi\|^2 \qquad (C.13)$$

defines a norm in $Q(a)$. If $Q(a)$ is complete with respect to this norm, we say that a is closed; if a has a closed extension, we say that it is *closable*. Note that the norm $\| \cdot \|_{+1,a}$ is associated with a scalar product in $Q(a)$. $Q(a)$ is pre-Hilbert; if a is closed, $Q(a)$ is a Hilbert space.

It is easily proved that the form associated with a given semi-bounded, self-adjoint operator is closed and semi-bounded. Conversely, we can prove the following fundamental theorem: *every symmetric, closed and semi-bounded a is a self-adjoint form*. The associated operator A is characterized by having as its domain the subspace of $Q(a)$ formed by those ψ_2 such that there exist $\psi_2' \in \mathcal{H}$ with the property

$$a(\psi_1, \psi_2) = \langle \psi_1, \psi_2' \rangle, \forall \psi_1 \in Q(a) , \qquad (C.14)$$

and by defining $\psi_2' = A\psi_2$. It is clear that a and A have the same lower bound.

Given two semi-bounded self-adjoint forms a and b such that $Q(a) \cap Q(b)$ is dense in $\mathcal{H}$, the form $a + b$, in $Q(a) \cap Q(b)$, is self-adjoint. The associated operator will be written as $A \overset{f}{+} B$ (sum in the sense of quadratic forms). Notice that it might well happen that $D(A) \cap D(B) = \{0\}$, so that the sum $A + B$, in the operator sense, would certainly have quite a few self-adjoint extensions. The sum $A \overset{f}{+} B$ selects one of these in a natural way. Of course, if $A + B$ is essentially self-adjoint in $D(A) \cap D(B)$, $A \overset{f}{+} B$ is then $\overline{A + B}$.

Finally, if A is symmetric and semi-bounded, the form a with domain $Q(a) = D(A)$, given by $a(\psi_1, \psi_2) = \langle \psi_1, A\psi_2 \rangle$ is (symmetric, semi-bounded, and) closable. Denoting its closure by $\bar{a}$, this closure is symmetric and semi-bounded, and thus self-adjoint. The associated operator A_F is the *Friedrichs extension* of A (see Sect. C.2). From among all the semi-bounded self-adjoint extensions of A, A_F is the one which leads to the form with the smallest domain; in addition A_F is the only semi-bounded, self-adjoint extension of A with $D(A_F) \subset Q(\bar{a})$. It is clear that A_F and A have the same lower bound.

C.7 Perturbation of Self-Adjoint Operators

Let H_0 be self-adjoint and V be symmetric. We say that V is H_0-*small* if:

1) $D(V) \supset D(H_0)$

2) $\exists a < 1, b > 0 : \|V\psi\| \leq a\|H_0\psi\| + b\|\psi\|, \forall \psi \in D(H_0)$.   (C.15)

If a can, in addition, be chosen as small as we wish, it is said that V is H_0-*infinitesimal*: $V \ll H_0$.

An important theorem by Kato–Rellich assures us that if V is H_0-small, then $H_0 + V$ is self-adjoint in $D(H_0)$. Similarly, it suffices that (2) be satisfied in a domain D in which H_0 is essentially self-adjoint in order to be certain that it is satisfied in $D(H_0)$ and that $H_0 + V$ is essentially self-adjoint in D. The condition $\|V(H_0 - i\alpha)^{-1}\| < 1$ for some real α is equivalent to the condition (2).

The Kato–Rellich theorem defines an interesting *criterion* for the *stability of self-adjointness*. Nevertheless, the spectral measure can undergo drastic changes upon changing H_0 into $H_0 + V$, even though V be very "small". A theorem due to Weyl and von Neumann asserts that, whatever the self-adjoint operator H_0 may be, there exists a self-adjoint and Hilbert–Schmidt V with $\|V\|_2 < \varepsilon, \varepsilon > 0$ but otherwise arbitrary, such that $H_0 + V = (H_0 + V)_{\text{p.p.}}$. [We recall that V is said to be *Hilbert–Schmidt* (H-S) if, for every orthonormal basis $\{\varphi_n\}$,

$$\|V\|_2^2 \equiv \sum \|V\varphi_n\|^2 < \infty .$$ (C.16)

In such a case, the sum of this series is independent of the chosen basis. The Hilbert–Schmidt class of operators is a subset $\mathcal{A}_2(\mathcal{H})$ of the family $\mathcal{A}_0(\mathcal{H})$ of *compact* operators, namely those bounded operators defined over the entire $\mathcal{H}$ which transform weakly convergent sequences into strongly convergent ones. Their spectral properties are very simple; thus, if A is self-adjoint and compact, and if $\dim(\mathcal{H}) = \aleph_0$, then $\sigma_{\mathrm{ess}}(A) = \{0\}$. In other words, $\sigma(A)$ is a finite or countable set, having no accumulation point different from zero, and if $\lambda \in \sigma(A)$ with $\lambda \neq 0$, then $\lambda \in \sigma_{\mathrm{disc}}(A)$. This is true for all compact, but not necessarily self-adjoint A. Compact operators, as well as H-S operators, form two-sided ideals in the algebra of bounded operators $\mathcal{A}(\mathcal{H})$; another important ideal, $\mathcal{A}_1(\mathcal{H})$, is that of the *trace-class* operators, which are products of two H-S operators. These are characterized by being those bounded operators (with $\mathcal{H}$ as their domain) for which there exists an orthonormal basis $\{\varphi_n\}$ such that

$$\|A\|_1 \equiv \sum \langle \varphi_n, |A|\varphi_n \rangle < \infty \,. \tag{C.17}$$

Again, the indicated sum is independent of the basis chosen. Both $\mathcal{A}_1$ and $\mathcal{A}_2$ are complete with respect to the norms $\| \cdot \|_1, \| \cdot \|_2$, respectively, and $\mathcal{A}_0(\mathcal{H})$ is complete with respect to the ordinary norm $\| \cdot \|$. The self-adjoint operators A belonging to $\mathcal{A}_1$ are characterized by being those compact self-adjoint operators for which $\sum_{\lambda \in \sigma(A)} m_\lambda |\lambda| (= \|A\|_1) < \infty$, and those of $\mathcal{A}_2$ by $\sum_{\lambda \in \sigma(A)} m_\lambda |\lambda|^2 (= \|A\|_2^2) < \infty$. Here m_λ stands for the multiplicity of the eigenvalue λ. Finally, in $L^2(\mathbf{R}^N, d\mu)$ the H-S operators A are defined via integral kernels $A(x, y)$ such that $A(\cdot, \cdot) \in L^2(\mathbf{R}^N \times \mathbf{R}^N, d\mu \otimes d\mu)$, and $\|A\|_2 = \|A(\cdot, \cdot)\|_2$.]

The Weyl-von Neumann theorem notwithstanding, there is something which is *stable* under addition of a compact term V. This is the *essential spectrum*: $\sigma_{\mathrm{ess}}(H_1) = \sigma_{\mathrm{ess}}(H_0)$, for those H_0, H_1 which are self-adjoint and for which $R_\lambda(H_1) - R_\lambda(H_0) \in \mathcal{A}_0(\mathcal{H})$, for some $\lambda \in \varrho(H_0) \cap \varrho(H_1)$; in particular, whenever $V \in \mathcal{A}_0(\mathcal{H})$, and is self-adjoint, we will have $\sigma_{\mathrm{ess}}(H_0 + V) = \sigma_{\mathrm{ess}}(H_0)$.

The perturbation V in the Weyl-von Neumann theorem was assumed to be H-S. The results of the theorem are valid for many other classes of ideals in $\mathcal{A}_0(\mathcal{H})$, with the exception of $\mathcal{A}_1(\mathcal{H})$. The reason it fails if $V \in \mathcal{A}_1(\mathcal{H})$ is the *stability of* $\sigma_{\mathrm{a.c.}}(H_0)$ under this type of perturbation: $\sigma_{\mathrm{a.c.}}(H_0 + V) = \sigma_{\mathrm{a.c.}}(H_0)$ whenever V is self-adjoint and a member of the trace class. This important result is a particular consequence of the *Kato-Birman theorem*: If H_0, H_1 are self-adjoint, and $R_\lambda(H_1) - R_\lambda(H_0) \in \mathcal{A}_1(\mathcal{H})$, for some non-real λ, then $H_{0,\mathrm{a.c.}}$ and $H_{1,\mathrm{a.c.}}$ are equivalent (isometrically isomorphic). This isomorphism is defined by any of the operators (called Möller operators)

$$\Omega_{\pm}(H_1, H_0) \equiv \lim_{t \to \mp\infty} \mathrm{e}^{itH_1} \mathrm{e}^{-itH_0} E_{\mathrm{a.c.}}(H_0) \,, \tag{C.18}$$

where $E_{\mathrm{a.c.}}(H_0)$ is the projection on the subspace $\mathcal{H}_{\mathrm{a.c.}}^{(0)}$ absolutely continuous with respect to H_0, and the limit is understood in the strong sense.

Examples:

1) Let $H_0 = -\Delta$ in $L^2(\mathbf{R}^N)$. It is essentially self-adjoint in the space $C_0^\infty(\mathbf{R}^N)$, and in $\mathcal{S}(\mathbf{R}^N)$. We will continue to use the symbol $-\Delta$ also for its closure. It is easy to show, making use of the Fourier transform of $D(-\Delta)$ (as in Example

4 of Sect. C.2), that if $\psi(\cdot) \in D(-\Delta)$, then $\|\psi\|_\infty \leq C\alpha^{-(2-N/2)}\|(-\Delta + \alpha^2)\psi\|$ for all α greater than zero, so long as $N \leq 3$. Moreover, $\psi(x)$ is uniformly continuous and Hölder continuous, with exponents $1, < 1, < 1/2$ for $N = 1, 2, 3$, respectively.

Now let $V(x) \in L^2 + L^\infty$, i.e., $V = V_1 + V_2$, with $V_1 \in L^2, V_2 \in L^\infty$. Given $\psi \in D(-\Delta)$, it is clear that $\|V\psi\| \leq \|V_1\psi\| + \|V_2\psi\| \leq \|V_1\|_2\|\psi\|_\infty + \|V_2\|_\infty\|\psi\|$; thus, $\psi \in D(V)$. On the other hand,

$$\|V_1\psi\| \leq C\|V_1\|_2\, \alpha^{-(2-N/2)}\|(-\Delta + \alpha^2)\psi\|$$
$$\leq C'[\alpha^{-(2-N/2)}\| - \Delta\psi\| + \alpha^{N/2}\|\psi\|]$$

(C.19)

and as consequence, $V \ll -\Delta$. Thus, $-\Delta + V$ is self-adjoint in $D(-\Delta)$. The "potential" operators $V \in L^2 + L^\infty$ constitute the so-called *Kato class*.

2) With the previous notation, suppose that $V \in L^2 + L^\infty_\varepsilon$, that is, given an arbitrary $\varepsilon > 0$, there exists a decomposition $V = V_{1,\varepsilon} + V_{2,\varepsilon}$, with $V_{1,\varepsilon} \in L^2$, $\|V_{2,\varepsilon}\|_\infty < \varepsilon$. Supposing that $N = 1, 3$, the operator $(-\Delta - k^2)^{-1}$ for $\text{Im}\, k > 0$ is given by the integral kernel

$$\frac{i}{2k}e^{ik|x-y|}, \qquad\qquad N = 1,$$
$$\frac{1}{4\pi}\frac{1}{|x-y|}e^{ik|x-y|}, \qquad N = 3.$$

(C.20)

Since $\lim_{\varepsilon \to 0} V_{2,\varepsilon} = 0$, in norm, and $V_{1,\varepsilon}(-\Delta - k^2)^{-1}$ has a Hilbert–Schmidt kernel, $V(-\Delta - k^2)^{-1}$ is the norm limit of H-S operators, and is therefore compact. Then, if k is sufficiently large, $(-\Delta + V - k^2)^{-1} - (-\Delta - k^2)^{-1} \in \mathcal{A}_0(\mathcal{H})$ and as a consequence $\sigma_{\text{ess}}(-\Delta + V) = \sigma_{\text{ess}}(-\Delta) = [0, \infty)$.

3) Continuing the above, if $V \in L^1 \cap L^2$, then $|V|^{1/2} \in L^2$, and thus $|V|^{1/2}(-\Delta - k^2)^{-1} \in \mathcal{A}_2(\mathcal{H})$. On the other hand, if k is sufficiently large, $(-\Delta - k^2)(-\Delta + V - k^2)^{-1} \in \mathcal{A}(\mathcal{H})$, and thus $V^{1/2}(-\Delta + V - k^2)^{-1} = V^{1/2}(-\Delta - k^2)^{-1}(-\Delta - k^2)(-\Delta + V - k^2)^{-1} \in \mathcal{A}_2(\mathcal{H})$, where $V^{1/2}(x) = |V|^{1/2}(x)\text{sgn}V(x)$. Then $(-\Delta + V - k^2)^{-1} - (-\Delta - k^2)^{-1} \in \mathcal{A}_1(\mathcal{H})$, as it is the product of two H-S operators. As a consequence, both $\Omega_\pm(-\Delta + V, -\Delta)$ exist, and $(-\Delta + V)_{\text{a.c.}}$ is equivalent to $(-\Delta)_{\text{a.c.}} = -\Delta$. In particular, $\sigma_{\text{a.c.}}(-\Delta + V) = [0, \infty)$.

Finally, the question could be asked: what happens to the isolated parts of the spectrum of H_0 under the addition of an H_0-small V? If the "magnitude" of V is not restricted, the effect of the perturbation can be to destroy the isolation. This is the origin of the utility of studying this problem for $H_0 + gV = H_g$, where g measures the strength of V. If V is H_0-*bounded* (i.e., it satisfies conditions 1 and 2 which define H_0-small, but without the restriction that $a < 1$), gV will be H_0-small for $|g| < g_0 = 1/a$, and H_g will then be a family of self-adjoint operators, all having the domain $D(H_0)$. Furthermore, for complex $g, |g| < g_0$, H_g is closed with domain $D(H_0)$, and defines what is called a *holomorphic self-adjoint family of type* (A): (1) All H_g with $g \in$ a neighborhood of 0 have the same domain in which they are closed; (2) H_g is self-adjoint for g real; (3) $H_g\psi$

is a holomorphic function of $g, \forall \psi \in D(H_0)$ [i.e., $\langle \varphi | H_g | \psi \rangle$ is a holomorphic function of $g, \forall \varphi \in \mathcal{H}, \forall \psi \in D(H_0)$].

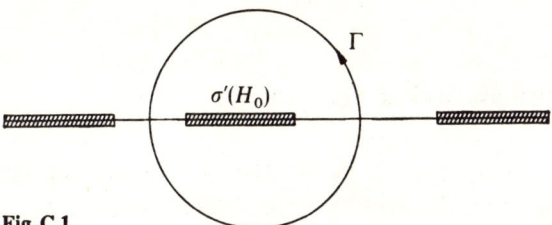

Fig. C.1

We can thus demonstrate that if $\sigma'(H_0)$ is an isolated part of $\sigma(H_0)$, and if Γ is a simple contour enclosing it, and separating it from the rest (Fig. C.1), then Γ continues to be in $\varrho(H_g)$ for sufficiently small $|g|$ [$\lambda \in \varrho(H_0) \Rightarrow R_\lambda(H_g)$ is holomorphic for sufficiently small $|g|$ depending upon λ], and

$$P_g \equiv (2\pi i)^{-1} \int_\Gamma \frac{d\lambda}{\lambda - H_g} \qquad (C.21)$$

is a projection (orthogonal if g is real). This projection is also holomorphic in g, and for real g permits the decomposition $\mathcal{H} = \mathcal{H}'_g \oplus \mathcal{H}''_g, \mathcal{H}'_g \equiv P_g\mathcal{H}, \mathcal{H}''_g \equiv (I - P_g)\mathcal{H}$. These subspaces $\mathcal{H}'_g, \mathcal{H}''_g$ are stable under H_g, and H_g has a spectrum $\sigma'(H_g)$ in $\mathcal{H}'_g$ interior to Γ and exterior in $\mathcal{H}''_g$. By continuity, $\dim(\mathcal{H}'_g) = \dim(\mathcal{H}'_0)$. The separation of the spectrum by Γ is thus maintained for sufficiently small $|g|$. Concretely, it suffices that

$$|g| < \min\left(a^{-1}, \min_{\lambda \in \Gamma}[a \sup_{\lambda' \in \sigma(H_0)} |\lambda'||\lambda - \lambda'|^{-1} + b \sup_{\lambda' \in \sigma(H_0)} |\lambda - \lambda'|^{-1}]^{-1}\right).$$
$$(C.22)$$

In particular, if $\sigma'(H_0) \subset \sigma_{\text{disc}}(H_0)$ [i.e., $\sigma'(H_0)$ contains only a finite number of points of finite multiplicity], the same will occur for $\sigma'(H_g)$ if g satisfies (C.22). When $\sigma'(H_0) = \{\lambda_0\}$, and $d \equiv$ distance $(\lambda_0, \sigma(H_0) - \{\lambda_0\})$, then (C.22) simplifies to

$$|g| < (1/2)[a + (a|\lambda_0| + b)/d]^{-1}, \qquad (C.23)$$

taking for Γ the circle of radius $d/2$ centered on λ_0. If $V \ll H_0$, a can be chosen arbitrarily small, and H_g is holomorphic and self-adjoint for all g. Finally, in the case $\sigma'(H_0) = \{\lambda_0\}$, and λ_0 is of finite multiplicity m, $\sigma'(H_g)$ will consist of a finite set of eigenvalues of finite multiplicity, so long as g satisfies (C.22), and they will be holomorphic functions of g. For real and sufficiently small g, there exists a set $\lambda_1(g), \ldots, \lambda_N(g)$ of analytic real functions such that $\lambda_i(g) \in \sigma_p(H_g)$, $\lambda_i(0) = \lambda_0$, and the corresponding spectral projections $P_i(g)$ are also analytic and self-adjoint with ranks $m_i(g)$ such that $\sum_i m_i(g) = m$. Each of these $\lambda_i(g), P_i(g)$ admit analytic extensions along the real axis so long as $\lambda_i(g)$ remains isolated in $\sigma(H_g)$ and of finite multiplicity (even though it may cross some other eigenvalue), yielding an eigenvalue of H_g and the projection onto the associated eigensubspace of H_g.

C.8 Perturbation of Semi-Bounded Self-Adjoint Forms

Analogous to the Kato–Rellich theorem, is the *KLMN theorem* (Kato–Lax–Milgram–Lions–Nelson) concerning the *stability of self-adjoint forms* under a perturbation:

"Let h_0 be a semi-bounded self-adjoint form associated with the operator H_0; and let v be a symmetric form, h_0-*small*; that is

1) $Q(v) \supset Q(h_0)$.

2) $\exists a < 1, b > 0 : |v(\psi, \psi)| \leq a h_0(\psi, \psi) + b\|\psi\|^2, \forall \psi \in Q(h_0)$. $\qquad$ (C.24)

Then $h = h_0 + v$ is a semi-bounded self-adjoint form in $Q(h_0)$."

In particular, if $H_0 \geq 0$, and $v(\psi_1, \psi_2) = \langle \psi_1, V\psi_2 \rangle$, V being a real "potential" operator $V(x)$, then (1) and (2) are satisfied if $|V|^{1/2}$ is $H_0^{1/2}$-small. Using H to denote the operator associated with h, it is clear that $H \supset H_0 + V$. But the domain $D(H_0) \cap D(V)$ of $H_0 + V$ can be very small, even $\{0\}$. We shall nevertheless often write $H_0 + V$ to denote such H.

Examples:

1) In $L^2(\mathbf{R})$, let the operator $H_0 = -\Delta$, and the symmetric form $\mu(\psi_1, \psi_2) = \int \psi_1^*(x)\psi_2(x)\, d\mu(x)$, where μ is a real Borel measure with finite negative and positive parts. [Such could be a real linear combination of "deltas", for example.] Just as in Example 1 of Sect. C.7, a simple estimate shows that $\|\psi\|_\infty \leq C\alpha^{-1}\{\langle\psi, -\Delta\psi\rangle + \alpha^2\|\psi\|^2\}, \forall \psi \in D(H_0^{1/2}), \forall \alpha > 0$, and since $|\mu(\psi, \psi)| \leq C'\|\psi\|_\infty^2$ there results $\mu \ll h_0$, i.e., μ is h_0-infinitesimal [a in (C.24) can be taken to be arbitrarily small]. The operator H associated with the self-adjoint form $h_0 + \mu$ in $Q(h_0)$ can be symbolically represented by $-\Delta + "d\mu/dx"$. The domain of H consists of those $\psi \in Q(h_0)$ such that the distribution

$$(-\Delta + "d\mu/dx")\psi : \phi \in C_0^\infty \to \langle\psi, -\Delta\phi\rangle + \int d\mu\, \psi^*\phi \qquad (C.25)$$

is realizable by an element of L^2, which we write as $H\psi$.

2) Let $V(x) \in L^1 + L^\infty$, in $\mathbf{R}$. With estimates similar to the ones in the previous case, $|V|^{1/2} \ll H_0^{1/2}$, and therefore, the KLMN theorem permits the definition of $H = H_0 + V$, a self-adjoint operator whose domain is $D(H) \subset D(H_0^{1/2})$, consisting of those $\psi \in D(H_0^{1/2})$ such that the distribution $H\psi \equiv (-\Delta + V)\psi \in L^2$.

3) In $L^2(\mathbb{R}^3)$, a potential $V(x)$ is said to belong to the *Rollnik class* if $V \in R + L^\infty$, R being the family of potentials V_1 with finite Rollnik norm $\| \cdot \|_R$:

$$\|V_1\|_R^2 = \int_{\mathbb{R}^3 \times \mathbb{R}^3} d^3x \, d^3y \frac{|V_1(x)| \, |V_1(y)|}{|x - y|^2} . \tag{C.26}$$

Thus, for example, one can prove that $L^p \cap L^q \subset R$, if $p \leq 3/2 \leq q$. In particular, $L^{3/2} \subset R, L^1 \cap L^2 \subset R \cap L^1$. Similarly, $L^p + L^\infty \subset R + L^\infty, \forall p \geq 3/2$; thus, the Kato class is in the Rollnik class. With the norm $\| \cdot \|_R$, R is complete. All $V \in R$ are L^1_{loc}.

Then, $V \in R + L^\infty \Rightarrow |V|^{1/2} \ll \sqrt{-\Delta} \equiv H_0^{1/2}$. It suffices to show this for $V \in R$, since the part in L^∞ gives rise to an operator in $\mathcal{A}(\mathcal{H})$, and thus is always $H_0^{1/2}$-infinitesimal. Assume that $V \in R$. Taking $\psi(x) \in C_0^\infty$, such $\psi \in D(|V|^{1/2})$, and using the integral kernel of $(-\Delta + \alpha^2)^{-1}, \alpha > 0$, it is easy to see that $\|(H_0 + \alpha^2)^{-1/2}|V|^{1/2}\psi\| \leq a\|\psi\|$, where $a \to 0$ if $\alpha \to \infty$. Therefore $(H_0 + \alpha^2)^{-1/2}|V|^{1/2}$ has a bounded extension to all of $\mathcal{H}$, with bound a; its adjoint $|V|^{1/2}(H_0 + \alpha^2)^{-1/2}$ has a similar property. Now taking $\psi \in D(H_0^{1/2})$, and writing $|V|^{1/2}\psi = |V|^{1/2}(H_0 + \alpha^2)^{-1/2}(H_0 + \alpha^2)^{1/2}\psi$, we have immediately

$$\| |V|^{1/2}\psi\|^2 \leq a^2(\|H_0^{1/2}\psi\|^2 + \alpha^2\|\psi\|^2) , \tag{C.27}$$

from which $|V|^{1/2} \ll H_0^{1/2}$.

Thus the KLMN theorem permits one to define $H = H_0 + V, \forall V \in R + L^\infty$. In the particular case for which $V = V_1 + V_2, V_1 \in R \cap L^1, V_2 \in L^\infty$, the domain of the self-adjoint operator H has a simple characterization: it consists of those $\psi \in D(H_0^{1/2})$ for which the distribution $-\Delta\psi + V\psi \in L^2$ (the condition $V_1 \in R \cap L^1$ ensures that $V\psi$ is a distribution).

In order to discuss the stability of $\sigma_{\text{ess}}(H_0)$ and $\sigma_{\text{a.c.}}(H_0)$, we use the Tiktopoulos formula:

$$(H - k^2)^{-1} = (H_0 - k^2)^{-1/2}$$
$$\times [1 + (H_0 - k^2)^{-1/2}V(H_0 - k^2)^{-1/2}]^{-1}(H_0 - k^2)^{-1/2} , \tag{C.28}$$

valid for Im $k > 0, |k|$ sufficiently large, $V \in L^1 + L^\infty$ [in $L^2(\mathbb{R})$], $V \in R + L^\infty$ [in $L^2(\mathbb{R}^3)$], $H_0 = -\Delta$, and where the left-hand member is the resolvent of H, with H defined by the method of forms.

Let us consider two situations:

a) $V \in L^1 + L_\epsilon^\infty$ (dimension 1)

    $V \in R + L_\epsilon^\infty$ (dimension 3)

b) $V \in L^1$ (dimension 1)

    $V \in R \cap L^1$ (dimension 3) .

For situation (a), $A \equiv (H_0 - k^2)^{-1/2}V(H_0 - k^2)^{-1/2}$ is the norm limit of the operators $A_n \equiv (H_0 - k^2)^{-1/2}V_n(H_0 - k^2)^{-1/2}$, with $V_n \in L^1$ (or R, depending on the dimension). Taking Fourier transforms, it is not difficult to establish that such operators A_n are H-S, and therefore A is compact. Thus, $(H - k^2)^{-1} - (H_0 - k^2)^{-1} \in \mathcal{A}_0(\mathcal{H})$, from which it follows that $\sigma_{\text{ess}}(H_0 + V) = [0, \infty)$.

In case (b), we rewrite Tiktopoulos' formula (C.28) in an equivalent form:

$$(H - k^2)^{-1} - (H_0 - k^2)^{-1} = -(H_0 - k^2)^{-1}$$
$$\times V^{1/2}[1 + |V|^{1/2}(H_0 - k^2)^{-1}V^{1/2}]^{-1}|V|^{1/2}(H_0 - k^2)^{-1}, \qquad (C.29)$$

and note that $(H_0 - k^2)^{-1}V^{1/2}$ and $|V|^{1/2}(H_0 - k^2)^{-1}$ are H-S. The difference in the resolvents, for $\text{Im }k > 0$, and sufficiently large $|k|$, is of trace class, so that an application of the Kato–Birman theorem suffices to conclude that for the conditions (b) $\exists \Omega_{\pm}(H_0 + V, H_0)$, which implement the equivalence between $(H_0 + V)_{\text{a.c.}}$ and $(H_0)_{\text{a.c.}}[= H_0]$. In particular, $\sigma_{\text{a.c.}}(H_0 + V) = [0, \infty)$.

Finally, the stability of the isolated parts of $\sigma(H_0)$, H_0 semi-bounded, under addition of an h_0-bounded symmetric form is investigated in the way suggested in Sect. C.7. The forms $h_0 + gv$ are self-adjoint and semi-bounded for sufficiently small g. Let H_g be the associated operator. Even for complex and small g, $h_0 + gv$ determines a densely defined, *sectorial* closed operator H [where sectorial refers to its numerical range $\langle \varphi | H_g | \varphi \rangle$, $\|\varphi\| = 1$, $\varphi \in D(H_g)$, being a subset of a sector $\{z \in \mathbb{C} : |\arg(z - \gamma)| \leq \theta < \pi/2, \gamma \text{ real}\}$] such that $(h_0 + gv)(\varphi_1, \varphi_2) = \langle \varphi_1 | H_g | \varphi_2 \rangle$, $\forall \varphi_1 \in Q(h_0), \forall \varphi_2 \in D(H_g)$, the domain $D(H_g)$ being determined by the requisite analogue to that for semi-bounded symmetric, closed forms. Thus, the family H_g, g complex and sufficiently small, constitutes what is known as a *self-adjoint holomorphic family of type (B)* [it is now the forms which have a common domain $Q(h_0)$]. Once again if $\lambda \in \varrho(H_0)$, $R_\lambda(H_g)$ is holomorphic in λ, g, and the isolation of $\sigma'(H_g)$ is maintained for sufficiently small $|g|$. An isolated eigenvalue of H_0 gives rise to one or more eigenvalues of H_g, which are analytic real functions of g, for g real, as well as to (self-adjoint) projections onto the corresponding eigensubspaces.

C.9 Min-Max Principle

The *min-max principle* (essentially due to Weyl) is very useful:

"Let A be self-adjoint and let

$$\lambda_n^{(1)}(A) \equiv \sup_{\mathcal{H}_{n-1}} [\inf_{\substack{\phi \in D(A); \phi \perp \mathcal{H}_{n-1} \\ \|\phi\|=1}} \langle \phi | A | \phi \rangle], \quad n = 1, 2, \dots, \qquad (C.30)$$

$$\lambda_n^{(2)}(A) \equiv \inf_{\mathcal{H}_n \subset D(A)} [\sup_{\phi \in \mathcal{H}_n; \|\phi\|=1} \langle \phi | A | \phi \rangle], \quad n = 1, 2, \dots, \qquad (C.31)$$

where $\mathcal{H}_n$ denotes an arbitrary n-dimensional subspace of $\mathcal{H}$. Then:

1) $\lambda_n^{(1)}(A) = \lambda_n^{(2)}(A) \equiv \lambda_n(A), \forall n$.
2) $\lambda_n(A)$ is non-decreasing.
3) If $\lambda_n(A) \neq -\infty$, for some n, it is so for all n; this occurs if and only if A is semi-bounded.

4) If $\lambda_\infty(A) = \lim_{n\to\infty} \lambda_n(A)$, then $\lambda_\infty(A) = \inf \sigma_{\mathrm{ess}}(A) [\equiv \infty$ if $\sigma_{\mathrm{ess}}(A) = \emptyset]$.

5) If $\lambda_n(A) < \lambda_\infty(A), \lambda_1(A), \dots \lambda_n(A)$ are the smallest eigenvalues of A (including multiplicities)."

The proof is simple using the spectral theorem.

One immediate consequence of this principle is: if there exists an $\mathcal{H}_n \subset D(A)$ such that $\langle \varphi, A\varphi \rangle \leq c\|\varphi\|^2, \forall \varphi \in \mathcal{H}_n$, then $\lambda_1(A), \lambda_2(A) \dots, \lambda_n(A) \leq c$. Thus, if A is semi-bounded, and $c < \inf \sigma_{\mathrm{ess}}(A)$, A will then have at least n independent eigenvectors.

Another useful application is the following: if H_1, H_2 are two self-adjoint operators such that $H_1 \leq H_2$ in $D(H_2)[\subset D(H_1)]$, then $\lambda_n(H_1) \leq \lambda_n(H_2), \forall n$. This frequently permits comparison of the discrete spectrum of two operators $H_i = -\Delta + V_i, i = 1, 2$; thus, for example, if $V_i \in L_2 + L_\varepsilon^\infty$, and $V_1(x) \leq V_2(x), \forall x$, then $\lambda_n(H_1) \leq \lambda_n(H_2), \forall n$.

If the operator A is semi-bounded, the min-max principle continues to be valid if we substitute $Q(A)$ for $D(A)$ in the expressions for $\lambda_n^{(1)}(A), \lambda_n^{(2)}(A)$. [Observe that if $A \geq -\alpha I, Q(A) = D((A + \alpha I)^{1/2})$, and it suffices to apply the principle to the operator $(A + \alpha I)^{1/2}$]. Thus, if $H_2 = H_1 \overset{f}{+} \delta V$, with $\delta V \geq 0, H_1$ being semi-bounded and $Q(H_1) \cap Q(\delta V)$ dense in $\mathcal{H}$, then also $\lambda_n(H_2) \geq \lambda_n(H_1), \forall n$. Such will be the case, for example, if $H_i = -\Delta + V_i, \delta V \equiv V_2 - V_1 \geq 0, V_i \in R + L^\infty$[in $L^2(\mathbf{R}^3)$] or $V_i \in L^1 + L^\infty$ [in $L^2(\mathbf{R})$].

C.10 Direct Integrals in Hilbert Spaces

Let $\mathcal{H}_1 \subset \mathcal{H}_2 \subset \dots \subset \mathcal{H}_\infty$ be a nested sequence of Hilbert spaces $\mathcal{H}_i$ of dimension i. Let X be a separable metric space, $X = B_1 \cup \dots \cup B_\infty$ be a partition of X in disjoint Borel sets, and μ be a regular, non-negative σ-finite Borel measure. We denote with the symbol

$$\int_X^\oplus d\mu(x)\mathcal{H}(x) \tag{C.32}$$

that collection of (equivalence classes) of functions $f: x \in X \to f(x) \in \mathcal{H}_n \subset \mathcal{H}_\infty$, if $x \in B_n$, such that

a) $f(x)$ is μ-measurable with values in $\mathcal{H}_\infty$;

b) $\int_X d\mu(x)\|f(x)\|^2 < \infty$.

and having a scalar product

c) $\langle f|g \rangle \equiv \int_X d\mu(x)\langle f(x)|g(x) \rangle$. $\tag{C.33}$

[The equivalence classes consist of functions which differ at most on μ-null sets; the μ-measurability of f means that $\langle \varphi | f(x) \rangle$ is measurable for all $\varphi \in \mathcal{H}_\infty$.] Then the set $\mathcal{H} \equiv \int_X^\oplus d\mu(x)\mathcal{H}(x)$ is a Hilbert space, called the *direct integral* of the $\mathcal{H}(x)$ (which is equal to $\mathcal{H}_n$ if $x \in B_n$) with measure μ, and with sets B_n of uniform dimension. The natural identification

$$\int_X^\oplus d\mu(x)\mathcal{H}(x) \equiv \bigoplus_1^\infty \int_{B_n}^\oplus d\mu(x)\mathcal{H}(x) = \bigoplus_1^\infty L^2(B_n, d\mu; \mathbb{C}^n) \tag{C.34}$$

is clear, where the last symbol expresses the Hilbert space of square integrable functions with respect to μ on B_n, with values in $\mathbb{C}^n$($\mathbb{C}^\infty$ being a Hilbert space of infinite dimension, and, as usual, separable).

Given a map $A : x \to A(x) \in \mathcal{A}(\mathcal{H}(x))$ which is μ-measurable [that is, $A(x)f(x)$ is μ-measurable for all $f \in \mathcal{H}$], such that $\|A(\cdot)\| \in L^\infty(X, d\mu)$, it is clear that $f(\cdot) \to A(\cdot)f(\cdot)$ defines an element $A \in \mathcal{A}(\mathcal{H})$, with norm $\|A\| = \|A(\cdot)\|_\infty$, called the direct integral of $A(\cdot)$:

$$A = \int_X^\oplus d\mu(x)A(x) . \tag{C.35}$$

The operators $\mathcal{A}(\mathcal{H})$ having this form are called *decomposable*. The set of such operators form a "*-algebra":

1) $A^\dagger = \int_X^\oplus d\mu(x)A^\dagger(x) .$

2) $\alpha_1 A_1 + \alpha_2 A_2 = \int_X^\oplus d\mu(x)(\alpha_1 A_1(x) + \alpha_2 A_2(x)) . \tag{C.36}$

3) $A_1 A_2 = \int_X^\oplus d\mu(x)A_1(x)A_2(x) .$

Further, it is a W^*-algebra: it is a *-algebra in $\mathcal{A}(\mathcal{H})$, closed under the weak operator topology. Such a W^*-algebra consists precisely of those operators in $\mathcal{A}(\mathcal{H})$ which commute with all *diagonal* operators $D = \int_X^\oplus d\mu(x)D(x)$, with $D(x) = d(x)I_{\mathcal{H}(x)}, d(\cdot)$ being a bounded Borel function. In other words, the *commutant* of the *abelian* W^*-algebra of the diagonal operators.

Reciprocally, every W^*-algebra whose commutant is abelian is isometrically isomorphic to that of the decomposable operators in a direct integral Hilbert space.

Appendix D: Elements of the Theory of Distributions

Given the importance of distributions in the theoretical formulation of physical problems, we collect here a few of the elementary concepts associated with them; the interested reader is urged to consult more complete and rigorous developments such as [SC 57,65], [GC 62], [ZE 65] and [BR 65], among others.

D.1 Spaces of Test Functions

Let Ω be a non-empty open set of $\mathbf{R}^n$, $p \equiv (p_1, \ldots, p_n)$ an n-tuple of integers $p_i \geq 0$, $|p| \equiv \sum p_i$, and $D^p = \partial^{p_1 + \cdots + p_n} / \partial x_1^{p_1} \ldots \partial x_n^{p_n}$. The set of complex-valued functions defined on Ω, having continuous derivatives D^p with $|p| \leq k$ [$\forall p$] will be denoted by $C^k(\Omega)[C^\infty(\Omega)]$. In the linear subspace $C_0^\infty(\Omega)[\equiv \mathcal{D}(\Omega)]$ of $C^\infty(\Omega)$ consisting of functions which in addition have compact support in Ω, we introduce the following *notion of convergence*:

$$\varphi_n \overset{\mathcal{D}(\Omega)}{\to} \varphi \overset{\text{def}}{\equiv} \exists \text{ compact } K \subset \Omega \text{ such that supp } \varphi_n \subset K,$$
$$\text{and } \|D^p \varphi_n - D^p \varphi\|_\infty \to 0, n \to \infty, \forall p.$$

The linear space $\mathcal{D}(\Omega), [\mathcal{D}(\mathbf{R}^n) \equiv \mathcal{D}$, if n is understood], is a *space of test functions*, of fundamental importance in the theory of distributions. All of these test functions are identically zero outside a finite region, and are infinitely differentiable. That $\mathcal{D}(\Omega)$ contains nontrivial functions is easily shown by providing an example:

If $x_0 \in \Omega, \exists$ a ball $\{x : |x - x_0| \leq r\} \subset \Omega$; with $r > 0$; and the function

$$\varrho_{x_0,r}(x) \equiv \begin{cases} 0, & |x - x_0| \geq r \\ \exp[r^2/(|x - x_0|^2 - r^2)], & |x - x_0| < r \end{cases} \tag{D.1}$$

belongs to $\mathcal{D}(\Omega)$.

The space $\mathcal{D}$ is not stable under Fourier transformation, but it can be extended to another one which is. This latter is the linear space $\mathcal{S}(\mathbf{R}^n) \equiv \mathcal{S}$ of functions C^∞ having *fast decrease*: $\varphi \in \mathcal{S}$ if and only if $\varphi \in C^\infty$ and $|x|^m D^k \varphi(x)$ is bounded in $\mathbf{R}^n, \forall m, k$. It is clear that $\mathcal{D} \subset \mathcal{S}$. The function $\exp(-x^2) \in \mathcal{S} - \mathcal{D}$. Another notion of convergence can be introduced in this space $\mathcal{S}$ of rapidly decreasing test functions:

$$\varphi_n \overset{S}{\to} \varphi \overset{\text{def}}{\equiv} \| |x|^m D^k \varphi_n - |x|^m D^k \varphi \|_\infty \to 0 , \quad n \to \infty , \quad \forall m, k .$$

Evidently, $\varphi_n \overset{D}{\to} \varphi \Rightarrow \varphi_n \overset{S}{\to} \varphi$. D is dense in S.

D.2 Concept of a Distribution or Generalized Function

Given a function $f \in L^1_{\text{loc}}$, it is clear that $\varphi \to \int f\varphi \, d^n x$ is a continuous linear map (with respect to the notion of convergence introduced above) of D in $\mathbb{C}$. Similarly, so is $\varphi \to \varphi(x_0)$ for fixed x_0. This latter cannot be realized as a L^1_{loc} function. Thus the concept of a continuous linear map of D into $\mathbb{C}$ generalizes the concept of a function. Such maps or functionals constitute the space D' of *distributions* on D:

$$T \in D' : \varphi \in D \to \langle T, \varphi \rangle \in \mathbb{C} \quad \text{such that} \tag{D.2}$$

$$\langle T, \alpha_1 \varphi_1 + \alpha_2 \varphi_2 \rangle = \alpha_1 \langle T, \varphi_1 \rangle + \alpha_2 \langle T, \varphi_2 \rangle , \quad \alpha_i \in \mathbb{C} ,$$
$$\varphi_n \overset{D}{\to} \varphi \Rightarrow \langle T, \varphi_n \rangle \to \langle T, \varphi \rangle . \tag{D.3}$$

The distributions on S, or *tempered distributions*, are defined in a completely analogous manner. The space S' of such distributions is the set of linear functionals on S, continuous in the sense of

$$\varphi_n \overset{S}{\to} \varphi \Rightarrow \langle T, \varphi_n \rangle \to \langle T, \varphi \rangle .$$

For $f \in L^1_{\text{loc}}$ to be a tempered distribution, it suffices that $f(x) = O(|x|^N)$, for some N, as $|x| \to \infty$; that is, that f increase at most polynomially at infinity. Finally, it is clear that $S' \subset D'$.

D.3 Operations with Distributions

It is evident that both D' and S' are linear spaces with the natural structure for linear combinations of linear functionals.

In the same fashion, if $\varrho(x) \in C^\infty$, and $\varphi_n \to \varphi$ (in D or S) $\Rightarrow \varrho\varphi_n \to \varrho\varphi$, it is clear that $\varrho T : \varphi \to \langle T, \varrho\varphi \rangle$ is an element of D' or S'. Such ϱ are called *multipliers* in this space. For D', all $\varrho \in C^\infty$ are multipliers. That is not true for S'. The multipliers ϱ of S' are the functions C^∞ with *slow growth*: $\forall m, \exists N_m$ such that $\| D^m \varrho / (1 + |x|^2)^{N_m} \|_\infty < \infty$.

Given a sequence $\varphi_n \to \varphi$ in D or in S, it is clear that $D^m \varphi_n \to D^m \varphi, \forall m$. Then

$$D^m T : \varphi \to \langle T, (-1)^{|m|} D^m \varphi \rangle \tag{D.4}$$

defines a distribution, called the mth *order derivative* of T. Every distribution, whether on D or S, is infinitely differentiable as a distribution.

Similarly, since the translation $\varphi(x) \to \varphi_a(x) \equiv \varphi(x - a)$ preserves convergence properties, the corresponding transform of T, namely $T_a : \varphi \to \langle T, \varphi_{-a} \rangle$ is also a distribution. Analogously, the image of T under parity $P : \varphi(x) \to \varphi_P(x) \equiv \varphi(-x)$, defined as $T_P : \varphi \to \langle T, \varphi_P \rangle$ is also a distribution.

All of these operations (linear combinations, product with a multiplier, differentiation, translation and parity) are *continuous* with respect to the following definition of convergence of distributions in $\mathcal{D}'$ (or $\mathcal{S}'$):

$$T_n \to T \overset{\text{def}}{\equiv} \langle T_n, \varphi \rangle \to \langle T, \varphi \rangle , \quad \forall \varphi \in \mathcal{D} \text{ (or } \mathcal{S}) .$$

Under this notion of convergence, $\mathcal{D}'$ and $\mathcal{S}'$ are complete and $\mathcal{S}'$ is dense in $\mathcal{D}'$.

D.4 Examples of Distributions

Other than the distributions which are L^1_{loc}, the most familiar of the tempered distributions is the Dirac "delta function":

$$\delta_{x_0} : \varphi \to \varphi(x_0) . \tag{D.5}$$

Frequently, distributions T are written as if they were functions $T(x)$ so that, symbolically

$$\langle T, \varphi \rangle = \int_{\mathbf{R}^n} d^n x \, T(x)\varphi(x) , \tag{D.6}$$

and its translated T_a by $T_a(x) = T(x - a)$. And thus, δ_{x_0} could be represented by $\delta(x - x_0)$, $\delta(x)$ being the symbolic representation of δ_0.

In one dimension, δ_{x_0} is the derivative of the Heaviside step function:

$$\delta_{x_0}(x) = \frac{d}{dx}\theta(x - x_0) , \quad \theta(x) \equiv \begin{cases} 1 , & x > 0 \\ 0 , & x < 0 , \end{cases} \tag{D.7}$$

as can be verified immediately. The delta function appears frequently in applications as the distribution limit of a sequence of functions. Thus, for example,

$$\delta(x) = \lim_{\lambda \to \infty} \sqrt{\frac{\lambda}{\pi}} e^{-\lambda x^2} = \lim_{\varepsilon \downarrow 0}(\pi i \varepsilon)^{-1/2} e^{ix^2/\varepsilon} = \lim_{\lambda \to \infty} \frac{\sin \lambda x}{\pi x} . \tag{D.8}$$

Another important tempered distribution is $\mathrm{PV}(1/x)$ ("principal value of $1/x$") which is defined to be

$$\left\langle \mathrm{PV}\frac{1}{x}, \varphi \right\rangle = \lim_{\varepsilon \downarrow 0} \int_{|x| \geq \varepsilon} dx \frac{\varphi(x)}{x} . \tag{D.9}$$

It is easy to see that

$$\mathrm{PV}\frac{1}{x} = \frac{d}{dx} \ln |x| . \tag{D.10}$$

On the other hand, if we choose the principal branch of the (complex valued) logarithm function:

$$\lim_{\varepsilon \downarrow 0} \ln(\varepsilon + ix) = \begin{cases} \ln|x| - i\frac{\pi}{2}, & x < 0 \\ \ln|x| + i\frac{\pi}{2}, & x > 0 \end{cases} = \ln|x| - i\frac{\pi}{2} + i\pi\theta(x) \qquad (D.11)$$

then differentiating (D.11) (a continuous operation for distributions) results in

$$\frac{1}{x - i0} \equiv \lim_{\varepsilon \downarrow 0} \frac{1}{x - i\varepsilon} = \mathrm{PV}\frac{1}{x} + i\pi\delta(x) , \qquad (D.12)$$

and complex conjugation gives

$$\frac{1}{x \pm i0} = \mathrm{PV}\frac{1}{x} \mp i\pi\delta(x) . \qquad (D.13)$$

In particular,

$$\frac{1}{x - i0} - \frac{1}{x + i0} = \lim_{\varepsilon \downarrow 0} \frac{2i\varepsilon}{x^2 + \varepsilon^2} = 2\pi i\delta(x) . \qquad (D.14)$$

Finally, if $x \to y = f(x)$ is a C^∞ one-to-one mapping of $\mathbf{R}$ onto $\mathbf{R}$, the formal change of integration variable suggests the definition

$$\langle T(f(\cdot)), \varphi(\cdot)\rangle \equiv \langle T, \varphi(f^{-1}(\cdot))|f^{-1}(\cdot)'|\rangle . \qquad (D.15)$$

In particular,

$$\delta(f(x)) = \frac{1}{|f'(x_0)|}\delta(x - x_0) , \quad f(x_0) = 0 ,$$

$$\delta(ax) = \frac{1}{|a|}\delta(x) , \quad a \text{ real } \neq 0 . \qquad (D.16)$$

When $x \to f(x)$ is not bijective, but all the zeros $x_1, \ldots, x_n$ of $f(x)$ are simple and finite in number, the first of equations (D.16) can be generalized to

$$\delta(f(x)) = \sum_{i=1}^{n} \frac{1}{|f'(x_i)|}\delta(x - x_i) , \qquad (D.17)$$

taking into account that the support of $\delta(f(\cdot))$ must be $\{x_1, \ldots, x_n\}$, and that locally, in the neighborhood of x_i, $x \to f(x)$ is a bijection. [Given a distribution T, the union of the open regions Ω such that $\langle T, \varphi\rangle = 0$ if $\mathrm{supp}\,\varphi \subset \Omega$ is called the *null set* of T. The complement of this null set is termed the *support* supp T of the *distribution T*.]

Thus, for example,

$$\delta(x^2 - a^2) = \frac{1}{2|a|}[\delta(x - a) + \delta(x + a)] , \qquad (D.18)$$

for all real, nonzero a.

D.5 Fourier Transformation

Given $f \in L^1(\mathbb{R}^n)$, it is known that its Fourier transform

$$(\mathcal{F}f)(k) \equiv \hat{f}(k) \equiv (2\pi)^{-n/2} \int d^n x \, f(x) e^{-ik \cdot x} , \tag{D.19}$$

where $k \cdot x = k_1 x_1 + \ldots + k_n x_n$, satisfies

a) $\hat{f}$ is a bounded, uniformly continuous function;
b) given $f, g \in L^1, \mathcal{F}f = \mathcal{F}g \Rightarrow f = g$ almost everywhere;
c) $f, g \in L^1 \Rightarrow \int d^n x \, f(x)\hat{g}(x) = \int d^n x \, g(x)\hat{f}(x)$. [Parseval]   (D.20)

In the linear subspace $S \subset L^1(\mathbb{R}^n)$ of test functions with fast fall-off, $\mathcal{F}$ has the following properties:

a) $\mathcal{F}S = S$;
b) $\mathcal{F}$ is a continuous one-to-one linear mapping of S onto S, as is $\mathcal{F}^{-1}$, whose explicit expression is

$$f(x) = (\mathcal{F}^{-1}\hat{f})(x) \equiv (2\pi)^{-n/2} \int d^n k \, \hat{f}(k) e^{ik \cdot x} ; \tag{D.21}$$

c) $(\mathcal{F}(D^r \varphi))(k) = (ik)^r(\mathcal{F}\varphi)(k)$,
$(\mathcal{F}((-ix)^r \varphi))(k) = (D^r \mathcal{F}\varphi)(k)$,
$(\mathcal{F}\varphi_a)(k) = e^{-ia \cdot k}(\mathcal{F}\varphi)(k)$,
$(\mathcal{F}(\mathcal{F}\varphi))(x) = \varphi(-x)$, $\mathcal{F}^4 = $ identity   (D.22)

where $(b)^r = b_1^{r_1} b_2^{r_2} \ldots b_n^{r_n}$.

The Parseval equality (D.20) with $f \in L^1, g \in S$, suggests the definition of the Fourier transform $\mathcal{F}T$ of $T \in S'$:

$$\langle \mathcal{F}T, \varphi \rangle \equiv \langle T, \mathcal{F}\varphi \rangle . \tag{D.23}$$

The continuity of $\mathcal{F}$ in S indicates that $\mathcal{F}T \in S'$. $\mathcal{F}$ is also a linear, continuous and bijective map from S' onto S', and its inverse

$$\langle \mathcal{F}^{-1}T, \varphi \rangle = \langle T, \mathcal{F}^{-1}\varphi \rangle \tag{D.24}$$

is also continuous. The formulae (D.22) continue to hold upon replacement of φ by T.

Examples:

1) $(\mathcal{F}\delta_{x_0})(k) = (2\pi)^{-n/2} e^{-ix_0 \cdot k}$. \tag{D.25}

2) $(\mathcal{F}D^r \delta_{x_0})(k) = (2\pi)^{-n/2} (ik)^r e^{-ix_0 \cdot k}$. \tag{D.26}

3) $(\mathcal{F}e^{ik_0 \cdot x})(k) = (2\pi)^{n/2} \delta_{k_0}(k)$. \tag{D.27}

4) $(\mathcal{F}((-ix)^r e^{ik_0 \cdot x}))(k) = (2\pi)^{n/2} D^r \delta_{k_0}(k)$.   (D.28)

5) $(\mathcal{F}\theta)(k) = -(2\pi)^{-1/2} i \dfrac{1}{k - i0}$.   (D.29)

6) $\left(\mathcal{F} \dfrac{1}{x + i0} \right)(k) = -(2\pi)^{1/2} i\theta(k)$.   (D.30)

The linear subspace $L^2(\mathbf{R}^n) \subset \mathcal{S}'$ is also stable under $\mathcal{F}$, and in it the Fourier transformation $\mathcal{F}$ is a *unitary operator*.

Finally, $\mathcal{F}$ relates the *convolution product* to the ordinary product: given $T_1 \in \mathcal{D}'(\mathbf{R}^{n_1}), T_2 \in \mathcal{D}'(\mathbf{R}^{n_2})$, there exists a distribution $T_1 \times T_2 \in \mathcal{D}'(\mathbf{R}^{n_1+n_2})$ such that, if $\varphi(x_1, x_2) = \varphi_1(x_1)\varphi_2(x_2), \varphi_{1,2} \in \mathcal{D}(\mathbf{R}^{n_{1,2}})$, then

$$\langle T_1 \times T_2, \varphi \rangle = \langle T_1, \varphi_1 \rangle \langle T_2, \varphi_2 \rangle .$$ (D.31)

Similar considerations apply for tempered distributions. Thus, for example,

$$\delta(x_1, x_2) = \delta(x_1) \times \delta(x_2) .$$ (D.32)

(The symbol $\times$ is frequently omitted). Given $f, g \in \mathcal{D}$, their convolution is defined by

$$(f * g)(x) \equiv \int d^n y \, f(x - y)g(y) .$$ (D.33)

It is easy to see that $f * g = g * f \in \mathcal{D}$, and that, given $\varphi \in \mathcal{D}$,

$$\langle f * g, \varphi \rangle = \langle f(x) \times g(y), \varphi(x + y) \rangle ,$$ (D.34)

a relation which suggests the definition, for $T, S \in \mathcal{D}'(\mathbf{R}^n)$:

$$\langle T * S, \varphi \rangle \equiv \langle T(x) \times S(y), \varphi(x + y) \rangle .$$ (D.35)

The r.h.s. is not always defined, since $\varphi(x+y) \notin \mathcal{D}(\mathbf{R}^{2n})$ unless $\varphi = 0$. It suffices, nevertheless, for T or S to have compact support in order for (D.35) to define $T * S \in \mathcal{D}'(\mathbf{R}^n)$. Similarly, it suffices that supp T and supp S both be "bounded from the left", or both be "bounded from the right".

For example

$$(D^r \delta_{x_0}) * T = D^r T_{x_0} , \quad \forall T \in \mathcal{D}' .$$ (D.36)

The convolution product, when it exists, is abelian, and depends linearly on its arguments. Associativity is not guaranteed:

$$1 * (D^1 \delta_0 * \theta) \neq (1 * D^1 \delta_0) * \theta .$$ (D.37)

It is ensured if all supports are compact, or if they are all bounded from the left (from the right).

Under differentiation,

$$D^t(T * S) = (D^r T) * (D^s S) , \tag{D.38}$$

so long as $r + s = t$.

Given $T, S \in \mathcal{S}'$, if one of these distributions (say S) has compact support, then

$$\mathcal{F}(T * S) = (\mathcal{F}T)(\mathcal{F}S) , \tag{D.39}$$

where the r.h.s. defines a tempered distribution since $\mathcal{F}S \in C^\infty$ and is a multiplier for $\mathcal{S}'$. [Actually, for (D.39) to hold, it suffices that either T or S be a distribution with "fast decrease".]

Appendix E: On the Measurement Problem in Quantum Mechanics

The postulates IV and V in Chap. 2 assign different changes to the states of a physical system, depending upon whether one is dealing with a measurement process or an evolution between consecutive measurements [see (2.106) and (2.134)]. The reconciliation of both of these types of changes constitutes the central problem of measurement in quantum mechanics. In this appendix we will sketch, albeit superficially, some of the solutions which have been proposed for this problem. We can anticipate that none of them have achieved general acceptance. There is a substantial literature on this subject; as basic sources we recommend [NE 32], [LB 39], [JA 68], [ES 71], [ES 71a], [JA 74], [PI 76], [WZ 83], and [KA 84].

E.1 Types of Evolution

Quantum mechanics postulates a *deterministic* (or *causal*) time evolution for states of isolated physical systems, or systems in interaction with external classical systems which they do not influence appreciably:

$$\varrho(t_0) \rightarrow \varrho(t) = U(t, t_0)\varrho(t_0)U^{-1}(t, t_0) . \tag{E.1}$$

In this evolution, a state which is known at an instant t_0 is uniquely determined at any other time t. In particular, (E.1) indicates that $\varrho(t)$ is unitarily equivalent to $\varrho(t_0)$, and thus it represents a pure state if the latter is pure.

On the other hand, the process of measurement (an interaction of the system being observed with an apparatus which reacts to its presence) of an observable A at the instant t_0, if it is an ideal measurement which selects values of A in a set Δ, is postulated to produce the change

$$\varrho(t_0) \rightarrow \varrho_{A,\Delta}(t_0) \equiv \frac{1}{\mathrm{Tr}\{\varrho(t_0)E_A(\Delta)\}} \sum_{a \in \Delta} E_A(\{a\})\varrho(t_0)E_A(\{a\}) , \tag{E.2}$$

known as the *collapse of the state* or *reduction of the wave packet*. In contrast to (E.1), this new change is *irreversible* [in general, $\varrho_{A,\Delta}(t_0)$ does not determine $\varrho(t_0)$] and *stochastic*, and can transform a $\varrho(t_0)$ which is pure into a state $\varrho_{A,\Delta}(t_0)$ which is strictly mixed.

With the definition

$$s(\varrho) \equiv -\mathrm{Tr}\{\varrho \ln \varrho\} \in [0, \infty] \qquad\qquad (E.3)$$

for the *entropy* of the state [if the operator $\varrho \ln \varrho$, bounded and compact, does not belong to the trace class, $s(\varrho)$ is taken to be ∞], the concavity of $-x \ln x$ allows us to prove [RU 71], [RO 71] that the entropy of a state is a concave function (i.e., it increases with the incoherence of the state)

$$s(\lambda \varrho_1 + (1 - \lambda)\varrho_2) \geq \lambda s(\varrho_1) + (1 - \lambda)s(\varrho_2), \quad 0 \leq \lambda \leq 1, \qquad (E.4)$$

and it grows under the change $\varrho \to \varrho_A \equiv \varrho_{A,\mathbf{R}}$:

$$s(\varrho_A) \geq s(\varrho). \qquad\qquad (E.5)$$

[Since the measurement of A which causes $\varrho \to \varrho_A$ extracts information, it necessarily provokes greater disorder in the system being observed, resulting in an increase of its entropy.] On the other hand, under the reversible and deterministic evolution (E.1), the entropy remains constant:

$$s(\varrho(t)) = s(\varrho(t_0)). \qquad\qquad (E.6)$$

E.2 Sketch of a Measurement Process

In order to make evident the difficulties generated by the comparison of (E.1) and (E.2), we must take into account that the measurement of an observable A for a physical system S requires the use of another physical system, namely the measurement apparatus A, and their interaction produces two effects: S acts on A, altering the state of A and producing in the apparatus the effect necessary to inform us of the value of the observable; but also, A acts on S, giving rise to the change (E.2) in its state. Let $\mathcal{H}_S$ and $\mathcal{H}_A$ be the Hilbert spaces associated with S and A. We will assume a discrete spectrum for A (in practice we measure only discretised approximations to any observable) with distinct eigenvalues $a_r, r = 1, 2, \ldots$, and denote the corresponding eigensubspaces in $\mathcal{H}_S$ by the symbol $\mathcal{E}_r$. The apparatus A should have a pointer or other indicator having positions $f_r, r = 1, 2, \ldots$, and such that, before the measurement, A should be in a neutral or "ready" state, and after the measurement the pointer should be at f_r if the observable A has the value a_r on the system S. If we denote by $\mathcal{F}_r$ those orthogonal subspaces of $\mathcal{H}_A$ which contain the pure states of A in which the pointer occupies the positions f_r, it is clear that the interaction between S and A should lead to a time evolution operator U for the entire system $S + A$ (assumed to be isolated) during the course of the measurement such that

$$U(\mathcal{E}_r \otimes \mathcal{F}_0) \subset \mathcal{H}_S \otimes \mathcal{F}_r, \qquad\qquad (E.7)$$

where $\mathcal{F}_0$ indicates the subspace of neutral pure states of the apparatus. In particular, when $U(\mathcal{E}_r \otimes \mathcal{F}_0) \subset \mathcal{E}_r \otimes \mathcal{F}_r$, the measurement will be of the first kind.

Example [EM 72]: Let $\mathcal{H}_S = \mathbb{C}^2$ (for a system of spin $1/2$ without any translational motion), $\mathcal{H}_A = L^2(\mathbb{R})$ (the apparatus consisting of, for example, a simple particle in one dimension), and an interaction between S and A such that the total Hamiltonian for $S + A$ is $H = \sigma_z \otimes (-\mathrm{i}d/dx)$. Clearly

$$|\uparrow\rangle \otimes \psi(x) \overset{U(t,0)}{\longrightarrow} |\uparrow\rangle \otimes \psi(x - t) ,$$
$$|\downarrow\rangle \otimes \psi(x) \overset{U(t,0)}{\longrightarrow} |\downarrow\rangle \otimes \psi(x + t) ,$$

$$(E.8)$$

and, thus, if $\mathcal{F}_0 \equiv \{\psi \in L^2 : \psi(x) = 0, |x| > a/2\}, \mathcal{F}_\pm = \{\psi \in L^2 : \psi(x) = 0, \pm x < 0\}, \mathcal{E}_+ = \{\alpha|\uparrow\rangle, \alpha \in \mathbb{C}\}, \mathcal{E}_- = \{\alpha|\downarrow\rangle, \alpha \in \mathbb{C}\}$, we have $U(\mathcal{E}_r \otimes \mathcal{F}_0) \subset \mathcal{E}_r \otimes \mathcal{F}_r$ for $U = U(t,0), t > a/2$. This ideal apparatus will measure the spin, correlating $|\uparrow\rangle$ with the position of the particle to the right of the origin, and $|\downarrow\rangle$ to the left. It is therefore a highly idealized version of a Stern–Gerlach measurement.

We simplify the above description (which was already schematic and superficial) of the measurement process even further by assuming that the subspaces $\mathcal{E}_r, \mathcal{F}_0$, and $\mathcal{F}_r$ are one-dimensional with basis normalized kets $|a_r\rangle, |f_0\rangle$, and $|f_r\rangle$, respectively. Since U is *unitary*, then for a measurement of the first kind

$$|a_r\rangle|f_0\rangle \overset{U}{\longrightarrow} |a_r\rangle|f_r\rangle ,$$

$$(E.9)$$

and the *linear* character of U implies that for any other normalized ket $\sum \alpha_r |a_r\rangle$ we will also have

$$(\textstyle\sum \alpha_r |a_r\rangle)|f_0\rangle \overset{U}{\longrightarrow} \sum \alpha_r |a_r\rangle|f_r\rangle .$$

$$(E.10)$$

The final state $|\psi\rangle \equiv \sum \alpha_r |a_r\rangle|f_r\rangle$ of $S + A$ is a pure state, since it is so initially. But its *reduction* to the system S (i.e., the state ϱ_S of S such that $\mathrm{Tr}\{\varrho_S B\} = \mathrm{Tr}\{|\psi\rangle\langle\psi|B \otimes 1_A\}$ for all observables B of S) is

$$|\psi\rangle\langle\psi| \overset{\text{reduction to } S}{\longrightarrow} \varrho_S = \sum |\alpha_r|^2 |a_r\rangle\langle a_r| ,$$

$$(E.11)$$

which coincides with the collapsed state ϱ_A of $\varrho \equiv (\sum \alpha_r |a_r\rangle)(\sum \alpha_s^* \langle a_s|)$ after the measurement of A. We thus see the "reason" for the collapse of ϱ to ϱ_A: as a consequence of the interaction with A, which establishes a strong correlation between S and A, the state of the subsystem S ceases to be pure. The relative phases of α_r and α_s are no longer observables in the subsystem S, since $|a_r\rangle$ and $|a_s\rangle$ are correlated to $|f_r\rangle$ and $|f_s\rangle$, which are orthogonal states of the other subsystem A.

But have we really *explained* the collapse? Certainly not, since we have forgotten something essential to the measurement: during each measurement on S, the pointer in A takes on a definite position f_r, while in the final state $\sum \alpha_r |a_r\rangle|f_r\rangle$ its position would in general be indeterminate. In other words, after the measurement we "see" that the state of $S + A$ is $\sum |\alpha_r|^2 |a_r f_r\rangle \langle a_r f_r|$ and not $\sum \alpha_r |a_r\rangle |f_r\rangle$. What do we mean here by "see"? This requires the observation of the position of the pointer, and thus another apparatus A', for example, a light source and the eyes of the observer. With this, the total system becomes $(S + A) + A'$. Proceeding as before, we would have

$$(\sum \alpha_r |a_r\rangle)|f_0\rangle|f_0'\rangle \xrightarrow{\text{interaction}} \sum \alpha_r |a_r\rangle |f_r\rangle |f_r'\rangle \,, \tag{E.12}$$

with the reduction to $S + A$ given by $\sum |\alpha_r|^2 |a_r f_r\rangle \langle a_r f_r|$, which in turn reduces to S as $\sum |\alpha_r|^2 |a_r\rangle \langle a_r|$. But now we have the same problem with $S + A + A'$: its state, on the one hand, is $\sum \alpha_r |a_r\rangle |f_r\rangle |f_r'\rangle$, and, on the other, it appears to be $\sum |\alpha_r|^2 |a_r f_r f_r'\rangle \langle a_r f_r f_r'|$. We could argue that our analysis is merely at the level of the eye retina, and that we have not yet "become aware" of the result, since the visual stimulation must still traverse the nervous system, arrive at the brain, …, and have the observer "become conscious" of the signal as a last step in an infinite regression. It is in this last step of becoming aware, according to von Neumann, and London and Bauer, that the collapse of the wave function takes place. This analysis has at least one positive aspect, that of establishing the objective character of the measurement process: the final state of S is independent of the choice of boundary between what is considered to be the observed system and the apparatus for observation.

For many physicists, this solution to the problem of the collapse is inadmissible, referring as it does, to something – the "consciousness" – which does not appear to be analyzable in physical terms. If it were, the collapse would remain unexplained: it suffices to introduce the observer into the system. If the mind of the observer were also subject to the laws of quantum mechanics, and we considered $S + A + \ldots + A^\infty$, the state of the total system would be a linear superposition and not the mixed state resulting from the measurement. We could consider yet another observer $\bar{A}^\infty$, a "friend" of A^∞, who in turn observes $S + A + \ldots + A^\infty$. His act of observation could consist of asking A^∞ "What have you observed?". A^∞ would respond "a_r with probability $|\alpha_r|^2$". According to the above, this verbal communication between A^∞ and $\bar{A}^\infty$ would then be the interaction responsible for the collapse of $S + A + \ldots + A^\infty$. But the second observer could have then asked "What were you observing before I asked you the first question?" and the inevitable response would be "I already told you: a_r with probability $|\alpha_r|^2$". Said another way, before the first interaction between A^∞ and $\bar{A}^\infty$, the first observer was already aware of the result and therefore the state of $S + A + \ldots + A^\infty$ was already a mixed one and not the pure state result of a deterministic evolution.

Summarizing, what seems to emerge from the preceding is that either we abandon attempts to explain the collapse, accepting it as a postulate, even knowing that it may be irreconciliable with the deterministic evolution of closed systems, or we accept the existence of entities ("consciousness") whose actions are outside the framework of applicability of the laws of physics. Herein lies the grand dilemma which confronts the theory of measurement.

E.3 Solutions to the Dilemma

The problem confronting the theory of the measurement process, which, when extrapolated to situations with living beings or consciousnesses, leads to such paradoxical situations as that of "Schrödinger's cat" [JA 68], [LE 84] or that of "Wigner's friend" ($\mathcal{A}^\infty$ versus $\bar{\mathcal{A}}^\infty$), has instigated some interesting controversies and given rise to numerous novel contributions, by physicists as well as philosophers. We will limit ourselves here to listing and discussing some of the arguments proposed to avoid this dilemma.

1) Probably the first consideration which occurs to everyone is that the measurement process has been simplified to an extreme. In fact, a measuring apparatus is microscopically very complicated and it is illusory to think that we can describe its state by a vector $|f_0\rangle$ or $|f_r\rangle$. It would be more appropriate to describe it through a density operator, which can reflect our ignorance about its initial state. Thus it has been maintained [HE 62] that this initial indeterminacy could give rise to $S + \mathcal{A}$ ending up in a "final reading" state without the necessity of inspection by an observer. Said in another way, the state of $S + \mathcal{A}$, after the interaction of S with $\mathcal{A}$, would be a mixed state in which the pointer takes on definite positions, albeit statistically distributed, with weights $|\alpha_r|^2$. *Wigner* [WI 63] was the first to prove that this idea was not feasible; later, *Stein* and *Shimony* [ES 71] gave a more general proof of the fact that no measurement, not even one of the second kind, is possible in which the linear laws of quantum mechanics can lead to definite positions (even if unknowable) of the pointer of the apparatus without the necessity for an observer.

2) Perhaps the problem has arisen because we have required mathematically discrete and discernable positions f_r, of the pointer, when it would suffice that the pointer be found "preponderantly" in a neighborhood of f_r. This also fails to eliminate the dilemma, as *d'Espagnat* [ES 71a] has proven.

3) Is it possible instead that the laws of quantum mechanics apply only to few-body systems, as is the case with the applicability of classical mechanics when the number of particles becomes very large? If this were the case, we might still not know the laws of interaction between microscopic and macroscopic systems, and these unknown laws could be the cause of the introduction of new statistical elements which might explain the collapse of the wave function during measurement. This is the viewpoint, at this time not refutable, taken by *Ludwig* [LU 61], which postpones the solution to our dilemma to experimental determination of these hypothetical physical laws for "mesoscopic" systems.

4) *Everett* [EV 57], [WG 73], has proposed an interpretation of quantum mechanics based on the hypothesis of a simultaneous realization of all possible alternatives in each measurement process, as a result of a supposed coexistence of multiple editions of parallel and mutually unobservable worlds which emanate branch-like from these processes. The defenders of this interpretation, who assign a state vector undergoing deterministic evolution to the universe as a whole, support the view that the mathematical formalism of quantum theory is capable

of structuring its own statistical interpretation and need not be imposed a priori. Even though the theoretical framework of this theory may be logically consistent, many find it difficult to accept the basic principle of an infinite multiplicity of these real worlds which the universe would be continuously shedding. For a critical discussion see [ES 71a].

5) Finally, the consideration of the classical and/or macroscopic nature of a measurement apparatus can lead to a *practical* reconciliation between a deterministic evolution of $S + A$ (which originates, for instance, the state $\sum \alpha_r |a_r\rangle |f_r\rangle$) and a stochastic evolution associated with a measurement (the state $\sum |\alpha_r|^2 |a_r f_r\rangle \langle a_r f_r|$). The reconciliation takes the following form: even though both states are in principle distinguishable from each other by means of some observable, in fact they are equivalent under a certain abelian set of observables (generated in our example by the projections $|a_r\rangle\langle a_r| \otimes 1_A$ and $1_S \otimes |f_r\rangle\langle f_r|$) which are exactly those observables we need to assign to $S + A$ as a classical system for the measurement to have an objective character [JA 68]. On the other hand, if we take into account the macroscopic nature of the measurement instruments (even though it should be noted that sometimes the essential part of the measurement process is completed at the microscopic level; for example the spin reversal of a magnetic ion in a crystal which is produced by the passage of a neutron, the role of the rest of the apparatus being limited to classically amplifying the signal up to the level of human perception), the relative phases among the various amplitudes α_r will depend on an enormous number of unknown microscopic parameters of the instrument, and therefore not measurable in practice. Much more elaborate considerations on the role of the classical and/or macroscopic character of the measurement apparatus and a critique of these arguments can be found in [JA 68], [DL 62], [ES 71] and [ES 71a]. (Some simple models [HE 72] in the framework of quantum systems with infinite numbers of degrees of freedom, for which the collapse is rigorously achieved with respect to all local observables in the $t \to \infty$ limit using deterministic evolution, also offer no fundamental solution to the problem of the reduction of a wave packet [BE 75].)

After this brief, superficial and in no way exhaustive presentation of the windmill of arguments and counter-arguments concerning the resolution of the problem posed by the quantum theory of measurement, something clearly shows up: despite the indubitable and impressive successes which quantum mechanics enjoys, we still do not know exactly how and when the evolution rule given by the Schrödinger equation gives way to the collapse, and therefore we lack a complete and unambiguous formulation of our most fundamental physical theory.

For a critical discussion of these issues, see [WZ 83] and [KA 84].

Appendix F: Models for Hidden Variables (A Summary)

The objective of this appendix is to introduce the interested reader to some of the alternative formulations of quantum mechanics tending, among other things, to reestablish in physics the determinism which the probabilistic interpretation of the Copenhagen school has displaced. As general sources we recommend [NE 32], [JA 66], [CF 70], [ES 71], [BE 73a], [JA 74], [FM 76], [WZ 83], and [KA 84], among others. The conclusions, as we shall see at the end, are at present not very optimistic for the defenders of determinism.

F.1 Motivation

At the 1927 Solvay Conference, Born and Heisenberg claimed: "Quantum mechanics leads to accurate results concerning average values, but gives no information as to the details of an individual event. The determinism, which so far has been regarded as the basis of the exact sciences, has to be given up." They ended with the following provocative affirmation: "We maintain that quantum mechanics is a complete theory; its basic physical and mathematical hypotheses are not further susceptible of modifications." Einstein (who the year before had written to his friend Ehrenfest: "I look upon quantum mechanics with admiration and suspicion.") was present. During this Solvay Conference, in a scientific debate among giants, he was defeated, but not persuaded, by Bohr. He had to admit that quantum mechanics, assisted by the principle of complementarity, formed a consistent logical edifice, but accepting the statistical basis of the quantum theory as a complete description of physical reality was contrary to his scientific instincts. The collapse of the virtual presence of a particle in a region of space occupied by its wave function ψ at the instant of detection implies action at a distance, which Einstein could only comprehend if ψ were merely a catalog of probabilities, describing not physical reality for an individual system, but instead the results of possible measurements on a statistical ensemble.

Half a century has now passed, and physicists are still playing with two sets of cards for describing natural phenomena: for the *classical world*, an *objective description*; for the *quantum world*, a *subjective one*. Where does one start and the other end? A substantial majority of us believe that measurement apparatus, counters, scales, etc., are all in the classical category. Some set the boundary as far out as the actual cognitive act; others have different boundary markers.

The majority prefer not to think about it. In fact, what difference does it make? It is quantitatively irrelevant, given the difference in scales between the two worlds. But even if minute, the differences are not null, and for that reason the displacement of the boundary between two descriptions is of limited validity. This brings us to consider the provisional character of this double description, and justifies thinking about its possible evolution. How? Where to?

It may still be worthwhile to search for the exact line of demarcation between classical and quantum regimes. If we find it in the middle of the road, we need to become used to living in two worlds simultaneously. This offends many. Thus, one feels obliged to adopt the extreme positions. But if everything is quantal, how can we imagine ourselves discoursing about a world without objective events to correlate, even if only at the level of our own consciousness? It is by contrast easier to see ourselves in a world in which classical reasoning is the rule. According to *Bell* [ES 71], this is possibly the *principal motivation* for the study of *hidden variables* (HV).

Another possible motivation is based on a fundamental, almost religious belief in a deterministic nature, in which everything is determined by initial conditions, even though our own recognition of these may not be as precise as to know the causes of events. It is thus assumed that the erratic nature of the quantum fluctuations could be made deterministic by a suitable choice of some HV, hidden because we only suspect their existence, but which lie outside our control. From a practical point of view, such variables would be useless, but they would satisfy an *emotional need* which permits belief in predetermining causes, even if unknown. We would thus be dealing with a *cryptodeterministic* attitude. There exists also an opposing faction, moved likewise by impulses of a religious nature, who do not believe in determinism, and view the occurrence of chance in quantum mechanics as a manifestation of the omnipresence and omnipotence of a Creator. Things which are not explainable deterministically are pure "acts of God". In this debate rational thought seems of little help, as indicated by *Belinfante* [BE 73a].

Polemical reasons are also responsible for the existence of scientists worried by this problem. These latter reject the "brainwashing" by the Copenhagen school, who by fiat deny the existence of HV which determine the course of events for individual systems. These people do not accept royal decrees, and maintain that the establishment should point out the difficulties with HV rather than simply ignore their existence.

Aside from philosophical, religious or polemical motivations mentioned above, there exists a more concrete motivation, based on the peculiar character of some particular quantum predictions, which practically cries out for an interpretation based on HV. This is the famous *Einstein–Podolsky–Rosen* (EPR) *argument* [ER 35], which was advanced to refute the completeness of quantum mechanics alleged by the Copenhagen school, and which some label as a *paradox*, as if it had a solution, or as a *fallacy*, when in fact it is a true *theorem*, paradoxical only in that its conclusion is unexpected:

The following two alternatives are incompatible:

1) (*Completeness*): ψ represents a complete, exhaustive description of an individual system.
2) (*Local action*): The real physical conditions of isolated subsystems, space-like separated, are independent of each other.

It is worth our while to summarize the EPR argument, in *Bohm*'s version (EPRB) [BA 57]: consider an unstable state of two distinguishable spin $1/2$ (in units such that $\hbar = 1$) fermions initially in a singlet configuration, which decays, the spin being conserved in the process. (Such a system might be an H atom which is being photo-dissociated in an E1 transition.) The pair separates and one of the two fermions, call it the first one, traverses a Stern–Gerlach apparatus (S-G), I_1, its wave function collapsing at a point on a screen and thus being forced to make a quantal binary decision. If, for example, the result corresponds to $s_1 \cdot \hat{n}_1 = +1/2$ ($\hat{n}_1 \equiv$ unit vector along the I_1-axis), quantum mechanics predicts that the result of a measurement of $s_2 \cdot \hat{n}_2$ for fermion 2, carried out immediately afterward with an S-G aligned along I_2, is certain to be $-1/2$ if $\hat{n}_1 = \hat{n}_2$. This foreknowledge is obtained without any local interaction with subsystem 2, without allowance for time for the information as to how subsystem 1 has responded to a measurement of its spin to reach 2. For Einstein, this endows fermion 2 with an *attribute of physical reality*: "its spin along $\hat{n}_2$ is $-1/2$", not only after the measurement on 1, but also immediately *before* it. And since quantum mechanics only assigns a probability of $1/2$ to the latter, it is obvious that if we adopt (2), we must reject (1).

In his reply, Bohr emphasized the essential importance which the process of measurement necessarily has on the conditions underlying the very definition of physical quantities. These conditions are inherent in any quantity to which we wish to attribute physical reality. In simple terms for our example, a measurement of I_1, performed on fermion 1, is also a measurement on $1 + 2$, and as a consequence on fermion 2. But this instantaneous non-causal influence was distasteful to Einstein, the architect of special relativity, who preferred the rejection of (1) and affirmation of (2).

EPR nevertheless do not suggest how one should "complete" the quantum theory to salvage the principle of *local action*. It is also strange that they should have ignored the 1932 theorem by *von Neumann* [NE 32], according to which HV do not exist in the following sense: the stochastic aspects of quantum mechanics *cannot* be reproduced under the hypothesis that there exist some parameters analogous to positions and velocities of the molecules of a gas, whose knowledge determines a *micro-state* of the system, *dispersion-free* for *every observable*, in terms of which an ordinary pure state ψ of the system is merely an incoherently superposed *macro-state* (similar to a thermodynamic one specified by variables such as p, V and T). It is probable that the formal reasoning of von Neumann had little overlap with the physical intuition of Einstein, and because of this the latter did not notice how damaging this theorem was to the program of the completion of quantum mechanics. Note that the simplest solution to the problem of ensuring local interactions consists of supposing that the individual responses of the fermions 1 and 2 to the magnets I_1 and I_2 in the *Gedankenexperiment* of EPRB are completely determined by the HV impressed in these fermions, and that the correlation between these responses is not causally propagated, but it was born when the fermions were freed.

F.2 Impossibility Theorems

If von Neumann has proven that HV cannot exist, why should we pursue this avenue of endeavor? What explains the fact that despite that rotund statement, many recognized and respected scientists, such as de Broglie, Bohm, Wiener and others have developed and elaborated on theories based on HV? In fact, the authority and respect for von Neumann was such that it halted all interesting developments on this problem for twenty years. The first critique of von Neumann's work on physical grounds was that of *Bohm* in 1952 [BO 52], where he indicated there was something unsatisfactory in the frame which that great mathematical physicist had laid for this problem. But it was not until 1966 that *Bell* [BE 66] was able to put his finger on the problem by identifying the specific premise of von Neumann's theorem which is a priori unacceptable in any realistic theory of hidden variables. We recall that observables in the ordinary context of quantum mechanics can be identified with self-adjoint operators in a complex, separable Hilbert space, and that the conventional quantum states are just positive linear functionals on the bounded observables. Von Neumann required this same property of the micro-states, i.e., of dispersion-free states. His theorem then follows from this hypothesis:

There exist no dispersion-free states. Equivalently, *there exist no microstates*; that is, *there exist no hidden variable descriptions.*

Indeed, let ξ be dispersion-free, and A be a bounded observable. We must have

$$\langle (A - \langle A \rangle_\xi)^2 \rangle_\xi = 0 \ . \tag{F.1}$$

If

$$(A - \langle A \rangle_\xi)^2 \geq \alpha > 0 \ , \quad \text{then} \tag{F.2}$$

$$\langle (A - \langle A \rangle_\xi)^2 \rangle_\xi \geq \langle \alpha \rangle_\xi = \alpha > 0 \ . \tag{F.3}$$

There being thus no such α, it follows that $\langle A \rangle_\xi \in \sigma(A)$. Since the existence of $\lambda_1 \in \sigma(A_1)$ and $\lambda_2 \in \sigma(A_2)$ does not in general imply that $\lambda_1 + \lambda_2 \in \sigma(A_1 + A_2)$, the theorem is proved.

However, does the measurement of $A_1 + A_2$ *not* require an experimental apparatus which may be incompatible with those needed for A_1 and A_2 separately? Paradoxically, von Neumann himself was aware of this, as indicated by his example of the hydrogen atom: the kinetic energy is determined if we measure the momentum of the electron, the potential energy upon measurement of the position. But the total energy requires a spectroscopic measurement. There is no a priori reason to require the linearity of the expectation values in this case. That such linearity holds experimentally on physically realizable macro-states is very surprising and in no way trivial and foreseeable.

Bell did not limit himself to a critique of von Neumann's impossibility theorem, but pointed out physically unacceptable premises in other more sophisticated theorems having the same negative conclusions for hidden variables: those of *Jauch–Piron* [JP 63] and *Gleason* [GL 57], for example. An important consequence of this analysis is the clarification of which conditions *cannot* be required of physical micro-states. We have already indicated one of these: the *linearity* of the expectation values of observables. Another less obvious one which also cannot be maintained as part of the price for salvaging determinism is the following: the value (= expectation value) $\langle T \rangle_\xi$ of the observable T on a micro-state ξ depends only on the observable T, and not on the particular apparatus used for its measurement. That this is not a supportable proposition is evidenced by the following example.

Consider a system with angular momentum $J = 1$. For this system, the operators $T_i \equiv 1 - J_i^2, i = 1, 2, 3$ corresponding to an orthonormal triad in $\mathbf{R}^3$, are orthogonal, pairwise commuting projections of rank 1 and which satisfy $T_1 + T_2 + T_3 = 1$. $\{T_1, T_2, T_3\}$ being abelian, a single apparatus $\bar{\mathcal{T}}$ can simultaneously measure all three. (For example, such a measurement can be made on an atom of spin 1 by immersing it in an octahedral electric field.) In other words, there exists in principle an apparatus $\bar{\mathcal{T}}$ which measures an observable $\bar{T}$ of which T_1, T_2, T_3 are functions, $T_i = f_i(\bar{T})$. Therefore the values $\langle T_i \rangle_{\xi, \bar{\mathcal{T}}}$ of T_i in ξ, measured with $\bar{\mathcal{T}}$, satisfy

$$\langle T_i \rangle_{\xi, \bar{\mathcal{T}}} = \langle f_i(\bar{T}) \rangle_{\xi, \bar{\mathcal{T}}} = f_i(\langle \bar{T} \rangle_{\xi, \bar{\mathcal{T}}}) , \tag{F.4}$$

the last equality being justified because ξ is dispersion-free. Then

$$\langle T_1 \rangle_{\xi, \bar{\mathcal{T}}} + \langle T_2 \rangle_{\xi, \bar{\mathcal{T}}} + \langle T_3 \rangle_{\xi, \bar{\mathcal{T}}} = (f_1 + f_2 + f_3)(\langle \bar{T} \rangle_{\xi, \bar{\mathcal{T}}}) = \langle 1 \rangle_{\xi, \bar{\mathcal{T}}} = 1 . \tag{F.5}$$

As a consequence, if the value of T_1 in ξ did not depend on $\bar{\mathcal{T}}, \langle T_1 \rangle_\xi = 1 \Rightarrow \langle T \rangle_\xi = 0, \forall$ projections $T \perp T_1$ of rank 1. Is this possible? Can one decompose the spherical surface S_2 into two zones R and N such that each one will be symmetric with respect to the origin and such that if $P \in R$, the equator $P_\perp$ corresponding to the pole P lies in N? *Kochen* and *Specker* [KS 67] showed that such a decomposition is not possible by strategically locating one hundred and seventeen points in S_2. (For more "economical" arguments see [BE 73a], [GA 76].)

Given a bounded observable T, we will denote by $\mathcal{T}$ a maximal abelian algebra (i.e., the set of all bounded operators which are functions of a complete set of compatible observables) which contains T. Let $\bar{T}(\bar{T}')$ be a generator of $\mathcal{T}$ (every $T \in \mathcal{T}$ is a function of $\bar{T}$ (or $\bar{T}'$)) and $\bar{\mathcal{T}}(\bar{\mathcal{T}}')$ an apparatus for measuring $\bar{T}(\bar{T}')$. In particular, $\bar{T}'(\bar{T})$ is a function of $\bar{T}(\bar{T}')$ and thus $\bar{\mathcal{T}}$ and $\bar{\mathcal{T}}'$ are essentially, aside from a scale change, the same apparatus. Thus, the micro-states, if they exist, are functionals:

$$\xi : T, \mathcal{T} \rightarrow \langle T \rangle_{\xi, \mathcal{T}} \in \sigma(T) , \tag{F.6}$$

which assign to each observable T a number from its spectrum depending, in principle, on a maximal abelian algebra $\mathcal{T}$ which contains T. Given $\mathcal{T}, \xi$ will be a homomorphism of $\mathcal{T}$ into $\mathbb{R}$.

F.3 Hidden Variables of the First Kind and of the Second Kind (or Local Hidden Variables)

From these considerations, there arise naturally two alternative objectives for proponents of hidden variables:

1) To maintain the stochastic predictions of quantum mechanics (HV of the first kind); or
2) To save the principle of local action [second kind or local hidden variables (LHV)].

(Due to the success of quantum mechanics, Einstein believed in the simultaneous achievement of both goals.)

Macro-states being convex mixtures of micro-states, the first alternative is achieved if it is possible to find, given the pure state ψ, a probability measure μ_ψ in the space X of micro-states such that the conventional quantum expectation values satisfy

$$\langle T \rangle_\psi = \int_X \langle T \rangle_{\xi, \mathcal{T}} d\mu_\psi(\xi) \tag{F.7}$$

for every observable T and every maximal abelian algebra $\mathcal{T}$ containing T. In 1966, *Bohm* and *Bub* [BB 66], following up on an investigation of Wiener and Siegel, introduced a set of micro-states $\bar{\xi} = (\psi, \xi)$, formed by pairs of normalized vectors in Hilbert space, and a (polycotomic) *algorithm* for assigning the values of $\langle T \rangle_{\xi, \mathcal{T}}$. One member of the pair (ψ) is just the macro-state; the prescription is such that ordinary quantum mechanics is recovered when ξ has a *uniform* distribution, say that for *thermodynamic equilibrium*, which is the restriction to the unit sphere in the Hilbert space of a *gaussian* distribution. No details will be presented here. It suffices to anticipate that every cryptodeterministic theory which leads to quantum mechanics will necessarily violate the principle of local action, something which not even Einstein, the most vigorous of the defenders of hidden variables, could accept without strong arguments in favor. Nevertheless, it is important to emphasize that this model automatically leads to a causal reduction of the wave packet, if one uses a *nonlinear* deterministic dynamics which by assumption controls the evolution of $\bar{\xi}$ during the process of measurement.

Papaliolios subjected the Bohm–Bub theory to experimental verification in 1967 [PA 67]. The schematic idea for his experiment was the following: a beam of white light, with a flux of $10^8 - 10^9$ photons cm^{-2} s^{-1} is incident on a polarizer 15 μm thick. The transmitted light, with intensity I_1, is linearly polarized; it is sent into a similar polarizer with its axis making an angle θ_{12} with respect to the axis of the previous one. We make the "Ansatz" that the HV ξ have had sufficient time to reach equilibrium in between the two polarizers. Thus, an intensity obeying Malus' law, $I_2 = I_1 \cos^2 \theta_{12}$, will emerge from polarizer 2. But if a polarizer 3 is immediately juxtaposed to polarizer 2 with its axis making an angle θ_{23} with respect to that of 2, such that in passing through the 15 μm the HV do not have time to relax to equilibrium, the photons leaving 2 will be incident on 3 with an HV distribution which is strongly biased, and the transmitted intensity will depart from the quantum prediction $I_3 = I_2 \cos^2 \theta_{23}$. The experimental result showed no such effect, and it was compatible with a relaxation time $\tau \lesssim 1.9 \times 10^{-14}$ s (see *Freedman* et al. [FM 76]), thereby justifying the Ansatz. However, given the smallness of τ, it is not easy to attribute it to a thermal relaxation, at least for the room temperatures at which the experiment was performed.

If the HV are used to validate the principle of local action instead of having as their prime objective the explanation of quantum mechanics, the requirement must be that the expectation value of a correlation observable $T_1 \times T_2$ for a state ψ of the system composed of subsystems 1 and 2, when these are space-like separated and not interacting, should be a statistical average over all micro-states of products of the individual values associated with these micro-states. Each of these individual values (here is where locality enters) should depend exclusively on the apparatus which serves to measure the corresponding individual observables, and upon the HV of the micro-state, but not on actions taken on the other subsystem. Mathematically.

$$\langle T_1 \otimes T_2 \rangle_\psi^{\text{LHV}} \equiv \int_X \langle T_1 \rangle_{\xi, T_1} \langle T_2 \rangle_{\xi, T_2} d\mu_\psi(\xi) , \tag{F.8}$$

with (F.7) being obeyed in the remaining cases. It is not difficult to recognize that this hypothesis leads to deviations from quantum mechanics; an elementary calculation using (F.8) suffices to prove that for any four observables T_1, T_2 (first subsystem), T_3, T_4 (second subsystem) with spectra in the interval $[-a, +a]$, we have the inequality (which is an easy generalization of the famous *Bell inequality* [BE 64])

$$|\langle T_1 \otimes T_3 \rangle_\psi^{\text{LHV}} - \langle T_1 \otimes T_4 \rangle_\psi^{\text{LHV}}|$$
$$+ |\langle T_2 \otimes T_3 \rangle_\psi^{\text{LHV}} + \langle T_2 \otimes T_4 \rangle_\psi^{\text{LHV}}| \le 2a^2 , \tag{F.9}$$

which conventional quantum mechanical expectation values do not in general satisfy. Thus for example, in the *Gedankenexperiment* of EPRB, using $T_1 = s_1 \cdot \hat{n}_1, T_2 = s_1 \cdot \hat{n}_2, T_3 = s_2 \cdot \hat{n}_3$ and $T_4 = s_2 \cdot \hat{n}_4$, we have $(i = 1, 2, j = 3, 4)$

$$\langle T_i \otimes T_j \rangle_\psi = -\tfrac{1}{4} \cos \theta_{ij} , \quad \theta_{ij} \equiv \sphericalangle(\hat{n}_i, \hat{n}_j) , \tag{F.10}$$

and there is no reason for the previous inequality (now $a = 1/2$)

$$|\cos\theta_{12} - \cos\theta_{14}| + |\cos\theta_{23} + \cos\theta_{24}| \leq 2 \qquad (F.11)$$

to be fulfilled: consider the planar configuration $\theta_{13} = \theta_{32} = \theta_{24} = \pi/3, \theta_{14} = \pi$.

There now exists the possibility of submitting the question of the existence of LHV to direct experimental verification. The decay of parapositronium ($\tau \sim 10^{-10}$ s) produces two gamma rays with correlated perpendicular polarizations, since the transition amplitude is $\propto (\epsilon_1 \times \epsilon_2) \cdot k$. The high energy of these photons means that there are no adequate ideal polarizers at our disposal to analyze the correlation between their polarizations; one must then resort to Compton diffusion of these photons as a means of analysis. This experiment has been carried out (see *Kasday* in [ES 71]). Although it produces a result compatible with quantum mechanics, it does not completely rule out LHV, since the theoretical analysis requires the description of the Compton scattering. If this is carried out using the Klein–Nishina formula, local hidden variables and quantum mechanics are "mixed"; if not, it can be shown that there exists an ad hoc LHV description which is in agreement with experiment.

In 1972, *Clauser* and *Freedman* [FC 72] completed a very important experiment: Ca atoms excited by ultraviolet light decay through a complicated cascade in a time $\tau \sim 10\,\mathrm{ns}$. The two photons γ_1 and γ_2 of the $6\,^1S_0 - 4\,^1P_1 - 4\,^1S_0$ subcascade, whose wavelengths are $\lambda_1 = 5513\,\text{Å}$ and $\lambda_2 = 4227\,\text{Å}$, respectively, are among those produced. A system of wavelength filters, polarizers, coincidence monitors and photodetectors permits the measurement of the coincidence rates $R(\hat{n}_i, \hat{n}_j)$ of pairs of photons γ_1, γ_2 , each member of which travels in predetermined opposite directions and traverses different polarizers with axes $\hat{n}_i, \hat{n}_j, i = 1, 2, j = 3, 4$. One can extract a combination S of these quantities:

$$S = \frac{[R(\hat{n}_1, \hat{n}_3) - R(\hat{n}_1, \hat{n}_4) + R(\hat{n}_2, \hat{n}_3) + R(\hat{n}_2, \hat{n}_4) - R_0]}{R_0}. \qquad (F.12)$$

Here R_0 is the coincidence rate in the absence of the polarizers. The generalized Bell inequality implies that S should be non-positive in every local hidden variable description: $S^{\mathrm{LHV}} \leq 0$. On the other hand, the conventional quantum value S^{QM} reaches its maximum positive theoretical value of $(\sqrt{2} - 1)/2$ when $\theta_{13} = \theta_{32} = \theta_{24} = \pi/8$. The results of the experiment were in beautiful agreement with quantum mechanics (when pertinent corrections to account for efficiencies and finiteness of solid angles are made), the discrepancy with LHV predictions being at least six standard deviations.

For a good discussion of the experiments which have been performed on these topics through the 1970s, see [PI 78]. During the past decade, *Aspect* and collaborators have carried out a series of crucial experiments at the Institute of Optics at Orsay, some of which display discrepancies of as much as 40 standard deviations with respect to LHV models ([AG 81], [AG 82], [AD 82]). Specially noteworthy is the last of these experiments, which, unlike the previous ones, is not static, in the sense that the orientations of the polarizers are changed during the flight time of the photons (delayed choice) thereby destroying the possible causal communication between the two detection regions. In that experiment, the predictions of LHV theories are violated by some 5 standard deviations.

The defenders of the EPR viewpoint, noting the low efficiency of the detectors, maintain that these experiments are not yet conclusive. Nevertheless, notwithstanding the possibility that experimental techniques of the future may bring us closer to the ideal experiment, the results to date are a demonstration of the impressive success of quantum mechanics and of the mounting difficulties of hidden variable theories.

F.4 Conclusions

These are the facts. Nature refuses to collaborate with the defenders of local determinism. Of course, new experimental confirmations are desirable. The general impression we perceive is that the negative conclusions of Freedman, Clauser and Aspect will persist, and become very likely more accentuated.

The partisans of HV of the first kind, in order to make their theory more agreeable to the palates of the majority of the scientific community, should: (1) *provide* that theory with very *specific predictive character*. It appears that the theory has at least in principle more than enough flexibility to accommodate itself to *any* experimental result; (2) liberate the theory from a series of internal afflicting paradoxes dealing with the measurement of observables with degenerate spectra and with the description of composite systems; (3) explain convincingly Papaliolios' negative result within a specific HV dynamical scheme, and finally (this may well be the "mission impossible") convince the physicists as to why the principle of local action, grounded in special relativity and therefore in the geometry of space-time, should fail to hold in HV theory and not fail in what is observable to date.

If the absolute rejection of determinism at the quantum level is truly inevitable, and if, as we said at the beginning, we do not wish to introduce a dichotomy in the description of nature which destroys scientific unity, we must properly displace the boundary between the classical and the quantum regimes to the other extreme, leaning heavily towards the view that all of physics is essentially quantal in principle. In fact, classical determinism is itself not a logical necessity, nor is it empirically established. Suffice it to say that it is not absurd to think of a universe in which the laws of physics are different, so that the preparation of exactly precise initial conditions for a system would require infinite time.

Of course, the rejection of determinism as a scientific paradigm poses serious and important problems: how is it possible to establish objective and unambiguous facts upon which to base the theoretical description of nature, in a universe in which everything is governed by probabilistic laws? And intimately related to this question, why is deterministic theory such a good approximation in a large part of physics? The magnitude of these questions is evident.

It appears to be a characteristic of the tremendous epistemological commotion provoked by quantum mechanics that when we contemplate its fundamentals, we seem driven to interpretational oscillations between extremes: one defending determinism at all costs; and the other, supporting the unicity of the individual as incompatible with whatever scientific affirmation can be made concerning it.

Appendix G: Properties of Certain Antiunitary Operators

The study of time reversal introduces antiunitary operators into quantum mechanics. We collect here a few properties of these operators. For a proper study of antiunitary operators and of the phenomenological distinction between unitary and antiunitary operators, see [WI 60].

G.1 Definitions and Basic Properties

Given a Hilbert space $\mathcal{H}$, every *antilinear bijective* map V from $\mathcal{H}$ onto $\mathcal{H}$,

$$V(\alpha|x\rangle + \beta|y\rangle) = \alpha^* V|x\rangle + \beta^* V|y\rangle , \tag{G.1}$$

which is *isometric*,

$$\|V|x\rangle\| = \| |x\rangle\| , \tag{G.2}$$

is called an *antiunitary operator*.

From (G.1) and (G.2) we have

$$\langle Vx|Vy\rangle = \langle x|y\rangle^* = \langle y|x\rangle . \tag{G.3}$$

V being antilinear, $V^\dagger$ is defined as

$$\langle x|V|y\rangle = \langle y|V^\dagger|x\rangle , \tag{G.4}$$

so that the antiunitary character of V implies $V^\dagger = V^{-1}$. Thus,

$$\langle Vx|Vy\rangle = \langle y|x\rangle, \ \langle x|V|y\rangle = \langle y|V^\dagger|x\rangle = \langle y|V^{-1}|x\rangle . \tag{G.5}$$

Similarly, if A is a linear operator,

$$\langle Vx|A|Vy\rangle = \langle x|V^{-1}AV|y\rangle^* . \tag{G.6}$$

It is evident that the product of two antiunitary operators is a unitary operator, and the product of a unitary and an antiunitary operator is antiunitary. Then every antiunitary V can be factorized as $V = UK$, where K is a given antiunitary operator and U is unitary. Thus, for example, if $\{|e_n\rangle\}$ is an orthonormal basis in $\mathcal{H}$,

$$K : \sum \alpha_n |e_n\rangle \rightarrow \sum \alpha_n^* |e_n\rangle \tag{G.7}$$

defines an antiunitary involutive operator: the *complex conjugation* associated with this basis, which can be chosen as a typical factor in the decomposition $V = UK$. If $\mathcal{H} = L^2(\mathbf{R}^n)$, another very common complex conjugation is

$$K : \varphi(x_1, \ldots, x_n) \rightarrow \varphi^*(x_1, \ldots, x_n) \, . \tag{G.8}$$

Just as in the linear case, we shall say that $|x\rangle (\neq 0)$ is an *eigenvector* of V if

$$V|x\rangle = \omega |x\rangle \, . \tag{G.9}$$

V being isometric, it is necessary that $|\omega| = 1$. On the other hand, from (G.9)

$$V^2 |x\rangle = V\omega |x\rangle = \omega^* V |x\rangle = |\omega|^2 |x\rangle = |x\rangle \, , \tag{G.10}$$

so that $|x\rangle$ is an eigenvector of V^2, with eigenvalue $+1$. Therefore, if $1 \notin \sigma_p(V^2)$, V has no eigenvectors. Such is the case, for example, in time reversal for an odd number of fermions, since then $V^2 = -I$.

In the linear case, if $|x\rangle$ is an eigenvector of A with eigenvalue a, then $\lambda |x\rangle$ is also an eigenvector with the same eigenvalue; by contrast, if (G.9) is satisfied, the ray $|x\rangle_R$ contains eigenvectors of V having all possible eigenvalues of unit modulus:

$$V(e^{i\alpha} |x\rangle) = e^{-i\alpha} V |x\rangle = e^{-i\alpha} \omega |x\rangle = (e^{-2i\alpha} \omega)(e^{i\alpha} |x\rangle) \, . \tag{G.11}$$

Therefore, the eigenvectors of V with different eigenvalues *are not necessarily orthogonal*.

G.2 Canonical Form of Antiunitary V with $(V^2)_{\text{p.p.}} = V^2$

Suppose that the unitary operator V^2 has a spectrum which is pure point; i.e., the eigenvectors of V^2 span the entire Hilbert space. Let $\omega_1, \omega_2, \ldots$ be the distinct eigenvalues of V^2, and $\mathcal{H}_{\omega_j}$ be the eigensubspace of V^2 corresponding to ω_j. Then

$$\mathcal{H} = \bigoplus_j \mathcal{H}_{\omega_j} \, . \tag{G.12}$$

Since

$$V^2 |x\rangle = \omega_j |x\rangle \Rightarrow V^2 V |x\rangle = VV^2 |x\rangle = \omega_j^* V |x\rangle \, , \tag{G.13}$$

then also $\omega_j^* \in \sigma(V^2)$, and $V\mathcal{H}_{\omega_j} = \mathcal{H}_{\omega_j^*}$. Therefore only if $\omega_j = \pm 1$ will $\mathcal{H}_{\omega_j}$ be stable under V.

We now examine the structure of V in $\mathcal{H}_{+1}, \mathcal{H}_{-1}$ and $\mathcal{H}_\omega \oplus H_{\omega^*}$:

1) Let $\mathcal{H}_{+1} \neq \{0\}$. Given $|x\rangle(\neq 0) \in \mathcal{H}_{+1}$, we have

$$V(|x\rangle + V|x\rangle) = |x\rangle + V|x\rangle , \quad V(i|x\rangle - iV|x\rangle)) = i|x\rangle - iV|x\rangle , \qquad (G.14)$$

so that $\mathcal{H}_{+1}$ contains some normalized vector $|e_1\rangle$ with $V|e_1\rangle = |e_1\rangle$. The subspace $\mathcal{H}_{+1} \ominus \{|e_1\rangle\}$ is also stable under V; repeating the argument, if it is $\neq \{0\}$, it will contain some normalized $|e_2\rangle$ such that $V|e_2\rangle = |e_2\rangle$. We can thus successively find an orthonormal basis for $\mathcal{H}_{+1}$ such that

$$V|e_n\rangle = |e_n\rangle , \qquad (G.15)$$

and, therefore, in $\mathcal{H}_{+1}$, V will be the complex conjugation operator associated with this basis as in (G.7).

2) Let $\mathcal{H}_{-1} \neq \{0\}$. Given $|x\rangle(\neq 0) \in \mathcal{H}_{-1}$, we have

$$\langle x|V|x\rangle = -\langle V^2 x|V x\rangle = -\langle x|V|x\rangle , \qquad (G.16)$$

and therefore, $|x\rangle \perp V|x\rangle$. In the same way, if $|y\rangle(\in \mathcal{H}_{-1}) \perp |x\rangle, V|x\rangle$, then also $V|y\rangle \perp |y\rangle, |x\rangle$ and $V|x\rangle$. Therefore, there exists an orthonormal basis $\{|v_n\rangle\}$ in $\mathcal{H}_{-1}$ such that

$$V|v_{2n-1}\rangle = |v_{2n}\rangle , \quad V|v_{2n}\rangle = -|v_{2n-1}\rangle , \qquad (G.17)$$

and in particular, $\dim(\mathcal{H}_{-1})$ will be *even* if it is finite.

3) Let $\mathcal{H}_\omega \oplus \mathcal{H}_{\omega^*} \neq \{0\}, \omega \neq \pm 1$. As in case (2), every $|x\rangle \in \mathcal{H}_{\omega,\omega^*}$ is $\perp V|x\rangle$, and an orthonormal basis $\{|w_n\rangle\}$ in $\mathcal{H}_\omega \oplus \mathcal{H}_{\omega^*}$ can be chosen in such a way that

$$V|w_{2n-1}\rangle = |w_{2n}\rangle , \quad V|w_{2n}\rangle = \omega|w_{2n-1}\rangle , \quad \mathrm{Im}\,\omega > 0 . \qquad (G.18)$$

Thus, in those cases where there exists an integer N such that $V^{2N} = -I$, we will have $\mathcal{H}_{+1} = \{0\}, V$ shall then fail to have eigenvectors and $\dim(\mathcal{H})$ (if it is finite) will be even. This fact constitutes the mathematical foundation of the *Kramers degeneracy theorem*.

Bibliography

AA 87   Aharonov, A.; Anandan, J.: Phys. Rev. Lett. **58**, 1593 (1987)

AB 59   Aharonov, Y.; Bohm, D.: Phys. Rev. **115**, 485 (1959)

AB 65   Akhiezer, A.I.; Berestetskii, V.B.: *Quantum Electrodynamics* (translation of the Russian edition). Wiley, New York (1965)

AB 65a   Arens, R.; Babbitt, D.: J. Math. Phys. **6**, 1071 (1965)

AC 72   Arecchi, F.; Courtens, E.; Gilmore, R.; Thomas, A.: Phys. Rev. A **6**, 2211 (1972)

AC 72a   Amaldi, E.; Cabibbo, N.: "On the Dirac Magnetic Poles", in *Aspects of Quantum Theory*, Cambridge Univ. Press (1972)

AC 76   Aragone, C.; Chalbaud, E.; Salamó, S.: J. Math. Phys. **17**, 1963 (1976)

AC 84   Aguilar–Benítez, M,; Cahn, R.N.; Crawford, R.L.; Frosch, R.; Gopal, G.P.; Hendrick, R.E.; Hernández, J.J.; Höhler, G.; Losty, M.J.; Montanet, L.; Porter, F.C.; Rittenberg, A.; Roos, M.: Roper, L.D.; Shimada, T.; Shrock, R.E.; Törnqvist, N.A.; Trippe, T.G.; Trower, W.P; Walck, C.; Wohl, C.G.; Yost, G.P.; Armstrong, B.: Rev. Mod. Phys. **56**, no. 2, part II (1984)

AD 82   Aspect, A.; Dalibard, J.; Roger, G.: Phys. Rev. Lett. **49**, 1804 (1982)

AD 85   Adler, S.L.: Phys. Rev. Lett. **55**, 783 (1985)

AG 61   Akhiezer, N.A.; Glazman, I.M.: *Theory of Linear Operators in Hilbert Space* (two volumes), Ungar, New York (1961)

AG 74   Aragone, C.; Guerri, G.; Salomó, S.; Tani, J.L.: J. Phys. A **7**, L149 (1974)

AG 81   Aspect, A.; Grangier, P.; Roger, G.: Phys. Rev. Lett. **47**, 460 (1981)

AG 82   Aspect, A.; Grangier, P.; Roger, G.: Phys. Rev. Lett. **49**, 91 (1982)

AG 84   Aspect, A.; Grangier, P.: In Proceedings of the International Symposium *Foundations of Quantum Mechanics in the Light of New Technology.* Eds. S. Kamefuchi et al., Phys. Soc. of Japan (1984)

AG 84a   Albeverio, S.; Gesztesy, F.; Høegh–Krohn, R.; Holden, H.: *Some exactly solvable models in Quantum Mechanics and the low energy expansions*, Preprint. University of Bielefeld (1984)

AG 88   Abellanas, L.; Galindo, A.: *Espacios de Hilbert (Geometría, Operadores, Espectros)*, Eudema, Madrid (1988)

AH 78   Avron, J.; Herbst, I.W.; Simon, B.: Duke Math. J. **45**, 847 (1978)

AJ 70   Ajzenberg–Selove, F.: Nucl. Phys. A **152**, 1 (1970)

AJ 77   Amrein, W.O.; Jauch, J.M.; Sinha, K.B.: *Scattering Theory in Quantum Mechanics.* W.A. Benjamin, Reading, Mass. (1977)

AL 59   Ajzenberg–Selove, F.; Lauritsen, T.: Nucl. Phys. **11**, 1 (1959)

AL 69   Allcock, G.R.: Ann. Phys. (N.Y.) **53**, 253, 286, 311 (1969)

AL 70   Ans, J.D'.; Lax, E.: *Taschenbuch für Chemiker und Physiker.* Springer, Berlin, Heidelberg (1970)

AL 72   Albeverio, S.: Ann. Phys. (N.Y.) **71**, 167 (1972)

AL 88   Alvarez, G.: Phys. Rev. A **37**, 4079 (1988)

AM 74   Amrein, W.O.: "Some Questions in Non-Relativistic Quantum Scattering Theory", in *Scattering Theory in Mathematical Physics.* J.A. La Vita and J.P. Marchand. Reidel, Dordrecht, Holland (1974)

AN 69   Anderson, R.F.V.: J. Functional Anal. **4**, 240 (1969)

AN 81   Andrews, M.: J. Phys. A **14**, 1123 (1981)

AR 65   Alfaro, V. de; Regge, T.: *Potential Scattering.* North-Holland, Amsterdam (1965)

AS 64   Abramowitz, M.; Stegun, I.A. (Eds.): *Handbook of Mathematical Functions with Formulas, Graphs and Mathematical Tables*. Dover, New York (1964)
AS 83   Austen, G.J.M.; Swart, J.J. de: Phys. Rev. Lett. **50**, 2039 (1983)
AS 87   Avron, J.E.; Seiler, R.; Yaffe, L.G.: Commun. Math. Phys. **110**, 33 (1987)

BA 49   Bargmann, V.: Rev. Mod. Phys. **21**, 488 (1949)
BA 49a   Bargmann, V.: Phys. Rev. **75**, 301 (1949)
BA 52   Bargmann, V.: Proc. Nat. Acad. Sci. U.S.A. **38**, 961 (1952)
BA 52a   Bargmann, V.: Ann. Math. **59**, 1 (1952)
BA 57   Bohm, D.; Aharonov, Y.: Phys. Rev. **108**, 1070 (1957)
BA 60   Bazley, N.W.: Phys. Rev. **120**, 144 (1960)
BA 64   Bargmann, V.: J. Math. Phys. **5**, 862 (1964)
BA 67   Bacry, H.: *Leçons sur la Théorie des Groups et les Symétries des Particules Elémentaires*. Gordon and Breach, Paris (1967)
BA 75   Baker, G.A.: *Essentials of Padé Approximants*. Academic Press, New York (1975)
BA 76   Banerjee, K.: Lett. Math. Phys. **1**, 323 (1976)
BB 66   Bohm, D.; Bub, J.: Rev. Mod. Phys. **38**, 453 (1966)
BB 78   Banerjee, K.; Bhatnagar, S.P.; Choudhry, V.; Kanwali, S.S.: Proc. Roy. Soc. London A **360**, 575 (1978)
BC 57   Bardeen, J.; Cooper, L.N.; Schrieffer, J.R.: Phys. Rev. **108**, 1175 (1957)
BD 71   Biswas, S.N.; Datta, K.; Saxena, R.P.; Srivastaka, P.K.; Varma, V.S.: Phys. Rev. D **4**, 3617 (1971)
BD 87   Bitter, T.; Dubbers, D.: Phys. Rev. Lett. **59**, 251 (1987)
BE 42   Bergmann, P.G.: *Introduction to the Theory of Relativity*. Prentice-Hall, Englewood Cliffs, N.J. (1942)
BE 64   Bell, J.S.: Physics **1**, 195 (1964)
BE 66   Bell, J.S.: Rev. Mod. Phys. **38**, 447 (1966)
BE 73   Bertrand, J.: C. R. Acad. Sci. **77**, 849 (1873)
BE 73a   Belinfante, F.J.: *A Survey of Hidden-Variables Theories*. Pergamon, Oxford (1973)
BE 75   Bell, J.S.: Helv. Phys. Acta **48**, 93 (1975)
BE 84   Berry, M.V.: Proc. Roy. Soc. London A **329**, 45 (1984)
BG 84   Baumgartner, B.; Grosse, H.; Martin, A.: Phys. Lett. **146B**, 363 (1984)
BG 85   Baumgartner, B.; Grosse, H.; Martin, A.: *"Inequalities for scattering phase shifts"*. U.W. Th. Ph.-1985-6 (1985)
BH 85   Bhattacharya, S.K.: Phys. Rev. A **31**, 1991 (1985)
BI 61   Birman, M.S.: Mat. Sb. **55**, 125 (1961)
BJ 68   Bethe, H.A.; Jackiw, R.: *Intermediate Quantum Mechanics*, 2nd ed., Benjamin, New York (1968)
BK 61   Bromley, D.A.; Kuehner, J.A.; Almqvist, E.: Phys. Rev. **123**, 878 (1961)
BL 46   Bloch, F.: Phys. Rev. **70**, 460 (1946)
BL 73   Blin–Stoyle, R.J.: *Fundamental Interactions and the Nucleus*. North-Holland, Amsterdam (1973)
BL 77   Brezin, E.; Le-Gillou, J.; Zinn-Justin, J.: Phys. Rev. D **15**, 1544 and 1558 (1977)
BL 81   Biedenharn, L.C.; Louck, J.D.: *Angular Momentum in Quantum Physics*, Addison-Wesley, Reading, Mass. (1981)
BM 69   Bohr, A.; Mottelson, B.R.: *Nuclear Structure* (two volumes). Benjamin, New York (1969)
BM 72   Berry, M.V.; Mount, K.E.: Rep. Prog. Phys. **35**, 315 (1972)
BO 13   Bohr, N.: Phil. Mag. **26**, 1, 476 and 875 (1913)
BO 18   Bohr, N.: Kgl. Danske Vid. Selsk. Skr., nat.-math. Afd., 8 Raekke IV, 1 (1918)
BO 22   Bohr, N.: Z. Phys. **13**, 117 (1922)
BO 26   Born, M.: Z. Phys. **37**, 863 (1926) and **38**, 803 (1926)
BO 28   Bohr, N.: "The Quantum Postulate and the Recent Development of Atomic Theory", in *Atti del Congresso Internazionale dei Fisici*. N. Zanichelli, Bologna (1928)
BO 52   Bohm, D.: Phys. Rev. **85**, 166, 180 (1952)

BO 57   Bogolyubov, N.N.: J. Phys. USSR **9**, 23 (1957)

BO 58   Bogolyubov, N.N.: J.E.P.T. **7**, 41 (1958)

BO 67   Bogolyubov, N.N.: *Lectures on Quantum Statistics* (two volumes). Gordon and Breach, New York (1967)

BO 68   Bogolyubov, N.N. (Ed.): *The Theory of Superconductivity*. Gordon and Breach, New York (1968)

BO 77   Bender, C.M.; Olaussen, K.; Wang, P.S.: Phys. Rev. D **16**, 1740 (1977)

BP 79   Balian, R.; Parisi, G.; Voros, A.: "Quartic Oscillator" in *Feynman Path Integrals*. Lecture Notes in Physics, Vol. 106, Springer, Berlin, Heidelberg (1979)

BP 83   Balsa, R.; Plo, M.; Esteve, J.G.; Pacheco, A.F.: Phys. Rev. D **28**, 1945 (1983)

BR 23   Broglie, L. de: C. R. Acad. Sci. **177**, 507, 548 and 630 (1923)

BR 26   Brillouin, L.: C. R. Acad. Sci. **183**, 24 (1926)

BR 33   Bohr, N.; Rosenfeld, L.: Klg. Danske Vid. Sels. Math.-Fys. Medd. **12**, no. 8 (1933)

BR 50   Bohr, N.; Rosenfeld, L.: Phys. Rev. **78**, 794 (1950)

BR 59   Brueckner, K.A.: "Theory of Nucleon Structure", in *The Many Body Problem*. Lectures given at l'École d'Été de Physique Théorique. Les Houches. Session 1958. Wiley, New York (1959)

BR 65   Bremermann, H.: *Distributions, Complex Variables, and Fourier Transforms*. Addison-Wesley, Reading, Mass. (1965)

BS 59   Bogolyubov, N.N.; Shirkov, D.V.: *Introduction to the Theory of Quantized Fields*. Wiley, New York (1959)

BW 65   Born, M.; Wolf, E.: *Principles of Optics*, 3rd ed., Pergamon, Oxford (1965)

BW 69   Bender, C.M.; Wu, T.T.: Phys. Rev. **184**, 1231 (1969)

CA 67   Calogero, F.: *Variable Phase Approach to Potential Scattering*. Academic Press, New York (1967)

CA 85   Caves, C.M.: Phys. Rev. Lett. **54**, 2465 (1985)

CB 85   Comtet, A.; Bandrauk, A.D.; Campbell, K.: Phys. Lett. **150B**, 159 (1985)

CD 73   Cohen–Tannoudji, C.; Diu, B.; Laloë, F.: *Mecanique Quantique*, Hermann, Paris (1973)

CE 68   Cerny, J.: Ann. Rev. Nuc. Sci. **18**, 27 (1968)

CF 70   Capasso, V.; Fortunato, D.; Selleri, F.: Riv. Nuovo Cimento **11**, 149 (1970)

CF 81   Creutz, M.; Freedman, B.: Ann. Phys. (N.Y.) **132**, 427 (1981)

CG 80   Caliceti, E.; Graffi, R.; Maioli, M.: Commun. Math. Phys. **75**, 51 (1980)

CH 39   Chandrasekhar, S.: *An Introduction to the Study of Stellar Structure*. Dover, New York (1939)

CH 50   Cherry, T.M.: Trans. Am. Math. Soc. **68**, 224 (1950)

CH 60   Chambers, R.G.: Phys. Rev. Lett. **5**, 3 (1960)

CH 68   Chadan, K.: Nuovo Cimento **58A**, 191 (1968)

CM 70   Calles, A.; Moshinsky, M.: Am. J. Phys. **38**, 456 (1970)

CM 82   Chiu, C.B.; Misra, B.; Sudarshan, E.C.G.: Phys. Lett. **117B**, 34 (1982)

CN 68   Carruthers, P.; Nieto, M.M.: Rev. Mod. Phys. **40**, 411 (1968)

CO 21   Compton, A.H.: Phil. Mag. **41**, 749 (1921); Phys. Rev. **18**, 96 (1921) and **19**, 267 (1922)

CO 23   Compton, A.H.: Phys. Rev. **21**, 483, 715 (1923); ibid. **22**, 409 (1923)

CO 65   Cohn, J.H.E.: J. London Math. Soc. **40**, 523 (1965)

CO 75   Colella, R.; Overhauser, A.W.; Werner, S.A.: Phys. Rev. Lett. **34**, 1472 (1975)

CO 77   Coleman, S.: *The Uses of Instantons*. Proc. Int. School of Physics. Erice (1977)

CO 83   Coleman, S.: In *The Unity of the Fundamental Interactions*. Ed. A. Zichichi. Plenum, London (1983)

CO 85   Common, A.K.: J. Phys., A.: Math. Gen. **18**, 2219 (1985)

CP 48   Casimir, H.G.R.; Polder, D.: Phys. Rev. **73**, 360 (1948)

CP 63   Cerny, J.; Pehl, R.H.; Rivet, E.; Harvey, B.G.: Phys. Lett. **7**, 67 (1963)

CR 66   Conley, C.C.; Rejto, P.A.: "Spectral Concentration II-General Theory", in *Perturbation Theory and Its Applications in Quantum Mechanics*. Ed. C.H. Wilcox. Wiley, New York (1966)

CS 57   Condon, E.U.; Shortley, G.H.: *The Theory of Atomic Spectra*, 4th ed., Cambridge University Press, London (1957)

CS 77   Chadan, K.; Sabatier, P.C.: *Inverse Problems in Quantum Scattering Theory*. Springer, New York (1977)

CS 78   Clauser, J.F.; Shimony, A.: Rep. Progr. Phys. **41**, 1881 (1978)

CW 66   Cotton, F.A.; Wilkinson, G.: *Advanced Inorganic Chemistry*, Wiley, New York (1966)

CW 77   Cwikel, M.: Ann. Math. **106**, 93 (1977)

DA 61   Dalgarno, A.: "Stationary Perturbation Theory", in *Quantum Theory* (three volumes). Ed. D.R. Bates. Academic Press, New York (1961)

DG 27   Davisson, C.; Germer, L.H.: Phys. Rev. **30**, 705 (1927)

DI 28   Dirac, P.A.M.: Proc. Roy. Soc. London A **117**, 610 and A **118**, 351 (1928)

DI 30   Dirac, P.A.M.: Proc. Cambridge Phil. Soc. **26**, 376 (1930)

DI 31   Dirac, P.A.M.: Proc. Roy. Soc. London A **133**, 60 (1931)

DI 58   Dirac, P.A.M.: *The Principles of Quantum Mechanics*, 4th ed., Clarendon Press, London (1958)

DI 61   Dicke, R.H.: Phys. Rev. Lett. **7**, 359 (1961)

DK 75   Detwiler, L.C.; Klauder, J.R.: Phys. Rev. D **11**, 1436 (1975)

DK 79   Duru, I.H.; Kleinert, H.: Phys. Lett. **84B**, 185 (1979)

DL 62   Danieri, A.; Loinger, A.; Prosperi, G.M.: Nucl. Phys. **33**, 297 (1962)

DO 64   Dollard, J.D.: J. Math. Phys. **5**, 729 (1964)

DO 69   Dollard, J.D.: Commun. Math. Phys. **12**, 193 (1969)

DO 76   Dowker, J.S.: J. Math. Phys. **17**, 1873 (1976)

DP 84   Damburg, R.; Propin, R.; Martyschenko, V.: J. Phys. A **17**, 3493 (1984)

DR 81   Drummond, J.E.: J. Phys. A **14**, 1651 (1981); A **15**, 2321 (1982)

DS 64   Dunford, N.; Schwartz, J.T.: *Linear Operators* (two volumes). Wiley, New York (1964)

DS 78   Davies, E.B.; Simon, B.: Commun. Math. Phys. **63**, 277 (1978)

DT 79   Deift, P.; Trubowitz, E.: Comm. Pure Appl. Math. **32**, 121 (1979)

DU 32   Dunham, J.L.: Phys. Rev. **41**, 713 (1932)

DY 66   Dyson, F.J.: *Symmetry Groups in Nuclear and Particle Physics*. Benjamin, New York (1966)

DY 72   Dyson, F.J.: "Fundamental constants and their time variation", in *Aspects of Quantum Theory*. Eds. A. Salam and E.P. Wigner. Cambridge Univ. Press (1972)

EC 30   Eckart, C.: Rev. Mod. Phys. **2**, 205 (1930)

EC 30a   Eckart, C.: Phys. Rev. **35**, 1303 (1930)

ED 57   Edmonds, A.R.: *Angular Momentum in Quantum Mechanics*. Princeton Univ. Press, Princeton, N.J. (1957)

EG 70   Eisenberg, J.M.; Greiner, W.: *Nuclear Theory* (three volumes). North-Holland, Amsterdam (1970)

EH 13   Ehrenfest, P.: Versl. Gewone Vergard. Wis. Natuurk. Afd. Kon. Akad. Wetensch. Amsterdam **22**, 586 (1913) and *Collected Scientific Papers*. Ed. M.J. Klein. North-Holland, Amsterdam (1959)

EI 05   Einstein, A.: Ann. Phys. **17**, 132 (1905)

EI 71   Einsenbud, L.: *The Conceptual Foundations of Quantum Mechanics*. Van Nostrand Reinhold, New York (1971)

EL 67   Endt, P.M.; Van der Leuen, C.: Nucl. Phys. A **105**, 1 (1967)

EM 63   Emch, G.: Helv. Phys. Acta **36**, 739 and 770 (1963)

EM 72   Emch, G.G.: Helv. Phys. Acta **45**, 1049 (1972)

EN 78   Enss, V.: Commun. Math. Phys. **61**, 285 (1978)

EN 79   Enss, V.: Ann. Phys. (N.Y.) **119**, 117 (1979)

EN 84   Enss, V.: Physica **124A**, 269 (1984)

EN 85   Enss, V.: "Quantum Scattering Theory for Two- and Three-Body Systems with Potentials of Short and Long Range", in *Schrödinger Operators*. Lecture Notes in Mathematics, Vol. 1159. Ed. S. Graffi. Springer, Berlin, Heidelberg (1985)

EP 16   Epstein, P.S.: Ann. Phys. (Leipzig) **50**, 489 (1916)
ER 35   Einstein, A.; Rosen, N.; Podolsky, B.: Phys. Rev. **47**, 777 (1935)
ER 53   Erdély, A. (Ed.): *Higher Transcendental Functions* (three volumes) and *Tables of Integral Transforms* (two volumes). McGraw-Hill, New York (1953)
ES 55   Elliot, J.P.; Skyrme, T.H.R.: Proc. Roy. Soc. London A **232**, 561 (1955)
ES 71   Espagnat, B.d' (Ed.): *Foundations of Quantum Mechanics*. Proceedings of the International School of Physics "Enrico Fermi". Curso IL. Academic Press, New York (1971)
ES 71a   Espagnat, B.d': *Conceptual Foundations of Quantum Mechanics*. Benjamin, Menlo Park, Cal. (1971)
ES 80   Enss, V.; Simon, B.: Commun. Math. Phys. **76**, 177 (1980)
ET 74   Eberhard, P.H.; Tripp, R.D.; Déclais, Y.; Séguinot, J.; Baillon, P.; Bricman, C.; Ferro-Luzzi, M.; Perreau, J.M.; Ypsilantis, T.: Phys. Lett. **53B**, 121 (1974)
EV 57   Everett, H.: Rev. Mod. Phys. **29**, 454 (1957)

FA 57   Fano, U.: Rev. Mod. Phys. **29**, 74 (1957)
FA 59   Faddeev, L.D.: Usp. Mat. Nauk **14**, 57 (1959).[English translation J. Math. Phys. **4**, 72 (1963)]
FA 65   Faddeev, L.D.: *Mathematical Aspects of the Three Body Problem in the Quantum Theory of Scattering*. Israel Program of Scientific Translations, Jerusalem (1965)
FA 67   Faddeev, L.D.: Am. Math. Soc. Trans. **65**, 139 (1967)
FA 67a   Faris, W.G.: Pac. J. Math. **22**, 47 (1967)
FA 75   Faris, W.G.: *Self-Adjoint Operators*. Springer, Berlin, Heidelberg (1975)
FA 78   Faris, W.G.: J. Math. Phys. **19**, 461 (1978)
FC 56   Feynman, R.P.; Cohen, M.: Phys. Rev. **102**, 1189 (1956)
FC 72   Freedman, S.J.; Clauser, J.F.: Phys. Rev. Lett. **28**, 938 (1972)
FD 70   Fröman, N.; Dammert, Ö.: Nucl. Phys. A **147**, 627 (1970)
FE 27   Fermi, E.: Rend. Accad. Naz. Lincei. **6**, 602 (1927); Z. Phys. **48**, 73 (1928)
FE 30   Fermi, E.: Z. Phys. **60**, 320 (1930)
FE 32   Feenberg, E.: Phys. Rev. **40**, 40 (1932)
FE 39   Feynman, R.P.: Phys. Rev. **56**, 340 (1939)
FE 48   Feynman, R.P.: Rev. Mod. Phys. **20**, 367 (1948)
FE 58   Feshbach, H.: Ann. Rev. Nucl. Sci. **8**, 49 (1958)
FE 69   Feenberg, E.: *Theory of Quantum Fluids*. Academic Press, New York (1969)
FF 65   Fröman, H.; Fröman, P.O.: *JWKB Approximation. Contribution to the Theory*. North-Holland, Amsterdam (1965)
FG 59   Feinberg, G.; Goldhaber, M.: Proc. Nat. Acad. Sci. U.S.A. **45**, 1301 (1959)
FG 70   Fonda, L.; Ghirardi, G.C.: *Symmetry Principles in Quantum Physics*. Marcel Dekker, New York (1970)
FH 14   Franck, J.; Hertz, G.: Verh. Deut. Phys. Ges. **16**, 457, 512 (1914); Phys. Z. **17**, 609 (1916); ibid. **20**, 132 (1919)
FH 65   Feynman, R.P.; Hibbs, A.R.: *Quantum Mechanics and Path Integrals*. McGraw-Hill, New York (1965)
FL 71a   Frank, W.M.; Land, D.J.; Spector, R.M.: Rev. Mod. Phys. **43**, 36 (1971)
FL 74   Faris, W.G.; Lavine, R.B.: Commun. Math. Phys. **35**, 39 (1974)
FL 77   Feynman, R.P.; Leighton, R.B.; Sands, M.: *The Feynman Lectures on Physics*, Vol. 3. Addison-Wesley, Reading, Mass. (1977)
FM 73   Fonte, G.; Mignani, R.; Schiffrer, G.: Commun. Math. Phys. **33**, 293 (1973)
FM 76   Flato, M.; Maric, Z.; Milojevic, A.; Sternheimer, D.; Vigier, J.P.: *Quantum Mechanics, Determinism, Causality and Particles*. Reidel, Dordrecht, Holland (1976)
FO 30   Fock, V.: Z. Phys. **61**, 126 (1930); ibid. **62**, 795 (1930)
FO 76   Fonda, L.: *Preprint Trieste IC/76/20*. Lectures presented at the XIII Winter School of Theoretical Physics. Karpacz, Poland (Feb. 1976)
FP 66   Frankowski, K.; Pekeris, C.L.: Phys. Rev. **146**, 46 (1966)
FR 26   Frenkel, J.: Z. Phys. **37**, 243 (1926)

FR 50   Fröhlich, H.: Phys. Rev. **79**, 845 (1950); Proc. Roy. Soc. London A **215**, 291 (1952)
FR 55   Friedrichs, K.O.: *On the Adiabatic Theorem in Quantum Theory*. Report IMM NYU-218, New York (1955)
FR 59   Fano, U.; Racah, G.: *Irreducible Tensorial Sets*. Academic Press, New York (1959)
FR 71   Friar, J.L.: Nucl. Phys. A **173**, 257 (1971)
FR 72   Friedman, C.N.: J. Funct. Anal. **10**, 346 (1972)
FS 70   Feinberg, G.; Sucher, J.: Phys. Rev. A **2**, 2395 (1970)

GA 68   Galindo, A.: Anales de Física **64**, 141 (1968)
GA 70   Gari, M.: Phys. Lett. **31B**, 627 (1970)
GA 76   Galindo, A.: "Another Proof of the Kochen-Specker Paradox", in *Algunas cuestiones de Física Teórica*. Ed. GIFT, Zaragoza (1976)
GA 85   Grangier, P.; Aspect, A.; Vigué, J.: Phys. Rev. Lett. **54**, 418 (1985)
GC 34   Gorter, C.J.; Casimir, H.: Phys. Z. **35**, 963 (1974); Z. Tech. Phys. **15**, 539 (1934)
GC 62   Guelfand, J.M.; Chilov, G.E.: *Les Distributions* (vol. III). Dunod, Paris (1962)
GF 66   Gorodetzky, S.; Freeman, R.M.; Gallmann, A.; Haas, F.: Phys. Rev. **149**, 801 (1966)
GG 66   Gilbey, D.M.; Goodman, F.O.: Am. J. Phys. **34**, 143 (1966)
GG 67   Gardner, C.S.; Greene, J.M.; Kruskal, M.D.; Miura, R.M.: Phys. Rev. Lett. **19**, 1095 (1967)
GG 70   Graffi, S.; Grecchi, V.; Simon, B.: Phys. Lett. **32B**, 631 (1970)
GG 78   Graffi, S.; Grecchi, V.: J. Math. Phys. **19**, 1002 (1978)
GG 78a   Graffi, S.; Grecchi, V.: Commun. Math. Phys. **62**, 834 (1978)
GH 78   Grosse, H.; Herjel, P.; Thirring, W.: Acta Phys. Austr. **49**, 89 (1978)
GL 51   Gelfand, I.M.; Levitan, B.M.: Izv. Akad. Nauk SSSR. Ser Mat. **15**, 309 (1951)
GL 57   Gleason, A.M.: J. Math. Mech. **6**, 885 (1957)
GM 52   Gamba, A.; Malvano, R.; Radicati, L.A.: Phys. Rev. **87**, 440 (1952)
GM 62   Galindo, A.; Morales, A.; Núñez-Lagos, R.: J. Math. Phys. **3**, 324 (1962)
GM 65   Greenberg, O.W.; Messiah, A.M.L.: Phys. Rev. **138B**, 1115 (1965)
GM 76   Glaser, V.; Martin, A.; Grosse, H.; Thirring, N.: "A family of optimal conditions for the absence of bound states in a potential", in *Studies in Mathematical Physics*. Eds. E.H. Lieb et al. Princeton Univ. Press, Princeton (1976)
GM 86   García-Alcaine, G.; Molera, J.M.: "Efectos de tipo Aharonov-Bohm", Tesina UCM
GN 64   Gell-Mann, M.; Ne'eman, Y.: *The Eightfold Way*. Benjamin, New York (1964)
GN 80   Gutschick, V.P.; Nieto, M.M.: Phys. Rev. D **22**, 403 (1980)
GO 56   Gombás, P.: *Handbuch der Physik*, Vol. 36 (1956)
GO 70   Goldstein, H.: *Classical Mechanics*. Addison-Wesley, Reading, Mass. (1970)
GP 67   Guardiola, R.; Pascual, P.: Anales de Física **63**, 279 (1967)
GP 71   Gudder, S.; Piron, C.: J. Math. Phys. **12**, 1583 (1971)
GP 75   Galindo, A.; Pascual, P.: Nuovo Cimento **30A**, 111 (1975)
GP 76   Galindo, A.; Pascual, P.: Nuovo Cimento **34B**, 155 (1976)
GP 77   Galindo, A.; Pascual, P.: Anales de Física **73**, 217 (1977)
GR 64   Greenberg, O.W.: Phys. Rev. **135B**, 1447 (1964)
GR 83   Greenberg, D.M.: Rev. Mod. Phys. **55**, 875 (1983)
GR 83a   Goy, P.; Raimond, J.M.;; Gross, M.; Haroche, S.: Phys. Rev. Lett. **50**, 1903 (1983)
GS 22   Gerlach, W.; Stern, O.: Z. Phys. **9**, 349 (1922)
GS 61   Galindo, A.; Sánchez del Río: Am. J. Phys. **29**, 582 (1961)
GS 85   Gislason, E.A.; Sabelli, N.H.; Wood, J.W.: Phys. Rev. A **31**, 2078 (1985)
GT 30   Gebauer, R.; Traubenberg, H.R.V.: Z. Phys. **62**, 289 (1930)
GU 66   Guardiola, R.: *Estudio Analítico de la Matriz S*. University of Valencia (1966)
GU 77   Guardiola, R.: *Recoil Corrections in Electric Gamma Transitions*. University of Granada (1977)
GW 55   Goldhaber, M.; Weneser, J.: Ann. Rev. Nucl. Sci. **5**, 1 (1955)
GW 64   Goldberger, M.L.; Watson, K.W.: *Collision Theory*. Wiley, New York (1964)
GY 69   Grotch, H.; Yennie, D.R.: Rev. Mod. Phys. **41**, 350 (1969)

HA 28   Hartree, D.R.: Proc. Cambridge Phil. Soc. **24**, 89, 111 (1928)
HA 57   Hartree, D.R.: *The Calculation of Atomic Structure*. Wiley, New York (1957)
HA 62   Hamermesh, M.: *Group Theory and Its Applications to Physical Problems*. Addison-Wesley, Reading, Mass. (1962)
HA 80   Harrell, E.M.: Commun. Math. Phys. **75**, 239 (1980)
HA 88   Hallwachs, W.: Ann. Phys. (Wiedemann) **33**, 301 (1888); ibid. **34**, 731 (1888)
HC 84   Hallin, A.L.; Calaprice, F.P.; MacArthur, D.W.; Piilonen, L.E.; Schneider, M.B.; Schreiber, D.F.: Phys. Rev. Lett. **52**, 337 (1984)
HE 25   Heisenberg, W.: Z. Phys. **33**, 879 (1925)
HE 27   Heisenberg, W.: Z. Phys. **43**, 172 (1927)
HE 30   Heisenberg, W.: *The Physical Principles of Quantum Theory*. University of Chicago Press, Chicago, Ill. (1930)
HE 32   Heisenberg, W.: Z. Phys. **77**, 1 (1932)
HE 35   Hellmann: Acta Physicochemica URSS **16**, 913 (1935); **IV2**, 225 (1936)
HE 54   Heitler, W.: *The Quantum Theory of Radiation*, 3rd ed. Oxford Univ. Press, Oxford (1954)
HE 62   Heisenberg, W.: *Physics and Philosophy*. New York (1962)
HE 69   Hepp, K.: Helv. Phys. Acta **42**, 425 (1969)
HE 69a   Henley, E.M.: Ann. Rev. Nucl. Sci. **19**, 367 (1969)
HE 69b   Henley, E.M.: "Charge Independence and Charge Symmetry of Nuclear Forces", in *Isospin in Nuclear Physics*. Ed. D.H. Wilkison. North-Holland, Amsterdam (1969)
HE 72   Hepp, K.: Helv. Phys. Acta **45**, 237 (1972)
HE 74   Herbst, I.W.: Commun. Math. Phys. **35**, 181 (1974)
HE 74a   Hepp, K.: Commun. Math. Phys. **35**, 265 (1974)
HE 81   Herbst, I.W.: In *Rigorous Atomic and Molecular Physics*. Eds. G. Velo and A.S. Wightman. Plenum, New York (1981)
HE 85   Hegerfeldt, G.C.: Phys. Rev. Lett. **54**, 2395 (1985)
HE 87   Hertz, H.: Ann. Phys. (Wiedemann) **31**, 982 (1887)
HH 70   Hättig, H.; Hünchen, K.; Wäffler, H.: Phys. Rev. Lett. **25**, 941 (1970)
HI 77   Hill, R.N.: Phys. Rev. Lett. **38**, 643 (1977)
HI 77a   Hill, R.N.: J. Math. Phys. **18**, 2316 (1977)
HI 82   Hirsbrunner, B.: Helv. Phys. Acta **55**, 295 (1982)
HK 60   Harting, D.; Kluyver, J.C.; Kusumegi, A.; Rigopoulos, R.; Sachs, A.M.; Tibell, G.; Vander-haeghe, G.; Weber, G.: Phys. Rev. **119**, 1716 (1960)
HK 64   Hohemberg, P.; Kohn, W.: Phys. Rev. **136B**, 864 (1964)
HL 51   Hulthén, L; Laurikainen, K.V.: Rev. Mod. Phys. **23**, 1 (1951)
HO 71   Hochstadt, H.: *The Functions of Mathematical Physics*. Wiley, New York (1971)
HO 85   Hansen, T.O.; Østgaard, E.: Can. J. Phys. **63**, 1022 (1985)
HR 60   Hughes, V.W.; Robinson, H.G.; Beltram-López, V.: Phys. Rev. Lett. **4**, 342 (1960)
HR 80   Hegerfeldt, G.C.; Ruijsenaars, N.M.: Phys. Rev. D **22**, 377 (1980)
HS 63   Herman, F.; Skillman, S.S.: *Atomic Structure Calculations*. Prentice Hall, Englewood Cliffs, N.J. (1963)
HS 78   Herbst, I.W.; Simon, B.: Phys. Lett. **78B**, 304 (1978)
HT 56   Hanbury-Brown, R.; Twiss, R.Q.: Nature **157**, 27 (1956); Proc. Roy. Soc. London A **242**, 300 (1957); A **243**, 291 (1957)
HU 25   Hund, F.: Z. Phys. **33**, 345 (1925)
HU 42   Hulthén, L.: Arkiv. Mat. Astron. Fys. **28A**, no. 5 (1942); ibid. **29B**, no. 1 (1942)
HU 66   Hunziker, W.: Helv. Phys. Acta **39**, 451 (1966)
HU 68   Hunziker, W.: "Mathematical Theory of Multi-particle Quantum Systems", in *Lectures in Theoretical Physics*. Vol. X-A. Eds. A.O. Barut and W.E. Brittin. Gordon and Breach, New York (1968)
HU 79   Hunziker, W.: *I.A.M.P. Proc. Lausanne*. Springer, Berlin, Heidelberg (1979)
HW 61   Henshaw, D.G.; Woods, A.D.B.: Phys. Rev. **121**, 1266 (1961)
HY 30   Hylleraas, E.: Z. Phys. **65**, 209 (1930)

HY 37   Hylleraas, E.A.: Z. Phys. **107**, 258 (1937)

IH 28   Ishida, Y.; Himaya, S.: *Inst. Phys. Chem. Res. Tokyo. Sci. Paper,9*. Vol. 152 (1928)
IH 51   Infeld, L.; Hull, T.E.: Rev. Mod. Phys. **23**, 21 (1951)
IK 60   Ikebe, T.: Arch. Rat. Mech. Anal. **5**, 1 (1960)
IK 65   Ikebe, T.: Pacific J. Math. **15**, 511 (1965)
IS 15   Ishiwara, J.: Tokyo Sugaku Buturigakkawi Kizi **8**, 106 (1915)
IZ 80   Itzykson, C; Zuber, J.B.: *Quantum Field Theory*. McGraw-Hill, New York (1980)

JA 62   Jacobson, N.: *Lie Algebras*. Wiley, New York (1962)
JA 66   Jammer, M.: *The Conceptual Development of Quantum Mechanics*. McGraw-Hill, New York (1966)
JA 68   Jauch, J.M.: *Foundations of Quantum Mechanics*. Addison-Wesley, Reading, Mass. (1968)
JA 73   Jauch, J.M.: In *The Physicist's Conception of Nature*. Ed. J. Mehra. Reidel, Dordrecht, Holland (1973)
JA 74   Jammer, M.: *The Philosophy of Quantum Mechanics*. Wiley, New York (1974)
JA 75   Jackson, J.D.: *Classical Electrodynamics*. 2nd ed., Wiley, New York (1975)
JE 05   Jeans, J.H.: Phil. Mag. **10**, 91 (1905)
JE 24   Jeffreys, H.: Proc. London Math. Soc. **23**, 428 (1924)
JK 52   Jost, R.; Kohn, W.: Phys. Rev. **87**, 979 (1952); ibid. **88**, 382 (1952)
JL 72   Jauch, J.M.; Lavine, R.; Newton, R.G.: Helv. Phys. Acta **45**, 325 (1972)
JO 75   Joachain, C.J.: *Quantum Collision Theory*. North-Holland, Amsterdam (1975)
JP 28   Jordan, P.; Pauli, W.: Z. Phys. **47**, 151 (1928)
JP 63   Jauch, J.M.; Piron, C.: Helv. Phys. Acta **36**, 827 (1963)
JR 55   Jauch, J.M.; Rohrlich, F.: *The Theory of Photons and Electrons*. Addison-Wesley, Reading, Mass. (1955)
JS 72   Jauch, J.M.; Sinha, K.: Helv. Phys. Acta **45**, 580 (1972)
JS 72a   Jauch, J.M.; Sinha, K.B.; Misra, B.N.: Helv. Phys. Acta **45**, 398 (1972)
JW 59   Jacob, M.; Wick, G.C.: Ann. Phys. (N.Y.) **7**, 404 (1959)

KA 11   Kamerlingh-Onnes, H.: *Commun. Phys. Lab. Univ. Leiden,* 119b, 120b, 133b (1911)
KA 38   Kapitza, P.L.: Nature **141**, 74 (1938)
KA 49   Kato, T.: Progr. Theor. Phys. **4**, 514 (1949)
KA 50   Kato, T.: J. Phys. Soc. Jap. **5**, 435 (1950)
KA 50a   Kac, M.: In *Proc. 2nd Berk. Symp. Math. Statist. Probability,* p. 189 (1950)
KA 51   Kato, T.: J. Fac. Sci. Univ. Tokio, Sect. I, **6**, 145 (1951)
KA 59   Kac, M.: *Probability and Related Topics in the Physical Sciences*. Wiley, New York (1959)
KA 80   Kato, T.: *Perturbation Theory for Linear Operators*. Springer, Berlin, Heidelberg (1980)
KA 84   Kamefuchi, S. et al. (Eds.): Proceedings of the International Symposium *Foundations of Quantum Mechanics in the Light of New Technology*, Phys. Soc. of Japan (1984)
KE 35   Kemble, E.C. Phys. Rev. **48**, 549 (1935)
KE 37   Kemble, F.: *The Fundamental Principles of Quantum Mechanics with Elementary Applications*. Dover, New York (1937)
KG 84   Kaiser, H.; George, E.A.; Werner, S.A.: Phys. Rev. A **29**, 2276 (1984)
KH 84   Kolbas, R.M.; Holonyak, N.: Am. J. Phys. **52**, 431 (1984)
KI 57   Kinoshita, T.: Phys. Rev. **105**, 1490 (1957)
KI 60   Kirchhoff, G.R.: Ann. Phys. (Poggendorf) **109**, 275 (1860)
KI 66   Kittel, C.: *Introduction to Solid State Physics*. Wiley, New York (1966)
KM 55   Kobayashi, S.; Matsukuma, T.; Nagai, S.; Umeda, K.: J. Phys. Soc. Japan **10**, 759 (1955)
KO 60   Köhler, H.S.: Phys. Rev. **118**, 1345 (1960)
KO 67   Kolos, W.: Int. J. Quantum Chem. **1**, 169 (1967)
KR 26   Kramers, H.A.: Z. Phys. **39**, 828 (1926)
KR 30   Kramers, H.A.: Proc. Acad. (Amsterdam) **33**, 959 (1930)
KS 65   Kohn, W.; Sham, J.L.: Phys. Rev. **140A**, 1133 (1965)
KS 67   Kochen, S.; Specker, E.P.: J. Math. Mech. **17**, 59 (1967)

KS 68   Klauder, J.R.; Sudarshan, E.C.G.: *Fundamentals of Quantum Optics*. Benjamin, New York (1968)

KU 25   Kuhn, W.: Z. Phys. **33**, 408 (1925)

KU 62   Kuhn, H.G.: *Atomic Spectra*. Longmans, Green and Co., London (1962)

KU 71   Kujawski, E.: Am. J. Phys. **39**, 1248 (1971)

LA 23   Landé, A.: Z. Phys. **15**, 189 (1923); ibid. **19**, 112 (1923)

LA 37   Langer, R.E.: Phys. Rev. **51**, 669 (1937)

LA 41   Landau, L.D.: J. Phys. USSR **5**, 71 (1941); ibid. **8**, 1 (1944)

LA 55   Latter, R.: Phys. Rev. **99**, 510 (1955)

LA 74   Labonté, G.: Commun. Math. Phys. **36**, 59 (1974)

LB 39   London, F.W.; Bauer, E.: *La théorie de l'observation en mécanique quantique*. Hermann, Paris (1939)

LE 26   Lewis, G.N.: Nature **118**, 874 (1926)

LE 49   Levinson, H.: Kgl. Danske. Videnskab. Selskab., Mat.-Fys. Medd. **25** (9) (1949)

LE 74   Levy-Leblond, J.M.: Riv. Nuovo Cimento **4**, 99 (1974)

LE 83   Leinfelder, H.: J. Operator Theory **9**, 163 (1983)

LE 84   Leggett, A.J.: Contemp. Phys. **25**, 6 (1984)

LE 99   Lenard, P.: Wien Ber. **108**, 1649 (1899); Ann. Phys. (Leipzig) **2**, 359 (1900); **8**, 149 (1902)

LI 58   Lipkin, H.J.: Phys. Rev. **110**, 1395 (1958)

LI 73   Lipkin, H.J.: *Quantum Mechanics. New Approaches to Selected Topics*. North-Holland, Amsterdam (1973)

LI 76   Lieb, E.: Bull. Amer. Math. Soc. **82**, 751 (1976)

LI 77   Lipatov, L.: Sov. Phys. JEPT **72**, 411 (1977)

LI 81   Lieb, E.H.: Rev. Mod. Phys. **53**, 603 (1981)

LI 84   Lieb, E.H.: Phys. Rev. A **29**, 3018 (1984)

LL 35   London, F.; London, H.: Proc. Roy. Soc. London A **149**, 71 (1935); Physica (Utrecht) **2**, 341 (1935)

LL 65   Landau, L.D.; Lifshitz, E.M.: *Quantum Mechanics: Non-Relativistic Theory*. Pergamon Press, Oxford (1965)

LL 80   Landau, L.D.; Lifshitz, E.M.: *Statistical Physics*. 3rd ed., Pergamon Press, London (1980)

LM 69   Loeffel, J.J.; Martin, A.; Simon, B.; Wightman, A.S.: Phys. Lett. **30B**, 656 (1969)

LM 70   Loeffel, J.J.; Martin, A.: Proc. R.C.P. 25 May (1970)

LO 30   London, F.: Z. Phys. **63**, 245 (1930)

LO 63   Lousiell, W.: Phys. Lett. **7**, 60 (1963)

LO 71   Loebel, E.M. (Ed.): *Group Theory and Its Applications*. Academic Press, New York (1968)

LO 97   Lorentz, H.A.: Ann. Phys. (Wiedemann) **63**, 278 (1897)

LP 72   Lautrup, B.E.; Peterman, A.; Rafael, E. de: Phys. Rep. **3C**, 193 (1972)

LP 97   Lummer, O.; Pringsheim, E.: Ann. Phys. (Wiedemann) **63**, 395 (1897); Verh. Deut. Phys. Ges. **2**, 163 (1900)

LS 73   Lieb, E.H.; Simon, B.: Phys. Rev. Lett. **31**, 681 (1973)

LS 74   Lieb, E.H.; Simon, B.: J. Chem. Phys. **61**, 735 (1974)

LS 77   Lieb, E.H.; Simon, B.: Commun. Math. Phys. **53**, 185 (1977)

LS 77a Lieb, E.H.; Simon, B.: Adv. Math. **23**, 22 (1977)

LS 81   Leinfelder, H.; Simader, C.G.: Math. Z. **176**, 1 (1981)

LT 85   Lieb, E.H.; Thirring, W.E.: *The Universal Nature of van der Waals Forces for Coulomb Systems*, preprint. Vienna (1985)

LU 61   Ludwig, G.: In *Werner Heisenberg und die Physik unserer Zeit*, Vieweg, Braunschweig (1961)

LY 56   Lee, T.D.; Yang, C.N.: Phys. Rev. **104**, 254 (1956)

LY 62   Lynton, E.A.: *Superconductivity*. Wiley, New York (1962)

MA 50   Marchenko, V.A.: Dokl. Akad. Nauk SSSR **72**, 457 (1950)

MA 55   Marchenko, V.A.: Dokl. Akad. Nauk SSSR **104**, 695 (1955)

MA 63   Mackey, G.W.: *The Mathematical Foundations of Quantum Mechanics*. Benjamin, New York (1963)
MA 68   Marchand, J.P.: "Rigorous Results in Scattering Theory", in *Quantum Theory and Statistical Physics*. Eds. A.O. Barut and W.E. Brittin. Gordon and Breach, New York (1968)
MA 72   Maslov, V.P.: *Théorie des Perturbations et Méthodes Asymptotiques*. Dunod, Paris (1972)
MA 75   Martin, A.: Ref. *TH 2085-CERN* (1975)
MA 79   Martin, A.: Commun. Math. Phys. **69**, 89 (1979)
MA 80   Martin, A.: Commun. Math. Phys. **73**, 79 (1980)
ME 59   Messiah, A.: *Mécanique Quantique* (two volumes). Dunod, Paris (1959)
ME 65   Meyer-Abich, K.M.: *Korrespondenz, Individualität und Komplementarität*. Vol. 5 of the series "Boethius-Texte und Abhandlungen zur Geschichte der exakten Wissenschaften". Eds. J.E. Hofmann, F. Klemm and B. Sticker. Franz Steiner, Wiesbaden (1965)
MG 53   Miller, S.C.; Good, R.H.: Phys. Rev. **91**, 174 (1953)
MI 16   Millikan, R.A.: Phys. Rev. **7**, 356 (1916)
MI 54   Miller, S.C.: Phys. Rev. **94**, 1345 (1954)
MI 64   Mikhlin, S.G.: *Variational Methods in Mathematical Physics*. Pergamon, Oxford (1964)
MI 65   Midtdal, J.: Phys. Rev. **138A**, 1010 (1965)
MI 87   Michelson, W.: J. Phys. **6**, 467 (1887)
MI 91   Michelson, A.A.: Phil. Mag. **31**, 338 (1891); ibid. **34**, 280 (1892)
MK 84   Meitzler, C.R.; Khalil, A.E.; Robbins, A.B.; Temmer, G.M.: Phys. Rev. C **30**, 1105 (1984)
ML 67   Mendt, P.; Van der Leon, C.: Nucl. Phys. A **105**, 1 (1967)
MM 65   Mott, N.F.; Massey, H.S.N.: *The Theory of Atomic Collisions*. 3rd ed., Clarendon Press, Oxford (1965)
MO 29   Morse, P.M.: Phys. Rev. **34**, 57 (1929)
MO 58   Morpurgo, G.: Phys. Rev. **110**, 721 (1958)
MO 68   Moshinsky, M.: Am. J. Phys. **36**, 52 and 763 (1968)
MP 79   Marshall, J.T.; Pell, J.L.: J. Math. Phys. **20**, 1297 (1979)
MR 69   Marshak, R.E.; Riazuddin; Ryan, C.P.: *Theory of Weak Interactions in Particle Physics*. Wiley, New York (1969)
MR 85   Manson, J.R.; Ritchie, R.H.: Phys. Rev. Lett. **54**, 785 (1985)
MT 45   Mandelstam, L.; Tamm, I.G.: J. Phys. USSR **9**, 249 (1945)
MZ 70   Mulherin, D.; Zinnes, I.I.: J. Math Phys. **11**, 1402 (1970)

NA 04   Nagaoka, H.: Bull. Math. Phys. Soc. Tokyo **2**, 140 (1904); Nature **69**, 392 (1904); Phil. Mag. **7**, 445 (1904)
NA 68   Naimark, M.A.: *Linear Differential Operators*. Ungar, New York (1968)
NE 32   Von Neumann, J.: *Mathematische Grundlagen der Quantenmechanik*. Springer, Berlin (1932)
NE 59   Nelson, E.: Ann. Math. **70**, 572 (1959)
NE 60   Newton, R.G.: J. Math. Phys. **1**, 319 (1960)
NE 64   Nelson, E.: J. Math. Phys. **5**, 332 (1964)
NE 76   Newton, R.G.: Am. J. Phys. **44**, 639 (1976)
NE 82   Newton, R.G.: *Scattering Theory of Waves and Particles*. 2nd ed., Springer, New York (1982)
NI 80   Nieto, M.M.: Phys. Rev. D **22**, 391 (1980)
NS 79   Nieto, M.M.; Simmons, L.M.: Phys. Rev. D **20**, 1321, 1332, 1342 (1979)
NW 29   Von Neumann, J.; Wigner, E.P.: Phys. Z. **30**, 465 (1929)

OP 28   Oppenheimer, J.R.: Phys. Rev. **31**, 66 (1928)
OP 85   Olariu, S.; Popescu, I.I.: Rev. Mod. Phys. **57**, 339 (1985)

PA 16   Paschen, F.: Ann. Phys. (Leipzig) **50**, 901 (1916)
PA 25   Pauli, W.: Z. Phys. **31**, 765 (1925)
PA 67   Papaliolios, C.: Phys. Rev. Lett. **18**, 622 (1967)
PA 79   Pascual, P.: Anales de Física **75**, 77 (1979)

PA 97   Paschen, F.: Ann. Phys. (Wiedemann) **60**, 662 (1897)
PD 88   Particle Data Group: Phys. Lett. **204B**, 1 (1988)
PE 01   Perrin, J.: Rev. Sci. **15**, 449 (1901)
PE 62   Pekeris, C.L.: Phys. Rev. **126**, 1470 (1962)
PE 72   Peterman, A.: Phys. Lett. **38B**, 330 (1972)
PE 75   Pearson, D.B.: Commun. Math. Phys. **40**, 125 (1975)
PI 53   Pippard, A.B.: Proc. Roy. Soc. London, **A216**, 547 (1953)
PI 76   Piron, C.: *Foundations of Quantum Physics*. Benjamin, Reading, Mass (1976)
PI 78   Pipkin, F.M.: *Advances in Atomic and Molecular Physics*. Eds. P.R. Bates and B. Bederson. Academic Press, New York (1978)
PL 00   Planck, M.: Verh. Deut. Phys. Ges. **2**, 202, 237 (1900)
PL 01   Planck, M.: Sitzungsber. K. Preuss. Akad. Wiss. Berlin **25**, 544 (1901)
PL 12   Planck, M.: "La Théorie du Rayonnement et les Quanta", in *Rapports et Discussion de la Réunion Tenue à Bruxelles 1911*. Eds. P. Langevin and M. de Broglie. Gauthier-Villars, Paris (1912)
PR 62   Preston, M.A.: *Physics of the Nucleus*. Addison-Wesley, Reading, Mass. (1962)
PR 67   Prugovečki, E.: Can. J. Phys. **45**, 2173 (1967)
PR 71   Prugovečki, E.: *Quantum Mechanics in Hilbert Space*. Academic Press, New York (1971)
PR 80   Privman, V.: Phys. Rev. A **22**, 1833 (1980)
PR 84   Preskill, J.: Ann. Rev. Nuc. Part. Sci. **34**, 461 (1984)
PR 97   Preston, T.: Sci. Trans. Roy. Dublin Soc. **6**, 385 (1897)
PS 68   Prepost, R.; Simons, R.M.; Wiik, B.H.: Phys. Rev. Lett. **21**, 1271 (1968)
PS 84   Pak, N.K.; Sökmen, I.: Phys. Lett. **100A**, 327 (1984)
PT 84   Pascual, P.; Tarrach, R.: *Q.C.D.: Renormalization for the Practitioner*. Springer, Berlin, Heidelberg (1984)
PU 56   Purcell, E.M.: Nature **178**, 1449 (1956)
PU 67   Putnam, C.R.: *Commutation Properties of Hilbert Space Operators and Related Topics*. Springer, Berlin, Heidelberg (1967)
PW 34   Paley, R.; Wiener, N.: *Fourier Transforms in the Complex Domain*. American Math. Soc. Providence, RI (1934)
PW 99   Paschen, F.; Wanner, H.: Sitzungsber. K. Preuss. Akad. Wiss. Berlin **2**, 5 (1899)

RA 00   Lord Rayleigh: Phil. Mag. **49**, 539 (1900)
RA 37   Rabi, I.I.: Phys. Rev. **51**, 652 (1937)
RA 42   Racah, G.: Phys. Rev. **62**, 438 (1942)
RA 43   Racah, G.: Phys. Rev. **63**, 367 (1943)
RA 75   Radin, C.: J. Math. Phys. **16**, 544 (1975)
RB 59   Rotenberg, M.; Bivins, R.; Metropolis, N.; Wooten, J.K.: *The 3-j and 6-j Symbols*. Crosby Lockwood, London (1959)
RG 70   Rogers, F.J.; Graboske, H.C.; Harwood, D.J.: Phys. Rev. A **1**, 1577 (1970)
RH 63   Robinson, P.D.; Hirschfelder, J.O.: J. Math. Phys. **4**, 338 (1963)
RI 08   Ritz, W.: Phys. Z. **9**, 521 (1908); Astrophys. J. **28**, 237 (1908)
RI 65   Rickayzen, G.: *Theory of Superconductivity*. Wiley, New York (1965)
RK 01   Rubens, H.; Kurlbaum, F.: Ann. Phys. (Leipzig) **4**, 649 (1901)
RK 68   Rosenzweig, C.; Krieger, J.B.: J. Math. Phys. **9**, 849 (1968)
RM 72   Rau, A.R.P.; Mueller, R.O.; Spruch, L.: Comm. Atom. Mol. Phys. **3**, 87 (1972)
RO 57   Rose, M.E.: *Elementary Theory of Angular Momentum*. Wiley, New York (1957)
RO 71   Robinson, D.W.: *The Thermodynamic Pressure in Quantum Statistical Mechanics*. Springer, Berlin, Heidelberg (1971)
RO 72   Rosenbljum, G.V.: Dokl. Akad. Nauk SSSR **202**, 1012 (1972)
RR 73   Rauch, J.; Reed, M.: Commun. Math. Phys. **29**, 105 (1973)
RS 25   Russell, H.N.; Saunders, F.A.: Astrophys. J. **61**, 38 (1925)
RS 55   Riesz, F.; Sz-Nagy, B.: *Functional Analysis*. Ungar, New York (1955)
RS 70   Robiscoe, R.T.; Shyn, T.W.: Phys. Rev. Lett. **24**, 559 (1970)

RS 72   Reed, M.; Simon, B.: *Functional Analysis*. Academic Press, New York (1972)

RS 75   Reed, M.; Simon, B.: *Fourier Analysis, Self-Adjointness*. Academic Press, New York (1975)

RS 78   Reed, M.; Simon, B.: *Analysis of Operators*. Academic Press, New York (1978)

RS 79   Reed, M.; Simon, B.: *Scattering Theory*. Academic Press, New York (1979)

RS 85   Radzig, A.A.; Smirnov, B.M.: *Reference Data on Atoms, Molecules and Ions*, Springer Series in Chemical Physics, Vol. 31. Springer, Berlin, Heidelberg (1985)

RT 25   Reiche, F.; Thomas, W.: Z. Phys. **34**, 510 (1925)

RU 00   Rubens, H.: Verh. Deut. Phys. Ges. **3**, 263 (1900)

RU 11   Rutherford, E.: Phil. Mag. **21**, 669 (1911)

RU 71   Ruelle, D.: "Equilibrium Statistical Mechanics of Infinite Systems", in *Statistical Mechanics of Quantum Field Theory*. Eds. C. de Witt and R. Stora. Gordon and Breach, New York (1971)

RZ 75   Rauch, H.; Zeilinger, A.; Badurek, G.; Wilfing, A.; Bauspiess, W.; Bonse, U.: Phys. Lett. **54A**, 425 (1975)

SA 64   Sakurai, J.J.: *Invariance Principles and Elementary Particles*. Princeton Univ. Press, Princeton, N.J. (1964)

SA 67   Sakurai, J.J.: *Advanced Quantum Mechanics*. Addison-Wesley, Reading, Mass. (1967)

SA 84   Sakaki, H.: Proceedings of the International Symposium *Foundations of Quantum Mechanics in the Light of New Technology*. Eds. S. Kamefuchi et al., Phys. Soc. of Japan (1984)

SB 82   Summhammer, J.; Badurek, G.; Rauch, H.; Kischko, U.: Phys. Lett. **90A**, 110 (1982)

SB 88   Samuel, J.; Bhandari, R.: Phys. Rev. Lett. **60**, 2339 (1988)

SC 14   Schwarzschild, K.: Verh. Deut. Phys. Ges. **16**, 20 (1914)

SC 26   Schrödinger, E.: Ann. Phys. (Leipzig) **79**, 361, 489 (1926); **80**, 437 (1926); **81**, 109 (1926)

SC 28   Schlapp, R.: Proc. Roy. Soc. London A **119**, 313 (1928)

SC 57   Schwartz, L.: *Théorie des Distributions* (two volumes). Hermann, Paris (1957)

SC 61   Schwinger, J.: Proc. Nat. Acad. Sci. **47**, 122 (1961)

SC 64   Schrieffer, J.R.: *Theory of Superconductivity*. Benjamin, New York (1964)

SC 65   Schwartz, L.: *Méthodes Mathématiques pour les Sciences Physiques*. 2nd ed., Hermann, Paris (1965)

SC 68   Schwartz, J.: W^*-*Algebras*. Nelson, London (1968)

SC 68a   Schiff, L.I.: *Quantum Mechanics*, 3rd ed., McGraw-Hill, New York (1968)

SC 81   Schechter, M.: *Operator Methods in Quantum Mechanics*. North-Holland, New York (1981)

SC 81a   Schulman, L.S.: *Techniques and Applications of Path Integration*. Wiley, New York (1981)

SG 61   Stueckelberg, E.C.G.; Guenin, M.: Helv. Phys. Acta **34**, 621 (1961)

SG 61a   Stueckelberg, E.C.G.; Guenin, M.; Piron, C.; Ruegg, H.: Helv. Phys. Acta **34**, 675 (1961)

SG 62   Stueckelberg, E.C.G.; Guenin, M.: Helv. Phys. Acta **35**, 673 (1962)

SH 59   Shewell, J.R.: Am. J. Phys. **27**, 16 (1959)

SI 68   Siegbahn, K. (Ed.): *Alpha-, Beta- and Gamma-Ray Spectroscopy*. North-Holland, Amsterdam (1968)

SI 70   Simon, B.: Helv. Phys. Acta **43**, 607 (1970)

SI 70a   Simon, B. (with an appendix by A. Dicke): Ann. Phys. (N.Y.) **58**, 76 (1970)

SI 71   Simon, B.: *Quantum Mechanics for Hamiltonians Defined as Quadratic Forms*. Princeton Univ. Press, Princeton N.J. (1971)

SI 71a   Simon, B.: Commun. Math. Phys. **23**, 37 (1971)

SI 72   Simon, B.: In *Mathematics of Contemporary Physics*. Academic Press, New York (1972)

SI 72a   Sinha, K.: Helv. Phys. Acta **45**, 619 (1972)

SI 73   Simon, B.: Ann. Math. **97**, 247 (1973)

SI 74   Simon, B.: Proc. Amer. Math. Soc. **42**, 395 (1974)

SI 76   Simon, B.: "On the Number of Bound States of Two Body Schrödinger Operators: A Review", in *Studies in Mathematical Physics*. Eds. E.H. Lieb, B. Simon and A.S. Wightman, Princeton Univ. Press, Princeton, N.J. (1976)

SI 76a   Simon, B.: Ann. Phys. **97**, 279 (1976)

SI 78   Sienert, C.E.: J. Math. Phys. **19**, 434 (1978)

SI 79   Simon, B.: *Functional Integration and Quantum Physics*. Academic Press, New York (1979)

SI 81   Silverman, J.N.: Phys. Rev. A **23**, 441 (1981)

SI 82   Simon, B.: Bull. Am. Math. Soc. **7**, 447 (1982)

SI 82a   Simon, B.: Int. J. Quant. Chem. **21**, 3 (1982)

SI 83   Simon, B.: Phys. Rev. Lett. **51**, 2167 (1983)

SI 84   Simon, B.: "Some Aspects of the Theory of Schrödinger Operators", in *C.I.M.E. Summer School on Schrödinger Operators*. Como, Italy (Sept. 1984)

SI 84a   Simon, B.: Ann. Math. **120**, 89 (1984)

SI 85   Silverstone, H.J.: Phys. Rev. Lett. **55**, 2523 (1985)

SL 29   Slater, J.C.: Phys. Rev. **34**, 1293 (1929)

SL 39   Slater, J.C.: *Introduction to Chemical Physics*. McGraw-Hill, New York (1939)

SL 60   Slater, J.C.: *Quantum Theory of Atomic Structure* (two volumes). McGraw-Hill, New York (1960)

SN 85   Silverstone, H.J.; Nakai, S.; Harris, J.G.: Phys. Rev. A **32**, 1341 (1985)

SO 16   Sommerfeld, A.: Ann. Phys. (Leipzig) **51**, 1, 125 (1916)

SO 19   Sommerfeld, A.: *Atombau und Spetrallinien*, 7th ed., Vieweg, Braunschweig (1951)

SP 67   Spector, R.M.: J. Math. Phys. **8**, 2357 (1967)

SS 87   Sigal, I.M.; Soffer, A.: Ann. Math. **126**, 35 (1987)

ST 13   Stark, J.: Sitzungsber. K. Preuss. Akad. Wiss. Berlin **47**, 932 (1913)

ST 60   Stueckelberg, E.C.G.: Helv. Phys. Acta **33**, 727 (1960)

ST 66   Stillinger, F.H.: J. Chem. Phys. **45**, 3623 (1966)

ST 68   Stakgold, I.: *Boundary Value Problems of Mathematical Physics*. Macmillan, New York (1968)

ST 81   Selleri, F.; Tarozzi, G.: Riv. Nuovo Cimento **4**, 1 (1981)

ST 88   Stoletow, A.: C. R. Acad. Sci. **106**, 1149, 1593 (1888); **107**, 81 (1888); **108**, 1241 (1889); J. Phys. **9**, 468 (1889)

SV 84   Seetharaman, M.; Vasan, S.S.: J. Phys. A **17**, 2485 (1984)

SW 64   Scadron, M.; Weinberg, S.: Phys. Rev. **133B**, 1589 (1964)

SW 64a   Scadron, M.; Weinberg, S.; Wright, J.: Phys. Rev. **135B**, 202 (1964)

SW 64b   Streater, R.F.; Wightman, A.S.: *PCT, Spin, Statistics, and All That*. Benjamin, New York (1964)

SW 68   Swanson, C.A.: *Comparison and Oscillation Theory of Linear Differential Equations*. Academic Press, New York (1968)

SZ 79   Seznec, R.; Zinn-Justin, J.: J. Math. Phys. **20**, 1398 (1979)

TA 72   Taylor, J.R.: *Scattering Theory*. Wiley, New York (1972)

TB 84   Trelle, R.P.; Birkhäuser, J.; Hinterberger, F.; Kuhn, S.; Prashun, D.; von Rossen, P.: Phys. Lett. **134B**, 34 (1984)

TC 86   Tomita, A.; Chiao, R.Y.: Phys. Rev. Lett. **57**, 937 (1986)

TE 28   Temple, G.: Proc. Roy. Soc. London A **119**, 276 (1928)

TG 29   Traubenberg, H.R.V.; Gebauer, R.: Z. Phys. **54**, 307 (1929); ibid. **56**, 254 (1929)

TG 30   Traubenberg, H.R.V.; Gebauer, R.; Lewin, G.: Naturwissenschaften **18**, 418 (1930)

TH 03   Thomson, J.J. Phil. Mag. **6**, 673 (1903), and **7**, 237 (1904)

TH 25   Thomas, W.: Naturwissenschaften **13**, 627 (1925)

TH 26   Thomas, L.H.: Nature **117**, 514 (1926)

TH 26a   Thomas, L.H.: Proc. Cambridge Phil. Soc. **23**, 542 (1926)

TH 55   Thirring, W.: *Einführung in die Quantenelektrodynamik*. Deuticke, Vienna (1955)

TH 81   Thirring, W.: *Quantum Mechanics of Atoms and Molecules*. Springer, New York (1981)

TH 83   Thirring, W.: *Quantum Mechanics of Large Systems*. Springer, New York (1983)

TI 63   Tisza, L.: Rev. Mod. Phys. **35**, 151 (1963)

TR 27   Thomson, G.P.; Reid, A.: Nature **119**, 890 (1927)

TR 52   Trainor, L.E.H.: Phys. Rev. **85**, 962 (1952)

UG 25   Uhlenbeck, G.E.; Goudsmit, S.: Naturwissenschaften **13**, 953 (1925)

VI 72   Vinh Mau, N.: *Non Conservation de la Parité dans les Noyaux*. Report GIFT 20/72 (1972)
VL 26   Van Vleck, J.H.: Proc. Nat. Acad. Sci. U.S.A. **12**, 662 (1926)
VP 66   Vessot, R.; Peters, H.; Vanier, J.; Bechler, R.; Halford, D.; Harrach, R.; Allan, D.; Glaze, D.; Snider, C.; Barnes, J.; Cutler, L.; Bodily, L.: IEEE Trans. Instr. Meas., IM-15, 165 (1966)
VW 81   Velo, G.; Wightman, A.S. (Eds.): *Rigorous Atomic and Molecular Problems*. Plenum, New York (1981)

WA 26   Waller, I.: Z. Phys. **38**, 635 (1926)
WA 57   Wu, C.S.; Ambler, E.; Hayward, R.W.; Hoppes, D.D.; Hudson, R.P.: Phys. Rev. **105**, 1413 (1957)
WA 72   Wagner, D.: *Introduction to the Theory of Magnetism*. Pergamon, Oxford (1972)
WC 75   Werner, S.A.; Colella, R.; Overhauser, A.W.; Eagen, C.F.: Phys. Rev. Lett. **35**, 1053 (1975)
WE 26   Wentzel, G.: Z. Phys. **38**, 518 (1926)
WE 63   Wei, J.: J. Math. Phys. **4**, 1337 (1963)
WE 63a   Weinberg, S.: Phys. Rev. **130**, 776 (1963)
WE 63b   Weinberg, S.: Phys. Rev. **131**, 440 (1963)
WE 64   Weinberg, S.: Phys. Rev. **133B**, 232 (1964)
WE 67   Weidmann, J.: Bull. Am. Math. Soc. **73**, 452 (1967)
WE 67a   Weidmann, J.: Math. Zeit. **98**, 268 (1967)
WE 70   Weisskopf, V.F.: CERN 70-8 (1970)
WG 73   Witt, B. de; Graham, N.: *The Many-Worlds Interpretation of Quantum Mechanics*. Princeton University Press, Princeton (1973)
WI 15   Wilson, W.: Phil. Mag. **29**, 795 (1915)
WI 23   Wiener, N.: J. Mathematical and Physical Sci. **2**, 132 (1923)
WI 31   Wigner, E.P.: *Gruppentheorie und ihre Anwendung auf die Quantenmechanik der Atomspektren*. Vieweg, Braunschweig (1931). English translation: *Group Theory and Its Application to the Quantum Mechanics of Atomic Spectra*. Academic Press, New York (1959)
WI 41   Wintner, A.: *The Analytical Foundations of Celestial Mechanics*. Princeton University Press, Princeton, N.J. (1941)
WI 55   Wigner, E.P.: Phys. Rev. **98**, 145 (1955)
WI 60   Wigner, E.P.: J. Math. Phys. **1**, 409 and 414 (1960)
WI 62   Wightman, A.S.: Rev. Mod. Phys. **34**, 845 (1962)
WI 63   Wigner, E.P.: Am. J. Phys. **31**, 6 (1963)
WI 69   Wilkinson, D.H. (Ed.): *Isospin in Nuclear Physics*. North-Holland, Amsterdam (1969)
WI 72   Wigner, E.P.: In *Aspects of Quantum Theory*. Eds. A. Salam and E.P. Wigner. Cambridge Univ. Press, London (1982)
WI 81   Witten, E.: Nucl. Phys. B **188**, 513 (1981)
WI 82   Wilczek, F.: Phys. Rev. Lett. **48**, 1144 (1982)
WI 84   Will, C.M.: Phys. Rep. **113C**, 345 (1984)
WI 94   Wien, W.: Ann. Phys. (Wiedemann) **52**, 132 (1894)
WI 96   Wien, W.: Ann. Phys. (Wiedemann) **58**, 662 (1896)
WR 68   Witsch, W. von; Richter, A.; Bretano, P. von: Phys. Rev. **169**, 923 (1968)
WS 55   Wightman, A.S.; Schweber, S.: Phys. Rev. **98**, 812 (1955)
WS 72   Weinstein, A.; Stenger, W.: *Intermediate Problems for Eigenvalues*. Academic Press, New York (1972)
WU 84   Wu, Y.S.: Phys. Rev. Lett. **53**, 111 (1984)
WW 30   Weisskopf, V.; Wigner, E.: Z. Phys. **63**, 54 (1930)
WW 52   Wick, G.C.; Wightman, A.S.; Wigner, E.P.: Phys. Rev. **88**, 101 (1952)
WW 69   Whittaker, E.T.; Watson, G.N.: *A Course of Modern Analysis*, 4th ed., Cambridge Univ. Press, Cambridge, Mass. (1969)
WZ 83   Wheeler, J.A.; Zurek, W.H.: *Quantum Theory and Measurement*. Princeton Univ. Press, Princeton (1983)

YB 78   Yaris, R.; Bendler, J.; Lovett, A.; Bender, C.M.; Fedders, P.A.: Phys. Rev. A **18**, 1816 (1978)
YU 84   Yuen, H.P.: Phys. Rev. Lett. **52**, 1730 (1984)

ZE 65   Zemanian, A.H.: *Distribution Theory and Transform Analysis*. McGraw-Hill, New York (1965)
ZE 97   Zeeman, P.: Phil. Mag. **43**, 226 (1897); **44**, 55 and 255 (1897); Versl. Gewone Vergad. Wis. Natuurk. Afd. Kom. Akad. Wetensch. Amsterdam **6**, 13, 99 and 260 (1897)
ZO 80   Zorbas, J.: J. Math. Phys. **21**, 840 (1980)

Subject Index

Contents of **Quantum Mechanics II**